Chemistry of Pesticides

CHEMISTRY OF PESTICIDES

Edited by
K. H. BÜCHEL

Contributors:

	R. A. Fuchs	H.-J. Riebel
	G. Jäger	K.-J. Schmidt
K. H. Büchel	W. Krämer	R. Schröder
W. Draber	W. Lunkenheimer	W. Sirrenberg
Ch. Fest	H.-J. Niessen	J. Stetter

Translated by GRAHAM HOLMWOOD

A Wiley-Interscience Publication

JOHN WILEY & SONS

New York   Chichester   Brisbane   Toronto   Singapore

Library of Congress Cataloging in Publication Data:

Pflanzenshutz und Schädlingsbekämpfung. English.
   Chemistry of pesticides.

   Translation of: Pflanzenschutz und Schädlingsbekämp-
fung.
   "A Wiley-Interscience publication."
   Bibliography: p.
   Includes indexes.
   1. Pesticides. I. Büchel, K. H. II. Title.

SB951.P4613   1982   668'.65   82-17340
ISBN 0-471-05682-0 ✓

Printed in the United States of America

10 9 8 7 6 5 4 3 2 1

Contributors

Prof. Dr. Karl-Heinz Büchel
Board of Management
Bayerwerk, 5090 Leverkusen, FRG

Dr. Wilfried Draber
Chemical Research Laboratory
Agrochemicals Division, Bayer AG
Research Center, 5600 Wuppertal, FRG

Dr. Christa Fest
Chemical Research Laboratory
Agrochemicals Division, Bayer AG
Research Center, 5600 Wuppertal, FRG

Dr. Rainer A. Fuchs
Chemical Research Laboratory
Agrochemicals Division, Bayer AG
Research Center, 5600 Wuppertal, FRG

Dr. Gerhard Jäger
Corporate Research Laboratory,
* Bayer AG*
Bayerwerk, 5090 Leverkusen, FRG

Dr. Wolfgang Krämer
Chemical Research Laboratory
Agrochemicals Division, Bayer AG
Research Center, 5600 Wuppertal, FRG

Dr. Wilfried Lunkenheimer
Chemical Research Laboratory
Agrochemicals Division, Bayer AG
Research Center, 5600 Wuppertal, FRG

Dr. Heinz-Josef Niessen
Institute for Formulation Technology
Agrochemicals Division, Bayer AG
Bayerwerk, 5090 Leverkusen, FRG

Dr. Hans-Jochen Riebel
Chemical Research Laboratory
Agrochemicals Division, Bayer AG
Research Center, 5600 Wuppertal, FRG

Dr. Karl-Julius Schmidt *(deceased)*
formerly Planning Commission
* (Agrochemicals)*
Staff Executive to the Board of
* Management*
Bayerwerk, 5090 Leverkusen, FRG

Dr. Rolf Schröder
Production Department
Agrochemicals Division, Bayer AG
Bayerwerk, 5600 Wuppertal, FRG

Dr. Wilhelm Sirrenberg
Chemical Research Laboratory
Agrochemicals Division, Bayer AG
Research Center, 5600 Wuppertal, FRG

Dr. Jörg Stetter
Chemical Research Laboratory
Agrochemicals Division, Bayer AG
Research Center, 5600 Wuppertal, FRG

TRANSLATOR

Dr. Graham Holmwood
Chemical Research Laboratory
Pharmaceuticals Division, Bayer AG
Research Center, 5600 Wuppertal, FRG

Preface

Of the 4 billion people inhabiting the world today, about 700 million are under-nourished and 1.3 billion have an unbalanced or inadequate diet. Statistics show that the world population increases by about 2% annually and by the year 2000 will be well beyond the 6-billion mark. At the present time one-third of the *potential* world harvest is lost to pests; without chemical crop protection this loss would be twice as high. Substantial food reserves could be mobilized by judiciously widening the use of pesticides.

Crops are exposed to attack from their various adversaries throughout the growing season. Those grown continuously as monocultures are especially prone to attack. Therefore, to maintain crop productivity at the highest possible level, intensive use of crop protection chemicals is essential. In highly developed agricultural systems, crop protection chemicals are also inputs that serve to reduce labor requirements, to facilitate harvesting, and thus to improve the economy of food production. Additionally, chemicals are invaluable aids post-harvest for the safe storage, transportation, distribution, preservation, and processing of agricultural produce. The following conclusion is thus inevitable: *without chemical products the already critical world food supply would collapse, there being little potential for expanding the area of land under cultivation.*

Another vital role for chemical products is in the control of disease vectors such as those for malaria, yellow fever, and Chagas' disease. In this function they protect millions of human lives, particularly in tropical and subtropical regions, and hence make a significant contribution to national economies and the maintenance of public health.

All these facts lead to one conclusion. *Chemical compounds such as pesticides are not luxury articles of a technical civilization but necessities for the survival of the world population.* For this reason chemical research and the crop protection industry have their special position and significance for the future.

Owing to the finiteness of fossil energy resources, raw materials based on annually renewable agricultural products will gain special importance in the coming decades. Here pesticides will have the new task of optimizing the production of biomass to be used as a source of chemical feedstocks.

This monograph gives a concise survey of the significance, structure, synthesis, and chemical and biological properties of the important commercial pesticides. The chapters are arranged according to the different major groups of target organisms, the content ranging from the classical chemical structures to the newer control methods, the so-called third-generation products, and to formulation aids.

This volume is intended not only for specialists in the pesticide field but also for a wide circle of readers in university, industry, agriculture, and government authority, who are offered a comprehensive survey of the most important facts in this discipline.

K. H. Büchel

Leverkusen, West Germany
January 1983

Contents

Chemistry of Pesticides

Chapter One

Introduction

K. H. BÜCHEL
Bayer AG, Leverkusen

1.1. Definition of Terms

The classical term *crop protection products* embraces all chemical compounds used for the control of arthropods, microorganisms, and weeds injurious to economic or crop plants. These chemicals thus serve to maintain crops in a healthy condition throughout their growth, as well as to reduce agricultural labor requirements and, by enabling crops to realize their full yield potential, to safeguard and maintain the food supply of man. The more comprehensive term *pesticides* relates to compounds used for the control of pests of all descriptions—including those directly hazardous to the health of man and animals. Notable among insect pests that act as vectors of human diseases are malaria-transmitting mosquito species, typhus-transmitting lice, plague-transmitting fleas, and various dipterous species recognized as vectors of filariasis, yellow fever, meningitis, and onchocerciasis. The term *pesticides* also embraces compounds for the control of rodents, slugs, snails, and household and stored-product pests. Products for the control of disease vectors are of especially great political and economic importance because they help to preserve millions of human lives, particularly in tropical and subtropical regions. The well-known insecticide *DDT*, for example, is now used mainly for malaria and yellow-fever control under programs arranged by the World Health Organization. The carbamate insecticide Baygon® is also used for malaria eradication and, among other things, for the control of reduviid bugs such as *Triatoma infestans*, vectors of the dreaded Chagas' disease (South American trypanosomiasis).

1.2. Organization of Chapters and Their Significance

This book presents in concise form a survey of the most important crop protection products currently on the market. The comprehensive work entitled *Chemie der*

Pflanzenschutz–und Schädlingsbekämpfungsmittel (edited by R. Wegler)[1] should be consulted for further information, particularly in relation to historical and non-commerical products, experimental compounds, and fringe areas such as metabolism and residues.

The arrangement of chapters in this book is oriented around the major groups of target organisms.

In Chapter 2 on "Agents for Control of Animal Pests," the individual classes of insecticides (organophosphates, carbamates, etc.) are discussed as well as nematicides, acaricides, molluscicides, and rodenticides. A further section in this chapter is devoted to new methods of control and the so-called third-generation pesticides.

Products for the control of microbial pathogens are described in Chapter 3 on "Fungicides and Bactericides." The next, Chapter 4, is devoted to chemical weed control (i.e., herbicides), and the following Chapter 5 is on plant growth regulators, an area still very largely in the investigational stage.

The book concludes with Chapter 6 on adjuvants used for the formulation of active ingredients.

The *insecticides* continue to be a group of products of immense agricultural and economic importance.[2] Notwithstanding the many highly effective products already available, there still remains much scope for new ones. This is due in part to the development of resistance, particularly in species that produce several generations in rapid succession and have high multiplication rates. Legislative measures such as the restrictions imposed on the use of chlorinated hydrocarbon insecticides also create a need for substitutes. Newly developed insecticides are expected to display ideal properties of a wide array including broad-spectrum activity concomitant with safety to beneficial insects, low mammalian and fish toxicity, and sufficiently long residual action yet with insignificant, harmless residue levels. These ideal properties should all be attained at the lowest possible cost. Such requirements, however, which are sometimes even mutually exclusive, make research and development in crop protection a very risky enterprise.

Of great topical interest are the new approaches to pest control using insect hormones and hormone mimics, attractants, virus toxins, chemosterilants, and the like, as well as biological control methods, although these have achieved relatively little practical importance as yet. Special problems accompany many of these product groups. The hormone mimics are expensive to produce and require the same outlay on toxicological and ecological studies as conventional insecticides. Various virus toxins also have teratogenic and allergenic side effects.[3]

One disadvantage of hormone-related products is their chemical instability under field conditions. Apart from this, their high specificity to only one or no more than a few pest species can create practical problems owing to the selection of uncontrolled pest species, which then occupy the resultantly vacated ecological niches. The use of attractants for the direct control of insect pests is a goal that still awaits realization; practical solutions have been found only for their utilization in monitoring the buildup of pest populations.

It follows from these comments that crop protection and pest control are the clear domain of chemical products today and will remain so in the near future.

The relative significance of *fungicides* and *bactericides* differs substantially throughout the world. Whereas fungicides have a 26% and 24% share of the pesticide market in Europe and Asia, respectively, their share in the United States is only 6%. Nevertheless, the fungicide market is a constantly growing one that holds great promise for new developments. Changes in cultural practices associated with closer cropping sequences, breeding of new, higher-yielding crop cultivars, and use of heavier fertilizer dressings all lead to an increased incidence of fungal disease. The restrictions imposed on the use of the old but highly effective mercurials will also encourage the occurrence of fungal diseases and accelerate the development of new compounds.[4]

The new systemic fungicides such as *benomyl, carbendazim*, and *thiophanate* had a stimulating influence on the development of the fungicide sector, although following their introduction the problem of resistance buildup among fungi arose on a major scale for the first time ever. The good systemic products nonetheless still continue to be of great practical value. Work directed toward developing compounds for the control of phytopathogenic diseases from among the microbial metabolites (antibiotics) is showing great promise. As such compounds are effective at very low concentrations, they are of ecological benefit.

Fungicidal and bactericidal antibiotics are already playing a significant role in disease control programs in rice crops in Japan. It must be pointed out, however, that because of the possible development of resistance to such antibiotics, which are used also in medicine, their utilization in crop protection is heavily restricted in many countries.

Usage of *herbicides* has expanded enormously during the last 10 to 15 years, a trend that shows no signs of diminishing. Herbicides now account for the biggest share of all pesticide usage in crop farming. The reasons for this trend are essentially twofold: Besides contributing largely to the yield improvements achieved, they have had a large sociological impact on agriculture. Herbicides are work-saving tools. They reduce labor requirements and eliminate the need for certain husbandry operations, which factors are of decisive importance especially in highly developed countries with their high wage costs. This also has a stabilizing influence on the price of agricultural produce. The sugar-beet-growing industry in Europe, for example, has remained in existence only by rationalizing crop husbandry and adopting the use of selective herbicides. Modern mechanical harvesting techniques also rely for their practicability and efficiency on the use of chemical weed killers.

Herbicides also have some advantages on the ecological and toxicological side. As they are usually applied long before harvest, the residue problem is relatively small. Many herbicides are selective inhibitors of photosynthesis—a mechanism specific to plants—and hence are not acutely toxic to mammals.

Against this background it is obvious that many opportunities for new developments are also open in the herbicide sector. In particular, compounds that display high selectivity to crop plants yet are fully effective against problem weeds will be a major focus of research activity in the future, as indeed they are now.

Plant growth regulators have specific effects on the physiological processes of plants. They modify plant composition, for example, by increasing sugar or protein content. They also inhibit growth, for example, by shortening internodes, an effect that in cereals reduces stem length, improves standing power, and prevents lodging even at high fertilizer rates.

Other uses of plant growth regulators include the suppression of sucker growth (in tobacco), induction of flowering, improvement of drought and frost resistance, promotion of fruit abscission to facilitate mechanical harvesting, acceleration of ripening and the suppression of growth of grasses, woody plants, and roadside vegetation. Plant growth regulators thus serve to improve quality and yields of crops and to reduce the work load in agriculture and horticulture.[5]

Despite these many potentialities, the development of this sector must be assessed conservatively and critically. Estimates for the U.S. market put the total turnover for plant growth regulators in 1980 at only $40 million, equivalent to a mere 1.7% of the whole U.S. pesticide market.[6] Research in this area is very precarious because of the great uncertainty whether uses of new compounds will prove practicable, economically feasible, and commercially viable. Work directed toward developing plant growth regulators for use as aids to husbandry incurs the least risk. Field evaluation of compounds that produce yield increases or regulate the content of plant constituents is very difficult since phytophysiological processes are greatly influenced by other factors such as soil, temperature, insolation, and rainfall, and the results thus vary considerably. Further changes in metabolism are also to be expected from intervention in the physiology of a plant. In view of the very wide range of factors involved, complicated and costly investigations of a long-term nature are necessary. It is because of all these reasons that progress in bringing plant growth regulators to the market has been modest to date.

1.3. Research and Development Costs of a Crop Protection Product

Research and development costs for crop protection products have increased out of all proportion during the course of the past 20 years. In 1956 the average bill for R&D costs was $1.2 million; in 1967, $3.4; in 1970, $5.5; and in 1974, $7.6.[7] An evaluation of the Bayer AG figures (G. Haug[4]) covers 153,000 compounds screened in the period 1950-1974. In the years 1970-1974 one single successful commercial product cost DM42.6 million (~$18 million) on average to develop. The detailed analysis[4] shows that the increase in costs is due very largely to registration requirements in the areas of toxicology, metabolism, residue behavior, and ecology.

The R&D costs of a crop protection product include synthesis, biological screening, field evaluation, formulations, toxicity trials, metabolic studies, residue studies, environmental safety studies, product and metabolite analysis, patent protection, and process development for the manufacture of the product. For the research-based companies, on whose work all commercial innovations rest, these costs amount to about 6-10% of turnover. The heavy investment for the building of the manufacturing plant is not included; this usually exceeds the R&D cost.

These figures show what vast sums of money are involved in the development of new products and how risky an involvement in this business has become. Only large, vigorous, research-intensive companies are able to bear this risk and thus find solutions for current problems. This, in turn, can only happen when such companies are able to make an adequate return on investment.

1.4.   The World Market for Crop Protection Products According to Geographical Region, Crops, and Product Types

The world market turnover volume for crop protection products was DM13.4 billion in 1973; in 1979 this had risen to ~DM22 billion (wholesale price base), mainly as a result of increases in energy and raw-material costs.
    This DM22 billion turnover was shared as follows:

Europe	35%	Africa	6%
North America	21%	Asia	21%
Latin America	15%	Australia/New Zealand	2%

West Germany accounted for about 3% of this sales figure. Considering that new crop protection products cost such huge R&D sums, their development will obviously be a sound economic venture only if world market demand for their proposed uses is large enough to justify it.
    The distribution of crop protection product usage in the major crops is very interesting; almost one quarter of the total market is shared by cotton and corn alone, as the following list shows:

Corn	13%	Soybeans	9%
Cotton	11%	Vegetables	6%
Cereals		Pome fruit	4%
(excluding corn and rice)	11%	Rest	36%
Rice	10%		

The percentage breakdown of world turnover for the different product groups reveals the following:[4]

Herbicides	43%	Household, vector control and	
Insecticides	35%	hygiene products, miscellaneous	3%
Fungicides	19%		

It must be borne in mind when considering these world figures that the percentage share of the different product groups can vary widely from one geographical region to another according to the situations in the particular countries. A comparison of Europe, North America, and Africa shows that the share of herbi-

TABLE 1. Breakdown of Crop Protection Product Use by Region

Region	Herbicides (%)	Insecticides (%)	Fungicides (%)
Western Europe	47	21	27
United States and Canada	65	28	6
Africa	16	60	17

cides in the total turnover ranges from 65% (in North America) to 16% (in Africa).[4] (See Table 1.)

The potential for development in overseas regions is reflected most clearly in these figures, both with respect to the use of crop protection products and to improvement in the food supply situation for the growing populations of these countries.

1.5. Crop Protection, World Food Production, Public Health, and Biomass Production

Statistics show that the world population increases by about 2% annually (see Table 2) and will have grown to approximately 6.4 billion by the year 2000.[8] Of the 4 billion people alive today, about 700 million are undernourished, and 1.3 billion have an unbalanced or inadequate diet.

At the present, food production is far below requirements, and it will remain so in the future according to predictions for up to the year 2000.[9,10] Achieving the necessary production is one of our greatest future tasks, and crop protection products have an essential part to play. In 1967 around 35% of the *potential* harvest was destroyed by insect pests, fungal diseases, and weeds, equivalent to a financial loss of approximately $75 billion.[8] The loss related to the value of the *actual* harvest was 54%.[10]

Depending on the agricultural region, losses in some crops are much higher than 35% of the potential harvest, for example, rice 46%, millet 38%, sugarcane 55%, coffee 44%, and copra 44%. Just by avoiding cereal losses it would be possible to provide 2 billion undernourished people each with almost 700 g of cereals per day.[10] H. H. Cramer has presented many examples to demonstrate that the use

TABLE 2. World Population Trends

Period	Population at End of Period
~7000 B.C.–A.D. 1830	1 billion
1830–1930	2 billion
1930–1960	3.3 billion
1960–1980	4.4 billion
1980–2000	6.4 billion

of crop protection products not only prevents losses but also makes for extraordinary increases in yield.[8,11] A most impressive example is provided by rice production in Asia. Rice is the staple food of the majority of mankind. Yields per hectare vary considerably from one rice-growing area to another. In Asia, Japan produces the highest yield of 62.5 decitonnes/hectare (1 decitonne/hectare = ∼0.8 cwt/acre), India only 19.8 dt/ha, while some African countries produce a mere 5 dt/ha. Between 1946 and 1950 the rice yield in Japan was little higher than it is in India today. After it had been demonstrated in 1952 that the rice stem borer could be controlled by organophosphate insecticides and the dreaded rice blast disease (*Pyricularia oryzae*) by new fungicides, however, yields increased tremendously.

Insecticides play a decisive role in disease vector control and hygiene programs for the protection of health and life.

The large-scale programs for the control of insect vectors, under the aegis of the World Health Organization (WHO), have led to a dramatic reduction of disease in many parts of the world.[12] The number of deaths from malaria in India has been reduced from an annual 750,000 to 1500 in the last 15 years by the use of insecticides. In Sri Lanka (Ceylon) extensive use of *DDT* reduced the number of malaria cases from 2.8 million in 1946 to 110 in 1961. After the control program was stopped in 1963, however, the number of malaria cases rose rapidly to more than 1.0 million by 1968. On the island of Mauritius where 1589 deaths due to malaria were registered in 1948, this disease has now been virtually eradicated thanks to anopheles control programs.

Pesticides have saved millions of human lives in all parts of the world and will continue to do so. They are one of the most important weapons in the fight against hunger.

Agricultural products (i.e., *renewable* raw materials) are most likely to become an important source of chemical feedstocks in the forthcoming decades as the availability of fossil resources declines. This increased biomass production will definitely need the support of crop protection chemicals to secure the economics of biomass use for feedstock purposes.[12a,12b]

1.6. Crop Protection and the Environment

The tremendous growth in industrial production and the consequent improvement in the standard of living have provoked a worldwide discussion on environmental quality. The question of crop protection and environment is only one aspect of this entire complex. In spite of this, tendentious publications such as Rachel Carson's *Silent Spring* have brought crop protection into the foreground of environmental discussions. This is most unjust because among the so-called anthropogenic chemicals there are none studied more intensively than crop protection products; any potential hazards presented by them are thus most readily identified. It is particularly the chlorohydrocarbons such as *DDT*, *lindane*, and *dieldrin* that are the subject of public discussion. Their persistence, that is, their high stability, is regarded as problematic. But this—and their low prices—is precisely the reason why they have been such useful pesticides in the last 35 years and the mainstay of vector-control programs.

In general, the residue problem is exaggerated in public, not the least owing to the highly developed trace analysis techniques, which, with residue levels expressed in ppb (parts per billion, i.e., 1 g in 1,000,000 kg), suggest larger quantities than are actually present.

All commercial crop protection products are subject to strict legislation in all industrialized countries. Before a product is released for sale, safe residue levels have to be established.

International organizations such as WHO and the FAO (Food and Agriculture Organization of the United Nations) set internationally valid maximum levels for residues (MRLs) that present *no* hazard to humans and animals. Research-based industry does its utmost to develop products that satisfy these regulations. It is the declared aim of crop protection research to find products that are safe both in toxicological and environmental terms—the successes in recent years give unequivocal confirmation.

Crop protection is only part of the agricultural economy and must be seen in context with it. It is agriculture itself that has led to the most fundamental changes in the human environment.[13] Crop protection is thus more a necessary consequence of than the cause of environmental changes.[14] In all areas of life one must weigh the desired advantages against possible disadvantages. As F. Coulston[15] expresses it again and again, the aim is to make the benefit–risk ratio calculable.

The American Norman E. Borlaug, who was awarded the 1970 Nobel Peace Prize for his breeding of high-yielding wheat cultivars, expressed himself very clearly on the theme "crop protection and environment" before a hearing of the U.S. Environmental Protection Agency (EPA).[16] He pointed out that the United States today produces sufficient food from an area of 290 million acres. Thirty years ago 600 million acres would have been needed to produce the same quantity of food because varieties and crop protection agents were not so far advanced. The increases in yield have thus preserved 310 million acres of forest and green land for nature. Here crop protection reveals itself as active environmental protection.[11]

Borlaug makes it absolutely clear. We must feed ourselves and protect ourselves from the dangers that threaten our health. To do this we need chemical products. Without them the people of the Earth will starve.

Chapter Two

Agents for Control of Animal Pests

2.1.   Naturally Occurring Insecticides and Synthetic Analogues

R. A. FUCHS AND R. SCHRÖDER
Bayer AG, Wuppertal-Elberfeld

2.1.1.   INTRODUCTION

A multitude of insecticidal agents are synthesized by plant and animal species. Approximately 2000 plant species are known which contain toxic principles effective against insects, but only a small number are used in practice. Economic and biological considerations such as expensive manufacture and low stability prevent most insecticides produced by animals and plants from competing effectively against synthetic agents. In spite of this, they are of interest in crop protection and pest control in that chemical modification of the natural product can open the way to novel, synthetic insecticides. A classical example of the successful derivatization of natural products is given by the pyrethrins and their synthetic analogues.

2.1.2.   PYRETHROIDS

2.1.2.1.   Naturally Occurring Pyrethroids (Pyrethrins)[17-21]

Pyrethrins are the insecticidal principles of a number of chrysanthemum species, the dried and powdered flowers of which have been used since the beginning of the 19th century for control of household pests. The first such products were imported more than 100 years ago from Dalmatia and Iran.

Chrysanthemum cinerariaefolium, of the numerous chrysanthemum species the one best suited for isolation of the pyrethrins, has been grown in Japan since about 1880 and in Kenya, the major producer today, since 1920. World production in 1972 was ~20,000 tonnes of dried flowers, more than half coming from Kenya.

In the process of isolation of the insecticidal principles, the concentration of which ranges between 0.3 and 2% depending on area and treatment, the flowers are harvested shortly after blooming and are then either dried and powdered or extracted with a combination of polar and unpolar solvents (e.g., methanol-kerosene) to yield a crude product of ~25% a.i.* content. The dusts and extracts so obtained are known as pyrethrum and are mainly used in the hygiene sector (for flies, mosquitos, etc.). Pyrethrum has only a limited use in crop protection because the stability of the active principles under field conditions is very poor. Pyrethrum is inactivated both by air and by sunlight. Addition of antioxidants such as hydro quinone or resorcinol improves its stability and storage properties.

The insecticidal components of pyrethrum are optically active esters derived from the acids **1** and **2** and the three keto alcohols **3**, **4**, and **5**:

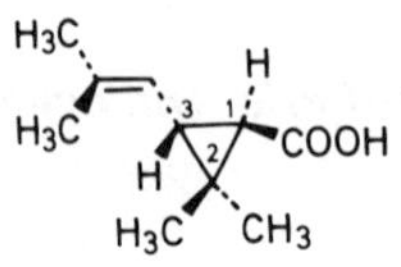

1; (+)-*trans*-Chrysanthemic acid

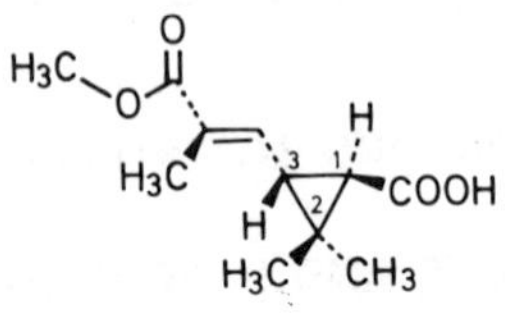

2; (+)-*trans*-Pyrethoic acid

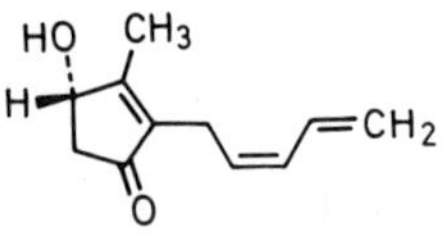

3; (+)-Pyrethrolone

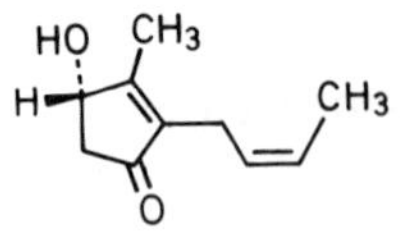

4; (+)-Cinerolone

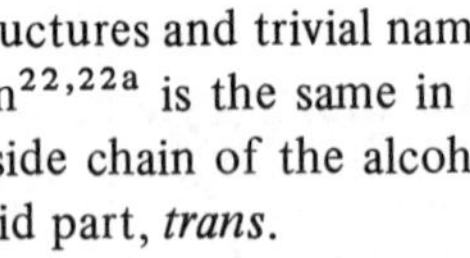

5; (+)-Jasmolone

The six esters, which are found in various proportions in the plants, are given in Table 3 together with their structures and trivial names.

The absolute configuration[22,22a] is the same in all the esters, namely, *1R, 3R, 4S*. The double bond in the side chain of the alcohol moiety has the *cis* configuration; that in the carboxylic acid part, *trans*.

The fundamental work on the structural elucidation of the pyrethrum active principles was carried out by Staudinger and Ruzicka[23] and by La Forge and Barthel.[24] Staudinger and Ruzicka also performed the first partial synthesis of pyrethrin I (**6**) by reaction of the acid chloride of chrysanthemic acid **1** with pyrethrolone **3**.

---

*Here and throughout, the abbreviation a.i. stands for active ingredient.

TABLE 3. Naturally Occurring Pyrethroids

No.	Structure	Name
6		Pyrethrin I
7		Pyrethrin II
8		Cinerin I
9		Cinerin II
10		Jasmolin I
11		Jasmolin II

2.1.2.2. Synthetic Analogues[24a]

In view of the high production costs of pyrethrum it was logical to search for compounds available by chemical modification of the natural structures that could be easily synthesized but that would possess the same properties as the natural product (e.g., low mammalian toxicity, good knockdown effect).

The first synthetic pyrethroid, *allethrin* (**12**, Table 4), appeared on the market as long ago as 1950. It is the allyl homologue of natural cinerin I (**8**, Table 3) but

TABLE 4. Summary of the Important Pyrethroids

No.	Common Name, *Chem. Abstr.* Name, Trade Names, and Other Designations	Structure	Stereochemistry (Absolute Configuration) in Acid Moiety	Toxicity LD_{50}(mg/kg) Rat[29,30]
6–11	*Pyrethrum* A plant extract with an a.i. content of ~30% Use: hygiene pests	Composition: 11.4% Pyrethrin I, 10.5% pyrethrin II, 2.2% cinerin I, 3.5% cinerin II, 1.2% jasmolin I, 2% jasmolin II (structures—see Table 3)		Oral: 200–2600 Pyrethrin I, oral: 260–420, IV: 5
12	*Allethrin*[31] 2-Methyl-4-oxo-3-(2-propenyl)-2-cyclo-penten-1-yl 2,2-dimethyl-3-(2-methyl-1-propenyl)cyclo-propanecarboxylate Pynamin®		(+)-*trans* (*1R , 3R*) Alcohol moiety: *S*	Oral: 680–1000
13	*Bioallethrin* D-*trans*-Allethrin		Mixture of 70% (±)-*trans*, 30% (±)-*cis* Alcohol moiety: (±)	
14	*Barthrin*[32] 6-Chloro-1,3-benzodioxol-5-yl)methyl 2,2-dimethyl-3-(2-methyl-1-propenyl)cyclo-propanecarboxylate		(+)-*trans* (*1R , 3R*) Alcohol moiety: (±) (±)-*cis, trans*	Oral: 1000, IV: 4 Oral: 23.6 ml/kg
15	*Tetramethrin*[33] (1,3,4,5,6,7-Hexahydro-1,3-dioxo-2H-isoindol-2-yl)methyl 2,2-dimethyl-3-(2-methyl-1-propenyl)cyclo-propanecarboxylate Phthalthrin, Neo-Pynamin® Use: hygiene sector, knockdown effect		(±)-*cis, trans*	Oral: >20,000

No.	Name	Stereochemistry	Toxicity
16	*Prothrin*[34] [5-(2-Propynyl)-2-furanyl] methyl 2,2-dimethyl-3-(2-methyl-l-propenyl)cyclo-propanecarboxylate Furamethrin	(±)-*cis, trans*	Mouse oral: 5900[34]
17	*Resmethrin*[35] [5-(Phenylmethyl)-3-furanyl] methyl 2,2-dimethyl-3-(2-methyl-l-propenyl)cyclo-propanecarboxylate Chryson®, Synthrin®, NRDC 104 Use: hygiene	Mixture of 70% (±)-*trans*, 30% (±)-*cis*	Oral: 1500, IV: 160
18	*Bioresmethrin*[35] NRDC 107 Use: hygiene	(+)-*trans (1R, 3R)*	Oral: >8000, IV: 340
19	*Cismethrin* NRDC 119 Use: hygiene	(+)-*cis (1R, 3S)*	Oral: 100, IV: 7
20	*Bioethanomethrin*[36] [5-(Phenylmethyl)-3-furanyl] methyl 3-(cyclopentylidenemethyl)-2,2-dimethyl-cyclopropanecarboxylate	(+)-*trans (1R, 3R)*	Oral: 100, IV: 5–10
20a	[5-(Phenylmethyl)-3-furanyl] methyl 3-[(dihydro-2-oxo-3 (2H)-thienylidene)-methyl] -2,2-dimethylcyclopropane-carboxylate[36a] Kadethrin® RU 15525 Use: hygiene, very good knockdown action	*cis (1R, 3S)*	Oral: 140–1300,[36c] IV: 0.5

TABLE 4. (Continued)

No.	Common Name, *Chem. Abstr.* Name, Trade Names, and Other Designations	Structure	Stereochemistry (Absolute Configuration) in Acid Moiety	Toxicity LD$_{50}$ (mg/kg) Rat[29,30]
20b	*Phenothrin*[36b] (3-Phenoxyphenyl)methyl 2,2-dimethyl-3-(2-methyl-l-propenyl)cyclo-propanecarboxylate		(±)-*cis, trans*	Oral: >10,000[36c]
20c	*Fenpropanate*[36d,36e] Cyano(3-phenoxyphenyl)methyl 2,2,3,3-tetramethylcyclopropanecarboxylate S 3206, SD 41706, WL 41706 Use: good acaricidal effect		No chiral center in acid moiety	—
21	*Permethrin*[30] (3-Phenoxyphenyl)methyl 3-(2,2-dichloroethenyl)-2,2-dimethyl-cyclopropanecarboxylate Pounce®, Ambush® NRDC 143, FMC 33297 Use: cotton, hygiene		Mixture of 70% (±)-*trans*, 30% (±)-*cis*	Oral: 1500–2000
21a	*Cypermethrin*[36f] Cyano (3-phenoxyphenyl)methyl 3-(2,2-dichloroethenyl)-2,2-dimethyl-cyclopropanecarboxylate Ripcord®, NRDC 149 Use: widely active[36e]–mainly in cotton		Mixture of 70% (±)-*trans*, 30% (±)-*cis* Alcohol moiety: (±)	Oral: 500,[36c] IV: 50

22 *Decamethrin*[37]
Cyano(3-phenoxyphenyl)methyl
3-(2,2-dibromoethenyl)-2,2-dimethyl-
cyclopropanecarboxylate
Decis®, NRDC 161, RU 22974
Use: widely active insecticide—mainly in
cotton, hygiene

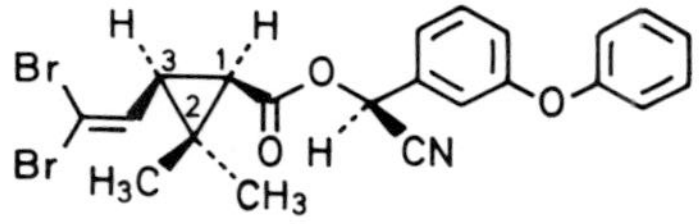

(+)-*cis* (*1R, 3R*)
Alcohol moiety: *S*

Oral: 25–60,[38] IV: 2.5

22a *Fenvalerate*[38a]
Cyano(3-phenoxyphenyl)methyl
4-chloro-α-(1-methylethyl)benzeneacetate
Sumicidin®, Belmark®, Pydrin®, S 5602,
SD 43775
Use: cotton, vegetables

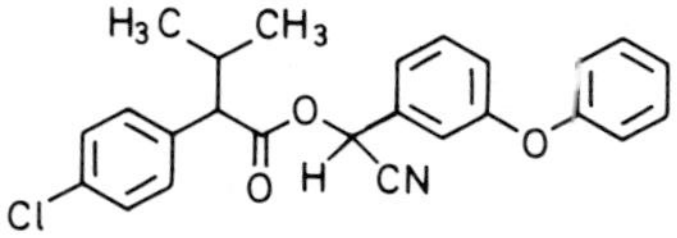

(±), *2RS*
Alcohol moiety: *RS*

Oral: 450,[36c] IV: 75

22b Cyano(3-phenoxyphenyl)methyl
3,3-dimethyl-spiro[cyclopropane-1,1'-
(1H)indene]-2-carboxylate[38b]
Use: veterinary medicine, ticks

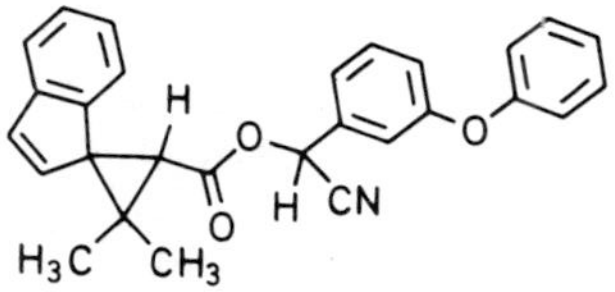

—

—

22c *Cyhalothrin*[38c]
Cyano(3-phenoxyphenyl)methyl
3-(2-chloro-3,3,3-trifluoro-1-propenyl)-
2,2-dimethylcyclopropanecarboxylate
Action: a very widely active insecticide

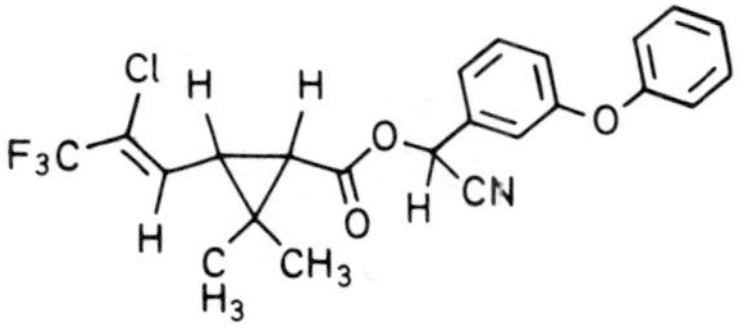

(±)-*cis*, (*1RS*)
Alcohol moiety: *RS*

—

—

TABLE 4. *(Continued)*

No.	Common Name, *Chem. Abstr.* Name, Trade Names, and Other Designations	Structure	Stereochemistry (Absolute Configuration) in Acid Moiety	Toxicity LD_{50}(mg/kg) Rat[29,30]
22d	(3-Phenoxyphenyl)methyl 2,2-dichloro-1-(4-ethoxyphenyl)cyclo-propanecarboxylate[38d]		(±), *1RS*	Mouse subcut.: >16,000[38d]
22e	Cyano(3-phenoxyphenyl)methyl 4-(difluoromethoxy)-α-(1-methylethyl)-benzeneacetate[38e] AC 222705[38f] Action: broad spectrum of activity with a good acaricidal and tickicidal effect		(±), *(2S)* Alcohol moiety: *RS*	Oral: 53–87

does not have the same broad spectrum of activity. Subsequently the natural alcohol components, available only with difficulty owing to their complicated stereochemistry, were replaced by a series of other, synthetic and readily available alcohols (see Table 4, structures **14–19**).

The activities of the chrysanthemates thus prepared are equal to, or even much higher than, those of the natural products. For example, the ester of (+)-*trans*-chrysanthemic acid and 5-(phenylmethyl)-3-furanmethanol (*bioresmethrin*, **18**), which matches the natural pyrethrin I (**6**) particularly well both in topological and electronic configuration, has an activity against *Musca domestica* 50 times higher than **6** (see Table 5).

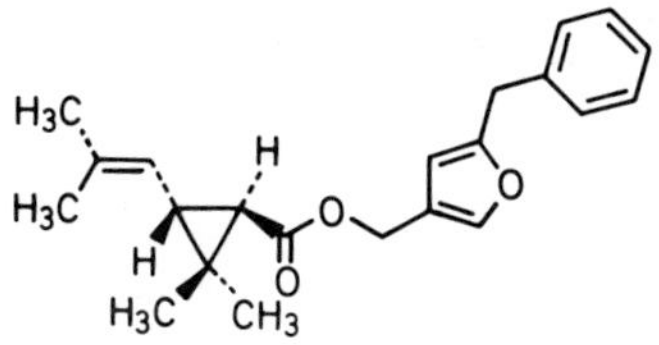

18, *Bioresmethrin*

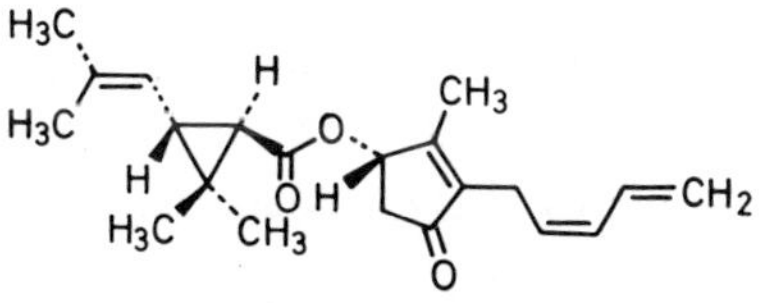

6, *Pyrethrin I*

Chrysanthemic acid was first synthesized by Staudinger and Ruzicka[23] by addition of diazoacetate onto 2,5-dimethyl-2,4-hexadiene. This process, improved by Campbell and Harper,[25] is still operated on a technical scale today.

The most important synthetic accesses to the cyclopropane ring are shown in the following scheme:[19,26]

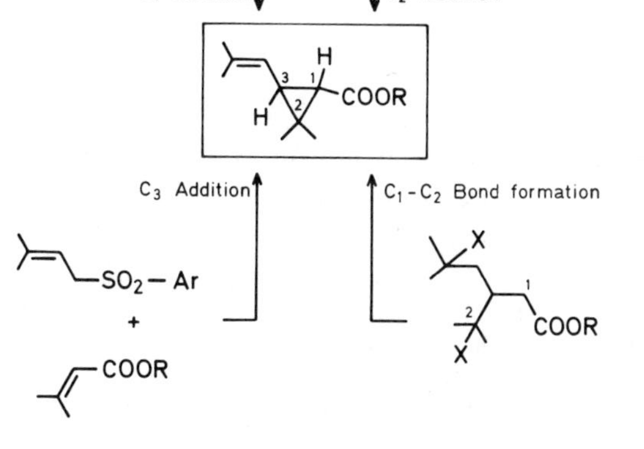

If the isobut-1-en-yl group of chrysanthemic acid is replaced by a 2,2-dihalovinyl unit, pyrethroids may be prepared that, in addition to improved stability toward light, have greatly increased potency.

For example, *permethrin* (**21**, Table 4) has a residual action of several weeks duration even under field conditions, in contrast to *resmethrin* (**17**, Table 4), which only has a short residual action.

Decamethrin (**22**, Table 4) is 1000 times more active than pyrethrin I (**6**, Table 3) against flies and 23 times more active than *bioresmethrin* (**18**, Table 4). This compound, first synthesized by Elliott et al.[27] in 1974, is one of the most active insecticides of all at the present time. The structural characteristics necessary for a high activity have been summarized by Elliott[27] as follows:

A   The geminal methyl groups on the cyclopropane ring are crucial for activity. Compounds without them are inactive.

B   The 5-hydroxy-3-oxo-cyclopentene ring can be replaced by structures of similar stereochemistry.

C   The alcohol component should bear a side chain. This must be unsaturated but can be a 1-alkenyl, 1-cycloalkenyl, or aromatic group.

D   The substituent at C-3 does not have to be isobut-1-en-yl, but can be 2,2-dihalovinyl, for example.

E   The configuration at C-1 and C-3 must be either *1R, 3S* or *1R, 3R*.

Based on the studies carried out up to that time, the general opinion was that the cyclopropanecarboxylate group was essential for the activity of the pyrethroids. However, Japanese chemists at Sumitomo[28] were able to synthesize new insecticidal

(1*R*,3*R*)-Chrysanthemic acid   (S)-α-(1-Methylethyl)benzene = acetic acid

esters by replacement of the cyclopropanecarboxylic acid by the structurally similar α-(1-methylethyl)benzeneacetic acid, for example:

These esters have a spectrum of activity analogous to that of the pyrethroids and, owing to their stability, are also suitable for use in agriculture (see *fenvalerate*, **22a** in Table 4).

The photostable pyrethroids (**21–22a**, Table 4) discovered through the work of Elliott et al. in Rothamsted, England and Ohno et al. at Sumitomo, Japan are the most interesting new products in the insecticide field at the present time. They are characterized by much lower application rates and lower mammalian toxicities compared to traditional insecticides.

Their main use is in cotton, where they have proved to be of special value for the control of *Heliothis*. The application rates here are about one-fifth to one-tenth those of the classical organophosphates and carbamates. Rates of 110–200 g a.i. per hectare have been recommended for *permethrin* (**21**, Table 4) and 12–25 g a.i. per hectare for *decamethrin* (**22**, Table 4).[38g,38h] Possibilities are opening up in many other areas, such as fruit and vegetable growing, hygiene, veterinary medicine, and materials preservation.[36c]

This successful modification of a natural product gave the impulse to a world-wide study of this class of insecticides. Some of the more recent developments are shown in Table 4, **22b–22e**.

2.1.2.3. Mechanism of Action and Toxicity[39–41]

The natural pyrethroids are purely contact poisons that penetrate rapidly into the nerve system and cause the characteristic symptoms in the insect. A phase of exceptional excitation is followed by disturbance to the coordination of movement, paralysis, and finally death. The initial effect is of such rapid onset that within a few minutes the insect is incapable of moving or flying away. This knockdown effect is attained by few insecticides, particularly against flies. The dose necessary for knockdown is usually insufficient to be lethal because the natural pyrethroids are rapidly detoxified in the insect by enzymatic action, and some of them recover. This enzymatic detoxification, which is particularly important in flies, may be delayed by the addition of synergists (see Section 2.1.3). In practice organophosphates or carbamates are frequently added to the pyrethroids to guarantee a lethal effect.

Among the synthetic analogues, *tetramethrin* (**15**, Table 4) in particular is characterized by a knockdown effect greater than that of pyrethrin I and is therefore preferentially used today in the hygiene sector (in aerosols).

In contrast, *resmethrin* (**17**, Table 4) and *permethrin* (**21**, Table 4) have a weaker knockdown effect but a much higher toxicity toward a variety of insects.

Toxicity toward mammals ranges between very wide limits. Even *cis* and *trans* isomers, with almost identical insecticidal properties, can differ tremendously in their toxicities. For example, *bioresmethrin* (**18**, Table 4) has an $LD_{50} \geqslant 8000$ mg/kg rat, oral, acute, and *cismethrin* (**19**, Table 4) has an $LD_{50} = 100$ mg/kg.

The usually low oral toxicity of pyrethroids is due to the fact that most of the dose administered is not absorbed; toxicities are much higher when the doses are given intravenously (see Table 4).

The biological action of the pyrethroids depends on a disturbance to axonic

TABLE 5. Relative Activities of Various Pyrethroids against Flies and Beetle Larvae[24a,30,37]

		Relative Activity	
		Flies	Beetle Larvae
No.	Compound	(*Musca domestica*)	(*Phaedon cochleariae*)
6	Pyrethrin I	2	160
6	Pyrethrin I + synergist	60	
7	Pyrethrin II	4	50
12	*Allethrin*	3	1
17	*Resmethrin*	42	37
18	*Bioresmethrin*	100	100
20	*Bioethanomethrin*	140	170
21	*Permethrin*	60	120
22	*Decamethrin*	2300	1600
22a	*Fenvalerate*	47	100

nerve impulse conduction. In contrast to the organophosphates and carbamates, the precise mechanism of action is unknown (cf. Sections 2.3.3 and 2.4.6).

Detoxification *in vivo* occurs through enzymatic oxidative attack on the *trans*-methyl group of the isobut-l-en-yl residue. The ester bond is very stable, no saponification fragments being found among the metabolites. (See Table 5.)

2.1.3. SYNERGISTS

Pyrethrum offers the best-known example of the possibility of potentiation of an insecticidal compound by a second component that itself has not necessarily any intrinsic activity. This so-called synergism, first observed with a mixture of pyrethrum and sesame oil,[42-44] may be brought about by a number of compounds, some natural and some synthetic, that are usually characterized by the presence of a methylenedioxy group.

That an extension of the knockdown effect of sublethal pyrethroid doses is obtained in the presence of synergists indicates that the synergist has an inhibitory influence on the detoxification process in the insect (see Section 2.1.2.3). The synergist competes with the insecticide and is oxidatively degraded by the enzyme in its stead. The methylenedioxy group is a favored point of attack.[45] The synergistic effect is highly dependent on the nature of the pyrethroid itself and is most marked among the natural pyrethroids. The most important synergists for pyrethroids are listed in Table 6.[45]

2.1.4. ROTENOIDS

Rotenoids are obtained by extraction of the roots of *Derris* and related plants. *Rotenone* (**23,** R = H) was isolated first as the most active insecticidal compound in

TABLE 6. Synergists for Pyrethroids

Structure	Name
	Sesamin
	Sesamolin
	Piperonyl-butoxid
	Tropital®
	Sesamex (ESA) *
	Propyl-isome (ESA)
	Safroxan
	Sulfoxide (ESA)
	Piperonylcyclonene (Mixture, R=H and COOC$_2$H$_5$)
	MGK 264®
	Synepirin 500®
	SKF 5254

*ESA = Entomological Society of America

this class. Other related compounds were found in the resinous extract left after crystallization of rotenone, for example, sumatrol (**23**, R = OH), deguelin (**24**, R = H), α-toxicarol (**24**, R = OH), and elliptone (**25**).[46] These compounds differ from *rotenone* in the substituent R and variations of the E ring.

23

24   25

Syntheses[47] of these compounds have no practical relevance. The mechanism of action of the rotenoids remained unclear for a long time. Today it is known that *rotenone* interferes in the respiratory chain in that it blocks the coupled oxidation of $NADH_2$ and reduction of cytochrome b on the pyruvate side.[48]

Rotenoids are marketed as dusts or resins. As the composition of the product varies according to nature and origin, the toxicological data vary accordingly (LD_{50}: 132–1500 mg/kg rat, oral, acute).

2.1.5. ALKALOIDS

The most well-known insecticide of this class, (−)-*nicotine* (**26**, LD_{50}: 50–60 mg/kg rat, oral, acute),[49] is a nerve poison absorbed dermally. Both its toxicological properties and its good water solubility create problems in the use of *nicotine* in open-field conditions.[50]

In addition to *nicotine*, *nornicotine* **27** and *anabasine* **28** also have achieved some significance.

An insecticide, which is mainly used in the United States as a selective contact and stomach poison against caterpillars and borers, is obtained from the roots and stems of *Ryania speciosa*, a plant native to Trinidad and the Amazon basin. The insecticidal principle is the alkaloid ryanodine **29**.[51]

26   27   28

29

2.1.6. UNSATURATED ISOBUTYLAMIDES[52]

This group of phytoinsecticides contains the N-isobutylamides of unbranched, unsaturated fatty acids as active principle.

$$H_3C-(CH_2)_x-(CH=CH)_y-(CH_2)_2-(CH=CH)_z-CO-NH-CH_2-\underset{\underset{CH_3}{|}}{\overset{\overset{CH_3}{|}}{CH}}$$

Although the natural products are as active as the natural pyrethroids, their synthetic analogues have only a weak activity.

2.1.7. PROTEIN INSECTICIDES

Apart from the pyrethroids, rotenoids, and nicotinoids, insecticides from microorganisms merit special attention. For example, four toxins have been isolated from culture broths of *Bacillus thuringiensis*[53] —designated α-, β-, and γ-exotoxins and δ-endotoxin—that are atoxic toward mammals.[54] These toxins are proteins, the structures of which have not yet been completely elucidated. The commercial products contain δ-endotoxin as the active substance, a highly specific stomach poison for insects that is particularly effective against caterpillars and causes primarily paralysis of the gut.

2.1.8. INSECTICIDES FROM ANIMALS

There are numerous animals that produce insecticidal compounds. The defense poisons of the ant species *Iridomyrmex humilis* (Argentine ant) and *Iridomyrmex nitidus* have been known for many years and are synthetically available,[55] but have not found any practical application.

Nereistoxin (N,N-dimethyl-1,2-dithiolan-4-amine, **30**), however, a product of the marine annelid worm *Lumbriconereis heteropoda*, has proved of promise. The structure was modified successfully to give *cartap* (Padan®, **31**), which is effective against lepidopterous and coleopterous insects and the like.[56] *Cartap* has gained a significant share of the market in Japan.

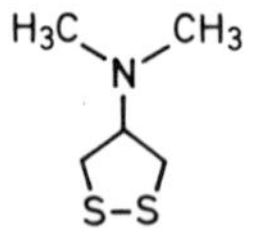

$$H_3C \diagdown N \diagup CH_3$$

30

$$H_3C \diagdown N \diagup CH_3 \qquad \cdot \; HCl$$

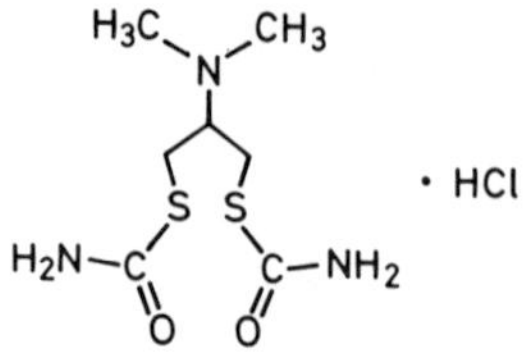

31

Numerous nitrogen-free natural products also have insecticidal properties, but remain however without practical significance.

2.2. *Insecticidal Chlorohydrocarbons*

J. STETTER
Bayer AG, Wuppertal-Elberfeld

2.2.1. GENERAL IMPORTANCE

Insecticidal chlorohydrocarbons have played an important role in the history of chemical crop protection.

DDT, probably the most well-known of all insecticides, was for many years the no. 1 pesticidal agent. The following factors are chiefly responsible for its phenomenal success in the years subsequent to its synthesis and the discovery of its insecticidal activity (Paul Müller, 1939).

1. High insecticidal activity.
2. Low acute mammalian toxicity.
3. Wide spectrum.
4. Simple manufacture and handling.
5. Low price.
6. Long duration of activity.

With certain qualifications these factors are valid for most other insecticidal chlorohydrocarbons.

The past and present importance of *DDT* lies in its suitability for the control of disease-carrying pests, for example, in the WHO programs for malaria control. It has been estimated that almost 1 billion people in all parts of the world have been saved from malaria by the use of *DDT* (for illustrative data, see Table 7). The heavy criticism leveled at *DDT*, and increasingly at most other chlorohydrocarbon insecticides, was sparked off by the behavior of residues from these agents. Their high persistence—usually a desirable property in relation to long duration of activity—led to a demonstrable accumulation in the environment and a concentration in food chains. The tendency of these residues to accumulate is regarded as a special cause for concern, particularly with respect to *DDT*, which was used worldwide on a huge scale.

Restrictions introduced by most Western industrialized countries at the start of the 1970s, particularly against *DDT*, have reduced use of chlorohydrocarbons to a fraction of the original quantities.

The special situation of the Third World countries, however, resulted in production peaks (on a worldwide basis) as late as the mid-1970s. Without sufficient quantities of *DDT* and *dieldrin*, the WHO is unable to fulfill its vector-control programs. (See the appropriate section for the present situation regarding an individual agent.)

The following sections have been divided up according to chemical considerations. Data on biological activity, toxicology, and metabolism of the agents listed are generally given. The literature sources quoted should be consulted for more detailed information.

2.2.2. DDT AND STRUCTURALLY RELATED AGENTS

2.2.2.1. DDT

1,1'-(2,2,2-Trichloroethylidene)bis(4-chlorobenzene); Gesarol®, Neocid®, Chlorphenothane®, Dicophane®, Gesapon®

2.2.2.1.1. History and Significance. First described by O. Zeidler,[57] 1,1'-(2,2,2-trichloroethylidene)bis(4-chlorobenzene) was resynthesized almost 70 years later by Paul Müller as part of a research program at J. R. Geigy A.G. This program was a search for a contact insecticide characterized by a long duration of activity. Compounds of type **1** (stomach poisons for moths) were forerunners on the way to *p,p'*-dichlorodiphenyltrichloroethane (*DDT*) **2**. Müller has written a detailed account of the discovery.[58]

1　　　　　　　　　　　　　　　　　　　　　　　　2

Y = O, S, SO₂, NH etc.

Following the discovery of the pronounced insecticidal properties of the new agent and the registration of the first patents in 1940,[59] the product, formulated in Switzerland, was introduced to the market in the spring of 1942 as Gesarol® for crop protection and Neocid® in the hygiene sector. The epidemic-promoting circumstances of World War II and the post-war years brought about the rapid triumph of *DDT* in the field of medicinal hygiene. Malaria, typhus, typhoid fever, and cholera were drastically reduced by the effective control of *Anopheles* mosquitoes, lice, and flies of all types or, as in the case of malaria, were virtually eradicated in many countries (Table 7).

Following this significant step forward in the control of epidemics, the conferment of the 1948 Nobel Prize in medicine on Paul Müller came as no surprise.

TABLE 7. Incidence of Malaria before and after Use of *DDT*[60]

Country	Year	No. of Cases
Cuba	1962	3,519
	1969	3
Jamaica	1954	4,417
	1969	0
Venezuela	1943	8,171,115
	1958	800
India	1935	>100 million
	1969	285,962
Italy	1945	411,602
	1968	37
Yugoslavia	1937	169,545
	1969	15
Taiwan	1945	>1 million
	1969	9
Ceylon	before 1950	>2 million
	1963	17
	1968[a]	1 million

[a]1963, control program stopped; 1968, restarted.

2.2.2.1.2. Synthesis and Physicochemical Properties. According to the reaction conditions, the condensation of chloral with chlorobenzene under the influence of sulfuric acid or oleum produces *DDT* in yields up to almost 100%.[61]

Technical grade DDT is obtained as a solid, yellowish-white to white mass or powder with a softening point of $\sim 90°C$. Normally it consists of $\sim 70\%$ *p,p'-DDT* (2) and $\sim 20\%$ the weakly insecticidal *o,p'-DDT* (3) as the main by-product.

DDT is reasonably stable to light, air, and acid. Under alkaline conditions, the noninsecticidal 1,1-(dichloroethylidene)bis(4-chlorobenzene) (DDE, 4), is formed readily by dehydrochlorination. Vigorous hydrolysis yields 4-chloro-α-(4-chloro-phenyl)-benzeneacetic acid (DDA, 5; see also Section 2.2.2.1.4).

2.2.2.1.3. Analysis. The Schechter–Haller method[62] is suitable for determination of *DDT* residues:

> Nitration of *DDT* to give the tetranitro derivative. Addition of methanolic sodium methoxide solution and colorimetric determination of the intense blue coloration at 596 nm. The electron-capture gas chromatography may be used for determination of nanogram quantities.[63]

2.2.2.1.4. Biological Properties.

Activity and use pattern. Against the background of a polytoxic activity embracing many kinds of insects, certain focuses of activity can be clearly distinguished. Diptera (flies, gnats) and chewing insects (beetles, caterpillars) are very susceptible to *DDT*, whereas aphids are only slightly susceptible and spider mites not at all. The pronounced contact activity of *DDT* is due to the highly lipophilic character of the compound, which enables it to penetrate the insect cuticle. *DDT* coatings on solid surfaces have a considerable duration of activity. A mixture of *DDT* with pyrethrum in household sprays guarantees a rapid knockdown effect. The great significance of *DDT* in vector control has already been discussed (see Section 2.2.2.1.1).

Toxicology. The acute mammalian toxicity is relatively insignificant: LD_{50}: 250–500 mg/kg rat, oral, acute. A problem is the tendency of *DDT* and its main metabolite DDE to accumulate in the fatty tissue of mammals. The legal restrictions mentioned earlier have already led to a considerable reduction of the residue quantities.[64] Little is known about the possibility of chronic poisoning by *DDT*.

Mode of action. Both Mullins[65] and Holan[66] have developed models of the mechanism of action of *DDT*. According to these models, a disturbance in the sodium balance of the nerve membranes, caused by the "fit" of the *DDT* molecule, is responsible for the poisoning of the insect (*DDT* acts as a nerve poison). Model compounds such as **6** synthesized on the basis of this hypothesis were in fact found to be highly insecticidal.[67] A comprehensive review on the mode of action has been published.[68]

6

Metabolism. The main metabolite of *DDT* is DDE (see also Section 2.2.2.1.2). DDE is also stored in fatty tissue. The hydrophilic DDA, the final step in the metabolic degradation, is excreted by mammals in the urine. 1,1′-(2,2-Dichloroethylidene)bis(4-chlorobenzene) (DDD) is a further main metabolite that has been detected in insect and mammalian organisms (see Section 2.2.2.4: DDD = *TDE* = Rothane®).

Resistance. Buildup of resistance to *DDT* is connected with the enzyme-catalyzed conversion into inactive DDE.[69,70] At least a partial resensitization can be achieved by addition of *N*-dibutyl-4-chloro-benzenesulfonamide.

2.2.2.1.5. ***Present Situation of DDT.*** Starting in Scandinavia, Canada, and the United States, the restrictions on the use of *DDT* have spread to almost all Western industrialized nations. As stated in the introduction, the initial effect was to bring about a shift of manufacture and use to the Third World countries, no significant reduction in the absolute quantities produced being achieved.

The major customer for *DDT* is the World Health Organisation (WHO), which, according to latest reports, is finding its antimalaria program increasingly endangered by a shortage of cheap insecticide.

2.2.2.2. Methoxychlor

$$H_3CO-\!\!\!\bigcirc\!\!\!-\overset{\overset{\displaystyle H}{|}}{\underset{\underset{\displaystyle CCl_3}{|}}{C}}-\!\!\!\bigcirc\!\!\!-OCH_3$$

1,1'-(2,2,2-Trichloroethylidene)bis(4-methoxybenzene); DMDT, Methoxy-DDT, Marlate®

Of the vast number of *DDT* analogues synthesized, *methoxychlor* is probably the most important. The future outlook for *methoxychlor* is very favorable, about 5000 tons being sold in the United States in 1972.[71]

Synthesis and physicochemical properties. Analogously to the manufacture of *DDT*, *methoxychlor* is made by condensation of anisole with chloral under the influence of sulfuric acid or boron trifluoride.[72] It is stable toward oxidation and less easily attacked by alcoholic alkali than *DDT*.

Biological properties. (See also Table 8.) It is not stored in fatty tissue. Owing to its favorable toxicological properties, *methoxychlor* is still used widely today. The estimated lethal oral dose for an adult human is ~450 g!

2.2.2.3. DFDT

$$F-\!\!\!\bigcirc\!\!\!-\overset{\overset{\displaystyle H}{|}}{\underset{\underset{\displaystyle CCl_3}{|}}{C}}-\!\!\!\bigcirc\!\!\!-F$$

1,1'-(2,2,2-Trichloroethylidene)bis(4-fluorobenzene); Gix®, Fluorogesarol®

The fluoro analogue of *DDT* was patented[73] by Hoechst in 1943 and was used in Germany during World War II under the name of Gix® for control of fleas and bedbugs. Synthesized from fluorobenzene and chloral, the product is only of historical interest.

Biological activity. See Table 8.

2.2.2.4. TDE

$$Cl-\!\!\!\bigcirc\!\!\!-\overset{\overset{\displaystyle H}{|}}{\underset{\underset{\displaystyle CHCl_2}{|}}{C}}-\!\!\!\bigcirc\!\!\!-Cl$$

1,1'-(2,2-Dichloroethylidene)bis(4-chlorobenzene) (developed by Geigy[74]); DDD, Rothane® (Rohm and Haas), D 3

TDE was synthesized at Geigy[74] and was introduced to the market by Rohm and Haas as Rothane®. It is synthesized from dichloroacetaldehyde and chlorobenzene. *TDE* is one of the major metabolites of *DDT* (see also Section 2.2.2.1.4).

Biological properties. See Table 8. *TDE* is no longer manufactured.

2.2.2.5. Perthane®

1,1'-(2,2-Dichloroethylidene)bis(4-ethylbenzene); Q 137

Perthane® (Rohm and Haas, 1950) is formed by condensation of dichloroacetaldehyde with ethylbenzene.

Biological properties. See Table 8.

2.2.2.6. Dilan®

1; 1,1'-(2-Nitrobutylidene)bis(4-chlorobenzene)
2; 1,1'-(2-Nitropropylidene)bis(4-chlorobenzene)

Dilan® (Commercial Solvents Corp., 1948)[75] is a mixture of Bulan® (**1**) and Prolan® (**2**) in the ratio 2:1.

Both components are labile toward alkali and sensitive to oxidation. The two-stage synthesis starts with the addition of nitroethane onto 4-chlorobenzaldehyde to give the carbinol **3**, which is then condensed with chlorobenzene to Prolan®. Bulan® is made by the analogous route from 1-nitropropane.

Dilan® is no longer manufactured.

Biological properties. See Table 8.

TABLE 8. DDT Analogues

Product	LD$_{50}$ (mg/kg) Rat, Oral, Acute	Pattern of Activity and Use
Methoxychlor	5000–7000	Ectoparasites (dairy cattle and fat stock), household, in fodder and vegetable crops,[76] for pest control in barns and stables, against cabbage seedpod weevil and cherry fruit fly
DFDT	900	Hygiene and veterinary sector, initial activity superior to *DDT*, not in crop protection (too phytotoxic)
TDE Rothane®	3400	On fruit and vegetable crops, more active than *DDT* against apple leaf roller, tobacco hornworm, caterpillars on brassicas
Perthane®	8170	Ectoparasites, pests in ornamental plant and vegetable growing, in dry cleaning and impregnation of textiles, against moths and carpet beetles
Dilan®	620–1320	Better activity than *DDT* against certain aphid species and the Mexican bean beetle

2.2.3. HEXACHLOROCYCLOHEXANE

A mixture of various stereoisomers of 1,2,3,4,5,6-hexachlorocyclohexane.
Other names: *BHC, HCH*

γ-1,2,3,4,5,6-Hexachlorocyclohexane.
Other names: γ-*BHC*, γ-*HCH*, *Lindane*, Gammexane®

The position of the chlorine atoms on the cyclohexane ring is a,a,a,e,e,e, where a = axial and e = equatorial.

2.2.3.1. History

In 1825 Faraday observed that benzene and chlorine react under the influence of sunlight to give a solid compound.[77] In 1912 van Linden showed the presence of four stereoisomers in the mixture.[78] Bender referred to the insecticidal activity of the hexachlorocyclohexane isomer mixture in 1935.[79] Independently of one another, the Frenchmen Dupire and Raucourt[80] and Slade et al.[81] of ICI discovered the valuable insecticidal properties of hexachlorocyclohexane. At ICI it was shown that the biological activity depended almost completely on the presence of the γ-isomer.

2.2.3.2. Synthesis and Physicochemical Properties

$$\text{benzene} \xrightarrow{3 \text{ Cl}_2, \text{ } h\nu} \text{hexachlorocyclohexane}$$

Chlorine is added onto benzene in a radical chain reaction under UV irradiation.[82] There are numerous variations on the process.[83–88] The isolation of the γ-isomer, which is used almost exclusively in pest control today, requires several concentration and purification steps (concentration of the γ-isomer in the raw product mixture is ~10–18%).

In contrast to the technical product, which has an intense musty smell, *lindane* (γ-HCH—at least 99% pure γ-isomer) is almost odorless. *Lindane* is stable under acidic or neutral conditions; in alkali, hydrogen chloride is eliminated to give 1,2,4-trichlorobenzene as final product.[89] *Lindane* is recovered unchanged after recrystallization from hot nitric acid. Stability toward light and oxidation is high.

Detailed surveys on the stereochemistry of the hexachlorocyclohexane isomers have been published.[90,91]

2.2.3.3. Analysis

A colorimetric method for determination of residual quantities is based on extraction with dichloromethane, dechlorination with zinc in acetic acid, nitration of the benzene so formed to give 1,3-dinitrobenzene, and determination of the colored complex with butanone at 565 nm.[92] Electron-capture gas chromatography is suitable for analysis of nanogram quantities.

An exhaustive description of analytical methods used also for the other insecticidal chlorohydrocarbons is contained in the CIPAC handbook.[93]

2.2.3.4. Biological Properties, Activity and Use Patterns

γ-HCH (*lindane*) is the only hexachlorocyclohexane isomer with pronounced insecticidal properties. The spectrum of activity is similar to that of *DDT*. Higher volatility (vapor pressure in millibars at 20°: *lindane*, 9.4×10^{-6};[94] *DDT*, 1.9×10^{-7}) causes a better initial activity of shorter duration. *Lindane*, a contact, stomach, and respiratory poison, is lethal to chewing and sucking insects but not to spider mites. The vapor pressure and relatively good water solubility (~10 ppm) make *lindane* an excellent soil insecticide. This method of use permits effective control of economically important soil pests (e.g., beetle larvae, wireworms, white grubs, corn rootworms, flea beetles, moth larvae, cutworms, dipterous larvae, Tipula, cabbage root fly, frit fly). The tainting property, even of highly pure *lindane*, prevents use on fruit and vegetable crops, but application in forest crops and cotton growing is wide. Under the name of Jacutin® it serves in veterinary medicine for control of ectoparasites such as ticks and mange mites.

The use of *lindane* is regarded as relatively harmless in view of the following favorable toxicological properties.

Toxicology. The acute mammalian toxicity of *lindane* is somewhat greater than that of *DDT* (LD$_{50}$, rat, oral 76-200 mg/kg). After administration, *lindane* is found in the milk, body fat, and kidneys but is excreted relatively quickly. The danger of an accumulation is very slight. For the technical product, hexachlorocyclohexane, the situation is different. The high chronic and cumulative toxicity of β-hexachlorocyclohexane (present to ~5-14% in the isomer mixture) make the use of the technical product very undesirable.

Metabolism. In mammals, *lindane* is mainly metabolized via various intermediate stages to 1,2,4-trichlorobenzene, which is further converted to isomeric trichlorophenols. Excretion occurs after formation of the glucuronate derivative. In insects degradation occurs mainly via pentachlorocyclohexene and its addition product with glutathione as far as isomeric dichlorothiophenols. Detailed information on the biological properties of *lindane* is contained in a monograph.[95]

2.2.4. TOXAPHENE

Chlorinated camphene with a chlorine content of 67-69%, Octachlorocamphene, *Polychlorocamphene* (USSR), *Camphechlor*, Phenacide®, Phenatox®, Strobane-T®, Toxakil®

Toxaphene is a mixture of chlorinated C$_{10}$ hydrocarbons with 5 to 12 chlorine atoms. The manufacture was patented in 1945.[96] Under the code name Hercules 3956®, it was introduced in 1948 by the Hercules Powder Company as an insecticide.[97]

2.2.4.1. Synthesis and Physicochemical Properties

Toxaphene is prepared by the photochlorination of camphene (the isomerization product of α-pinene from turpentine) in carbon tetrachloride.[96] According to latest investigations the technical product contains about 180(!) different components (GLC-MS analysis).[98] To date the structures of only four components have been elucidated.[99]

The technical product is a yellowish wax with a mild terpene odor. Irreversible decomposition with loss of insecticidal activity occurs under the influence of alkali, UV light, and higher temperatures (>150°C).

2.2.4.2. Analysis

A colorimetric method for determination of microgram quantities requires reaction of the cleaned-up material with diphenylamine–zinc chloride at 205° and measurement of the colored complex at 640 nm (ref. 100, see also ref. 93).

2.2.4.3. Biological Properties, Activity and Use Patterns

Toxaphene has a good activity against biting insects. Its lack of toxicity toward bees permits use in flowering crops. In terms of quantity, *toxaphene*, on its own or as a mixture with other insecticides (e.g., *DDT*), was the most important insecticide in cotton growing in the mid-1970s. *Toxaphene* is still used widely in tropical countries against ectoparasites on sheep and cattle. For a time its rodenticidal property was put to use for controlling field mice and voles. Since May 1971 the use of *toxaphene* has no longer been permitted in Germany for control of agricultural and stored product pests.

    Toxicology. (LD$_{50}$: 40–120 mg/kg rat, oral, acute). The product is stored in the body fat of mammals (review[101]). Skeletal damage in fish has been reported recently (broken backbone syndrome).[102]

2.2.5. DIMERIZATION PRODUCTS OF HEXACHLOROCYCLOPENTADIENE

Hexachlorocyclopentadiene (HCCP) is the key substance for the synthesis of numerous highly active insecticides, which are dealt with in this and the subsequent section.

2.2.5.1. Synthesis of HCCP

Reaction of cyclopentadiene with sodium hypochlorite[103] or the direct chlorination[104] are of little significance today. The current technical synthesis is based on straight-, branched-chain, or cyclic $C_5H_{12}(C_5H_{10})$ hydrocarbons, which are subjected to high-temperature chlorination on suitable contacts (fuller's earth) to give HCCP via various intermediates.[105]

$$n, \textit{iso}-C_5H_{12}\,(C_5H_{10}) \xrightarrow[\text{[various intermediates]}]{Cl_2}$$

2.2.5.2. Chlordecone

Decachlorooctahydro-1,3,4-metheno-2H-cyclobuta[c,d]pentalen-2-one; Kepone®, GC-1189
(IUPAC name: decachloro-5-oxo-pentacyclo[5.3.0.0^{2,6}.0^{3,9}.0^{4,8}]decane)

Synthesized in 1950 at the Allied Chemical Corp. and patented[106] under an incorrect structure, *chlordecone* was introduced as an insecticide and acaricide under the trade name Kepone® in 1958. The correct structure was published by McBee.[107]

2.2.5.2.1. Synthesis and Physicochemical Properties.
Under sulfonating conditions (fuming sulfuric acid, chlorosulfonic acid, SO_3), HCCP forms a dimerization product containing sulfonic acid groups, which is hydrolyzed to the stable Kepone hydrate.[106] Chlorination with phosphorous(V) chloride yields the perchlorinated cage compound identical with *mirex* (see also Section 2.2.5.4).

The carbonyl group in Kepone® readily forms solvates with water, alcohols, and amines.

2.2.5.2.2. Biological Properties, Activity and Use Patterns.
Kepone® has a weak contact activity but is a highly effective stomach poison. It is mainly used in bait formulations against cockroaches and ants.[108] Kepone® is also effective as a larvicide and an acaricide.

Toxicology. (LD_{50}: 95–140 mg/kg rat, oral, acute). The only manufacturer of Kepone® (Life Science Products) ceased production in 1975 after examinations of plant workers showed blood levels of 2–25 ppm Kepone®.[109]

2.2.5.3. Kelevan

Ethyl 1,1a,3,3a,4,5,5,5a,5b,6-decachloroocatahydro-2-hydroxy-γ-oxo-1,3,4-metheno-1H-cyclobuta[c,d]pentalene-2-pentanoate; Despirol®, GC-9160 {IUPAC name: decachloro-5-(4-ethoxycarbonyl-2-oxo-butyl)-5-hydroxypentacyclo[5.3.0.0.^{2,6}.0^{3,9}.0^{4,8}] decane}

At the beginning of the 1970s a derivative of Kepone® was introduced by Allied Chemical Corporation as an insecticide under the trade name Despirol® (patents[110,111]).

2.2.5.3.1. Synthesis and Physicochemical Properties. Addition of ethyl laevulinate to anhydrous Kepone® under reflux conditions in toluene or xylene gives Despirol® in very good yield.[110]

$$H_3C-CO-CH_2-CH_2-COOC_2H_5 \longrightarrow$$

$$-CH_2-CO-(CH_2)_2-COOC_2H_5$$

The technical product has a brownish color and suffers hydrolysis with relative ease under acidic or basic conditions.

2.2.5.3.2. Biological Properties. Despirol® is a stomach poison mainly used for control of Colorado beetle. It also has acaricidal properties but is atoxic toward bees (review[112]).

Toxicology. (LD$_{50}$: 240–290 mg/kg rat, oral, acute).

2.2.5.4. Mirex

Dodecachlorooctahydro-1,3,4-metheno-1H-cyclobuta[*c,d*]pentalene: GC 1283, Dechlorane®, (IUPAC name: dodecachloropentacyclo[5.3.0.0^{2,6}.0^{3,9}.0^{4,8}]decane).

First described by Prins,[113] *mirex* was patented[114] by the Allied Chemical Corporation in 1954 and introduced in 1959 as an insecticide.

2.2.5.4.1. Synthesis and Physicochemical Properties. Manufacture of *mirex* is based on the smooth dimerization of HCCP in dichloromethane or carbon tetrachloride under the influence of aluminum chloride.

This cage compound is characterized by an incredibly high melting point (485°) and exceptional chemical stability.

2.2.5.4.2. Biological Properties, Activity and Use Patterns. *Mirex* has weak contact activity but is a highly effective stomach poison. It is mainly used in bait formulations against various ant species, chiefly against the fire ant.

Toxicology. *Mirex* has a medium acute toxicity (LD_{50}: 235–702 mg/kg rat, oral, acute). It is relatively safe for birds, fish, and crustaceans. A distribution phenomenon that serves to emphasize the extreme persistence of *mirex* has been observed in Canada.[115]

2.2.6. CYCLODIENE INSECTICIDES

The term *cyclodiene insecticide* is used for an important group of chlorohydrocarbons having a basic structure constructed according to the principle of the Diels–Alder reaction. In honor of K. Alder and O. Diels, two of the most important representatives bear the names Aldrin and Dieldrin.

With one exception (*endrin*), hexachlorocyclopentadiene (HCCP) serves as the diene component. Reaction partners are acyclic, monocyclic, and bicyclic dienophiles.

Although reaching market maturity somewhat later than *DDT* or *lindane*, the agents discussed in the following represent a group of extremely active insecticides of considerable commercial significance. Residue problems have, however, tended to restrict their use in recent times.

2.2.6.1. Adducts of Hexachlorocyclopentadiene with Acyclic and Monocyclic Dienophiles

2.2.6.1.1. Chlordane. Common name: AG-Chlordane; trade name: Belt plus®.[116]

1,2,4,5,6,7,8,8-Octachloro-2,3,3*a*,4,7,7*a*-hexahydro-4,7-methano-1H-indene; Velsicol 1068®, Octachlor®, Belt®, Chlor-Kil®, Synchlor®

Chlordane was synthesized in 1944 by J. Hyman of Velsicol Chemical Corporation.[117] The insecticidal properties were first described by Kearns, Ingle, and Metcalf.[118] Pioneer work in the area of cyclodiene chlorohydrocarbons also originates from Riemschneider's group.[119]

2.2.6.1.1.1. Synthesis and Physicochemical Properties. HCCP **1** and cyclopentadiene **2** combine in an exothermic reaction to the endo Diels–Alder adduct **3** (chlordene), which is converted to *chlordane* **4** by chlorination either with elemental chlorine in refluxing carbon tetrachloride[117] or with sulfuryl chloride in the presence of iron(III) chloride.[120]

Technical *chlordane* is a brown, viscous liquid with a campherlike odor. Almost

Cl_2 or SO_2Cl_2

insoluble in water, it is, however, infinitely miscible with most organic solvents including mineral oils. In terms of volatility, *chlordane* lies between *DDT* and *lindane* (see also Table 9). *Chlordane* is stable toward acid but loses hydrogen chloride under the influence of alkali, forming insecticidally inactive products.

2.2.6.1.1.2. Composition of Technical Chlordane. Technical *chlordane* is a mixture of various compounds. At least 14 components can be detected by gas liquid chromatography (Polen,[121] Saha and Lee[122]).

Riemschneider,[123] Vogelbach,[124] and March[125] have done pioneering work on the isolation and identification of the compounds in the mixture. Two octachloro isomers ($C_{10}H_6Cl_8$), a heptachloro derivative ($C_{10}H_5Cl_7$), and a nonachloro derivative ($C_{10}H_5Cl_9$) have been determined as the main components.

The technical product consists of a maximum of 45% of the two octachloro isomers together. α-*Chlordane* **1** is the *trans* isomer with 1-exo, 2-endo positioning of the chlorine atoms; in β-*chlordane* **2** the chlorine atoms have the *cis* (1-exo, 2-endo) orientation. It has been shown that by specific *trans* chlorination of chlordene with trichloromethanosulfenyl chloride, pure α-*chlordane* **1** is formed (Büchel et al.).[126] The observation that β-*chlordane* is formed both by chlorination of α-dihydroheptachlor **3** and of β-dihydroheptachlor **4** proves the configuration of the chlorine atoms in β-*chlordane* **2**.

1

The same conclusion is drawn from the results of dehydrohalogenation with alcoholic sodium hydroxide (Christol).[127] The third main component (~10%) in technical *chlordane* is *heptachlor*.

In nonachlor (~7%) the three chlorine atoms in the cyclopentane ring have a *cis*-exo configuration (Cochrane).[128]

Continuous improvements in the process have recently led to a product containing 70% β- and 25% α-*chlordane*.

2.2.6.1.1.3. Analysis. Residues may be determined colorimetrically by reaction of *chlordane* with methanolic potash and diethanolamine and measurement of the absorption maximum at 521 nm.[129] Gas chromatography with electron-capture detector is also suitable.[130]

2.2.6.1.1.4. Biological Properties, Activity and Use Patterns. Chlordane is a broadly active contact, stomach, and respiratory poison especially suitable for control of soil pests (white grubs, wireworms, Tipula larvae). In Germany, where use of *chlordane* in agricultural crops is no longer permitted (since May 1971), it was mainly used in bait formulations against mole crickets. In the United States the main use is in cotton growing and for control of grasshoppers.

Toxicology. (LD_{50}: 283–590 mg/kg rat, oral, acute). There are clear indications of a chronic and cumulative toxicity (review[131]). *Chlordane* is stored in body fat and lipoid-containing organs.

Metabolism. Via dehydrohalogenation and hydroxylation steps *chlordane* is converted by mammals and insects into hydrophilic metabolites, which are then excreted.

2.2.6.1.2. Heptachlor. Hyman's original patent on the synthesis of *chlordane*[117] contains an experiment on the radical chlorination of *chlordane* with sulfuryl chloride in which the main product is *heptachlor*.

1,4,5,6,7,8,8-Heptachloro-3*a*,4,7,7*a*-tetrahydro-4,7-methano-1H-indene; Velsicol 104®, Hepta-Klor®, Drinox®, Heptamul®

2.2.6.1.2.1. Synthesis and Physicochemical Properties. Apart from the process used today for the manufacture of *heptachlor*–chlorination of chlordene under exclusion of light in benzene solution in the presence of fuller's earth[132]–there are

various multistage processes described in the patent literature that mostly yield pure *heptachlor*.[133,134]

The technical product, a waxy, brown solid, contains ~72% *heptachlor*, the remainder being mostly *α-chlordane*. *Heptachlor* is exceptionally stable to light, air, acid, and base.[135]

2.2.6.1.2.2. Analysis. Treatment of a benzene solution of *heptachlor* with ethanolamine and potassium hydroxide in glycol monobutylether gives a violet coloration, the absorption maximum of which is measured at 560 nm.[136] Gas chromatographic determination is also possible.[137]

2.2.6.1.2.3. Biological Properties, Activity and Use Patterns. *Heptachlor* is a broadly active stomach and contact poison with an activity superior to that of technical *chlordane* in many cases (e.g., flies, cockroaches). It is mainly used as a soil insecticide against wireworms, corn rootworms, cutworms, vegetable flies, white grubs, and pygmy mangold beetles. In contrast to *chlordane, heptachlor* is also used for seed dressing and coating. *Heptachlor* is inactive against scale insects and spider mites.

Toxicology. (LD_{50}: 90–135 mg/kg rat, oral, acute). *Heptachlor* is stored in the body fat and excreted in the milk in the form of an epoxide (review[130]).

Metabolism. The *in vivo* oxidation of *heptachlor* to the epoxide **1** was first found in rats and dogs[138]. The epoxide is more insecticidal than *heptachlor* itself. Poisoning symptoms in insects appear parallel with the formation of the epoxide.[139]

1

2

The hydrophilic metabolite **2** seems to be the last step in the breakdown of *heptachlor* in the mammalian organism.[140] Soil microorganisms degrade *heptachlor* principally to 1-hydroxychlordene.[141]

2.2.6.1.3. Compounds Similar to Heptachlor. Apart from *heptachlor* only the monofluorinated analogue possesses comparable insecticidal activity. Owing to favorable toxicological properties (LD_{50}: 5000 mg/kg rat, oral, acute) β-dihydro-heptachlor **3**, made by addition of hydrogen chloride to chlordene,[142] was under development at Shell for a long time as an interesting experimental insecticide.

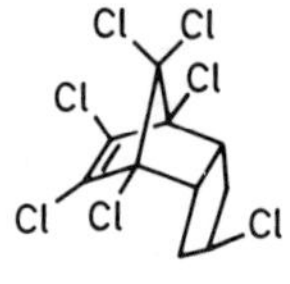

3

2.2.6.1.4. Isobenzan.[143] Problems associated with a high toxicity toward mammals were the reason for cessation of production in 1965.

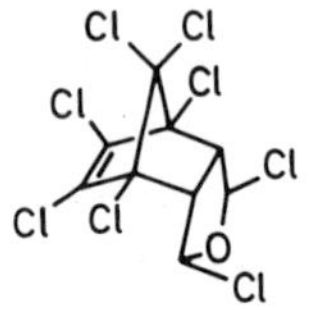

1,3,4,5,6,7,8,8-Octachloro-1,3,3a,4,7,7a-hexahydro-4,7-menthanoisobenzofuran; Telodrin® (Shell), R6700, SD 4402

2.2.6.1.4.1. Synthesis and Physicochemical Properties. Continuous Diels–Alder addition of HCCP **1** to 2,5-dihydrofuran in a high boiling solvent (C_{12}–C_{18} hydrocarbon) at temperatures of 120–180° affords the adduct **2** in very high yield.[143,144]

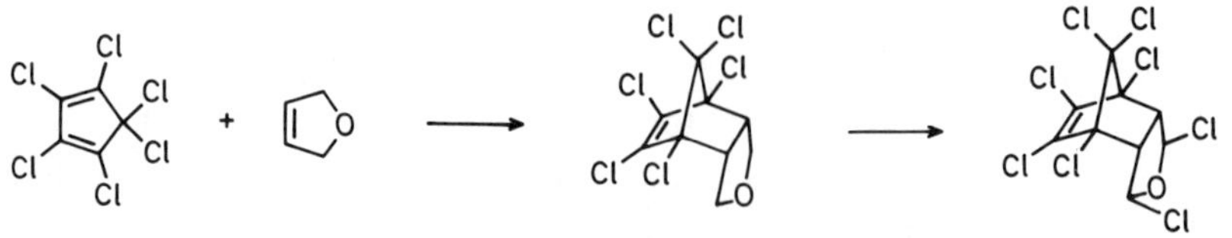

1   **2**

Conversion of structure **2** into *isobenzan* by chlorination under UV irradiation occurs almost quantitatively.[145,146] *Isobenzan* is unstable under alkaline conditions (see also Table 9).

2.2.6.1.4.2. Biological Properties, Activity and Use Patterns. Telodrin® is the most active of the cyclodiene insecticides. It has outstanding properties as a soil insecticide (duration of activity approximately 5 months). Major areas of use include control of insect pests in cotton, cane sugar, corn (maize), coffee, cocoa, and tobacco plantations.

Toxicology and metabolism. (LD_{50}: 7–8 mg/kg rat, oral, acute). The product is converted by the mammalian organism (rat) to the lactone **1** via a hydrophilic intermediate (LD_{50} of **1**: 306 mg/kg rat, oral, acute).

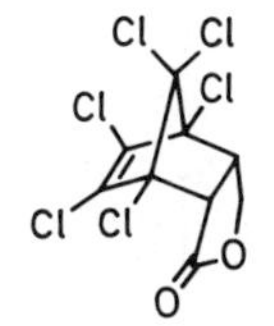

1

2.2.6.1.5. Endosulfan.

2.2.6.1.5. Endosulfan. The insecticidal properties were first described by Finken-brink.[150] Since 1950 *endosulfan* has been on the market under the name Thiodan®. Favorable toxicological findings and a low persistence in comparison to most other chlorohydrocarbon insecticides make the use of *endosulfan* still look interesting.

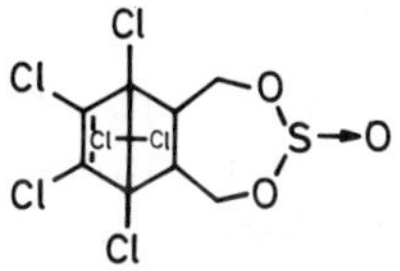

6,7,8,9,10,10-Hexachloro-1,5,5*a*,6,9,9*a*-hexahydro-6,9-methano-2,4,3-benzodioxathiepin-3-oxide (Hoechst 1954)[147-149]

Thiodan®, Malix®, HOE 2671, Cyclodan®, Thimul®, Thifor®

2.2.6.1.5.1. Synthesis and Physicochemical Properties. "Thiodandiol" **4**, obtained by saponification of the Diels–Alder adduct **3** from HCCP **1** and *cis*-1,4-diacetoxybutene-2, is converted into technical *endosulfan* **5** in high yield by heating with thionyl chloride in xylene.[151]

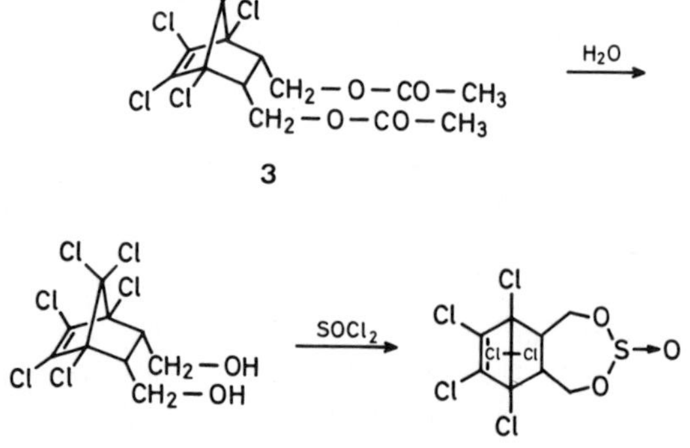

1

3

4

5

Technical *endosulfan* is a mixture of two isomers differing from one another only in the position of the sulfite group. *α-Endosulfan* (70%) has the structure **6**; *β-endosulfan* (30%) the structure **7**, the final structural elucidation being achieved with the aid of IR and NMR spectroscopy.[152]

6   7

Both *endosulfan* isomers yield the corresponding cyclic sulfate upon oxidation. The sulfate also appears as a metabolite of *endosulfan*.

Endosulfan is slowly hydrolyzed back to the "thiodandiol" **4** by the action of aqueous acid or base.

2.2.6.1.5.2. Analysis. For a review of macro- and microanalytical methods see the literature in ref. 153. Gas chromatography with electron-capture detector is suitable for residue analysis.

2.2.6.1.5.3. Biological Properties, Activity and Use Patterns. *Endosulfan* is a broadly active contact and stomach poison and has found wide application in ornamental plant growing, agriculture, and forestry. Its lack of toxicity toward beneficials is of special significance. In the correct application quantities it is innocuous toward bees, wood ants, spiders, lady bugs, and chalcids. Owing to its lack of toxicity toward bees, *endosulfan* is very suitable for use in flowering crops such as rape, brassicas, clover, and lucerne, as well as on fruit crops and for the control of cockchafer.

Toxicology. [LD_{50}: 100 mg/kg rat, oral, acute (*α-endosulfan*: 76 mg/kg; *β-endosulfan* 240 mg/kg[154]).] *Endosulfan* is rapidly broken down in the organism. There is no danger of accumulation. Long-term feeding trials gave no indications of a chronic toxicity.

Metabolism. At least some of the agent ingested by mammals is excreted unchanged. The corresponding cyclic sulfate and "thiodandiol" **4** are isolated as the main metabolites. Insects also metabolize *endosulfan* via hydrophilic intermediates to the lactone **1** above.[155]

2.2.6.1.6. Alodan®. Alodan® is formed by Diels–Alder reaction of HCCP and *cis*-1,4-dichlorobutene-2.[156] The extremely low mammalian toxicity (LD_{50}: 15,000 mg/kg rat, oral, acute) made Alodan suitable for use in the veterinary sector against ectoparasites. It was replaced by Bromodan®.

5,6-bis[Chloromethyl]-1,2,3,4,7,7-hexachloro-bicyclo[2.2.1]hept-2-ene; Chlorocyclen (ISO common name) (Hoechst, 1956)

2.2.6.1.7. Bromodan®. Bromodan® is synthesized by Diels–Alder reaction of HCCP and allyl bromide.[157] As successor to Alodan® it is used against ectoparasites and as an insecticide in stored-product protection. The low mammalian toxicity (LD_{50}: 12,900 mg/kg rat, oral, acute) and the fact that it is very well tolerated are worthy of note.

5-(Bromomethyl)-1,2,3,4,7,7-hexachloro-bicyclo[2.2.1]hept-2-ene; Alugan®, Bromocyclen (ISO common name)

2.2.6.2. Adducts of Hexachlorocyclopentadiene with Bicyclic Dienophiles

2.2.6.2.1. Aldrin. Aldrin is the common name for the product of Diels–Alder reaction between hexachlorocyclopentadiene (HCCP) and bicyclo[2.2.1]hepta-2,5-diene (norbornadiene) introduced to the market in 1948 as an insecticide under the trade name Octalene® (J. Hyman Co.). For a review of insecticidal properties see the literature in ref. 158.

1,2,3,4,10,10-Hexachloro-1α,4α,4aβ,5α,8α,8aβ-hexahydro-1,4:5,8-dimethanonaphthalene; HHDN (pure product), Compound 118, Octalene®, Drinox®, Aldrite®, Aldrosol®

2.2.6.2.1.1. Synthesis and Physicochemical Properties. Norbornadiene (**1**) is obtained in good yield by diene synthesis from cyclopentadiene and excess acetylene at pressures of 1–20 atm and temperatures of 150–400°.[159]

Thermal Diels–Alder reaction of excess norbornadiene—to avoid formation of a bis adduct—and HCCP in refluxing toluene[160] yields the technical product with a content of at least 95% **2** (abbreviation: *HHDN*) (for physical data see also Table 9).

Aldrin is stable toward higher temperatures, alkali, and dilute acid. Concentrated acid and oxidizing agents attack the double bond in the unchlorinated ring. *Aldrin* has a reasonable stability on storage and is miscible with most common crop protection agents and fertilizers.

Whereas the chlorinated part of the molecule is chemically almost inert, numerous reactions are possible at the double bond of the unchlorinated ring. The addition of halogen acids, halogens, epoxidation (see Section 2.2.6.2.2), Diels–Alder reaction, and dipolar cycloaddition reactions serve as examples.

2.2.6.2.1.2. Analysis. Electron-capture GLC[161] or a colorimetric method[162] are used for residue analysis. The latter involves coupling of the reaction product from *aldrin* and phenyl azide with diazotized 2,4-dinitroaniline and measurement of the maximum at 515 nm.

2.2.6.2.1.3. Biological Properties, Activity and Use Patterns. Owing to its relatively high vapor pressure (see also Table 9), *aldrin* is preferentially used as a soil insecticide in granulate and fertilizer formulations. The agent has excellent contact, stomach, and respiratory toxic properties. *Aldrin* is only used as a foliar insecticide in those cases where a short duration of activity is desired. Some of the crops in which *aldrin* guarantees effective pest control include corn (maize), potatoes, sugar beet, cane sugar, and bananas.

Toxicology. (LD_{50}: 67 mg/kg rat, oral, acute). Administration to rats of 0.5 ppm in the daily fodder over a period of 2 years led to liver enlargement. *Aldrin* acts primarily as a nerve poison, as do the other cyclodiene insecticides. A high solubility in blood ensures distribution into all tissues, particularly into fatty tissue.[163]

Metabolism. The activation of *aldrin* to *dieldrin* by enzymatic oxidation has been shown in mammals, insects, microorganisms, and on plants. Hydrophilic metabolites are also formed, and are excreted in the urine as glucuronic acid adducts. (For an exhaustive survey of the metabolism of cyclodiene insecticides see the literature refs. 164 and 165.)

2.2.6.2.2. Dieldrin. Numerous experiments to reduce the volatility of *aldrin* by chemical variation and thus improve the duration of activity led in 1948 to epox-

3,4,5,6,9,9-Hexachloro-1aα,2β,2aα,3β,6β,6aα,7β,7aα-octahydro-2,7:3,6-dimethanonapth[2,3-*b*]oxirene; HEOD (pure product), Alvit®, Dieldrite®, Octalox®

idation and hence to the synthesis of *dieldrin*.[166] A short time later *dieldrin* was brought onto the market as "Compound 497" under the trade name Octalox®.

2.2.6.2.2.1. Synthesis and Physicochemical Properties. Of the two possible synthetic paths to *dieldrin*, A and B, the latter has succeeded in practice.

Hydrogen peroxide–acetic anhydride or peracids such as peracetic or perbenzoic acids serve as epoxidizing agents.

The technical product thus obtained in about 90% yield contains at least 85% pure compound (*HEOD*). The vapor pressure of *dieldrin* is significantly lower than that of *aldrin* (see Table 9).

Dieldrin is stable toward alkali and dilute acid. Concentrated acid cleaves the epoxide ring.

2.2.6.2.2.2. Analysis. As in the case of *aldrin*, electron-capture GLC and a modified colorimetric method using phenyl azide can be used for determination of residues. Detailed information on analytical methods for *aldrin*, *dieldrin*, and *endrin* is contained in a handbook from the Shell Chemical Corporation.[167]

2.2.6.2.2.3. Biological Properties, Activity and Use Patterns. Low volatility and chemical stability make *dieldrin* an insecticide with a very good duration of activity (see Table 9). In comparison to *DDT*, the higher insecticidal potency of *dieldrin* permits far lower application quantities. In agriculture its use as a seed dressing predominates. Dieldrin is also used in combination with other insecticides in forest crops. Owing to its persistence *dieldrin* is especially suited for the control of crawling pests (crickets, cockroaches, etc.). *Dieldrin* was, and still is, used in vast quantities in vector programs for the control of *Anopheles* mosquitoes and tsetse flies. As in the case of *DDT*, the possibilities for application are limited by buildup of resistance.

Toxicology. (LD_{50}: 40–87 mg/kg rat, oral, acute). As with most other cyclodienes, *dieldrin* is absorbed easily through the skin and stored in organs rich in lipids. Excretion in the milk of mammals has been proven. Two-year feeding trials on rats produced liver damage.

Metabolism. A *trans*-dihydroxydihydroaldrin appears as a major metabolite in the mammalian organism. The structures of other hydrophilic metabolites are mostly unknown (see also refs. 164 and 165).

2.2.6.2.3. Endrin. *Endrin* was introduced to the market in 1951 by the J. Hyman Company as "Experimental Insecticide 269." A brief description of its insecticidal properties is to be found in the original patent.[166] Later *endrin* was manufactured and sold by two independent companies (Shell Development Co. and Velsicol Chemical Corp.).

3,4,5,6,9,9-Hexachloro-1α,2β,2aβ,3α,6α,6aβ,7β,7aα-octahydro-2,7:3,6-dimethanonaphth [2,3-*b*]oxirene; Compound 269, Hexadrin®

2.2.6.2.3.1. Synthesis and Physicochemical Properties. Epoxidation of the *aldrin* isomer, *isodrin* (2), leads to *endrin* (3). Isodrin is formed by the Diels–Alder reaction of hexachloronorbornadiene (1) (from "hex" and vinyl chloride or acetylene[169]) and cyclopentadiene.[168]

Endrin is a stereo-isomer of *dieldrin* with similar physicochemical properties (see also Table 9) and comparable stability toward alkali and dilute acid. An interesting difference that originates in the endo-endo structure of *endrin* is the possibility

TABLE 9. Physical Data of the Cyclodiene Insecticides

	mp of Tech. Product (°C)	Vapor Press. (mmHg/25°C)	Stability
Chlordan	Viscous oil	10^{-5}	Alkali labile
Heptachlor	46–74	4×10^{-4}	Stable to alkali at 27°C
Isobenzan	120–125	3×10^{-6}	Alkali labile
Endosulfan	70–100	Not measurable	Acid and alkali labile
Aldrin	~60	6×10^{-6}	Largely stable toward acid and alkali
Dieldrin	~150	1.8×10^{-7}	Similar to aldrin
Endrin	200 (dec.)	2×10^{-7}	Similar to aldrin, skeletal rearrangement above 200°C

of a thermal or photochemical rearrangement into the "half-cage ketone" **4** and the pentacyclic aldehyde **5**.[170,171] Both products are insecticidally inactive.

4   5

2.2.6.2.3.2. Analysis. Electron-capture GLC and the colorimetric determination of the 2,4-dinitro-phenylhydrazone of the rearrangement product **4** are useful methods for residue analysis.[167]

2.2.6.2.3.3. Biological Properties, Activity and Use Patterns. Endrin is mainly used alone or in combination with other insecticides (*DDT, toxaphen, mono-*, and *dicrotophos*) throughout the world in cotton growing. Approximately 70% of the quantity produced is applied in cotton; ~12% in rice growing.[172] As a foliar insecticide, *endrin* is definitely more active than *dieldrin* against aphids, caterpillars, and sucking insects. Some spider mite species are also killed. *Endrin* is also used as a rodenticide (field mice and voles).

Toxicology. Endrin, in company with *isobenzan*, is the cyclodiene insecticide with the highest mammalian toxicity (LD_{50}: 7.5–17.5 mg/kg rat, oral, acute). It is excreted much faster than *aldrin* or *dieldrin* by mammals, and thus storage in fatty tissue only takes place to a slight extent.

Metabolism. Analogous to *aldrin* and *dieldrin*, *endrin* is also converted mainly to hydrophilic metabolites by the mammalian or insect organism. The "half-cage ketone" **4** seems to appear as a metabolite in houseflies.[173]

2.2.6.2.4. Present Situation of the Cyclodiene Insecticides Aldrin, Dieldrin, and Endrin. In Germany the use of *aldrin* and *dieldrin* has been severely restricted since 1971. Use of *dieldrin* for agricultural purposes is no longer permitted. *Aldrin* may no longer be applied to plant parts. Use of *endrin* is completely forbidden. Owing to studies that indicated a relationship between *dieldrin* intake and an increased incidence of tumors in rats and mice, the manufacture and sale of *dieldrin* was temporarily forbidden in the United States.[174] In general, the importance of the cyclodiene insecticides is diminishing rapidly, particularly with the advent of the synthetic pyrethroids. The latter are regarded much more favorably in ecological terms and in view of their extreme potency.

2.3. Organophosphorus Insecticides

Ch. FEST AND K.-J. SCHMIDT
Bayer AG, Wuppertal-Elberfeld

2.3.1. INTRODUCTION

There are many instructive facets to the organophosphate insecticides. Following Lassaigne's first report in 1820 on the reaction of phosphoric acid with alcohol, an "incubation time" of more than 100 years, during which these compounds were studied almost exclusively under the terms of reference of pure science, ended with Schrader's discovery in the 1930s of their insecticidal applications.[176]

Schrader's discovery came at a time when the ever-increasing needs of a rapidly expanding agricultural industry were exceeding the production of natural insecticides (e.g., nicotine, rotenone, pyrethrum). In the subsequent period of applied research, there was a stormy development in organophosphates, which also induced new basic research in areas such as biology, biochemistry, physiology, toxicology, and entomology. In almost 35 years an annual production of ~150,000 tonnes has been reached. On the one hand, this shows that a differentiation between pure and applied research is not very fruitful, and on the other, that the time must be ripe for great developments, that research targets must be relevant to the needs of mankind.

In the case of crop protection these needs are easily defined. They are nothing other than ensuring the nourishment of a rapidly increasing world population that has passed the 4 billion mark! A withdrawal of crop protection no longer means just a small forfeit to competing pests but the loss of whole harvests and corresponding famine catastrophes. In spite of these consequences, one condition must always be maintained—the means used to safeguard crops should be kept in proportion.

Owing to their ester nature, the organophosphates offer fundamental advantages in this respect. Normally they can be easily degraded hydrolytically, enzymatically, or biologically.

A further advantage lies in the very low application quantities necessary for the desired insecticidal activity in the field. As a rule, only a few hundred grams per hectare and per application are recommended, which thus considerably reduces the danger of undesirable residues in the harvested product.

From a chemist's viewpoint, the most valuable property of the organophosphates is the sheer diversity of the combinations of substituents possible at the central phosphorus atom. This permits precise variation of the biological activity and toxicological, physical, and chemical properties within certain limits.

In the meantime, it is known that insecticides that interfere with central mechanisms in order to exert their effect must contain "weaknesses" to enable the mammalian organism to detoxify them in case of involuntary ingestion.

These aspects demonstrate that research cannot be wound up with an ideal product but, rather, that the state of the art must be continuously improved in the interest of safer products. The variability of the insecticidal organophosphates offers great opportunities for the future in this respect.

2.3.2. GENERAL STRUCTURE

Organophosphates having insecticidal activity may, almost without exception, be traced back to a basic structure that is defined by Schrader's acyl formula and has entered the literature under this name.

$$R^1 \diagdown \atop R^2 \diagup P \diagup{\diagup}^{O\,(S)} \diagdown_{Acyl}$$

This formula was the first attempt to demonstrate a relationship between structure and activity. Experience gained over the last 40 years has shown that it is still valid today. It has often been modified, for example, by the P-XYZ scheme, which did not, however, signify any progress. Acidity and pK values have also been related to activity (see Section 2.3.2.3.3) without decisive predictions resulting.

Schrader predicted an active phosphorus derivative in the following way:[176]

It is likely that a biologically active phosphate will be obtained when the following conditions are fulfilled: Either sulfur or oxygen must be directly bound to the pentavalent phosphorus, R^1 and R^2 may be alkoxy, alkyl, or amino residues; while "acyl" represents the anion of an organic or inorganic acid such as fluorine, cyanate, thiocyanate, or of other acidic residues (enol residues, mercapto, etc.).

The following types of compounds are obtained depending on the nature of the acyl function (R^1, R^2 = alkoxy or amino groups).

$$① \quad Acyl = -O- \overset{\overset{\textstyle O(S)}{\|}}{\underset{\underset{\textstyle R^2}{|}}{P}} -R^1 \quad \longrightarrow \quad R^1 - \overset{\overset{\textstyle O(S)}{\|}}{\underset{\underset{\textstyle R^2}{|}}{P}} - O - \overset{\overset{\textstyle O(S)}{\|}}{\underset{\underset{\textstyle R^2}{|}}{P}} -R^1$$

Pyrophosphoro(amid)ates (see Section 2.3.2.3.1)

$$\text{Hal} \longrightarrow R^1-\overset{\overset{\displaystyle O\,(S)}{\|}}{\underset{\underset{\displaystyle R^2}{|}}{P}}-\text{Hal}$$

Phosphorohalidates, phosphoro(di)aminohalidates (see Section 2.3.2.2.4)

$$\text{②   Acyl} = -S-CH_2-R^3 \longrightarrow R^1-\overset{\overset{\displaystyle O\,(S)}{\|}}{\underset{\underset{\displaystyle R^2}{|}}{P}}-S-CH_2-R^3$$

Substituted *S*-alkyl phosphorothioates (see Section 2.3.2.3.2)
(R^3 = H, alkyl, COOH, heterocycle)

③   Acyl = $-O-\text{(aryl)}^{(R^3)_n}$ $\longrightarrow$

$$R^1-\overset{\overset{\displaystyle O\,(S)}{\|}}{\underset{\underset{\displaystyle R^2}{|}}{P}}-O-\text{(aryl)}^{(R^3)_n}$$

O-Phenyl phosphoroates and phosphonates (see Section 2.3.2.3.3)

$$\text{Acyl} = -O-\text{Heteroaryl} \longrightarrow R^1-\overset{\overset{\displaystyle O\,(S)}{\|}}{\underset{\underset{\displaystyle R^2}{|}}{P}}-O-\text{Heteroaryl}$$

O-Heteroaryl phosphoroates and phosphonates (see Section 2.3.2.3.4)

$$\text{④   Acyl} = -O-N= \longrightarrow R^1-\overset{\overset{\displaystyle O\,(S)}{\|}}{\underset{\underset{\displaystyle R^2}{|}}{P}}-O-N=$$

O-Phosphorylated hydroxamates and hydroximates (see Section 2.3.2.3.5)

$$\text{⑤   Acyl} = -N=C\overset{\diagup}{\diagdown} \longrightarrow R^1-\overset{\overset{\displaystyle O\,(S)}{\|}}{\underset{\underset{\displaystyle R^2}{|}}{P}}-N=C\overset{\diagup}{\diagdown}$$

N-Phosphorylated imines (see Section 2.3.2.3.6)

$$\text{⑥   Acyl} = -O-C=C\overset{\diagup}{\diagdown} \longrightarrow R^1-\overset{\overset{\displaystyle O\,(S)}{\|}}{\underset{\underset{\displaystyle R^2}{|}}{P}}-O-C=C\overset{\diagup}{\diagdown}$$

Enol phosphates (see Section 2.3.2.3.7)

Phosphinates have not achieved any significance as insecticides and are not dealt with here. The P-XYZ scheme[177] has been formulated for phosphorylating agents as a modification of Schrader's acyl rule. Here X, Y, Z normally signify H, C, N, O, S, or halogen. For good phosphorylating properties the P-X bond must be as weak as possible, while the Z group should be strongly electron attracting or readily capable of becoming so under the influence of electrophilic agents. That is, the Z group must be able to accept electrons from the P-X bond.

$$-\overset{\overset{\text{\tiny ||}}{}}{\underset{|}{P}}-\overset{\ominus}{X}-Y=Z \quad \longleftrightarrow \quad -\overset{\overset{\text{\tiny ||}}{}}{\underset{|}{P}}-\overset{\oplus}{X}=Y-\overset{\ominus}{Z}|$$

The enol ester *dichlorvos* serves as an example for this scheme.

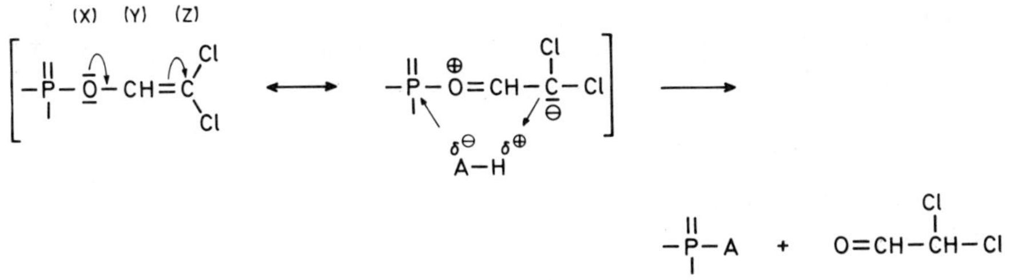

$$(X)\quad(Y)\quad(Z)$$

In cases where Y and Z are joined by a single bond the following fragmentation may be formulated.

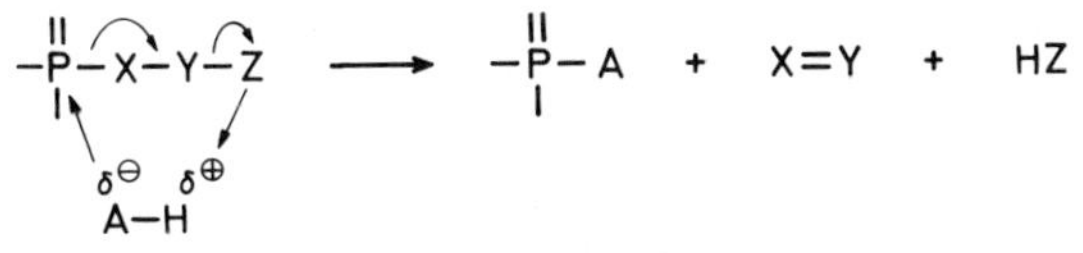

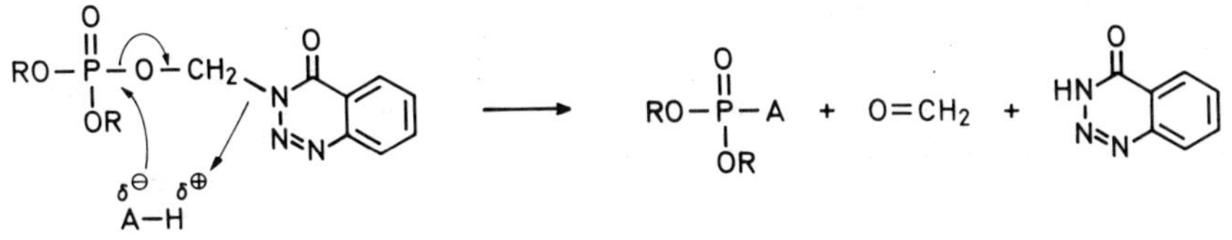

If X-Y-Z is a mesomerically stabilized anion such as phenolate then the following reaction occurs.

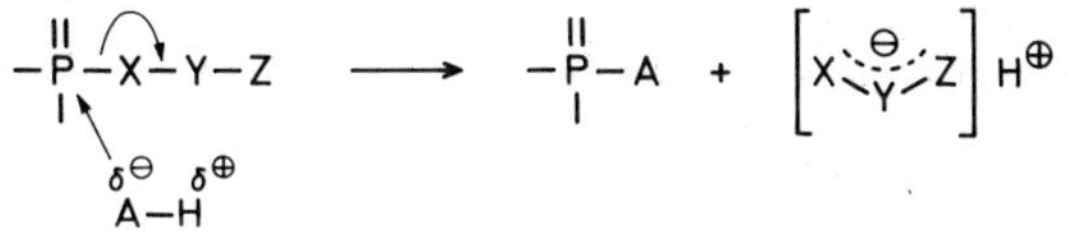

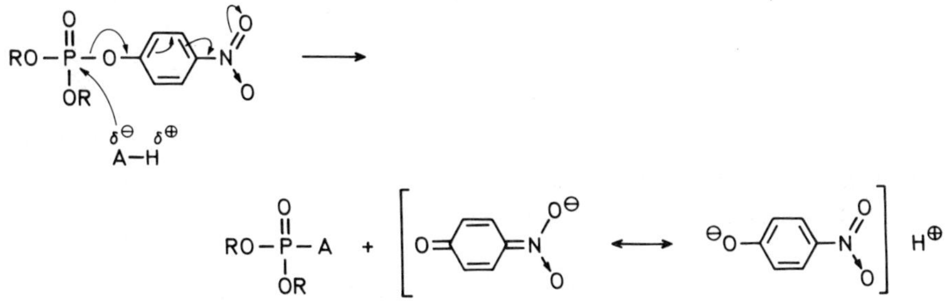

2.3.2.1. Nomenclature

It is a special problem to try to bring clarity into the sheer variety of nomenclature schemes. Every country has its own (e.g., the American, British, Scandinavian, Russian) that differs fundamentally from the others. In Germany there is the nomenclature according to Beilstein and to Houben–Weyl. Widespread and perhaps most useful is the nomenclature according to IUPAC, which is customary in English-speaking countries. Unfortunately, this system is not easily translated into the German language.[178]

In this section defined compounds are generally named according to IUPAC. To facilitate access to German literature the following list contrasts the Beilstein and IUPAC systems.

Dialkylphosphit
IUPAC: *O,O*-dialkyl phosphorous acid

Trialkylphosphit
IUPAC: trialkyl phosphoroite

O,O,O-Trialkylphosphorsäureester
IUPAC: *O,O,O*-trialkyl phosphoroate

O,O,O-Trialkylthionophosphorsäureester
IUPAC: *O,O,O*-trialkyl phosphorothioate

O,O-Dialkyl-*S*-alkylthiophosphorsäureester
IUPAC: *O,O*-dialkyl *S*-alkyl phosphorothioate

O,O-Dialkyl-*S*-alkyldithiophosphorsäureester
IUPAC: *O,O*-dialkyl *S*-alkyl phosphorodithioate

O,O-Dialkylphosphorsäureesteramid
IUPAC: *O,O*-dialkyl phosphoroamidate

O,O-Dialkylthionophosphorsäureesteramid
IUPAC: *O,O*-dialkyl phosphoroamidothioate

O,O-Dialkylphosphorsäureesterchlorid
IUPAC: *O,O*-dialkyl phosphorochloridate

O,O-Dialkylthionophosphorsäureesterchlorid
IUPAC: *O,O*-dialkyl phosphorochlorothioate

O,O-Dialkylphosphonsäureester
IUPAC: *O,O*-dialkyl phosphonate

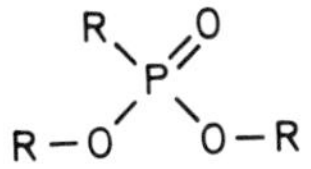

O,O-Dialkylthionophosphonsäureester
IUPAC: *O,O*-dialkyl phosphonothioate

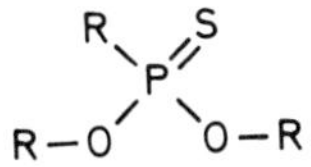

O-Alkylphosphonsäureesterchlorid
IUPAC: *O*-alkyl phosphonochloridate

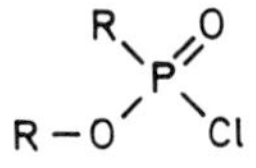

O-Alkylthionophosphonsäureesterchlorid
IUPAC: *O*-alkyl phosphonochlorothioate

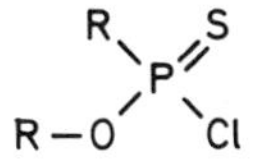

O-Alkylphosphinsäureester
IUPAC: *O*-alkyl phosphinate

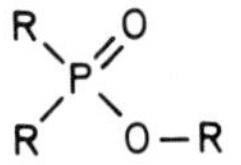

O-Alkylthionophosphinsäureester
IUPAC: *O*-alkyl phosphinothioate

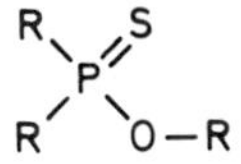

2.3.2.2. Preparation of the Necessary Precursors

For the synthesis of the various phosphates and phosphonates listed above, di- and trialkyl phosphites, salts of *O,O*-dialkyl phosphoro(di)thioic acids, and phosphor(*n*)o(thio)ic acid *O*-ester halogenides are used as "phosphorylating agents."

2.3.2.2.1. Di- and Trialkyl Phosphites and Their Reactions. The dialkyl phosphites are obtained by reaction of phosphorus(III) chloride with three molecules of alcohol in the absence of base.[179,180]

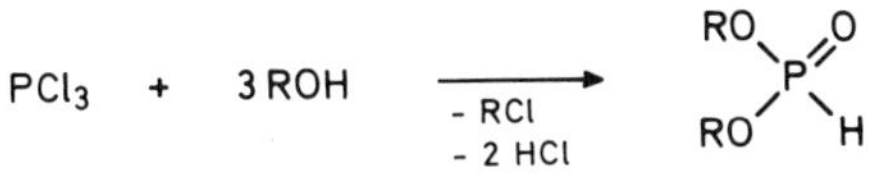

$$PCl_3 + 3\,ROH \xrightarrow[-2\,HCl]{-RCl} \quad \begin{array}{c} RO \\ RO \end{array}\!\! P \!\! \begin{array}{c} O \\ H \end{array}$$

1

Trialkyl phosphites are formed when the reaction is carried out in the presence of the equivalent amount of base.[181]

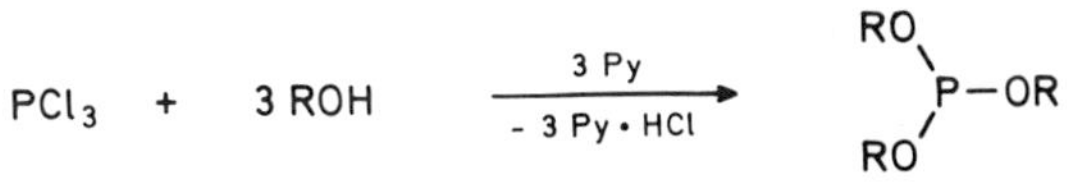

$$PCl_3 + 3\,ROH \xrightarrow[-3\,Py \cdot HCl]{3\,Py} \quad \begin{array}{c} RO \\ RO \end{array}\!\! P - OR$$

2

This method is uneconomic in that intensive cooling is necessary and the separation of the hydrochloride is difficult.

The esters are obtained almost quantitatively by reaction of phosphorus acid tris-amides with alcohols provided that the sec. amine formed has a lower boiling point than the alcohol used.[182]

$$[(R^1)_2N]_3P \quad + \quad 3\,R^2OH \quad \xrightarrow[-\,3\,(R^1)_2NH]{} \quad (R^2O)_3P$$

The dialkyl phosphites are reactive compounds. For example, they add easily onto carbonyl compounds (Dipterex®). They are themselves precursors in the preparation of phosphorochloridic acid diesters (Section 2.3.2.2.3) and phosphorothioic acid *O,O*-diesters (Section 2.3.2.2.2).

It has been deduced from kinetic data[183] that phosphorous acid diester exists in a trivalent form.

$$\underset{RO}{\overset{RO}{\diagup}}P-OH \quad \underset{?}{\rightleftharpoons} \quad \underset{RO}{\overset{RO}{\diagup}}P\underset{H}{\overset{O}{\diagdown}}$$

In contrast, a pentavalent form with a P-H bond has been found by NMR spectroscopy.

Trialkyl phosphites are important starting materials for the enol phosphates (see Section 2.3.2.3.7), which are synthesized according to the Michaelis–Arbuzov reaction.

Michaelis–Arbuzov and Perkow Reactions. The Michaelis–Arbuzov and Perkow reactions play a fundamental role in the chemistry of *O*-alkyl esters of trivalent phosphorus. There are no analogous reactions in carbon chemistry. They are a special feature of phosphorus and thus the subject of frequent investigations. Their origin lies in the desire of phosphorus to form P=O bonds. The formation of phosphonates from alkyl phosphites and alkyl halides is a long-known example of these reactions.

$$\underset{R^1O}{\overset{R^1O}{\diagup}}P-OR^1 \quad + \quad R^2X \quad \longrightarrow$$

$$\left[\underset{R^1O}{\overset{R^1O}{\diagup}}\overset{\oplus}{P}\underset{R^2}{\overset{OR^1}{\diagdown}}\right]X^{\ominus} \quad \xrightarrow{-\,RX} \quad \underset{R^1O}{\overset{R^1O}{\diagup}}P\underset{R^2}{\overset{O}{\diagdown}}$$

The intermediate is a quasiphosphonium salt, which stabilizes by loss of alkyl halide and formation of the phosphonate. This is called the Michaelis–Arbuzov reaction (R ≠ R′). The special case of this reaction in which R = R′ is the Arbuzov rearrangement. Here only isomerization occurs, and catalytic amounts of alkyl halide are sufficient. Dialkyl phosphites, in the form of their sodium salts, also react with alkyl halides to give the same final product (Michaelis–Becker–Nylén reaction).

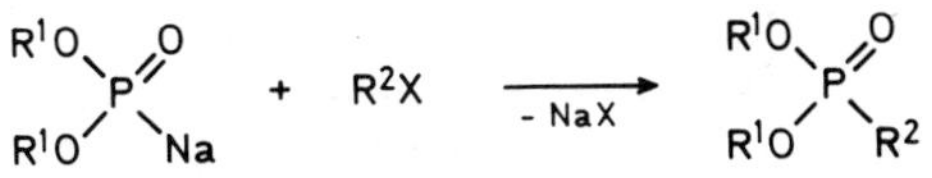

$$\begin{array}{c} R^1O \\ \diagdown P{=}O \\ R^1O \diagup \diagdown Na \end{array} + R^2X \xrightarrow[-NaX]{} \begin{array}{c} R^1O \\ \diagdown P{=}O \\ R^1O \diagup \diagdown R^2 \end{array}$$

In the reaction of trialkyl phosphites with α-halocarbonyl compounds, the so-called Perkow reaction occurs as well as or instead of the Michaelis–Arbuzov. *O,O*-Dialkyl *O*-vinyl phosphates are formed.

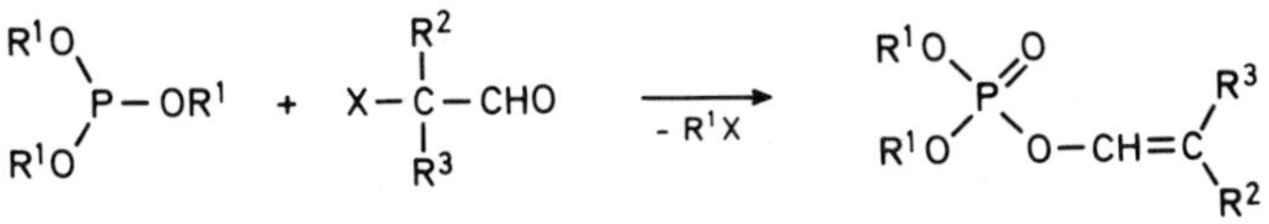

$$\begin{array}{c} R^1O \\ \diagdown P{-}OR^1 \\ R^1O \diagup \end{array} + \begin{array}{c} R^2 \\ | \\ X{-}C{-}CHO \\ | \\ R^3 \end{array} \xrightarrow[-R^1X]{} \begin{array}{c} R^1O \\ \diagdown P{=}O \\ R^1O \diagup \diagdown O{-}CH{=}C{\diagup R^3}_{\diagdown R^2} \end{array}$$

In the Michaelis–Arbuzov reaction, ketophosphonates are formed from α-halocarbonyl compounds. The structure of the latter and the reaction conditions determine the course of the reaction.[184,185] The following scheme serves to illustrate the relationship between the two reactions.

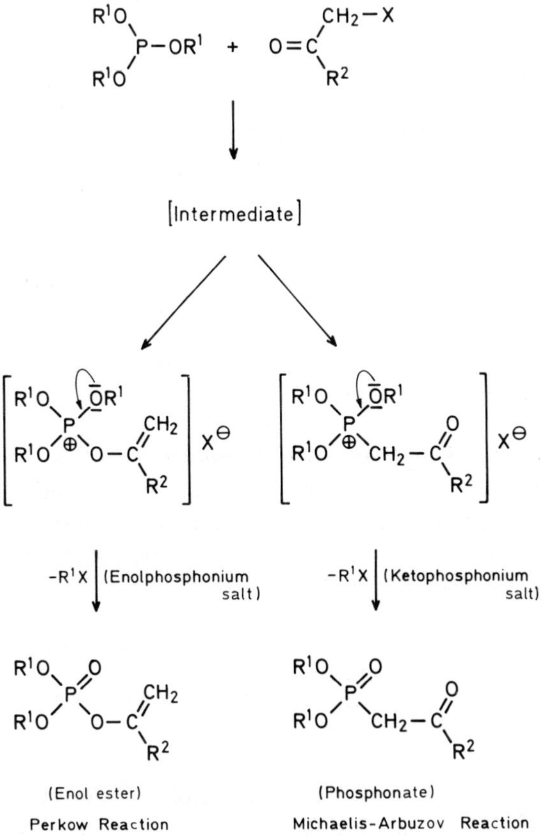

$$\begin{array}{c} R^1O \\ \diagdown P{-}OR^1 \\ R^1O \diagup \end{array} + \begin{array}{c} CH_2{-}X \\ O{=}C \diagup \\ \diagdown R^2 \end{array}$$

[Intermediate]

$$\left[\begin{array}{c} R^1O \\ \diagdown P \\ R^1O \diagup {}^{\oplus} \diagdown O{-}C{\diagup CH_2}_{\diagdown R^2} \end{array} \right] X^{\ominus} \qquad \left[\begin{array}{c} R^1O \\ \diagdown P \\ R^1O \diagup {}^{\oplus} \diagdown CH_2{-}C{\diagup O} \diagdown R^2 \end{array} \right] X^{\ominus}$$

$-R^1X$ | (Enolphosphonium salt) $-R^1X$ | (Ketophosphonium salt)

$$\begin{array}{c} R^1O \\ \diagdown P{=}O \\ R^1O \diagup \diagdown O{-}C{\diagup CH_2}_{\diagdown R^2} \end{array} \qquad \begin{array}{c} R^1O \\ \diagdown P{=}O \\ R^1O \diagup \diagdown CH_2{-}C{\diagup O}_{\diagdown R^2} \end{array}$$

(Enol ester)   (Phosphonate)

Perkow Reaction   Michaelis–Arbuzov Reaction

Either the enol ester or the phosphonate is formed from a common intermediate.[184,185] The Michaelis–Arbuzov reaction yields starting materials for the synthesis of phosphonochloridates (see Section 2.3.2.2.4), whereas the Perkow reaction is valuable for the preparation of the insecticidal enol phosphates (see Section 2.3.2.3.7).

2.3.2.2.2. O,O-Dialkyl Phosphorodithioic Acids and Their Salts.

The preparation of the frequently used ammonium salts of O,O-dialkyl phosphorodithioic acids starts with the reaction of phosphorus(V) sulfide with methanol or ethanol. The ammonium salt crystallizes out when ammonia is passed through the reaction mixture.

$$P_4S_{10} \quad + \quad 8\ ROH \quad \xrightarrow[-\ 2\ H_2S]{} \quad 4 \quad \begin{array}{c} RO \\ \end{array}\!\!P\!\!\begin{array}{c} S \\ SH \end{array}$$

$$\mathbf{1}$$

$$\xrightarrow{NH_3} \quad \begin{array}{c} RO \\ \end{array}\!\!P\!\!\begin{array}{c} S \\ S^{\ominus}\ NH_4^{\oplus} \end{array}$$

$$\mathbf{2}$$

The ammonium salts of O,O-dialkyl phosphorothioic acids are obtained by treatment of an alcoholic solution of dialkyl phosphite with sulfur in the presence of ammonia.[186]

$$\begin{array}{c} RO \\ \end{array}\!\!P\!\!\begin{array}{c} O \\ H \end{array} \quad + \ S\ +\ NH_3 \quad \xrightarrow{\sim 20°} \quad \begin{array}{c} RO \\ \end{array}\!\!P\!\!\begin{array}{c} O \\ :^{\ominus}\ NH_4^{\oplus} \\ S \end{array}$$

$$\mathbf{3}$$

The O,O-dialkyl phosphorothioic acids or their salts may also be made by hydrolysis of the corresponding ester chlorides.

2.3.2.2.3. O,O-Dialkyl Phosphoro(thio)chloridates.

These are the most important starting materials for the synthesis of insecticidal phosphates. Of the many existing syntheses only the technically most relevant are discussed here.

Starting from phosphorus oxychloride or sulfochloride, reaction with alcohols leads to O,O-dialkyl phosphoro(thio)chloridates.[187,188]

$$\begin{array}{c} (O) \\ PSCl_3 \end{array} \quad + \quad 2\ ROH \quad \xrightarrow{cool} \quad \begin{array}{c} RO \\ \end{array}\!\!P\!\!\begin{array}{c} S\ (O) \\ Cl \end{array}$$

excess

$$\mathbf{1}$$

The chlorination of O,O-dialkyl phosphorodithioic acids or their alkali metal salts to give O,O-dialkyl phosphorochlorothioates is technically very simple.[189]

$$2 \quad \underset{RO}{\overset{RO}{\diagup}} P \underset{SH}{\overset{S}{\diagdown}} \quad + \quad 3\,Cl_2 \quad \xrightarrow[\substack{-\ 2\,HCl \\ -\ S_2Cl_2}]{} \quad 2 \quad \underset{RO}{\overset{RO}{\diagup}} P \underset{Cl}{\overset{S}{\diagdown}}$$

2

Here it is possible to start directly from phosphorus(V) sulfide and chlorinate to give the chlorothioate without isolation of the intermediate dithioic acid.[190]

A further synthesis is the chlorination of dialkyl(thio)phosphites.[179]

$$\underset{RO}{\overset{RO}{\diagup}} P \underset{H}{\overset{O\,(S)}{\diagdown}} \quad + \quad Cl_2 \quad \xrightarrow[-\ HCl]{} \quad \underset{RO}{\overset{RO}{\diagup}} P \underset{Cl}{\overset{O\,(S)}{\diagdown}}$$

1

The chlorination can also be carried out on a large scale using sulfuryl chloride. Dialkyl phosphites may be chlorinated with carbon tetrachloride under mild conditions in the presence of a tertiary amine (see Section 2.3.2.3.4).[191]

$$\underset{R^1O}{\overset{R^1O}{\diagup}} P \underset{H}{\overset{O}{\diagdown}} \quad + \quad CCl_4 \quad \xrightarrow[-\ HCCl_3]{(R^2)_3N} \quad \underset{R^1O}{\overset{R^1O}{\diagup}} P \underset{Cl}{\overset{O}{\diagdown}}$$

3

Tetramethyl phosphorodiamidochloridate is made from phosphorus oxychloride and dimethylamine. To achieve higher yields only 3.5 moles amine per mole phosphorus oxychloride is used instead of the stoichiometric 4 moles.[192]

$$POCl_3 \quad + \quad 4\,(H_3C)_2NH$$

$$\xrightarrow[\substack{-\ 2\,(H_3C)_2NH \\ -\ 2\,HCl}]{cool} \quad \underset{(H_3C)_2N}{\overset{(H_3C)_2N}{\diagup}} P \underset{Cl}{\overset{O}{\diagdown}}$$

4

The thiochloridates are made by the corresponding route.

2.3.2.2.4. O-Alkyl Phosphono(thio)chloridates. Phosphonochloridates can be made from alkyl or aryl phosphonodichloridites (6). Some of the numerous methods for their synthesis use aluminum chloride as catalyst. A liquid complex (5) is formed:

$$R-Cl \quad + \quad 2\,AlCl_3 \quad + \quad PCl_3 \quad \longrightarrow$$

$$[R-PCl_3]^{\oplus}\,[Al_2Cl_7]^{\ominus}$$

5

R = CH_3, C_2H_5

which is reduced with aluminum. The aluminum chloride thus formed is bound by addition of alkali metal chloride and the alkyl phosphonodichloridite distilled out of the reaction mixture[193] (path A). (The reduction can also be carried out with phosphorus.[194]) If the aluminum complex (5) is decomposed with water,[195] the phosphonodichloridate is obtained, which can be converted to the desired phosphonochloridate by reaction with alcohol (path B, the Kinnear-Perren synthesis). Decomposition of the aluminum complex with hydrogen sulfide instead of water leads to the analogous phosphonodichlorothioate (9), which can be reacted further to the alkyl or aryl phosphonochlorothioate (10) (path C).

One interesting method of preparation is based on the Michaelis–Arbuzov reaction (see Section 2.3.2.2.1). Trialkyl phosphites are rearranged to O,O-dialkyl phosphonates with alkyl iodide and are then chlorinated with phosgene.[196]

Methyl phosphonodichloridate is available by pyrolysis of dimethyl phosphite followed by chlorination[197] and may be converted to the chloridate in the usual manner.

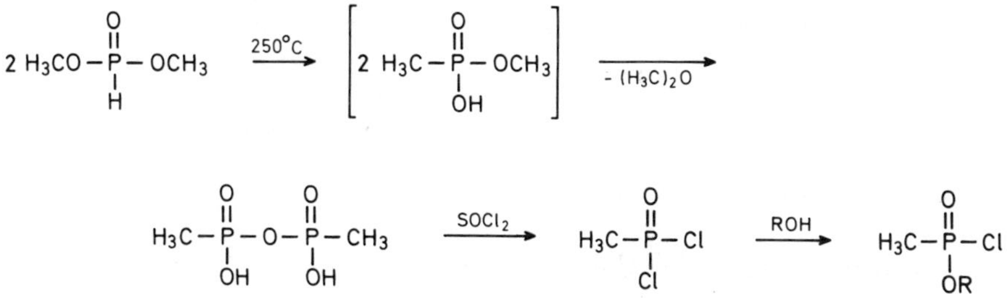

" Sumpf "

12

Phenylphosphonodichloridite ("phosphenyl chloride"), available from the reaction of benzene and phosphorus(III) chloride in the presence of aluminum chloride, can be reacted with sulfur without prior isolation.[198]

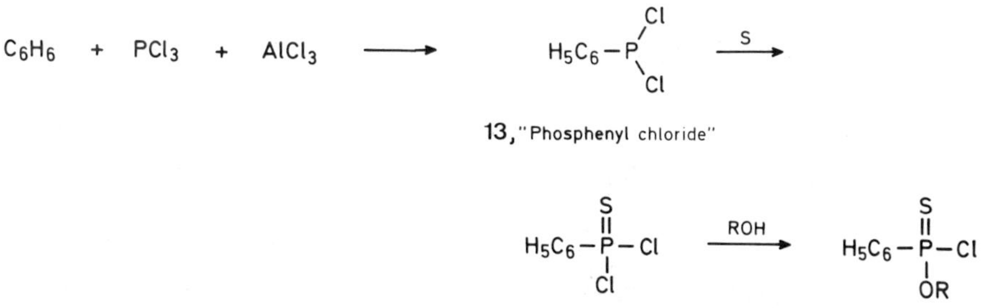

13, "Phosphenyl chloride"

A general method for obtaining the phosphonodichlorothioates is the thionation of the corresponding oxygen compounds. Alternatively, as above, the trivalent intermediate can be thionated directly.

2.3.2.3. Individual Compounds

The commercial products are obtained directly by reaction of the precursors discussed in the previous section, with the appropriate acyl compounds. Compounds in this section are grouped according to the various ACYL residues corresponding to the list in Section 2.3.2.

2.3.2.3.1. Phosphate Anhydrides (Pyrophosphates) and Phosphorohalidates (Acyl =

$$-O-\underset{\underset{R^2}{|}}{\overset{\overset{O(S)}{||}}{P}}-R^1$$

, *Hal).* Tetraethylpyrophosphate (*TEPP*) was first synthesized in 1854 by Ph. de Clermont, but its insecticidal activity remained unrecognized for almost a century.[199] The classical synthesis involved reaction of the silver salt of pyrophosphoric acid with alkyl iodide.

$$AgO-\overset{\overset{O}{\|}}{\underset{\underset{OAg}{|}}{P}}-O-\overset{\overset{O}{\|}}{\underset{\underset{OAg}{|}}{P}}-OAg \quad + \quad C_2H_5I \quad \longrightarrow \quad H_5C_2O-\overset{\overset{O}{\|}}{\underset{\underset{H_5C_2O}{|}}{P}}-O-\overset{\overset{O}{\|}}{\underset{\underset{OC_2H_5}{|}}{P}}-OC_2H_5$$

1; TEPP

Technical processes are as follows.

1. Partial hydrolysis of *O,O*-dialkyl phosphorochloridates.[200]

$$2\ RO-\overset{\overset{O}{\|}}{\underset{\underset{OR}{|}}{P}}-Cl \quad + \quad H_2O \quad + \quad 2\ R_3N \quad \xrightarrow[-\ 2\ R_3N\cdot HCl]{} \quad RO-\overset{\overset{O}{\|}}{\underset{\underset{OR}{|}}{P}}-O-\overset{\overset{O}{\|}}{\underset{\underset{OR}{|}}{P}}-OR$$

2. Chlorination of trialkyl phosphates with thionyl chloride.[201]

$$2\ RO-\overset{\overset{O}{\|}}{\underset{\underset{OR}{|}}{P}}-OR \quad + \quad SOCl_2 \quad \xrightarrow[-\ 2\ RCl]{-\ SO_2} \quad RO-\overset{\overset{O}{\|}}{\underset{\underset{OR}{|}}{P}}-O-\overset{\overset{O}{\|}}{\underset{\underset{OR}{|}}{P}}-OR$$

3. Reaction of trialkyl phosphate with dialkyl phosphorochloridate.[202]

$$RO-\overset{\overset{O}{\|}}{\underset{\underset{OR}{|}}{P}}-OR \quad + \quad Cl-\overset{\overset{O}{\|}}{\underset{\underset{OR}{|}}{P}}-OR \quad \xrightarrow[-\ RCl]{} \quad RO-\overset{\overset{O}{\|}}{\underset{\underset{OR}{|}}{P}}-O-\overset{\overset{O}{\|}}{\underset{\underset{OR}{|}}{P}}-OR$$

TEPP is an extremely toxic compound with an oral LD$_{50}$ for the rat of 1.12 mg/kg. In spite of its contact insecticidal and acaricidal properties, *TEPP* is seldom used today owing to its lack of stability toward hydrolysis.

The sulfur analogue *sulfotep* (*O,O,O',O'*-tetraethyl dithiopyrophosphate) is prepared, like *TEPP*, by partial hydrolysis of the phosphorochlorothioate.[203]

$$H_5C_2O-\overset{\overset{S}{\|}}{\underset{\underset{H_5C_2O}{|}}{P}}-O-\overset{\overset{S}{\|}}{\underset{\underset{OC_2H_5}{|}}{P}}-OC_2H_5$$

2; Sulfotep

The compound is relatively stable toward hydrolysis and is, under the trade name Bladafum®, recommended for control of aphids, mealybugs, whiteflies, thrips, and spider mites.

OMPA (octamethyl pyrophosphoroamidate) (common name: *schradan*) can also be synthesized by partial hydrolysis of the phosphorodiamidochloridate. It was first synthesized by reaction of *O*-alkyl-N,N,N',N'-tetramethyl phosphorodiamidate with N,N,N',N'-tetramethyl phosphorodiamidochloridate.[204]

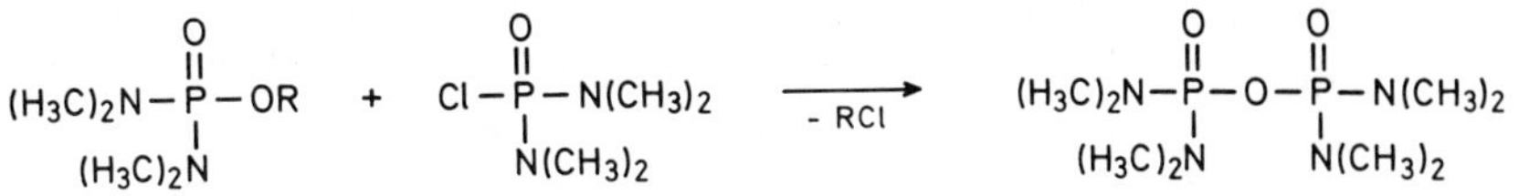

$$ 3;\ OMPA $$

A systemic insecticide, *OMPA* is today only of historical interest.

An example in which the acyl residue is fluorine is the systemic insecticide and acaricide *dimefox* (N,N,N',N'-tetramethyl phosphorodiamidofluoridate) **(4)**.

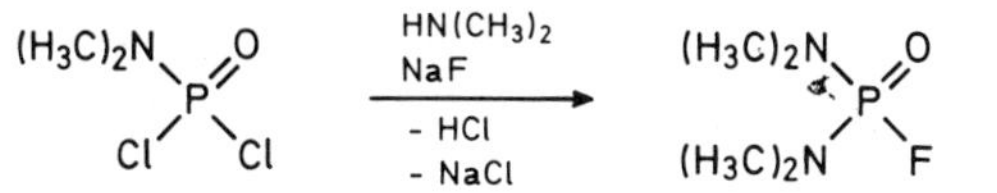

4; *Dimefox*;
LD_{50}: 5 mg/kg rat,
oral, acute

2.3.2.3.2. S-Alkyl Phosphorothioates (Acyl = S-CH₂—R³).

2.3.2.3.2. S-Alkyl Phosphorothioates ($Acyl = S\text{-}CH_2—R^3$). A large number of insecticidal phosphates are derived from *S*-alkyl phosphorothioates. According to the residue R^3 they may be divided into six basic types.

1. *S*-Methyl phosphorothioates (acyl = S—CH_3).
2. *S*-Alkyl(aryl)thiomethyl phosphoro(di)thioates
    [acyl = S—CH_2—S-alkyl(aryl)].
3. *S*-2-Ethyl(aryl)thioethyl phosphorothioates
    (acyl = S—CH_2—CH_2—S—C_2H_5).
4. *S*-Heteroarylmethyl phosphorothioates (acyl = S—CH_2-heterocycle).
5. *S*-Alkoxycarbonyl(aminocarbonyl)methyl phosphorothioates
    (acyl = S—CH_2—COOAlkyl; S—CH_2—$CONH_2$).
6. Special *S*-substituted phosphorothioates (acyl = *S*-substituted alkyl).

1. *S-Methyl phosphorothioates (acyl = S—CH_3).* The simplest compound in this series is Tamaron® (*O,S*-dimethyl phosphoroamidothioate[205]), also known as Monitor®.[206] It is remarkable in that three different residues are bound to the phosphorus atom. The starting material is *O,O*-dimethyl phosphorochlorothioate or the corresponding amidothioate, which is hydrolyzed by alkali and then alkylated.

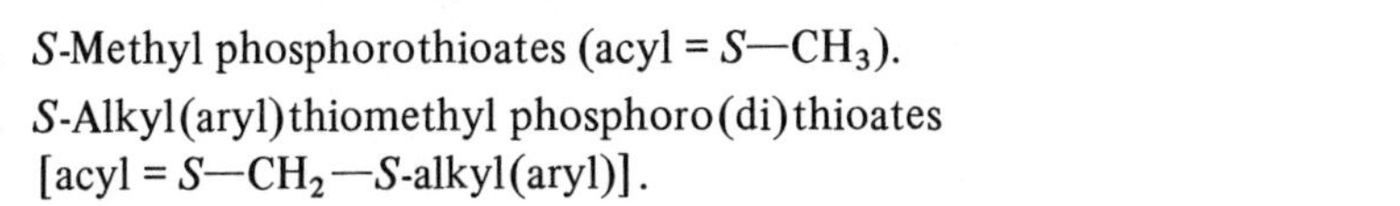

1; *Methamidophos*,
Tamaron®, Monitor®;
LD_{50}: 30 mg/kg rat,
oral, acute

Tamaron is used as an insecticide and an acaricide. In connection with this synthesis, attention is drawn to some rearrangement reactions of sulfur-containing phosphates.[207] The course of rearrangement is dependent on the nature of the attacking nucleophile.

a. *The realkylation.* A phosphorothioate anion, formed by dealkylation, is itself a nucleophilic agent and can also react with an alkylated product. The alkylation takes place preferentially at sulfur with the formation of a thiolate.

$$B = I^{\ominus}, (H_3C)_2S$$

b. *Alkylation of diester anion by triester.* When triethylamine is the nucleophile, no realkylation can take place owing to the strength of the bond between the alkyl group and the amine. The diester anion is thus alkylated by the triester (step 1). The reaction then proceeds according to the following equation (step 2).

c. *The Pistschimuka reaction, in which the thiono/thiol rearrangement is accelerated by alkyl iodide.*

This reaction is dependent on the substituents A and B. The ease of reaction increases in the following series.

$$Cl < RS < RO < R < R_2N$$

d. *In the absence of a nucleophilic agent, the sulfur atom of a phosphoro-thionate can be alkylated intermolecularly by another ester molecule (self-alkylation).* This thiono/thiol rearrangement takes place at around 120–180°C (e.g., with *parathion*).

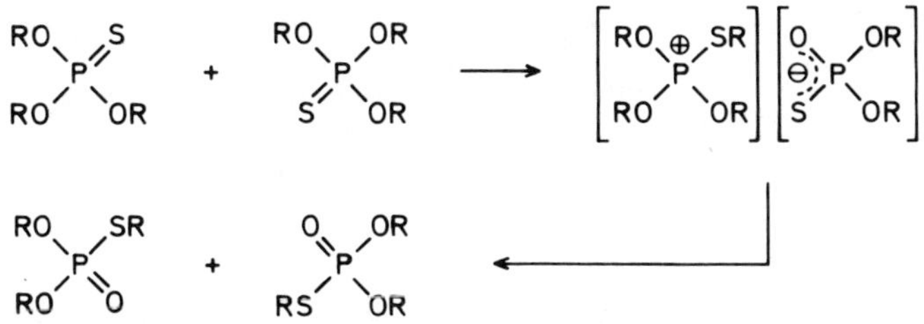

Of the alkylation reactions listed here, the Pistschimuka and the realkylation reactions are involved in the synthesis of Tamaron.

The *N*-acetyl derivative of Tamaron is *acephate* (*O,S*-dimethyl-*N*-acetyl phosphoramidothioate), which is simply made by acetylation of Tamaron with acetic anhydride. Remarkably, the toxicity is considerably reduced with respect to Tamaron (LD_{50}: 945 mg/kg rat, oral, acute). Like Tamaron, *acephate* is also a systemic insecticide and acaricide.

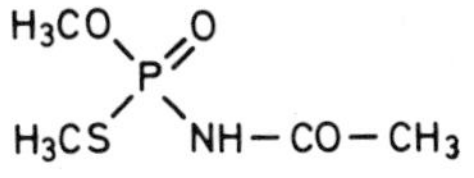

2; *Acephate*, ORTHO 12 420

2. *S-Alkyl(aryl)thiomethyl phosphoro(di)thioates [acyl = S—CH$_2$—S-alkyl (aryl)].* Phorate (*O,O*-diethyl-*S*-ethylthiomethyl phosphorodithioate) is obtained either by addition of *O,O*-diethyl phosphorodithioate onto formaldehyde and condensation with ethanethiol without isolation of the intermediate (A),[208] or from reaction of *O,O*-diethyl phosphorodithioic acid with ethylthiochloromethane.[209]

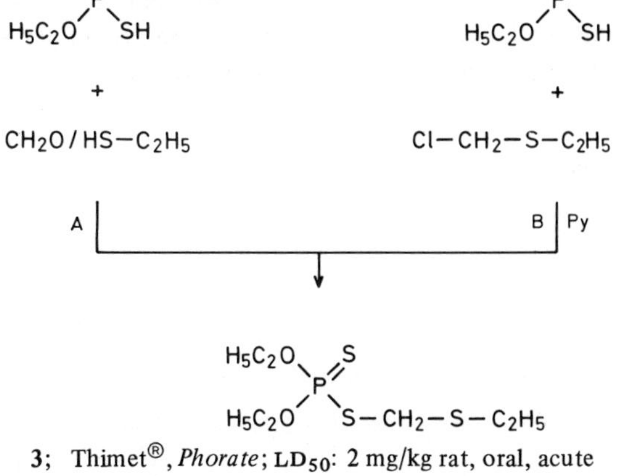

$CH_2O / HS—C_2H_5$

$Cl—CH_2—S—C_2H_5$

3; Thimet®, *Phorate*; LD_{50}: 2 mg/kg rat, oral, acute

It is relatively toxic and has contact insecticidal and systemic activity. Replacement of the alkyl group by an aryl residue leads to the contact insecticidal and acaricidal compound Trithion® [*carbophenothion, O,O*-diethyl *S*-(4-chlorophenylthio)-methyl phosphorodithioate] and the corresponding methyl ester Methyltrithion®.

4; Trithion®; LD_{50}: 28–100 mg/kg rat, oral, acute

5; Methyltrithion®; LD_{50}: 180–200 mg/kg rat, oral, acute

The aromatic thioether is prepared by reaction of 4-chlorothiophenol with aqueous formaldehyde and concentrated hydrochloric acid.[210]

$$4 \; ; R = C_2H_5$$
$$5 \; ; R = CH_3$$

If two molecules of *O,O*-diethyl phosphorodithioic acid are connected by a methylene bridge, the acaricidal and insecticidal (predominantly contact insecticidal) compound *ethion* [*O,O,O',O'*-tetraethyl *S,S'*-methylene di(phosphorodithioate)] is obtained. Several methods of synthesizing this compound are known. The simplest, the reaction of *O,O*-diethyl phosphorodithioic acid with formaldehyde, is given here.

6; *Ethion*; LD_{50}: 96 mg/kg rat, oral, acute

3. *S-2-Ethyl(aryl)thioethyl phosphorothioates [acyl = —S—CH$_2$—CH$_2$—S— C$_2$H$_5$ (aryl)]*. The most well-known compound from the series of systemically active insecticides is Systox® (*demeton*), a mixture of the isomeric thiono (**7a**) and thiolo esters (**7b**) in the ratio 2:1.

$$H_5C_2O\diagdown \underset{H_5C_2O\diagup}{P}\diagup^{\textstyle S} \diagdown O{-}CH_2{-}CH_2{-}S{-}C_2H_5$$

7a; *O,O*-Diethyl *O*-(2-ethylthioethyl) phosphorothioate

$$H_5C_2O\diagdown \underset{H_5C_2O\diagup}{P}\diagup^{\textstyle O} \diagdown S{-}CH_2{-}CH_2{-}S{-}C_2H_5$$

7b; *O,O*-Diethyl *S*-(2-ethylthioethyl)phosphorothioate

This isomer mixture is made technically by reaction of *O,O*-diethyl phosphorochlorothioate with 2-(ethylthio)ethanol in the presence of an acid acceptor.[211]

$$H_5C_2O\diagdown \underset{H_5C_2O\diagup}{P}\diagup^{\textstyle S} \diagdown Cl \quad + \quad HO{-}CH_2{-}CH_2{-}S{-}C_2H_5 \xrightarrow{K_2CO_3} \quad \textbf{7a} \quad + \quad \textbf{7b}$$

The following routes can also be employed for the synthesis of the thiolo ester **7b**. The starting materials are the sodium salt of diethyl phosphite (**8**)[212] or triethyl phosphite,[213] sodium *O,O*-diethyl phosphorothioate (**9**),[214] or the *S*-2-bromoethyl ester (**10**).[215]

$$H_5C_2O\diagdown \underset{H_5C_2O\diagup}{P}\diagup^{\textstyle S} \diagdown ONa$$

9

$$\updownarrow$$

$$H_5C_2O\diagdown \underset{H_5C_2O\diagup}{P}\diagup^{\textstyle O} \diagdown Na \qquad H_5C_2O\diagdown \underset{H_5C_2O\diagup}{P}\diagup^{\textstyle O} \diagdown SNa \qquad H_5C_2O\diagdown \underset{H_5C_2O\diagup}{P}\diagup^{\textstyle O} \diagdown S{-}R^1$$

8   **10**

NCS—R^2 Cl—R^2 | - NaCl $\overset{Na}{\underset{20-40°}{\overset{|}{S}{-}C_2H_5}}$ - NaBr

$$H_5C_2O\diagdown \underset{H_5C_2O\diagup}{P}\diagup^{\textstyle O} \diagdown S{-}R^2$$

7b

R^1 = (CH$_2$)$_2$—Br
R^2 = (CH$_2$)$_2$—S—C$_2$H$_5$

The thiolo ester is held responsible for the insecticidal activity. The acute oral LD_{50} of the isomer mixture is 6–12 mg/kg for the rat, the pure thiolo ester being 10 times more toxic than the thiono ester. At high rates of application, *demeton* is active against many pests, particularly against aphids and other sucking insects. It is a systemic insecticide.

An ionic mechanism has been suggested for the isomerization of the thiono form into the thiolo form.[216]

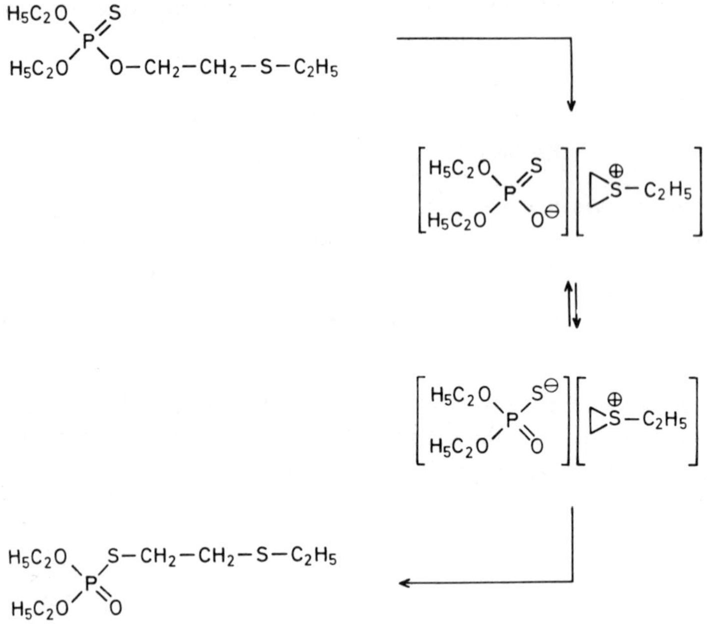

A radical mechanism is excluded because the polymerization of acrylonitrile is not initiated by the isomer mixture. The fact that rearrangement does not take place when the thioether is "blocked" by oxidation to sulfoxide or sulfone is supporting evidence for formation of the sulfonium cation. Metasystox® (*demeton-O-methyl*) (**11a**) and Metasystox(i)® (*demeton-S-methyl*) (**11b**) are the methyl esters analogous to Systox®.

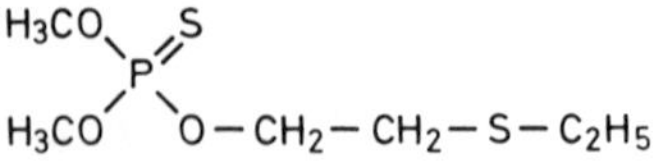

11a

11a; LD_{50}: 180 mg/kg rat, oral, acute

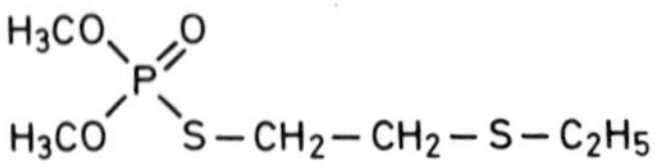

11b

11b; LD_{50}: 40–60 mg/kg rat, oral, acute

Metasystox(i)® is a systemic insecticide with a longer duration of activity than, for example, *parathion*. This is explained by the fact that metabolites more stable to hydrolysis, such as Metasystox R[®], are formed by oxidation in the plant (see Section 2.3.4.1). This oxidation in the plant can be achieved *in vitro* either with hydrogen peroxide[217,218] or with halogen[218] (with potassium permanganate, the corresponding sulfone is obtained[218]).

$$
\begin{array}{c}
H_3CO \\
\diagdown \\
P \\
H_3CO \diagup \diagdown S \\
| \\
(CH_2)_2 \\
| \\
S-C_2H_5
\end{array}
$$

(with P=O)

11b

| H_2O_2 $-H_2O$ | $\begin{array}{c}Br_2\\H_2O\\Na_2CO_3\end{array}$ $-2\,NaBr$ | equiv. $KMnO_4$ |

$$
\begin{array}{c}
H_3CO \\
\diagdown \\
P \\
H_3CO \diagup \diagdown S \\
| \\
(CH_2)_2 \\
| \\
SO-C_2H_5
\end{array}
\qquad
\begin{array}{c}
H_3CO \\
\diagdown \\
P \\
H_3CO \diagup \diagdown S \\
| \\
(CH_2)_2 \\
| \\
SO_2-C_2H_5
\end{array}
$$

12 **13**

The sulfoxide derivative Metasystox R[®] [*oxydemeton-methyl*: *O,O*-dimethyl *S*-(2-ethylsulfinylethyl) phosphorothioate] is not just a mere metabolite of Metasystox(i) (**11b**); it is itself a commercial product.[219] The syntheses are based on well-known processes.

Metasystox R (LD_{50}: 65 mg/kg rat, oral, acute) is also systemically active and compatible with all insecticides and fungicides apart from those of an alkaline character. It is specifically active against sucking insects and spider mites as well as sawflies.

The dithio derivative of Metasystox[®], *thiometon* [*O,O*-dimethyl *S*-(2-ethylthioethyl) phosphorodithioate], is a systemic and contact insecticide. One method of production is the alkylation of the dithio salt **14** with the *p*-toluenesulfonate of 2-(ethylthio)ethanol.[220]

$$
\begin{array}{c}
H_3CO \\
\diagdown \\
P \\
H_3CO \diagup \diagdown SNa
\end{array}
\quad + \quad H_3C-\langle\!\!\!\bigcirc\!\!\!\rangle-SO_2-O-(CH_2)_2-S-C_2H_5 \quad \longrightarrow
$$

(with P=S)

14

$$
\begin{array}{c}
H_3CO \\
\diagdown \\
P \\
H_3CO \diagup \diagdown S-(CH_2)_2-S-C_2H_5
\end{array}
$$

(with P=S)

15

15; *Thiometon*; LD_{50}: 85 mg/kg rat, oral, acute

Disyston® (*disulfoton*) is the diethyl ester corresponding to *thiometon*.[221]

$$H_5C_2O-\underset{H_5C_2O}{\overset{S}{P}}-S-CH_2-CH_2-S-C_2H_5$$

16; Disyston®; LD$_{50}$: 12 mg/kg rat, oral, acute

Like Systox® it is a systemic insecticide and is particularly active against sucking insects.

The isopropyl ether *isothioate* (Hosdon®) (*S*-2-isopropylthioethyl *O,O*-dimethyl phosphorodithioate) is a systemic insecticide particularly active against aphids when applied as a foliar spray.[221a]

$$H_3CO-\underset{H_3CO}{\overset{S}{P}}-S-CH_2-CH_2-S-CH(CH_3)_2$$

16a; *Isothioate*; LD$_{50}$: 150–170 mg/kg rat, oral, acute

Another systemic insecticide used for control of sucking pests is *vamidothion* {*O,O*-dimethyl *S*-[2-(1-methylcarbamoyl)ethylthio]ethyl phosphorothioate}.[222]

$$H_3CO-\underset{H_3CO}{\overset{O}{P}}-S-CH_2-CH_2-S-\underset{\underset{CH_3}{|}}{CH}-CO-NH-CH_3$$

17; *Vamidothion*; LD$_{50}$: 64–100 mg/kg rat, oral, acute

4. *S-Heteroarylmethyl phosphorothioates (acyl = S—CH$_2$-heterocycle).* The most well-known phosphate derived from an *N*-heterocycle is *azinphos-methyl* (Gusathion®, **18**) [*O,O*-dimethyl *S*-(3,4-dihydro-4-oxobenzo[*d*]-(1,2,3)triazin-3-ylmethyl) phosphorodithioate]. It is a contact and stomach insecticide and acaricide (LD$_{50}$: 10–18 mg/kg rat, oral, acute).

The spectrum of activity embraces species of Coleoptera, Diptera, Homoptera, and Lepidoptera. The residual activity is significantly longer than that of most other nonsystemic organophosphorus insecticides.

The ethyl ester (*azinphos-ethyl*) is also commercially available. It is effective against resistant spider mites. Starting from anthranilic amide, *azinphos-methyl* is obtained by the following route.[223,224]

Benzazimide

18, Azinphos-methyl (Gusathion®)

Another contact insecticide and acaricide with high activity from this class is Imidan® [*phosmet*: O,O-dimethyl S-(N-phthalimidomethyl) phosphorodithioate].

19; Imidan®; LD$_{50}$: 147 mg/kg rat, oral, acute

In Japan it is used for control of citrus pests. It is also active against ectoparasites on cattle, sheep, and pigs. The synthesis is carried out according to the general methods for the alkylation of phosphoric salts.

Closely related to Imidan® is *dialifos* [O,O-diethyl S-(2-chloro-1-phthalimido-ethyl) phosphorodithioate].

20; *Dialifos*; LD$_{50}$: 43–71 mg/kg rat, oral, acute

It is a contact and stomach insecticide and acaricide. The compound is synthesized by chlorination of N-vinylphthalimide and subsequent reaction with O,O-diethyl phosphorodithioic acid.[225]

A 1,3-benzoxazolone serves as starting material for the synthesis of *phosalone* [O,O-diethyl S-(6-chloro-2,3-dihydro-2-oxobenzoxazol-3-yl)methyl phosphoro-dithioate—often incorrectly described as the 5-chloro derivative in the literature].

21; Zolone®
Phosalone; LD$_{50}$: 135 mg/kg rat, oral, acute

Phosalone is a systemic insecticide and acaricide used in citrus and orchard fruit.[226]

Another compound worthy of mention in this class is Supracide® {*meth-idathion*: O,O-dimethyl S-[(2,3-dihydro-5-methoxy-2-oxo-1,3,4-thiadiazol-3-yl)-

methyl] phosphorodithioate}, which is highly active, particularly as an acaricide. It is used in arable and fruit crops (LD$_{50}$: 25–48 mg/kg rat, oral, acute).[227]

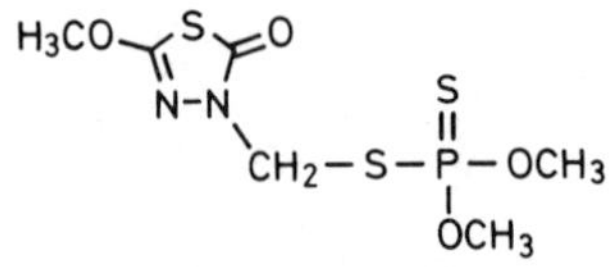

22; Supracide®

5. *S-Alkoxycarbonyl(aminocarbonyl)methyl phosphorothioates [acyl = —S— CH$_2$—COO(alkyl); —S—CH$_2$—CONR^{3}R^4].* α-Halocarboxylic acid derivatives represent important starting materials for a series of significant insecticides. *Acethion (O,O*-diethyl *S*-carbethoxymethyl phosphorodithioate)[228] is formed by alkylation of sodium *O,O*-diethyl phosphorodithioate with ethyl α-chloroacetate:

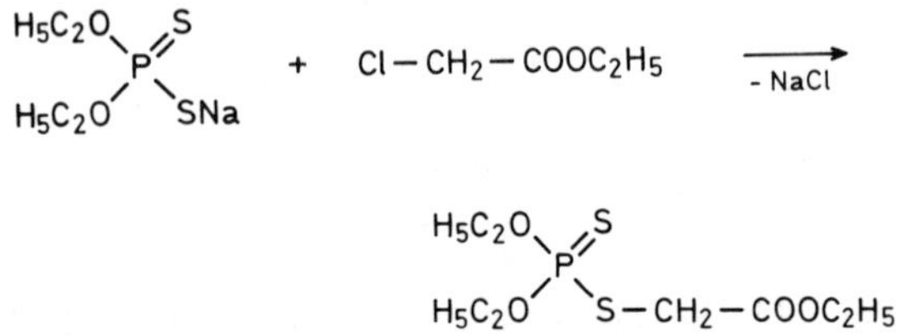

23; *Acethion*; LD$_{50}$: 1050–1100 mg/kg rat, oral, acute

Acethion has not achieved any practical significance but represents the first link in an important chain of insecticides.

Closely related to *acethion* is Cidial® **(24)** [*O,O*-dimethyl *S*-(α-carbethoxy)-benzyl phosphorodithioate]. It is used as a contact insecticide, particularly in Japan and is especially recommended for control of scale insects. Two syntheses have been selected here from the numerous patented processes.[229,230]

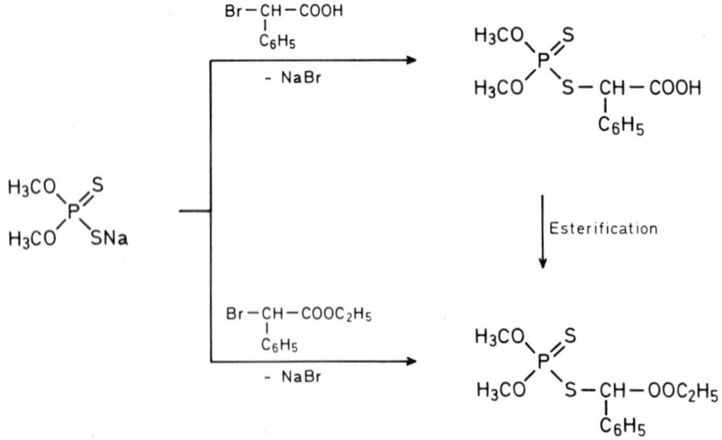

24,® Cidial, Phenthoat

LD$_{50}$: 250 mg/kg rat, oral, acute

Demethoate **(25)** [*O,O*-dimethyl *S*-(*N*-methylcarbamoyl)methyl phosphoro-dithioate], a well-known product from the phosphorylated acetate class, is a systemic insecticide with contact toxic activity.

Owing to the importance of this product, many companies have sought independent synthetic processes. Some of these have the mixed anhydride as a common intermediate (for details, see refs. 288, 231-235). The most important are given here*.

An oxidation product of *dimethoate*, found as a metabolite of the same, is *omethoate* **(26)** [*O,O*-dimethyl *S*-(*N*-methylcarbamoyl)methyl phosphorothioate], which is also a significant commercial product in its own right. It is a systemic acaricide and insecticide.[236]

$$\begin{array}{c} H_3CO \diagdown \quad \diagup O \\ \qquad P \\ H_3CO \diagup \quad \diagdown S-CH_2-CO-NH-CH_3 \end{array}$$

26; Folimat®, *Omethoate*; LD$_{50}$: 50 mg/kg rat, oral, acute

Omethoate is used for control of sucking insects, in particular against resistant spider mites, and has high activity against biting pests.

One compound obtained by variation of the amide side chain of *dimethoate* is *morphothion* **(27)** [*O,O*-dimethyl *S*-(morpholinocarbonyl)methyl phosphorodithioate], which has systemic and contact activity.[237]

$$\begin{array}{c} H_3CO \diagdown \quad \diagup S \\ \qquad P \\ H_3CO \diagup \quad \diagdown S-CH_2-CO-N \diagup \diagdown O \end{array}$$

27; *Morphothion*

The following are other variations on the dimethoate theme that have achieved significance:

Medithionat [*O,O*-dimethyl *S*-(*N*-methoxyethylcarbamoyl)methyl phosphoro-dithioate];[238] LD$_{50}$: 420-650 mg/kg rat, oral, acute.

$$\begin{array}{c} H_3CO \diagdown \quad \diagup S \\ \qquad P \\ H_3CO \diagup \quad \diagdown S-CH_2-CO-NH-CH_2-CH_2-OCH_3 \end{array}$$

Formothion [*O,O*-dimethyl *S*-(*N*-formyl-*N*-methylcarbamoyl)methyl phosphoro-dithioate];[239] LD$_{50}$: 353 mg/kg rat, oral, acute.

$$\begin{array}{c} H_3CO \diagdown \quad \diagup S \\ \qquad P \qquad\qquad\qquad CH_3 \\ H_3CO \diagup \quad \diagdown S-CH_2-CO-N \diagdown \\ \qquad\qquad\qquad\qquad\qquad CHO \end{array}$$

Cyanthoate {*O,O*-diethyl *S*-[*N*-(1-cyano-1-methylethyl)carbamoyl]methyl phosphorothioate};[240] LD$_{50}$: 3 mg/kg rat, oral, acute.

*See page 72.

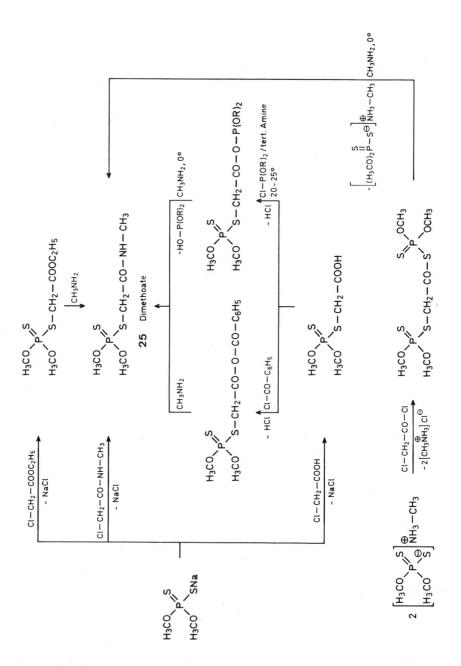

$$H_5C_2O\diagdown P\diagup^O_{\diagdown S-CH_2-CO-NH-\underset{\underset{CH_3}{|}}{\overset{\overset{CN}{|}}{C}}-CH_3}$$

Mecarbam [*O,O*-diethyl *S*-(*N*-ethoxycarbonyl-*N*-methylcarbamoylmethyl) phosphorodithioate];[214] LD$_{50}$: 31–35 mg/kg rat, oral, acute (aphicide and acaricide with ovicidal activity).

$$H_5C_2O\diagdown P\diagup^S_{\diagdown S-CH_2-CO-N\diagup^{CH_3}_{\diagdown COOC_2H_5}}$$

6. *Special S-substituted phosphorothioates (acyl = —S—subst. alkyl).* The heading "special phosphorothioates" refers to the class of compounds obtained by addition of phosphorodithioic acid derivatives onto activated C=C double bonds. For example, addition of *O,O*-dimethyl phosphorodithioic acid to diethyl maleate yields *malathion* (28) [*O,O*-dimethyl *S*-(1,2-dicarbethoxy)ethyl phosphorodithioate].

$$H_3CO\diagdown P\diagup^S_{\diagdown SH} \;+\; \overset{HC-COOC_2H_5}{\underset{HC-COOC_2H_5}{||}} \xrightarrow{\text{cat. amount alkali}} H_3CO\diagdown P\diagup^S_{\diagdown S-\underset{\underset{CH_2-COOC_2H_5}{|}}{CH}-COOC_2H_5}$$

28; *Malathion*

Owing to its good insecticidal and acaricidal properties combined with a low mammalian toxicity (LD$_{50}$: 1200 mg/kg rat, oral, acute), *malathion* is used on arable crops. In malaria zones it is also employed for the eradication of the *Anopheles* mosquito.[242]

Addition of the disulfan **29** to 1,4-dioxen leads to Delnav® [*dioxathion; S,S'*-(1,4-dioxane-2,3-diyl) *O,O,O',O'*-tetraethyl di(phosphorodithioate)],[243] mainly used in the veterinary sector.

$$\begin{array}{c}\text{O}\\[-2pt]\text{dioxene}\\[-2pt]\text{O}\end{array} + \; H_5C_2O\diagdown P\diagup^S_{\diagdown S-S}\diagdown P\diagup^{OC_2H_5}_{...} \xrightarrow{\text{Iodine (trace)}} \text{30}$$

29

30; *Dioxathion*, Delnav®

The reaction of *O,O*-diethyl phosphorodithioic acid with 2,3-dichloro-1,4-dioxen also yields Delnav®.[244] The *cis* and *trans* derivatives differ from each other in their toxicity (LD$_{50}$: *cis* form 65 mg/kg, *trans* form 240 mg/kg rat, oral, acute).

2.3.2.3.3. *O-Phenyl Phosphoroates and Phosphonates (Acyl = O-aryl).*

Most of the organophosphates—constructed according to the acyl rule—that have achieved a lasting significance in an economic, biological, and historical sense belong to the class of phenyl esters of phosphoric and phosphonic acids. They are synthesized by reaction of phenoxides with phosphorochloridates and phosphonochloridates.

The acylation of 4-nitrophenol with *O,O*-diethyl phosphorochloridate thus leads to *paraoxon* (diethyl 4-nitrophenyl phosphate).[245]

$$H_5C_2O{-}\underset{H_5C_2O}{\overset{O}{P}}{-}Cl \quad + \quad NaO{-}\langle\!\langle\ \rangle\!\rangle{-}NO_2 \quad \xrightarrow{-\ NaCl} \quad H_5C_2O{-}\underset{H_5C_2O}{\overset{O}{P}}{-}O{-}\langle\!\langle\ \rangle\!\rangle{-}NO_2$$

1; *Paraoxon*;
LD$_{50}$: 3 mg/kg rat, oral, acute

Owing to its high toxicity, this exceptionally active insecticide did not achieve any practical significance except in the field of ophthalmology as a miotic agent (pupil contractant) under the name Mintacol®.

Even more toxic (LD$_{50}$: 1 mg/kg rat, oral, acute) is Armin® —the corresponding phosphonate (ethyl 4-nitrophenyl ethylphosphonate), which is also used in opthalmology and in gynaecoiogy.[246]

$$H_5C_2{-}\underset{H_5C_2O}{\overset{O}{P}}{-}O{-}\langle\!\langle\ \rangle\!\rangle{-}NO_2$$

2; Armin®

The most important insecticide in this class is *parathion** (E 605®), for which there are various manufacturing processes that differ from one another only in the preparation of the phosphorochlorothioate.

$$H_5C_2O{-}\underset{H_5C_2O}{\overset{S}{P}}{-}Cl \quad + \quad NaO{-}\langle\!\langle\ \rangle\!\rangle{-}NO_2 \quad \xrightarrow{-\ NaCl} \quad H_5C_2O{-}\underset{H_5C_2O}{\overset{S}{P}}{-}O{-}\langle\!\langle\ \rangle\!\rangle{-}NO_2$$

3; *Parathion*; E 605®,
LD$_{50}$: 6.4 mg/kg rat, oral, acute

Parathion has a wide spectrum of activity embracing both biting and sucking pests and is a contact, stomach, and respiratory poison. It has a rapid initial activity, but the duration is variable and highly dependent on the substrate. The insecticidal activity is summarized in Table 10.[247]

The dimethyl ester *parathion-methyl* [*O,O*-dimethyl *O*-(4-nitrophenyl) phosphorothioate] is practically equivalent to *parathion* in activity and is used at least as widely.

---

*The registration of patents on this product was prevented in 1945 by the general circumstances of the last year of war. After the confiscation of the German documents by the Allies, parathion was developed in the United States and marketed under the name of Thiopor and Parathion. Only in 1948 was it possible to obtain patent protection for the German development E 605®.[245]

TABLE 10. Biological Activity of Parathion

	LC_{50}
Yellow-fever mosquito larvae (*Aedes aegypti*)	0.000005%
Potato aphid (*Macrosiphon solanifolii*)	0.0005%
Red spider mite (*Tetranychus urticae*)	0.001%
Citrus red mite (*Paratetranychus citri*)	0.0001%
Greenhouse thrips (*Heliothrips haemorrhoidalis*)	0.0001%
House fly (*Musca domestica*)	0.0001%

$$\begin{array}{c} H_3CO \\ \diagdown \\ \diagup P \diagdown \\ H_3CO \end{array} \!\!\!\! \overset{S}{=} \!\!\!\! O - \!\!\! \langle \!\!\! \bigcirc \!\!\! \rangle \!\!\! - NO_2$$

4; *Parathion-methyl*

As a dimethyl ester, *parathion-methyl* is more easily biodegraded than *parathion* and thus has a lower mammalian toxicity (LD_{50}: 15–20 mg/kg rat, oral, acute).

Parathion-methyl rearranges either thermally or in the presence of catalytic amounts of alkali to the thiolo ester.

$$\begin{array}{c} H_3CS \\ \diagdown \\ \diagup P \diagdown \\ H_3CO \end{array} \!\!\!\! \overset{O}{=} \!\!\!\! O - \!\!\! \langle \!\!\! \bigcirc \!\!\! \rangle \!\!\! - NO_2$$

5

The intermediate is an ambidentate anion that may be alkylated at sulfur (see Section 2.3.2.3.2).[248]

The phosphonothioate *EPN* [*O*-ethyl *O*-(4-nitrophenyl) phenylphosphonothioate] is a commercial product[249] of relatively low toxicity with well-proven control of spider mites, grasshoppers, and mosquito larvae.

$$\begin{array}{c} H_5C_2O \\ \diagdown \\ \diagup P \diagdown \\ H_5C_6 \end{array} \!\!\!\! \overset{S}{=} \!\!\!\! O - \!\!\! \langle \!\!\! \bigcirc \!\!\! \rangle \!\!\! - NO_2$$

6; *EPN*; LD_{50}: 36 mg/kg rat, oral, acute

The introduction of a methyl group into the phenyl ring reduces the toxicity without fundamentally changing the activity. Folithion® (Sumithion®) [*0,0*-dimethyl *O*-(3-methyl-4-nitrophenyl) phosphorothioate] is a contact and stomach insecticide.[250-252]

$$\begin{array}{c} H_3CO \\ \diagdown \\ \diagup P \diagdown \\ H_3CO \end{array} \!\!\!\! \overset{S}{=} \!\!\!\! O - \!\!\! \langle \!\!\! \bigcirc \!\!\! \rangle \!\!\! - NO_2$$
$$CH_3$$

7; *Fenitrothion*, Folithion®; LD_{50}: 500 mg/kg rat, oral, acute

Substituents in the *meta* position to the oxygen function (e.g., CH_3, Cl) reduce the toxicity drastically. In many cases the activity remains unaffected.

Thus Chlorthion® [*O,O*-dimethyl *O*-(3-chloro-4-nitrophenyl) phosphorothioate] is an insecticide with contact, stomach, and respiratory toxic activity but a lower mammalian toxicity.[253]

8; Chlorthion®; LD_{50}: 880 mg/kg rat, oral, acute

In contrast, the *ortho* isomer *dicapthon* [*O,O*-dimethyl *O*-(2-chloro-4-nitrophenyl) phosphorothioate] has a favorable toxicity but unfavorable insecticidal properties.[254]

9; *Dicapthon*; LD_{50}: 400 mg/kg rat, oral, acute

To improve the clarity of the following list, all phosphates and phosphonates derived from otherwise unmodified halophenols have been arranged according to the number of halogen atoms.

a. VC-13-Nemacide® (*dichlofenthion*) [*O,O*-diethyl *O*-(2,4-dichlorophenyl) phosphorothioate];[255] against soil nematodes (LD_{50}: 270 mg/kg rat, oral acute).

b. *S*-Seven® [*O*-ethyl *O*-(2,4-dichlorophenyl) phenylphosphonothioate];[256] acaricide, soil insecticide.

c. Nankor® (*fenchlorphos, ronnel*) [*O,O*-dimethyl *O*-(2,4,5-trichlorophenyl) phosphorothioate];[257] against ectoparasites.

d. Agritox®, Phytosol® (*trichloronat*) [*O*-ethyl *O*-(2,4,5-trichlorophenyl) ethylphosphonothioate];[258] for control of vegetable fly larvae and soil pests in meadows (LD$_{50}$: 37.5 mg/kg rat, oral, acute).

e. Nexion® (*bromophos*) [*O,O*-dimethyl *O*-(4-bromo-2,5-dichlorophenyl) phosphorothioate];[259] for control of flies and of ectoparasites on cattle, etc.; LD$_{50}$: 3000 mg/kg rat, oral, acute.

f. Nuvanol N® (*iodfenphos*) [*O,O*-dimethyl *O*-(4-iodo-2,5-dichlorophenyl) phosphorothioate].[259]

g. Phosvel® (*leptophos*) [*O*-methyl *O*-(4-bromo-2,5-dichlorophenyl) phenyl-phosphonothioate].[260]

h. Ruelene® (*crufomate*) [*O*-methyl *O*-(2-chloro-4-tert.butyl-phenyl) *N*-methyl phosphoroamidate][261] for control of intestinal parasites (as a substitute for phenothiazine) (LD$_{50}$: 100 mg/kg rat, oral, acute). It is synthesized by acylation of the corresponding phenol with phosphorus oxychloride, followed by stepwise replacement of the chlorines by methoxy and methylamino groups.

It is noticeable that all the *O*-phenyl phosphates and phosphonates just mentioned are derivatives of relatively acidic phenols.[262] Metcalf and Fukuto have found a correlation between the inhibition of fly-brain AChE by a phenyl ester and the influence of the phenyl ring substituents on the lability of the P–O–phenyl bond. The lability was related to the Hammett σ value. The logarithm of I_{50} (the molar concentration necessary for 50% inhibition of the fly-brain AChE), when plotted against the σ value of the substituents, gives almost a straight line. The outstanding insecticidal properties of the aklylmercaptophenol esters are thus, at first glance, even more suprising because the phenols are only weakly acidic (see also ref. 263).

The key to this contradiction lies in the fact that the thioether compound is not the active agent but merely a transport form. In mammalian and insect organisms the thioether is oxidized to the sulfoxide, which is regarded as the genuine active form (see also Section 2.3.4.1).

The sulfoxide is a derivative of 4-(methylsulfinyl)phenol, which is about as acidic as 4-nitrophenol. The 4-(methylsulfinyl)phenyl ester in fact has the same insecticidal properties as the mercapto compound.

The simplest compound of this nature is the dimethyl phosphorothioate of 4-(methylmercapto)phenol (**10**).

10

10; LD$_{50}$: 10 mg/kg rat, oral, acute

The principle that the toxicity of a phosphate can be reduced by the introduction of a methyl group into the position *ortho* to the substituent has also been employed here. *Fenthion* (Lebaycid®) {*O,O*-dimethyl *O*-[3-methyl-4-(methylthio)-phenyl] phosphorothioate} is such an example.

11; *Fenthion*; LD$_{50}$: 250 mg/kg rat, oral, acute

Lebaycid® has many applications. It is a contact and stomach insecticide with a long residual action. It is miscible with most insecticides and fungicides and has an exceptionally high activity against, for example, fruit flies, leafhoppers, cereal bugs, and rice stem borers. In the hygiene sector, it bears the name Baytex® and gives effective control of mosquitoes and flies with activity also against strains resistant to chlorohydrocarbons. Under the name Tiguvon® it finds application in veterinary medicine.[264]

The following technical synthesis of 3-methyl-4-(methylmercapto)phenol is worthy of mention.[265]

12

Fenamiphos [*O*-ethyl *O*-(3-methyl-4-methylthiophenyl) *N*-isopropyl phosphoroamidate], a phosphoroamidate of methylmercaptophenol, is a systemic nematicide. A point of interest is the synthesis, which deviates from the usual phosphorylation reactions[266] (compare also the list in Section 2.3.2.3.3, item h):

13; *Fenamiphos*; LD_{50}: 15–20 mg/kg, rat, oral, acute

New developments in the insecticide field are Tokuthion® (*prothiofos*) [*O*-ethyl *S*-propyl *O*-(2,4-dichlorophenyl) phosphorodithioate] and Bolstar® (*sulprofos*) [*O*-ethyl *O*-(4-methylthiophenyl) *S*-propyl phosphorodithioate].[267]

14; Tokuthion®; **15;** Bolstar®, *Sulprofos*;
LD_{50}: 1134 mg/kg rat, oral, acute LD_{50}: 227 mg/kg rat, oral, acute

Both may be synthesized by basically the same reaction sequence.

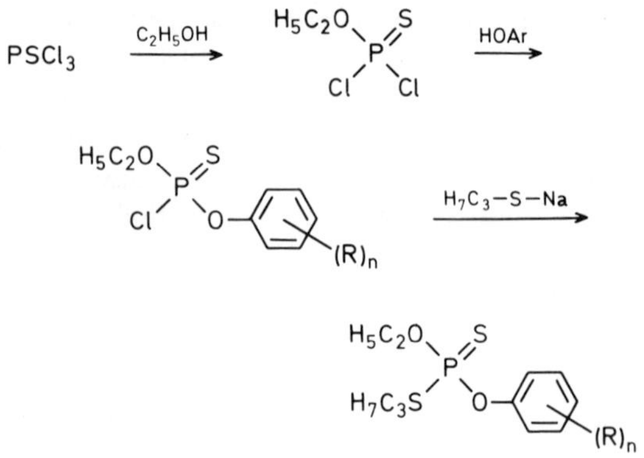

$$PSCl_3 \xrightarrow{C_2H_5OH} \underset{Cl}{\overset{H_5C_2O}{\underset{Cl}{\diagdown}}}\overset{S}{\underset{}{P}} \xrightarrow{HOAr}$$

$$\underset{Cl}{\overset{H_5C_2O}{\diagdown}}\overset{S}{P}\diagdown O-\text{(R)}_n \xrightarrow{H_7C_3-S-Na}$$

$$\underset{H_7C_3S}{\overset{H_5C_2O}{\diagdown}}\overset{S}{P}\diagdown O-\text{(R)}_n$$

Tokuthion® is an insecticide with high activity against leaf-eating caterpillars and other chewing insects, and also useful activity against spider mites and aphids.

Bolstar® is used against *Heliothis zea* on cotton. It has a low phytotoxicity and is especially active against pests such as *Spodoptera* and *Pectinophora*.

To conclude the mercaptophenyl derivatives, Abate® *(temephos)* is worthy of mention. Abate® [O,O,O',O'-tetramethyl O,O-thiodi-p-phenylene bis (phosphorothioate)] has a low acute toxicity (LD_{50}: 2000 mg/kg albino rat) and is principally used as a mosquito larvicide.[268]

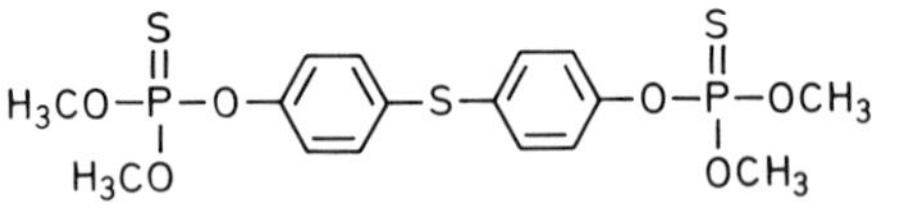

$$H_3CO-\overset{\overset{S}{\|}}{\underset{\underset{H_3CO}{|}}{P}}-O-\bigcirc-S-\bigcirc-O-\overset{\overset{S}{\|}}{\underset{\underset{OCH_3}{|}}{P}}-OCH_3$$

16; Abate®, *Temephos*

Owing to the strongly electron-withdrawing cyano group, the 4-cyanophenyl esters have good insecticidal properties, for example:

a. Cyanox® *(cyanophos)* [O,O-dimethyl O-(4-cyanophenyl) phosphorothioate];[269] against rice stem borers and house flies.

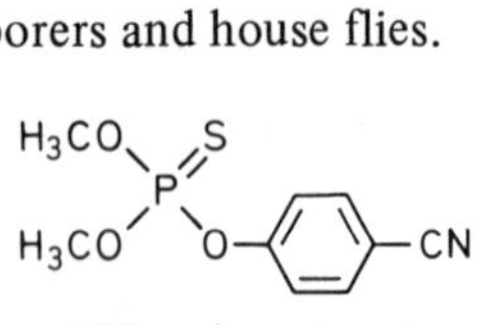

$$\underset{H_3CO}{\overset{H_3CO}{\diagdown}}\overset{S}{P}\diagdown O-\bigcirc-CN$$

LD_{50}: 920 mg/kg rat, oral, acute

b. Surecide® *(cyanofenphos)* [O-ethyl O-(4-cyanophenyl) phenylphosphonothioate];[270] activity spectrum analogous to Cyanox®, see above.

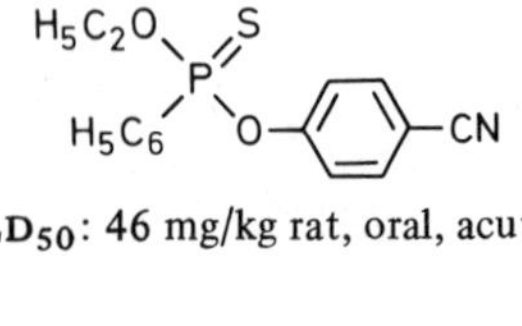

$$\underset{H_5C_6}{\overset{H_5C_2O}{\diagdown}}\overset{S}{P}\diagdown O-\bigcirc-CN$$

LD_{50}: 46 mg/kg rat, oral, acute

A recently introduced product for control of soil pests is Oftanol® (*isofenphos*) {*O*-ethyl *O*-[2-(isopropoxycarbonyl)phenyl] *N*-isopropyl phosphoroamidothio-ate}.[271]

17; Oftanol®; LD$_{50}$: 38.7 mg/kg rat, oral, acute

Even the weakly electron-withdrawing aminosulfonyl group suffices for insecticidal activity. For example, *famphur* {*O,O*-dimethyl *O*-[4-(dimethylsulfamoyl)-phenyl] phosphorothioate} is a systemic insecticide with nematicidal and anthelmintic properties.[272]

18; Famophos®, *Famphur*; LD$_{50}$: 35 mg/kg rat, oral, acute

Among the cyclic phosphates there are also insecticidal derivatives. Salioxon® (2-methoxy-4H-1,3,2-benzodioxa-phosphoran-2-one) and its thiono analogue, Salithion® are made by phosphorylation of salicylalcohol.[273]

19; Salioxon®; LD$_{50}$: 0.5–1 mg/kg rat, oral, acute

20; Salithion®; LD$_{50}$: 102 mg/kg rat, oral, acute

These esters were discovered in connection with studies on the metabolism of tri-*o*-cresyl phosphate, which is itself not insecticidal but causes paralysis in mammals by damaging the nerve system (delayed neurotoxicity).[274]

The *o*-tolyl ester "SM-1" was found to be a toxic metabolite of tri-*o*-cresyl phosphate. The first step is the hydroxylation of a methyl group, followed by an intramolecular transphosphorylation with expulsion of one *o*-cresol molecule.

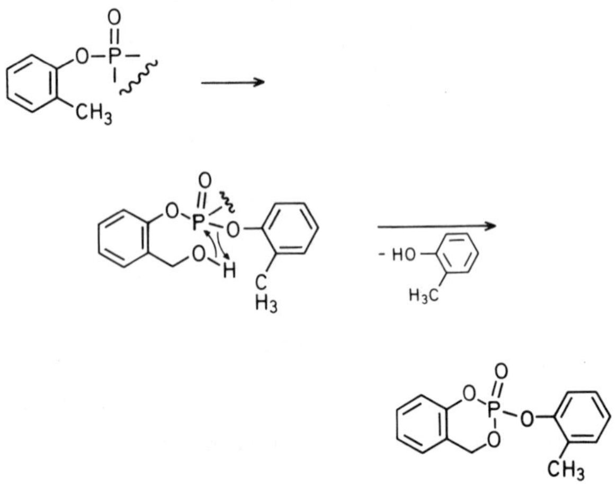

"SM – 1"

Dyfonate® (*fonofos*) (*O*-ethyl *S*-phenyl ethylphosphonodithioate) (acyl = —S—C$_6$H$_5$) is a soil insecticide with a good stability toward hydrolysis.[275]

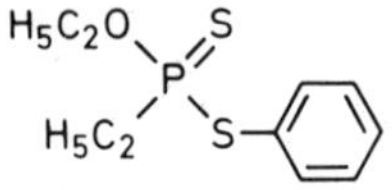

Dyfonate®; LD$_{50}$: 8–17 mg/kg rat, oral, acute

2.3.2.3.4. O-Heteroaryl Phosphoroates and Phosphonates (Acyl = O-Heteroaryl).

In this section, phosphoric and phosphonic esters of heterocyclic hydroxy compounds are described in which the hydroxy group is attached either to an annelated benzene ring or to the heterocycle itself.

In the post-war years, Potasan® [*O,O*-diethyl *O*-(4-methyl-2-oxo-2H-1-benzopyran-7-yl) phosphorothioate] was used for control of Colorado potato beetle.[276]

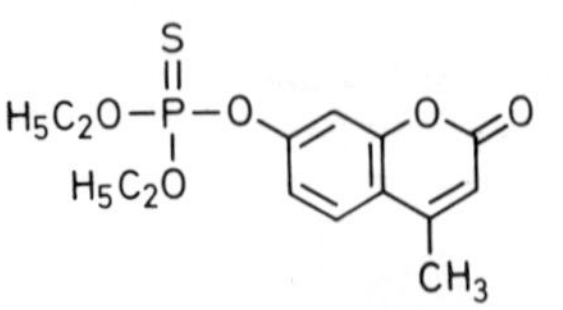

1; Potasan®;
LD$_{50}$: 19 mg/kg rat, oral, acute

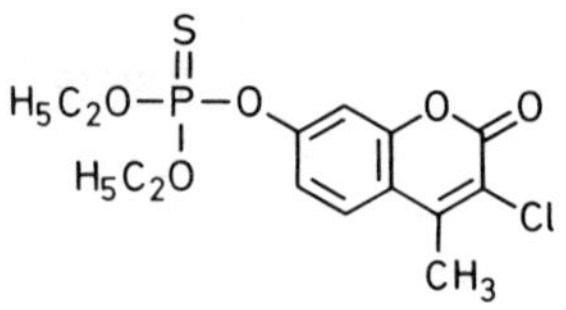

2; *Coumaphos*

This coumarin derivative is synthesized by condensation of resorcinol with acetoacetate. *Coumaphos* [*O,O*-diethyl *O*-(3-chloro-4-methyl-2-oxo-2H-1-benzopyran-7-yl) phosphorothioate] is the chloro derivative of Potasan®. Introduction of the chlorine atom causes a reduction in toxicity (LD_{50}: 100 mg/kg rat, oral, acute), on the one hand, and a thousandfold increase in activity against mosquito larvae, on the other.

This compound is used in veterinary medicine under the names Co-Ral® and Resitox® for control of ectoparasites in livestock. Good resistance to hydrolysis permits use in dips.[277]

Haloxon® [*O,O*-di(2-chloroethyl) *O*-(3-chloro-4-methyl-2-oxo-2H-1-benzopyran-7-yl) phosphorothioate] is a well-known anthelminthic.[278] Steinberg's synthesis is given here (see Section 2.3.2.2.3).

3; Haloxon®; LD_{50}: 950 mg/kg rat, oral, acute

Quintiofos (Bacdip®) [*O*-ethyl *O*-(quinolyl-8) benzenephosphonothioate] is an effective tickicide.[279]

4; *Quintiofos*

Quintiofos has a relatively low mammalian toxicity (LD_{50}: 150 mg/kg rat, oral, acute) and high resistance to hydrolysis, which permits use in dips.[279]

Dursban® (*chlorpyrifos*) [*O,O*-diethyl *O*-(3,5,6-trichloropyridine-2-yl) phosphorothioate] is used for mosquito control but may also be employed against ectoparasites on domestic animals.[280]

5; Dursban®; LD_{50}: 135–163 mg/kg rat, oral, acute

Pyridaphenthion (Ofunack®, Ofnack®) [*O,O*-diethyl *O*-(3-oxo-2-phenyl-2H-pyridazine-6-yl) phosphorothioate] is a promising, less-toxic insecticide for control of *Heliothis* sp. and the rice stem borer.[280a]

5a; *Pyridaphenthion*; LD$_{50}$: 769 mg/kg rat, oral, acute

The most well-known heterocyclic phosphate derivative is *diazinon* {*O,O*-diethyl *O*-[2-isopropyl-4-(methyl)pyrimidine-6-yl] phosphorothioate}, which has widespread application as a contact insecticide and acaricide.

6; *Diazinon*; LD$_{50}$: 108 mg/kg rat, oral, acute

The pyrimidine ring is formed by condensation of isobutyramidine with ethyl acetoacetate.[281]

Another pyrimidine derivative, *etrimfos*[281a] (Ekamet®) {*O,O*-dimethyl *O*-[6-ethoxy-2-(ethyl)pyrimidine-4-yl] phosphorothioate}, acts as a contact and stomach poison and has a low phytotoxicity.[281b]

6a; *Etrimfos*; LD$_{50}$: 1800 mg/kg rat, oral, acute

It is used in vegetables, fruit, and grapes for control of biting and sucking insects, and it is also effective against some aphid species.

Pirimiphos-methyl (Actellic®) [*O,O*-dimethyl *O*-(2-diethylamino-6-methyl-4-pyrimidinyl) phosphorothioate], also a pyrimidine derivative, is an insecticide and acaricide that acts as a contact and respiratory poison and is used mainly in cereal growing and stored-product protection.[281c]

6b; *Pirimiphos-methyl*; LD$_{50}$: 2050 mg/kg rat, oral, acute

Zinophos® [*O,O*-diethyl *O*-(2-pyrazinyl) phosphorothioate], a systemic nemati-cide, is very toxic.[282]

7; Zinophos®; LD$_{50}$: 5 mg/kg rat, oral, acute

The heterocycle is made by condensation of 1,2-diketones with aminoacetamide.

Quinalphos (Bayrusil®, Ekalux®) [*O,O*-diethyl *O*-(2-quinoxalinyl) phosphoro-thioate] is obtained by condensation of *o*-phenylenediamine with the hemi-acetal of glyoxylate.[283]

8; *Quinalphos*

The compound is highly active against biting and sucking insects and has an LD$_{50}$ of 70 mg/kg (rat, oral, acute).[283]

Pyrazothion® [*O,O*-diethyl *O*-(3-methylpyrazol-5-yl) phosphorothioate], a sys-temic insecticide and acaricide with relatively low mammalian toxicity, has high activity against aphids and spider mites.[281]

9; Pyrazothion®; LD$_{50}$: 36 mg/kg rat, oral, acute

In conclusion, a triazole derivative is worthy of mention. *Triazophos* (Hosta-thion®) [*O,O*-diethyl *O*-(1-phenyl-1,2,4-triazol-3-yl) phosphorothioate], although not systemic, has a good penetration action, long duration of activity, and is effec-tive against spider mites.[283a]

9a; *Triazophos*; LD$_{50}$: 83 mg/kg rat, oral, acute

The triazole moiety is produced by ring closure of phenyl semicarbazide with triethylformate.[283b]

2.3.2.3.5. *O-Phosphorylated Hydroxamates and Hydroximates (Acyl = —O—N=).*

Maretin® (*O,O*-diethyl *O*-naphthaloximido phosphate), a phosphorylated derivative of hydroxamic acid, is an anthelmintic used for the eradication of stomach and intestinal worms.[284]

$$\begin{array}{c}H_5C_2O \\ H_5C_2O\end{array} P \begin{array}{c}O \\ \end{array}$$

Maretin®; LD$_{50}$: 75 mg/kg rat, oral, acute

The starting material is obtained from naphthalic anhydride and hydroxylamine.

Volaton® (*O,O*-diethyl α-cyanobenzylideneamino-oxyphosphonothioate) has been introduced recently for control of *Heliothis* spp. on cotton. It has an extraordinarily low mammalian toxicity.

$$\begin{array}{c}H_5C_2O \\ H_5C_2O\end{array} P \begin{array}{c}S \\ O-N=C\end{array} \begin{array}{c}CN \\ \end{array}$$

Phoxim; LD$_{50}$: 2500 mg/kg rat, oral, acute

Volaton® (*phoxim*) is highly suitable for the control of older larva stages.[285] Under the trade name Baythion®, it is used for control of stored-product pests such as grain weevils in empty silos, granaries, warehouses, vessels, and port facilities.

2.3.2.3.6. N-Phosphorylated Imines (Acyl = $-N=C\underset{<}{}$). An interesting phosphoroimidate is Cyolane® (*phosfolan*) (diethyl-1,3-dithiolan-2-ylidenephosphoroamidate), which has high activity against chewing and sucking insects and spider mites. The synthesis of the required starting material begins with ethane-1,2-dithiol, which is reacted with cyanogen chloride in a nonpolar solvent in the presence of hydrogen chloride.[286,287]

$$\begin{array}{ccc}\text{SH} & & \\ | & + \text{ CICN} & \xrightarrow{\text{HCl}} \\ \text{SH} & & \end{array} \begin{bmatrix} S \\ | \\ S \end{bmatrix}=\overset{\oplus}{N}H_2 \end{array} Cl^{\ominus} \xrightarrow[\text{(dry)}]{\underset{OC_2H_5}{\overset{O}{Cl-P-OC_2H_5}}/N(C_2H_5)_3} \begin{array}{c} S \\ | \\ S \end{array}=N-\overset{O}{\underset{OC_2H_5}{P}}-OC_2H_5$$

The systemic soil insecticide and nematicide Nem-A-Tak® ("*fosthietan*" proposed) (diethyl 1,3-dithietan-2-ylidenephosphoroamidate) also belongs to this class.[287a]

$$H_5C_2O-\underset{\underset{H_5C_2O}{|}}{\overset{\overset{O}{||}}{P}}-N=\begin{array}{c} S \\ S \end{array}$$

Nem-A-Tak® ("*Fosthietan*" proposed), AC 64475; LD$_{50}$: 5 mg/kg rat, oral, acute

2.3.2.3.7. Enol Phosphates (Acyl = $-O-\overset{|}{C}=C\!\!<$). The enol phosphates are obtained simply and economically by the Perkow reaction. The simplest and most well known is *DDVP* (*dichlorvos*) [*O,O*-dimethyl *O*-(2,2-dichlorovinyl) phosphate], which is obtained by reaction of chloral with trimethyl phosphite[288] or with dimethyl phosphite (followed by alkaline rearrangement of the addition product and loss of hydrogen chloride[289]).

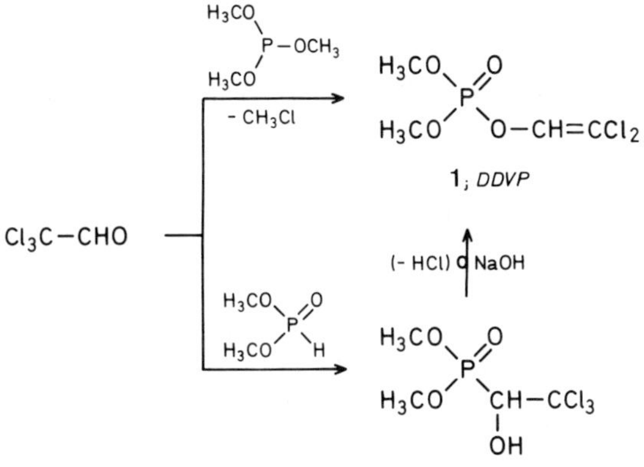

1; *DDVP*

2; *Trichlorfon*

It is an insecticide and acaricide of relatively high volatility and with contact and respiratory toxic activities. *DDVP* is used in the hygiene sector and in agriculture for control of biting and sucking pests and leaf miners (LD$_{50}$: 62 mg/kg rat, oral, acute).

The addition product **2** is also a commercially important compound called Dipterex® (*trichlorfon*) [*O,O*-dimethyl (1-hydroxy-2,2,2-trichloro)ethylphosphonate]. This is a contact and stomach insecticide used mainly against dipterous and lepidopterous pests (LD$_{50}$: 630 mg/kg rat, oral, acute) and is relatively nontoxic toward bees. The low toxicity also permits use in veterinary medicine for control of ecto- and endoparasites.

Naled [*O,O*-dimethyl *O*-(1,2-dibromo-2,2-dichloroethyl) phosphate], obtained by addition of bromine onto the C=C double bond of *DDVP*, is used in the hygiene sector.

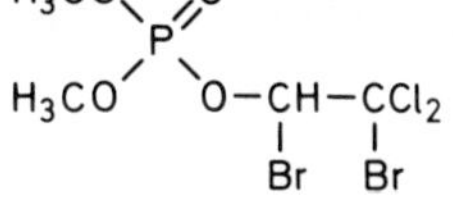

3; Dibrom®, *Naled*; LD$_{50}$: 450 mg/kg rat, oral, acute

Trialkyl phosphites are the starting materials for other halogenated vinyl esters:

a. *Forstenon* {*O,O*-diethyl *O*-[2,2-dichloro-1-(2-chloroethoxy)vinyl] phosphate};[290] contact insecticide; LD$_{50}$: 7–10 mg/kg rat, oral, acute.

$$H_5C_2O \diagdown P-OC_2H_5 \quad + \quad Cl_3C-CO-O-CH_2-CH_2-Cl \quad \longrightarrow$$

$$H_5C_2O \diagup$$

$$H_5C_2O \diagdown \overset{O}{\underset{\diagup}{P}} \diagdown O-C=CCl_2$$
$$H_5C_2O \diagup \qquad O-CH_2-CH_2-Cl$$

b. *Chlorfenvinphos*, Dermaton® {*O,O*-diethyl *O*-[2-chloro-1-(2,4-dichloro-phenyl)vinyl] phosphate};[291] soil insecticide; LD$_{50}$: 10–39 mg/kg rat, oral, acute.

$$H_5C_2O \diagdown P-OC_2H_5 \quad + \quad Cl- \underset{Cl}{\overset{Cl}{\bigcirc}} -CO-CHCl_2 \quad \longrightarrow$$
$$H_5C_2O \diagup$$

$$H_5C_2O \diagdown \overset{O}{\underset{\diagup}{P}} \diagdown O-C=CH-Cl$$
$$H_5C_2O \diagup$$

c. *Tetrachlorvinphos*, Gardona® {*O,O*-dimethyl *O*-[2-chloro-1-(2,4,5-trichlor-phenyl)vinyl] phosphate};[292] contact insecticide against *Lepidoptera* and *Diptera*; LD$_{50}$: 4000–5000 mg/kg rat, oral, acute.

$$H_3CO \diagdown \overset{O}{\underset{\diagup}{P}} \diagdown O-C=CH-Cl$$
$$H_3CO \diagup$$

Of great interest is the reaction of acetoacetate derivatives with trialkyl phosphites. A series of important commercial products stems from this, of which one is *mevinphos* (2-methoxycarbonyl-1-methylvinyl dimethyl phosphate).[292]

$$H_3CO \diagdown P-OCH_3 \quad + \quad H_3C-CO-\underset{Cl}{\overset{|}{CH}}-COOCH_3 \quad \longrightarrow \quad H_3CO \diagdown \overset{O}{\underset{\diagup}{P}} \diagdown O-C=CH-COOCH_3$$
$$H_3CO \diagup \qquad\qquad\qquad\qquad\qquad\qquad\qquad\qquad H_3CO \diagup \qquad\qquad \underset{CH_3}{|}$$

Phosdrin®, *Mevinphos*; LD$_{50}$: 3.7 mg/kg rat, oral, acute

In principle, the acylation of the sodium salt of methyl acetoacetate with phosphorochloridate is also possible, but this has not been realized technically.

Mevinphos is a mixture of *cis* and *trans* isomers of which the *cis*-form is ~100 times more active. *Mevinphos* is a systemic insecticide and acaricide with contact and respiratory toxic properties. It is remarkable in that it is rapidly degraded in the plant, thus permitting use shortly before harvesting. After 24 hr the agent is almost completely degraded.

The less-toxic Ciodrin® (*crotoxyphos*) [dimethyl 1-methyl-2-(1-phenylethoxy-carbonyl)vinyl phosphate][293] is closely related to *mevinphos* but is used for control of ectoparasites on cattle, horses, sheep, and swine.

$$H_3CO \diagdown \underset{\displaystyle H_3CO \diagup}{\overset{\displaystyle O}{P}} \diagdown O-C=CH-CO-O-CH \diagdown \\ | |$$

$$ CH_3 CH_3$$

Ciodrin®; LD$_{50}$: 140–200 mg/kg rat, oral, acute

Further systemic insecticides and acaricides are Bidrin® (dimethyl 2-dimethyl-carbamoyl-1-methylvinyl phosphate):

$$H_3CO \diagdown \overset{O}{P} \diagdown O-C=CH-CO-N(CH_3)_2 \\ H_3CO \diagup | \\ CH_3$$

Bidrin®, *Dicrotophos*; LD$_{50}$: 22 mg/kg rat, oral, acute

and the almost identically active Azodrin® (dimethyl 1-methyl-2-methylcarbamoyl-vinyl phosphate):

$$H_3CO \diagdown \overset{O}{P} \diagdown O-C=CH-CO-NH-CH_3 \\ H_3CO \diagup | \\ CH_3$$

Azodrin®, *Monocrotophos*; LD$_{50}$: 21 mg/kg rat, oral, acute

Phosphamidon (2-chloro-2-diethylcarbamyl-1-methylvinyl dimethyl phosphate), very similar to *mevinphos* in toxicity, has a broad spectrum of activity against biting and sucking pests and spider mites.[294]

$$H_3CO \diagdown \overset{O}{P} \diagdown O-C=C-CO-N(C_2H_5)_2 \\ H_3CO \diagup | | \\ H_3C Cl$$

Phosphamidon; LD$_{50}$: 10 mg/kg rat, oral, acute

A more recent enol phosphate is *heptenophos* (Hostaquick®) (7-chlorobicyclo [3.2.0] hepta-2,6-dien-6-yl dimethyl phosphate).[294a]

$$H_3CO-\overset{\overset{\displaystyle O}{\|}}{\underset{\displaystyle H_3CO}{P}}-O$$

Heptenophos; LD$_{50}$: 98–117 mg/kg rat, oral, acute

Manufacture is according to the Perkow reaction:

$$P(OCH_3)_3 \quad + \quad \cdots \longrightarrow \quad H_3CO-\overset{\overset{\displaystyle O}{\|}}{\underset{\displaystyle H_3CO}{P}}-O$$

Heptenophos is a systemic insecticide and acaricide with a short residual effect.

2.3.3. STRUCTURE AND ACTIVITY RELATIONSHIPS

The ecological and biochemical behavior of the organophosphorus pesticides is largely determined by their ester character. It is useful to distinguish the following aspects.

    a. Ease of chemical hydrolysis.
    b. Transesterification, phosphorylation.
    c. Dealkylation of triesters to diesters.
    d. Structure and toxic properties.
    e. Resistance phenomena.

    The first three modes of reaction may be combined with Schrader's rule (Section 2.3.2) so that it is possible to build up a qualitative picture of the properties of the phosphates.[176,178,295-297]

2.3.3.1. Hydrolysis Reactions

According to Schrader's rule (see Section 2.3.2), insecticidal phosphates must possess an electron-withdrawing substituent at phosphorus. Such compounds are susceptible to nucleophilic attack at the phosphorus atom, for instance:

$$\underset{R^1O}{\overset{R^1O}{>}}\!\!\overset{\delta^+}{P}\!\!\overset{O}{<}\!\!\underset{O-R^2}{} \quad \xrightarrow{OH^{\ominus}} \quad \underset{OR^1}{\overset{R^1O}{\underset{|}{HO\cdots P\cdots OR^2}}} \quad \xrightarrow{-R^2O^{\ominus}}$$

$$\underset{HO}{\overset{R^1O}{>}}P\!\!\overset{O}{<}\!\!\underset{OR^1}{} \quad \xrightarrow[-R^2OH]{R^2O^{\ominus}} \quad \underset{R^1O}{\overset{R^1O}{>}}P\!\!\overset{O}{<}\!\!\underset{O}{\ominus}$$

This reaction, analogous to S_N2 attack in carbon chemistry, only occurs in alkaline media. In neutral or acidic solution the course of the reaction is not so clear, since prior protonation equilibria play a part. Furthermore, in strongly acidic solution, direct attack of a water molecule on the phosphorus atom is possible. Whereas phenyl esters of the *parathion* type (see Section 2.3.2.3.3) are hydrolyzed relatively rapidly in alkaline solution but are relatively stable in neutral or acidic media, the speed of hydrolysis of the amidates **1**, for example, can be much higher in acidic media (in alkali the nitrogen lone electron pair is involved in the π-bonding system of the ester molecule and thus inhibits nucleophilic attack). In acidic media the formation of an ammonium phosphate **3** is favored by the prior protonation **2**.

In contrast, *N*-monoalkyl amidates are also susceptible to hydrolysis in alkaline media because the hydroxide ion attacks at the amino hydrogen and aids the expulsion of an ester group.

As would be expected, the enol phosphates are hydrolyzed relatively rapidly in alkali but have a half-life in acidic media a tenth that of the trialkyl phosphates. The cause of this is again a prior protonation, which favors the formation of the carbonyl compound corresponding to the enol.

For the thiono esters (**4**) the effective charge on phosphorus would be expected to be reduced as a result of the lower electronegativity of sulfur and the increased polarizability of sulfur in comparison with oxygen. This would mean increased stability toward hydrolysis, which is in fact found to be the case, as is demonstrated by the examples in Table 11.

$$
\begin{array}{ccc}
\text{4} & \text{5} & \text{6}
\end{array}
$$

Similar considerations also apply to the thiolo esters (**5**). Although the contribution of the sulfur atom to the π-bonding system in the molecule is lower than that

TABLE 11. Hydrolytic Half-Lives of Various Phosphates at 70°C in an Ethanol–Buffer Solution (1:4, pH 6)[298]

Structure	Name	Half-life (h)
$(H_5C_2O)_2\overset{\overset{O}{\|\|}}{P}-O-\langle\ \rangle-NO_2$	Paraoxon	28
$(H_5C_2O)_2\overset{\overset{S}{\|\|}}{P}-O-\langle\ \rangle-NO_2$	Parathion	43
$(H_3CO)_2\overset{\overset{S}{\|\|}}{P}-O-\langle\ \rangle-NO_2$	Methylparathion	8.4
$(H_3CO)_2\overset{\overset{S}{\|\|}}{P}-O-\langle\ \rangle-NO_2$, CH_3	Fenitrothion	11.2
$(H_5C_2O)_2\overset{\overset{S}{\|\|}}{P}-S-CH_2-S-C_2H_5$	Phorate	1.75
$(H_5C_2O)_2\overset{\overset{O}{\|\|}}{P}-S-CH_2-S-C_2H_5$	Phorate-Oxon	0.5
$(H_3CO)_2\overset{\overset{O}{\|\|}}{P}-O-CH=C\overset{Cl}{\underset{Cl}{}}$	Dichlorvos	1.3
$(H_3C)_2N\,\,O \\ \,\,\,\,\,\,\,P \\ (H_3C)_2N\,\,F$	Dimefox	212

of the oxygen analogues as a result of the lower electronegativity of sulfur, the effective charge on the phosphorus in thiolo esters should be somewhat higher. As expected, the thiolo compounds hydrolyze more rapidly than their oxygen counterparts in a given series (6).

Alkyl groups directly bound to phosphorus, as in the phosphonates (8) and the phosphinates (9), also contribute little toward compensating the partial positive charge on phosphorus when compared to the oxygens of the corresponding phosphates (7). The ease of hydrolysis in alkaline solution increases correspondingly from the phosphates over the analogous phosphonates to the phosphinates, whereas the stability under acidic conditions increases.

$$
\begin{array}{ccc}
\underset{\textbf{7}}{
\begin{array}{c}
R{-}CH_2{-}O \\
R{-}CH_2{-}O
\end{array}
\!\!\! P \!\!\!
\begin{array}{c}
O \\
OR^1
\end{array}
}
&
\underset{H^{\oplus}}{\overset{OH^{\ominus}}{\lessgtr}}
&
\underset{\textbf{8}}{
\begin{array}{c}
R{-}CH_2{-}O \\
R{-}CH_2
\end{array}
\!\!\! P \!\!\!
\begin{array}{c}
O \\
OR^1
\end{array}
}
\end{array}
$$

$$
\underset{H^{\oplus}}{\overset{OH^{\ominus}}{\lessgtr}}
\qquad
\underset{\textbf{9}}{
\begin{array}{c}
R{-}CH_2 \\
R{-}CH_2
\end{array}
\!\!\! P \!\!\!
\begin{array}{c}
O \\
OR^1
\end{array}
}
$$

In practice, phosphinates are not used owing to their lack of stability toward hydrolysis. Phosphonates are used occasionally (when the stability under acidic conditions is needed).

The hydrolytic degradation of phosphates is both qualitatively and quantitatively the most important detoxification reaction *in vivo*, that is, in living organisms. This is true both of the target organisms—insects—and of mammals. Water-soluble products result from this degradation, and although they are normally excreted rapidly, they are, at the very least, no longer capable of penetrating lipophilic membranes or surfaces. The insecticidal activity of a phosphate is thus a function of the ratio of the quantity of insecticide taken up to the amount hydrolyzed in a given period of time. This ratio is influenced by other parameters. If thiono esters are used as insecticidal agents then these are first metabolized to P-O esters by oxidase systems in the liver of mammals or in the lipoid tissue of insects. Hereafter the rate of hydrolysis, that is, the susceptibility toward detoxification, is considerably increased. On the other hand, since the P-S esters are more soluble in fat than the P-O esters, the former are better capable of penetrating lipophilic barriers.

The relationship between oxidative activation of P-S esters, their penetration, and the resulting toxicity, is shown in Figure 1. It is therefore possible for an intrinsically weak insecticide to penetrate so rapidly into the insect that the enzymatic hydrolysis capacity of the insect is insufficient and a lethal total toxicity results. One example of such cases is given in Figure 2. The insecticidal activity of *dimethoate* results mainly from the extraordinarily rapid penetration and consequential lethal concentrations at the sites of action.

The sensitivity toward hydrolysis, typical of the phosphate insecticides, is a decisive factor in their behavior in the environment. Under field conditions the phos-

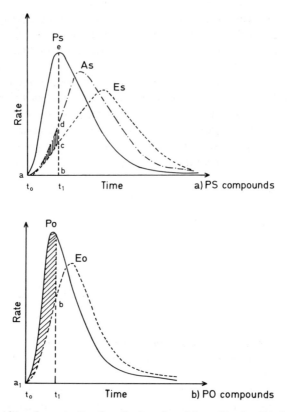

FIGURE 1. Rapidity of penetration P, activation A, and detoxification E in insects, according to Sun.[299] The resulting toxicity up to the point in time t_1 is hatched. In case (a) it is lower than in (b) by the area ade, since in the latter case the activation does not apply.

phates are exposed to hydrolytic influences, either on the plant or the soil surface, which can be of a purely chemical nature but which in the majority of cases are mediated by enzymes, for example, on the surface of leaves or after penetration into various parts of the plant. As a rule, the duration of activity of the organophosphorus insecticides lies between several days and a few weeks. In contrast, insecticides from other classes persist for months or even years. This means that accumulation phenomena are not to be feared with the phosphates. From a purely economic viewpoint of insecticide applications, however, the farmer is most interested in the longest duration of activity, by which the plant is protected from sowing until harvest. One can therefore go so far as to say that a more pressing problem in the development of new phosphate insecticides is sufficient stability in the field. Increasing the sensitivity toward hydrolysis influences means that degradation in the insect is so rapid that there is no chance for lethal concentrations to build up at the site of action *in vivo*.

Experience has shown that it is possible to estimate to a first approximation the lability of a phosphate toward hydrolysis from the pK value of the electron-with-

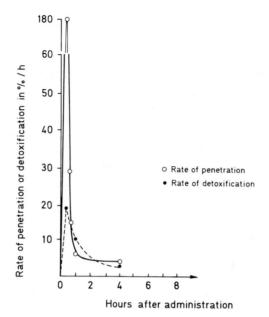

FIGURE 2. Rate of penetration and detoxification of dimethoate after topical application to house flies.[299]

drawing "acyl" substituent in Schrader's rule (Section 2.3.2). Three cases may be differentiated.[295]

a.   If the molecule to be phosphorylated has a pK value >8, then the phosphate resulting is too stable toward hydrolysis and, as will be discussed in the next section, too insensitive to transesterification so that no biological activity is observed. The Schrader rule, therefore, requires electron acceptor and not donor properties.

b.   In general, an optimum is to be found in a pK range of 6–8 with respect to Schrader's rule. That is, the insecticidal activity and instability toward hydrolysis are in a reasonable relation.

c.   Hydroxy compounds with strongly electron-withdrawing groups having a pK value below 6 are so susceptible to hydrolysis that they are rapidly detoxified and thus biologically inactive.

2.3.3.2.   Transesterification/Phosphorylation

In the same way as the hydrolysis, transesterifications are also governed by the partial charge on the phosphorus atom. Esters conforming to Schrader's rule may be transesterified with suitable alcohols. This is the basis of the mechanism of insecticidal action. The transesterification occurs with the primary alcohol group of a serine molecule in the enzyme acetylcholinesterase [3.1.1.7]. Seen from the viewpoint of the enzyme, this is a phosphorylation of the active site.

$$\begin{array}{c} RO \\ \diagdown \\ P \diagup \diagup O,\ S \\ RO \diagup \quad Y \end{array} + \text{Enzyme} - OH \quad \xrightarrow[- H-Y]{} \quad \begin{array}{c} RO \\ \diagdown \\ P \diagup \diagup O,\ S \\ RO \diagup \quad O - \text{Enzyme} \end{array}$$

Acetylcholinesterase plays an essential role in the transmission of nerve impulses. The electrical impulse built up in the nerve by fluctuations in the concentration of sodium and potassium ions reaches the so-called presynaptic membrane, where it must be transmitted chemically over the synapse to the postsynaptic membrane or a myoneural junction. The chemical transfer of the action potential is carried out by the transmitter, acetylcholine (10), which induces another potential on the postsynaptic side (receptor). After the impulse has been transferred, the transmitter compound must be removed, otherwise a permanent stimulation would result.

$$\underset{H_3C}{\overset{O}{\underset{\|}{C}}} - O - CH_2 - CH_2 - \underset{\overset{|}{CH_3}}{\overset{\overset{CH_3}{|}}{N}} \overset{\oplus}{-} CH_3$$

10

This is effected by hydrolysis to acetate and choline under the influence of the enzyme AChE [3.1.1.7]. If the AChE is blocked by suitable inhibitors, an endogenic acetylcholine intoxication results, the symptoms of which are similar either to the stimulation of the sympathetic system by nicotine or to the stimulation of the parasympathetic system by muscarine. This makes it clear that the site of action of the phosphates is to be found, on the one hand, at the myoneural junctions of the motor system, at the ganglia of the cholinergic system and, on the other hand, at the end plates of the parasympathetic branch in smooth muscle and glandular tissue. Further proof is furnished by the fact that the antagonist atropine corresponding to the agonist muscarine is also the most active antidote in cases of organophosphate intoxication.

All indications are such that this mechanism also holds true for the insecticidal action of the organophosphates, although the transmitter compound at the myoneural junctions in insects is most certainly not acetylcholine but an aminodicarboxylic acid. On the other hand, numerous experiments have demonstrated the presence of acetylcholine in insects.

The following equation due to Main[300] governs the kinetics of transesterification or phosphorylation at the site of action.

$$E + I \underset{k_2}{\overset{k_1}{\rightleftharpoons}} E - I_{rev} \overset{k_p}{\longrightarrow} EI_{irrev}$$

The enzyme E reversibly forms a complex E-I with the inhibitor I, the stability of which depends on the affinity of the inhibitor for the active site. The ratio $k_2 : k_1$ is therefore called the affinity constant K_a. The irreversible step leading to the phosphorylated enzyme is governed by the phosphorylation constant k_p. If Main's analysis[300] is carried out on different organophosphates, widely differing values are obtained (see p. 178 in ref. 178). This is explained by the fact that K_a is, among other things, a measure of the structural relationship of an inhibitor toward the natural substrate, acetylcholine, or toward the site of action itself, whereas k_p represents the binding relationships implied in Schrader's rule. The conclusion to be drawn from such kinetic considerations is that the activity profiles of biologically active phosphates can vary widely although a common mechanism of action exists. If, in addition, the different solubilities, that is, penetration and distribution properties in the insect organism, and the molecular changes possible as a result of metabolic degradation *in vivo* are taken into account, then it is perfectly understandable that one ester can be highly active against spider mites, another against leaf-eating caterpillars or against aphids, and a third can have an extremely wide spectrum. It is also understandable that, from a summation of all these variables, vast differences can arise between the insecticidal activity of an ester and its mammalian toxicity, which thus make the practical application of such compounds possible.

The construction of the active site in acetylcholinesterase is such that the transfer amino acid serine lies in spatial proximity to a polar group, the basic center, which, via an ionic interaction with the ammonium ion of acetylcholine, holds the substrate to the esterase site serine. It is not clear from pK considerations whether this polar group is a carboxylate or phosphorus ion. In the majority of cases the alcohol group of serine does not react directly with the phosphate and it is necessary to postulate a general base catalysis by the imidazole ring of a histidine residue lying near the serine. According to Leuzinger (see p. 166 in ref. 178), AChE [3.1.1.7] has a molecular weight of 260,000 ± 10,000 and is composed of two protomers, each with an α-chain that contains the active site and a β-chain with an as-yet-unknown function. One can imagine the situation for acetylcholinesterase to be somewhat like that described by Lehninger[301] for α-chymotrypsin.

The possibility that other enzymes apart from AChE are influenced by organophosphates must be taken into account.[302] It is difficult, however, to determine experimentally whether such additional inhibitory reactions are part of the primary mechanism of action or just a consequence of the original cholinesterase inhibition.

2.3.3.3. Structure and Toxicity

In view of the similar mechanism of action of the organophosphates for mammals and insects it is not surprising that, particularly with the first products from this class, the extremely high insecticidal activity was in some cases coupled with a relatively high toxicity (the acute toxicity is expressed as the LD_{50}, i.e., the dosage that is lethal to 50% of the test animals). In the meantime, one has learned to exploit differences between mammalian and insect metabolisms more effectively with the structures of new products—that is, to improve the safety margins.[295] For example, a switch from PO to PS analogues brings a reduction in the acute toxicity, partly by reducing the phosphorylating activity in accordance with Schrader's rule and partly because in mammals a metabolic oxidative activation of the PS compound to PO must first take place, which leaves more time for the hydrolytic detoxification processes (see Figure 1). An additional reduction of toxicity is achieved by switching from diethyl phosphates to the dimethyl analogues. The cause here is the additive effect of enzymatic demethylation by glutathione, which is only present in insects in insignificant quantities. Steric effects can be exploited in *O*-phenyl phosphates. For example, the introduction of a chlorine atom in the position *meta* to the oxygen function (the change from *parathion-methyl* to Chlorthion®) raises the acute oral LD_{50} in the rat from 14 mg/kg to 625 mg/kg. The analogous introduction of a methyl group leads to *fenitrothion*, with an LD_{50} of 250 mg/kg (rat, oral, acute).

The substituent in the *meta* position alters the affinity for the enzyme. Expressed more precisely, it increases the affinity toward insect acetylcholinesterase and reduces the affinity toward mammalian acetylcholinesterase. The phosphorylating activity falls slightly for both types of enzyme. Steric hindrance and a reduction of the mesomeric effect, owing to the nitro group being prevented from adopting a position planar to the ring, are probably responsible. The affinity toward the enzyme is mainly dependent on the steric effects of the phosphate molecule. The formation of an enzyme-inhibitor complex with insect AChE is apparently promoted by an interaction of the *meta*-alkyl group with the "anionic site" of the enzyme, whereas this complex formation with mammalian AChE is inhibited. As the following diagram shows, the distance between the "anionic" and "esteratic" sites in insect AChE (5.0-5.5 Å) is different from that in mammalian AChE (4.3-4.7 Å). The *m*-alkyl group, 5.2-6.5 Å apart from the phosphorus atom, fits well into the "anionic position" of the insect enzyme but badly into that of the mammalian enzyme.[303]

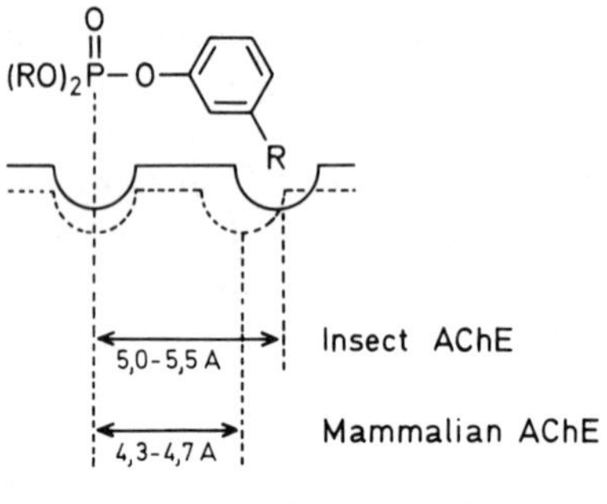

This explanation has the disadvantage that it is only valid for the *O,O*-dimethyl phosphates. A comparison of the *O,O*-diethyl analogues shows that a methyl group in the *meta* position is practically without effect.

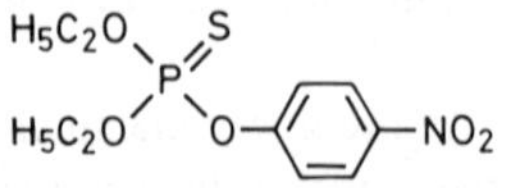

Parathion; LD_{50}: ~7 mg/kg rat, oral, acute 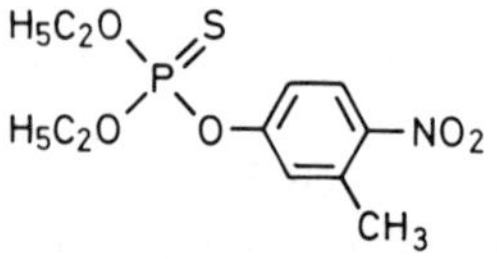 LD_{50}: ~10 mg/kg rat, oral, acute

It is inexplicable why the sterically controlled interaction between the benzene ring and the enzyme should not apply to the *O,O*-diethyl phosphates.

The introduction of so-called selectophoric groups is a further possibility for reducing toxicity. These are groups that are susceptible to specific enzymes. An example is carboxylesterase, which hydrolyzes, and thus detoxifies, *malathion* to "malathionic acid." This mechanism of detoxification is only available to insects on a minor scale. Other selectophoric groups are the alkylaminocarbonyl and dialkylaminocarbonyl, which are subject to the selective attack of amidases in mammals. Selective, oxidative dealkylation also plays a role here.

The principle of selectophoric groups, however, brings with it the danger of a relatively rapid buildup of resistance phenomena because of the selection of insect strains with increased ability to carry out the same degradation reaction.

Recent examples of an improvement in the safety margin are the *O*-ethyl *S*-propyl phosphorothioates, which have a much higher LD$_{50}$. This may, of course, be explained by a greater sensitivity toward hydrolysis, but additional, as yet unknown, degradation mechanisms cannot be excluded.

Some organophosphates such as tri-*O*-cresyl phosphate (TOCP) lead to characteristic symptoms of poisoning in which 1–2 weeks later paralysis of the extremities occurs. In contrast to the usual inhibition of the acetylcholine cleavage by blocking of the AChE, the axon itself is injured. The myelin sheath detaches itself and a "dying back" of the nerve occurs. This neurotoxic action was at first difficult to explain until the structure of the primary metabolite was taken into account (see ref. 297, p. 219). In the case of TOCP, a methyl group is hydroxylated enzymatically. The second step is an internal transesterification with elimination of one *o*-cresol molecule. This secondary ester is the real neurotoxic substance, not TOCP.

Another such group is the fluorides of phosphates and phosphonates, which are almost without exception neurotoxic. Here the special properties of the fluoride ion are responsible. Fluoride is a lipophilic, highly basic group with a comparatively high resistance toward hydrolysis. The phosphorus derivative resulting is a small molecule easily capable of penetrating the lipophilic barriers of the nerve system and thus reaching other sites of action not accessible to normal phosphates. As practical insecticides they have not been able to gain acceptance on the market. It is therefore essential that an additional "instability" be built into the ester molecule. This is the reason why—apart from the fluorides—no neurotoxic action has been found when at least one *O*-methyl, or mono- or dimethylamino, group is attached to the phosphorus atom. Apart from their favorable acute toxicities, di-*O*-methyl esters are thus found among the safest phosphates also with respect to neurotoxicity.

2.3.3.4. Resistance Phenomena

When arthropod pests are controlled for long periods of time with the same insecticide, then the formation of a new population can occur. This requires increasing concentrations of insecticide for a lethal effect to be seen. The appearance of resistance phenomena is most often seen in species with a high reproduction rate and rapid succession of generations, the control of which requires correspondingly more frequent spray treatments. The mechanisms of resistance may be of a varied nature, but genetically fixed tolerance probably plays a greater role than, for example, physiological mechanisms. In spite of their identical mechanism of action in principle, individual phosphates vary widely in their ability to initiate resistance phenomena. This allows the conclusion to be made that in the total mechanism of action, other factors must be involved apart from the AChE inhibition. A corollary is that it must then be possible to overcome organophosphate resistance with new phosphates, as has been demonstrated many times in practice. Resistance phenomena are not a property specific to the phosphates. For the purpose of survival of the species, all species defend themselves against all lethal factors, the so-called third-generation pesticides being no exception.

2.3.4. METABOLISM

All organisms are forced by their environment to take up foreign compounds and try to protect themselves from the same by endeavoring to convert them to innocuous substances, which are then excreted. Animal and plant organisms have their own degradation routes. Plants modify foreign compounds to a form that can react with peptides, proteins, or carbohydrates. Animal organisms convert foreign compounds into polar, water-soluble substances that are excreted as such or in the form of conjugates. Not all metabolic conversions are in fact detoxifications. For example, the phosphorothionates are converted into the phosphates, which are the actual AChE inhibitors. This activation is thus also the first step in the degradation. The phosphates are detoxified to inactive metabolites by hydrolysis and so on. Basically, there is a competition between activation and detoxification, the activity being determined by the more rapid reaction. Simple chemical reactions are involved in these conversions.

1. Oxidation	Thiono $\rightarrow$ oxo-form	
	Oxidative dealkylation	
	Thioether $\rightarrow$ sulfoxide $\rightarrow$ sulfone	
	Oxidation of aliphatic substituents	
	Hydroxylation of an aromatic ring	
2. Reduction	Nitro $\rightarrow$ amino group	
	Other reductions	
3. Isomerization		
4. Hydrolysis	Enzymatic	
	Nonenzymatic	
5. Dealkylation	Triester $\rightarrow$ diester	
6. Degradation at the carboxy group	"Saponification"	ester $\longrightarrow$ acid amide $\longrightarrow$ acid
7. Conjugation	Hydroxy compounds with glucuronic acid	
	Hydroxy compounds with sulfate	

As a rule, insecticides are not degraded by just one of the above mechanisms. Several reactions occur simultaneously.

In mammals the detoxification of the insecticides is mainly carried out in the liver owing to its high esterase content, the toxic oxygen compounds being hydrolyzed.

2.3.4.1. Oxidation

Thiono → *oxo-form*. Among the oxidation processes—which belong to the most important metabolic processes—one must differentiate according to the site of oxidative attack. The oxidation of sulfur converting phosphorothionates to phosphates does not signify detoxification (see Section 2.3.3.1) but activation, that is, an effect enhancing the toxic activity. In vertebrates oxidation takes place in the liver microsomes in the presence of $NADPH_2$ and oxygen,[304] in insects in the fat body.

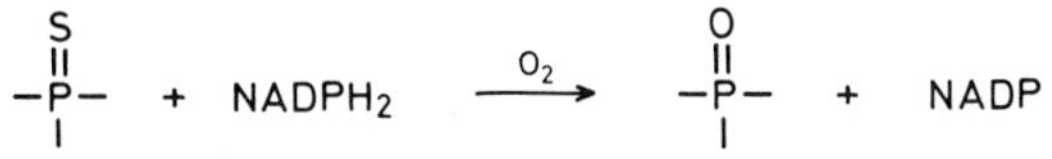

$$\underset{\underset{|}{\overset{\overset{S}{\|}}{-P-}}}{} + \quad NADPH_2 \quad \xrightarrow{O_2} \quad \underset{\underset{|}{\overset{\overset{O}{\|}}{-P-}}}{} + \quad NADP$$

This oxidation, which partly occurs in muscle tissue, is fundamental for thiono esters.[305] The enzyme, whose activity maximum lies at pH 7.5, is itself inactivated during the activation process. Possibly the oxygen molecule is reduced by the peroxidase-hydrogen donor system to hydroperoxy radicals,[306] which then react with the thiono sulfur. This is an attempt at explaining the apparently paradoxical fact that, in an oxidation process, a reducing agent is required. The oxidation product (P=O) is generally more water soluble and more toxic.

Oxidative dealkylation. Alkyl groups bound to nitrogen are removed by oxidative N-dealkylation,[307] that is, the substituted amides such as *dimethoate, dicrotophos, OMPA*, and Azodrin® are degraded to the unsubstituted amides,[307] for example:

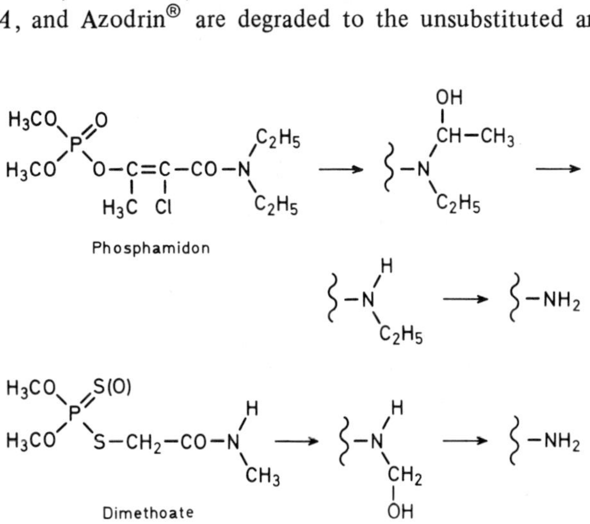

Phosphamidon

Dimethoate

The hydroxyalkyl intermediates have been isolated.

In the degradation of *phosphamidon* in bean plants, oxidative dealkylation ceases at the monoalkyl stage; the hydrolysis reaction participates simultaneously.[308]

$$\underset{HO}{\overset{H_3CO}{\diagdown}}\overset{O}{\underset{\diagup}{P}}{\diagdown}\underset{\underset{\underset{Cl}{|}}{H_3C}}{O-C}=\underset{|}{C}-CO-\underset{\diagdown}{\overset{\diagup R}{N}}R$$

$$\underset{H_3CO}{\overset{H_3CO}{\diagdown}}\overset{O}{\underset{\diagup}{P}}{\diagdown}\underset{\underset{\underset{Cl}{|}}{H_3C}}{O-C}=\underset{|}{C}-CO-\underset{\diagdown R}{\overset{\diagup R}{N}}$$

Phosphamidon

$$\underset{H_3CO}{\overset{H_3CO}{\diagdown}}\overset{O}{\underset{\diagup}{P}}{\diagdown}\underset{\underset{\underset{Cl}{|}}{H_3C}}{O-C}=\underset{|}{C}-CO-\underset{\diagdown R}{\overset{\diagup H}{N}} \longrightarrow \underset{H_3CO}{\overset{H_3CO}{\diagdown}}\overset{O}{\underset{\diagup}{P}}{\diagdown}OH \quad + \quad H_3C-CO-\underset{\underset{Cl}{|}}{CH}-CO-\underset{\diagdown R}{\overset{\diagup H}{N}}$$

$$\underset{H_3CO}{\overset{H_3CO}{\diagdown}}\overset{O}{\underset{\diagup}{P}}{\diagdown}OH \quad + \quad H_3C-CO-\underset{\underset{Cl}{|}}{CH}-CO-\underset{\diagdown R}{\overset{\diagup R}{N}} \longrightarrow \text{Smaller degradation products}$$

R = C₂H₅

Azodrin® is converted to the *N*-hydroxymethyl derivatives in insects and mammals, but simple hydrolysis is the main reaction in cotton plants.[309]*

The degradation of Bidrin® follows a similar path.[310] *O,O*-Dimethyl phosphoric acid and desmethylbidrin are found in plants. In rats hydrolysis occurs at the vinyl ester bond, leading to dimethyl phosphoric acid. The hydrolysis is accelerated by hydroxymethylation. Related compounds like Azodrin® and Bidrin® can be interconverted whereby the common intermediate is the *N*-hydroxymethyl compound.

The oxidative demethylation of *OMPA*[311] has been known for the longest time.

$$(H_3C)_2N-\overset{\overset{O}{\|}}{\underset{\underset{(H_3C)_2N}{|}}{P}}-O-\overset{\overset{O}{\|}}{\underset{\underset{N(CH_3)_2}{|}}{P}}-N(CH_3)_2$$

OMPA

$$\left[(H_3C)_2N-\overset{\overset{O}{\|}}{\underset{\underset{(H_3C)_2N}{|}}{P}}-O-\overset{\overset{O}{\|}\,\overset{O}{\uparrow}}{\underset{\underset{N(CH_3)_2}{|}}{P}}-N(CH_3)_2\right]$$

$$\left[(H_3C)_2N-\overset{\overset{O}{\|}}{\underset{\underset{(H_3C)_2N}{|}}{P}}-O-\overset{\overset{O}{\|}}{\underset{\underset{N(CH_3)_2}{|}}{P}}-\overset{\overset{CH_2-OH}{|}}{\underset{}{N}}-CH_3\right]$$

$-CH_2O$

$$(H_3C)_2N-\overset{\overset{O}{\|}}{\underset{\underset{(H_3C)_2N}{|}}{P}}-O-\overset{\overset{O}{\|}}{\underset{\underset{N(CH_3)_2}{|}}{P}}-NH-CH_3$$

*See scheme on p. 103.

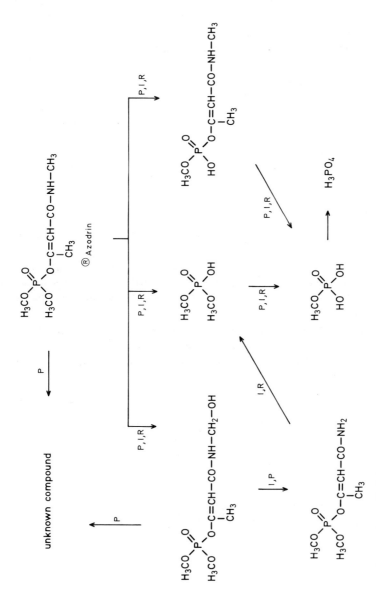

P = Plant; I = Insect; R = Rat

103

Intermediates that have not yet been identified are held responsible for the anti-cholinesterase activity.

Thioether → sulfoxide → sulfone. An important degradative route is the oxidation to the corresponding sulfoxides and sulfones observed in aliphatic and aromatic thioethers. This oxidation occurs before the oxidation of the phosphoro-thionate sulfur and is illustrated with a few examples.

Phorate forms sulfoxides and sulfones in plants.[312]

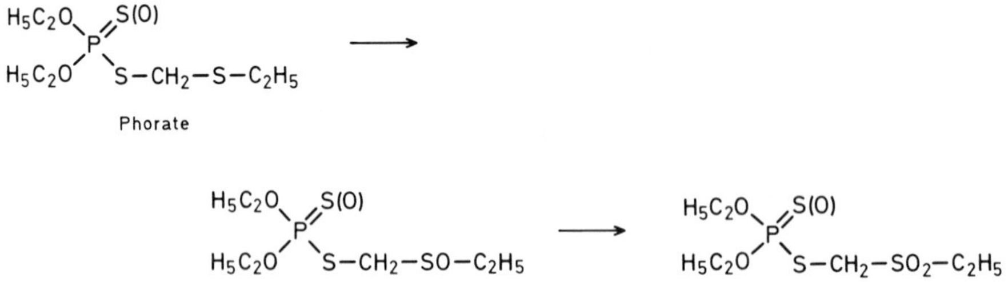

$$
\begin{array}{c}
H_5C_2O \diagdown \diagup S(O) \\
P \\
H_5C_2O \diagup \diagdown S-CH_2-S-C_2H_5
\end{array}
\qquad \longrightarrow
$$

Phorate

$$
\begin{array}{c}
H_5C_2O \diagdown \diagup S(O) \\
P \\
H_5C_2O \diagup \diagdown S-CH_2-SO-C_2H_5
\end{array}
\quad \longrightarrow \quad
\begin{array}{c}
H_5C_2O \diagdown \diagup S(O) \\
P \\
H_5C_2O \diagup \diagdown S-CH_2-SO_2-C_2H_5
\end{array}
$$

The same mechanism holds true for Trithion®, Systox®, Disyston®, *fenthion*, *phenamiphos*, and so on.

In the case of Lebaycid® the sulfoxide is the actual active form, the initial compound being merely the transport form. The toxicity and anticholinesterase activity increase on oxidation but not to the degree found on oxidation of phos-phorathionate to phosphate (cf. ref. 313).

The degradation of Lebaycid® under the influence of light, temperature, and plant enzymes in beans is shown in the following scheme.[314] *

Abate® is also metabolized to the corresponding sulfoxide and sulfone in mos-quito larvae.[315]

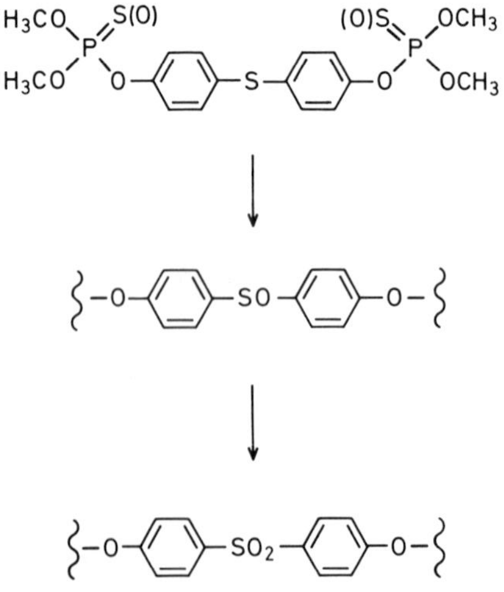

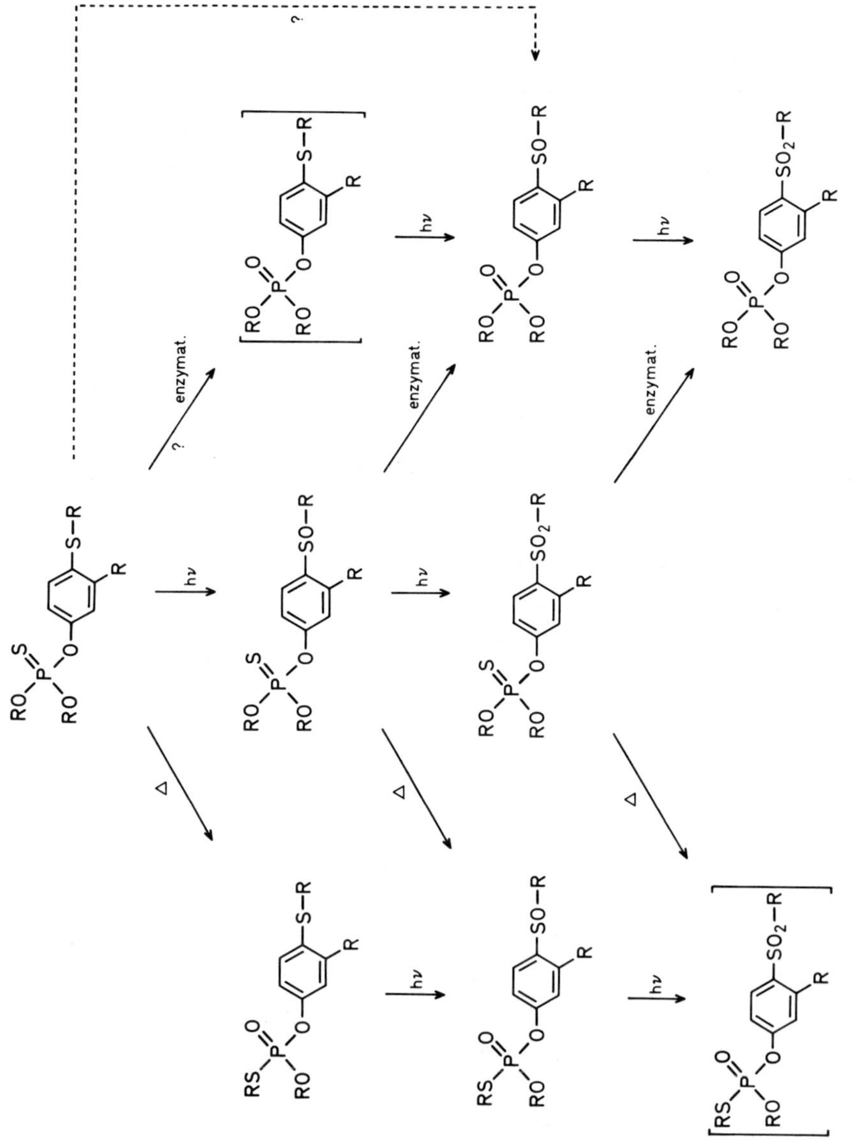
?
enzymat.
hν
Δ

Oxidation of thioethers also plays a role in the metabolism of the degradation products of Dyfonate® and *phosalone*.

These insecticides are first oxidized enzymatically. Thiophenol (from Dyfonate®) and the *N*-methylmercapto derivative **1** (from *phosalone*) are formed by hydrolysis, methylated by enzymes to the thioethers, and then further oxidized to sulfoxides and sulfones. It may be assumed that a mixed oxidase system, which is also responsible for the oxidation of phosphorothionate to phosphate, participates; for instance, the degradation of Dyfonate® in the rat.

The methylsulfonylbenzene formed from Dyfonate® is then hydroxylated in the 3- or 4-positions on the ring. The intermediacy of the arene-epoxide, which has been observed in other cases, has been assumed from the nature of the end products.[316] *In vitro* experiments on the thioether oxidation have not been carried out.

In the case of *phosalone*, other degradation pathways are found apart from the sulfone formation observed only in rats. In plants and in soil, cleavage of the C–S and P–S bonds occurs to yield 6-chloro-1,3-benzoxazolone (2), formaldehyde, and *O,O*-diethyl phosphorodithioic acid. The benzoxazolone 2 is converted either to 2-amino-7-chloro-3-oxo-3H-phenoxazine 3 as shown or, in plants, to the *N*-gluco-side 4. The phenoxazine derivative 3 and the sulfone are found side by side in the rat.[317]*

Oxidation of aliphatic substituents. The oxidation of aliphatic substituents may be illustrated with the example of *diazinon*. Apart from the oxidation to *diazoxon*, the mammalian organism (e.g., sheep) hydroxylates the aliphatic side chain in the α-position.[318] Some of the metabolites also inhibit cholinesterase.

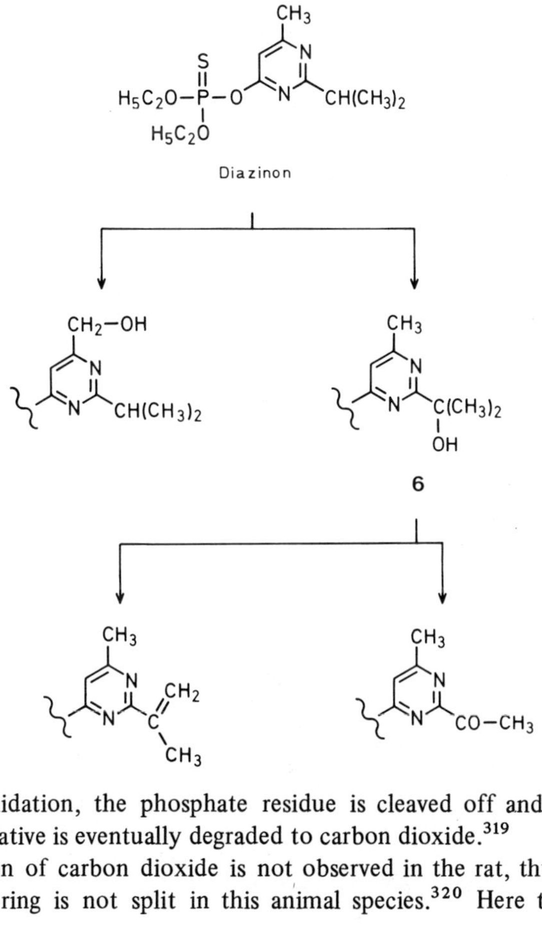

$$H_5C_2O-\overset{\overset{S}{\|}}{\underset{\underset{H_5C_2O}{|}}{P}}-O-\text{[pyrimidine ring: CH}_3\text{, N, N, CH(CH}_3)_2\text{]}$$

Diazinon

CH₂—OH pyrimidine with N, N–CH(CH₃)₂

CH₃ pyrimidine with N, N–C(CH₃)₂, OH

6

CH₃ pyrimidine with N, N–C(=CH₂)–CH₃

CH₃ pyrimidine with N, N–CO—CH₃

After the oxidation, the phosphate residue is cleaved off and the remaining pyrimidine derivative is eventually degraded to carbon dioxide.[319]

The formation of carbon dioxide is not observed in the rat, thus proving that the pyrimidine ring is not split in this animal species.[320] Here the pyrimidone

*See scheme on p. 108.

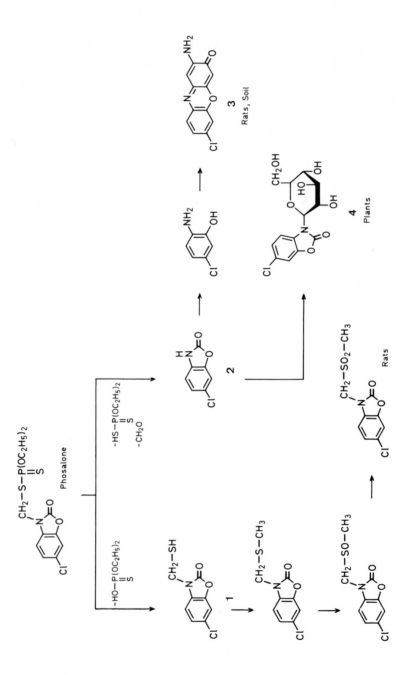

derivative **5** appears as an additional metabolite. In plants the α-hydroxy compound **6** has been detected as well as *diazoxon*.

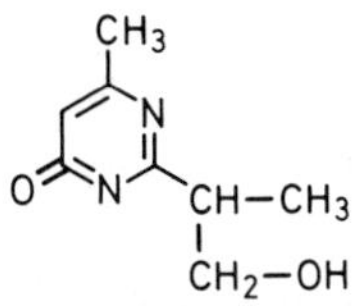

5

Hydroxylation of an aromatic ring. Although Hinosan® and Kitazin P® are fungicides and therefore, strictly speaking, have no place in this section on insecticides, they serve as useful examples for the enzymatic hydroxylation of an aromatic ring. This metabolic step has been detected in rats, cockroaches, and fungi.

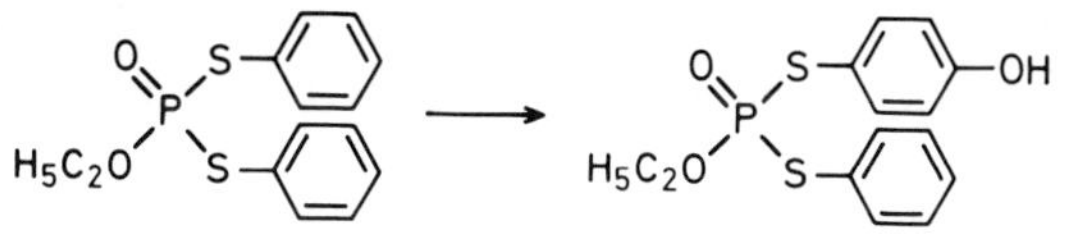

Hinosan®

Kitazin P®

2.3.4.2. Reduction

Parathion is the best example of the enzymatic reduction of nitro groups. The first step is the direct reduction of *parathion* by rumen fluid.[321] Oxidation of the thiono sulfur and hydrolysis also occur. The reduction requires NADPH as cofactor.

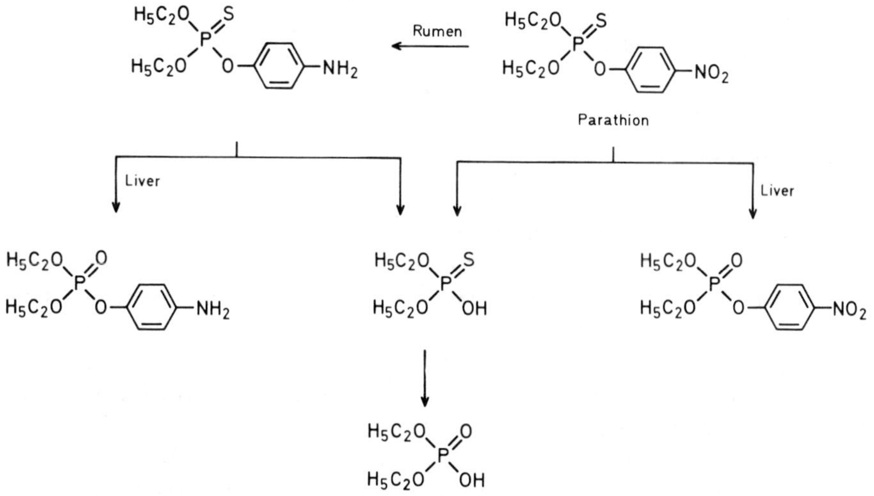

Parathion

Parathion, paraoxon, and the corresponding amino derivatives circulate in the blood and also appear in the milk. The 4-aminophenol released is excreted in the form of a glucuronide or sulfonate.

This reduction also occurs in all compounds related to *parathion* such as *parathion-methyl, EPN-O,* and Chlorthion®. The biological activity of the *p*-amino derivatives is not as high as that of the *p*-nitro compounds, but they are more resistant toward hydrolysis.

The reduction of *fensulfothion* to the corresponding thioether in soil and plants has also been reported.[322]

Fensulfothion

A reductive dechlorination of chlorinated enol phosphates is observed in plants and mammals. 2-(2,4,5-Trichlorophenyl)ethanol is formed from *tetrachlorvinphos.*[323]

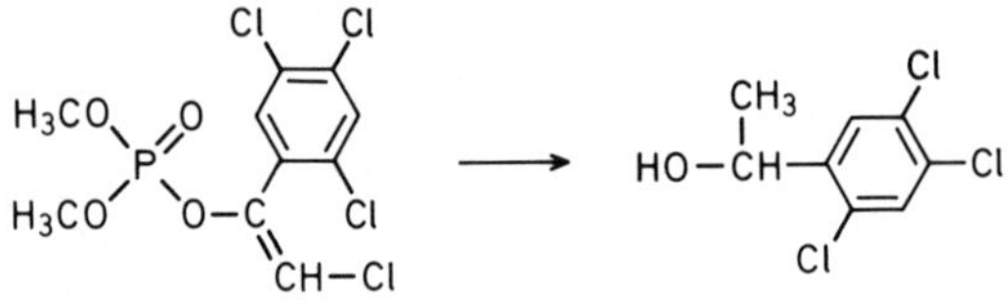

Tetrachlorvinphos

2.3.4.3. Isomerization

In the case of thiono phosphates, the possibility exists of isomerization to the thiolo phosphates. This isomerization is only observed with *parathion* at higher temperatures but has been found, for example, with *fensulfothion* in plant tissue.[324]

Fensulfothion

2.3.4.4. Hydrolysis

The most important reaction apart from the oxidation of thiono sulfur is hydrolysis, which accompanies almost all degradative reactions. The enzymes that cleave phosphates are generally referred to as phosphatases but are frequently named after the substrate that is cleaved, for example, paraoxonase. Enzymatic hydrolysis is the most significant detoxification reaction of phosphates. In the case of detoxification the phosphoryl group is transferred to water. Intoxication occurs when the phosphoryl is transferred to the serine–alcohol group in cholinesterase. The two reactions compete with one another. Maximum insecticidal activity is favored when the agent is taken up and transported in a form less sensitive to degradation reactions. Activation of the "transport form" to the "active form" should occur at the site of action. In the case of phosphorothionates the activity is determined by the following equation.[325]

$$\text{Toxicity} \sim \frac{\text{Enzymatic oxidation}}{\text{Enzymatic hydrolysis}}$$

The enzymatic hydrolysis of phosphates occurs primarily with loss of the acyl substituent or an *O*-alkyl residue. The primary metabolites are normally hydrolyzed further, as is shown by some examples.

Cyanox® can be hydrolyzed in animal organisms by cleavage either of methanol or 4-cyanophenol. The cyano group is not attacked.[326] The following degradative pathway has been suggested.

Paraoxon, O,O-diethyl phosphoric acid, *O,O*-diethyl phosphorothioic acid, and 4-nitrophenol have been detected in the *in vitro* metabolic breakdown of *parathion*. This degradation is inhibited under an atmosphere of nitrogen or carbon dioxide. The breakdown also requires NADPH and molecular oxygen. Three possible explanations are available for the formation of these products.[327]

    a. *Parathion* is bound to an enzyme in such a manner that sulfur and 4-nitrophenol can be cleaved off. In a smooth reaction the sulfur is released at the rate at which an enzyme-bound hydroxy group or hydroxy radical approaches, *paraoxon* being formed. Alternatively, when 4-nitrophenol is released, *O,O*-diethyl phosphorothioic acid is formed.

b. Two binding sites are present on the same enzyme. At one site *parathion* is bound such that sulfur is easily removed in a smooth reaction, and at the other site 4-nitrophenol is split off thus forming *paraoxon* and *O,O*-diethyl phosphorothioic acid.

c. There are two different enzymes, one forming *paraoxon* and the other 4-nitrophenol and *O,O*-diethyl phosphorothioic acid.

Mechanism b is less probable, and mechanism c is regarded as preferable.

Dursban® is oxidized to the phosphate in rats and then hydrolyzed, *O*-ethyl *O*-(3,5,6-trichloro-pyridine-2-yl) phosphoro(thio)ic acid and 2-hydroxy-3,5,6-trichloropyridine being identified as metabolites. Fish absorb Dursban® slowly but metabolize it rapidly. The metabolites enter plants from the water and are further metabolized. The following breakdown products have been identified.

The principal metabolite is 2-hydroxy-3,5,6-trichloro-pyridine, which is presumably broken down further by dehalogenation and ring cleavage.[328,329] In plants the main degradation product is also 2-hydroxy-3,5,6-trichloro-pyridine, which is dehalogenated to di- and triols, which can then be ring cleaved, leading eventually to carbon dioxide. In soil it is hydrolyzed to hydroxypyridine, which forms water-soluble salts at alkaline pH and is taken up by the plants. Ester acids are also formed.

In the case of Zinophos® the primary breakdown in soil begins with hydrolysis of the ester bond to the heterocycle and subsequent degradation of the latter to carbon dioxide.

Imidan[®], which is easily absorbed via the leaves, also suffers oxidation and hydrolysis with ring cleavage. The following metabolites have been detected.

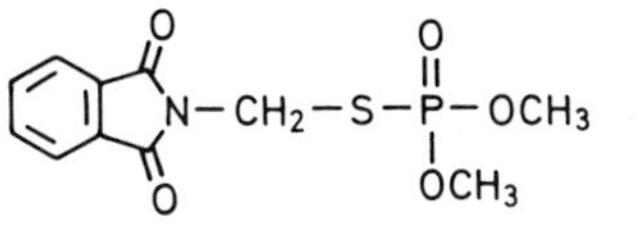

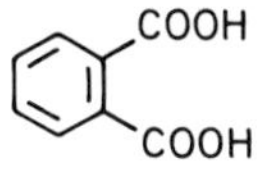

Imidan[®]

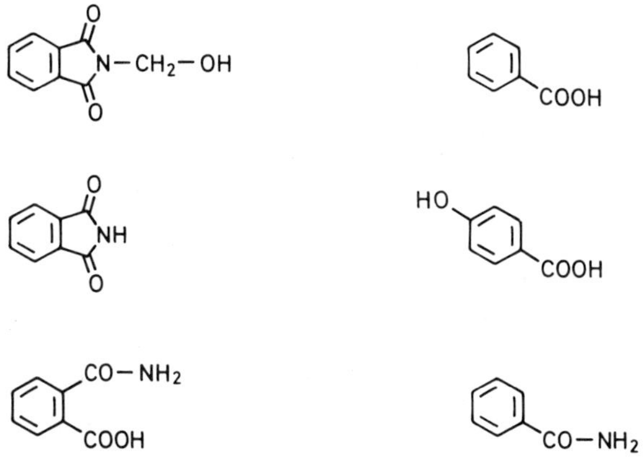

Hydrolysis predominates over oxidation in cotton plants. The intermediacy of phthalic and phthalamic acids is deduced from the formation of 4-hydroxybenzoic acid and carbon dioxide.[330] In soil Imidan[®] is degraded hydrolytically by microorganisms. If the soil is sterilized, Imidan[®] is stable for a longer period.

O,S-Dimethyl phosphorothioic acid, phosphoric acid, and two unidentified compounds have been detected in the metabolism of Monitor[®] by pine seedlings.[331]

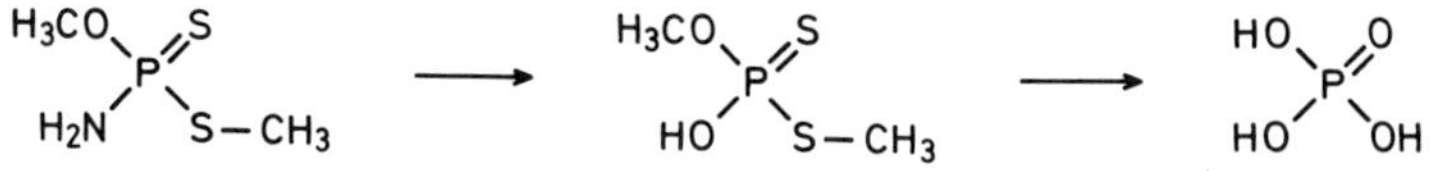

Monitor[®]

Loss of hydrogen chloride and rearrangement to *dichlorvos* occurs spontaneously with *trichlorfon* in weakly acidic, neutral, and basic media, that is, under physiological conditions. This reaction does not require the mediation of an enzyme system. Several authors have come to the conclusion that, as *dichlorvos* is found in all organisms treated with *trichlorfon*, *trichlorfon* itself is not toxic and is activated by rearrangement to *dichlorvos*. *Dichlorvos* is in fact more toxic than *trichlorfon*. Others, however, are of the opinion that *trichlorfon* is the inhibitor of cholinesterase. The problem remains unresolved. Hydrolysis and dealkylation take place in acidic media.[332,333]

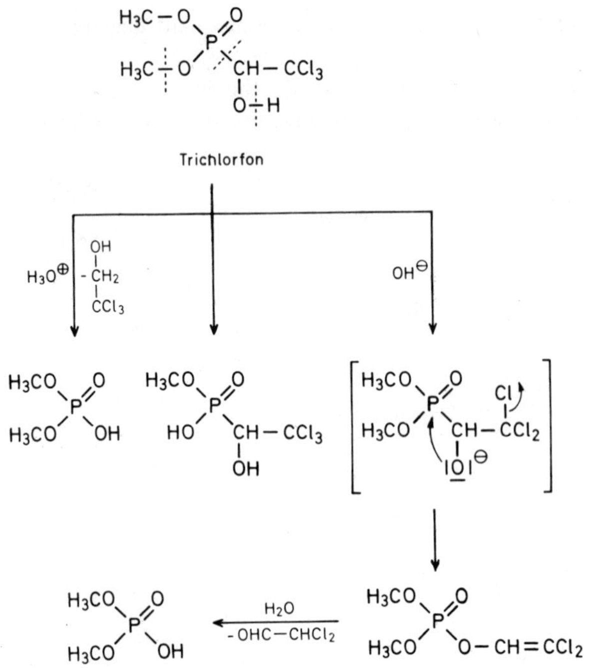

$$H_3C-O \diagdown \diagup O$$
$$P$$
$$H_3C \div O \diagup \diagdown CH-CCl_3$$
$$| \\ O \div H$$

Trichlorfon

$$\begin{array}{c} OH \\ | \\ H_3O^\oplus | \; - CH_2 \\ | \\ CCl_3 \end{array}$$

$$OH^\ominus$$

$$H_3CO \diagdown \diagup O \qquad H_3CO \diagdown \diagup O \qquad \left[H_3CO \diagdown \diagup O \diagdown \; Cl \right]$$
$$P \qquad\qquad P \qquad\qquad P \diagdown \; |$$
$$H_3CO \diagup \diagdown OH \qquad HO \diagup \diagdown CH-CCl_3 \qquad H_3CO \diagup (\; CH-CCl_2$$
$$\qquad\qquad\qquad\qquad | \qquad\qquad\qquad | \\ \qquad\qquad\qquad\qquad OH \qquad\qquad \;\; |O|^\ominus \qquad$$

$$H_3CO \diagdown \diagup O \qquad\quad \xleftarrow[\;-OHC-CHCl_2\;]{H_2O} \qquad H_3CO \diagdown \diagup O$$
$$P \qquad\qquad\qquad\qquad\qquad\qquad P$$
$$H_3CO \diagup \diagdown OH \qquad\qquad\qquad\qquad H_3CO \diagup \diagdown O-CH=CCl_2$$

The low mammalian toxicity of *trichlorfon* is ascribed to the hydrolysis of the phosphonate and subsequent conjugation of the 2,2,2-trichlorethanol (see Section 2.3.4.7).

Dichlorvos is probably broken down according to the following scheme.[334]

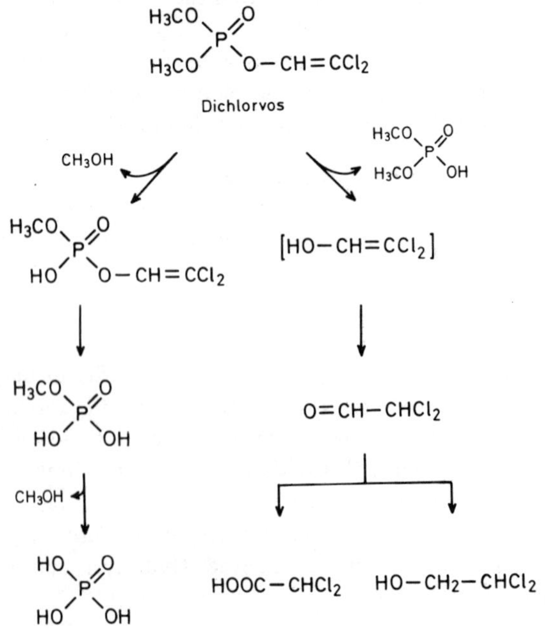

$$H_3CO \diagdown \diagup O$$
$$P$$
$$H_3CO \diagup \diagdown O-CH=CCl_2$$

Dichlorvos

$$CH_3OH \qquad\qquad\qquad\qquad H_3CO \diagdown \diagup O$$
$$\qquad\qquad\qquad\qquad\qquad\qquad\qquad P$$
$$\qquad\qquad\qquad\qquad\qquad\qquad H_3CO \diagup \diagdown OH$$

$$H_3CO \diagdown \diagup O \qquad\qquad\qquad [HO-CH=CCl_2]$$
$$P$$
$$HO \diagup \diagdown O-CH=CCl_2$$

$$H_3CO \diagdown \diagup O \qquad\qquad\qquad O=CH-CHCl_2$$
$$P$$
$$HO \diagup \diagdown OH$$

$$CH_3OH \blacktriangleleft$$

$$HO \diagdown \diagup O$$
$$P \qquad\qquad HOOC-CHCl_2 \qquad HO-CH_2-CHCl_2$$
$$HO \diagup \diagdown OH$$

2,2-Dichloroethanol has been detected as a conjugate in urine.

Studies on rats and mice have shown that the hydrolysis of *naled* begins at the H₃C-O-P group. Among other metabolites, dichlorobromoacetaldehyde is formed.[335] *Naled* can react with thiol groups whereby *dichlorvos* is re-formed. For example, cysteine reacts instantaneously with *naled*.

Naled

The degradation of *fenchlorphos* (*ronnel*) takes the same course both in rats and cattle such that it cannot be decided whether hydrolysis precedes oxidation or vice versa.[336]

The degradation pathways varying from animal to animal are seen in the scheme. They could be the result of different rates of hydrolysis at the methyl and phenyl groups, which could possibly account for the selective toxicity.

The degradation of *malathion* may be summarized as follows.

Malathion

Malathion is also degraded by an additional hydrolytic mechanism, the so-called malathionase. Liver microsomes cleave the molecule at the carboxy group to yield the water-soluble "malathionic acid." Nonresistant insects are unable to metabolize *malathion* as vertebrates do, and this explains the selective toxicity.

2.3.4.5. Dealkylation of Triesters to Diester

As expected, mono- and dimethyl phosphates react with suitable nucleophilic agents by transfer of a methyl group. A survey of the experimental literature shows that only relatively easily polarized nucleophiles such as mercaptans, thioethers, iodide, and basic tertiary amines come into consideration. In contrast, it is difficult to methylate hydroxy groups or weakly basic nitrogens. For detailed surveys on methylation reactions of phosphates, the reader should consult ref. 295, pp. 99ff. The essential facets are given here.

1. Dimethyl phosphates are moderate to weak methylating agents when compared to dimethyl sulfate, for example.

2. The methyl group is preferentially transferred to soft nucleophiles, that is, bases with a low charge concentration and high polarizability. Typical examples are iodide ion, mercaptans, and thioethers. Strongly basic species with a high charge concentration such as hydroxide ion and strongly basic amines attack preferentially at the phosphorus atom.

3. Weakly basic nitrogens in heterocyclic rings are only methylated with difficulty, if at all. It did not prove possible, for example, to obtain even traces of 7-methylguanine when guanine was warmed with a 10-fold excess of *DDVP* for one week at 37°C.[337]

The *O*-dealkylation of triesters to give water-soluble and rapidly excretable diester acids is the third most important mechanism of detoxification *in vivo*. This mechanism plays an important part in selectivity and resistance both in plant and animal organisms. At first this cleavage was thought to be an exclusively hydrolytic process catalyzed by a phosphorotriesterase enzyme system. Today processes of nonhydrolytic nature are known to be involved and possibly are more important. There are three biochemical mechanisms whereby an *O*-dealkylation can be brought about.

1. The oxidative process that can only take place in the presence of oxygen and NADPH. The enzyme system has the properties of microsomal hydroxylases. An unstable hydroxylated compound is probably the intermediate that then forms the *O*-dealkylated derivative and an aldehyde (e.g., *chlorfenvinphos*).
2. A soluble glutathione-dependent alkyl-transferase system is necessary in the second mechanism. The products are *O*-dealkylated derivatives and *S*-alkyl glutathione (e.g., *parathion-methyl*).[297,338,339]
3. The hydrolytic cleavage of the alkyl phosphate bond with formation of the *O*-dealkylated product and an alcohol (e.g., *paraoxon*).

The oxidative *O*-dealkylation has been proven in the case of *chlorfenvinphos. In vitro* tests confirmed the *O*-desethylation by enzyme preparations from rabbit liver. NADPH and oxygen are required for the reaction, the end products being *O*-desethyl chlorfenvinphos and acetaldehyde.

An unstable intermediate, formed by hydroxylation of the α-C atom of an *O*-ethyl group, has been postulated but not detected.

Chlorfenvinphos

In the *O*-dealkylation of *parathion-methyl* and *paraoxon-methyl*, the involvement of soluble enzyme fractions and reduced glutathione has been demonstrated. Studies on various organophosphates have proven that:

1. The enzyme involved acts as a phosphoric triester–glutathione–alkyl–transferase.
2. Methyl esters are favored, the products being *O*-dealkyl derivatives and *S*-alkyl glutathione.
3. The enzyme catalyzes the *O*-dealkylation of both phosphates and phosphorothionates.
4. The transferase activity is highest in the liver of mammals and in the fat body of insects.

In general, it has been observed that at first one alkyl group is split off, the second being attacked after the acyl bond in the molecule has been cleaved.[340] For the third mechanism the hydrolysis could be catalyzed by a phosphorotriesterase enzyme system. From the small amount of information available it seems that this reaction could play a part in the *in vivo* O-dealkylation of *paraoxon* in certain species of fly. One main metabolite detected was ethanol.

One must differentiate between *cis-* and *trans*-Phosdrin in the degradation of Phosdrin®. *cis*-Phosdrin is metabolized to *cis*-demethylphosdrin and *S*-methyl glutathione by an enzyme system requiring reduced glutathione, whereas *trans*-Phosdrin and both Bomyl isomers are metabolized to *O,O*-dimethyl phosphate; that is, they are cleaved at the P—O—C=C bond.

The significance of glutathione-dependent enzymatic detoxification is particularly well illustrated by *parathion-methyl* and Sumithion®, which are demethylated only by this mechanism, whereas the corresponding ethyl derivatives are only degraded very slowly. It must be concluded from the fact that *cis-* and *trans*-Phosdrin are degraded differently that steric factors have an influence. In the other compounds the methyl groups are relatively unhindered but in *trans*-Phosdrin are partly screened by the carboxylate group. The breakdown of Phosdrin isomers in mouse liver is shown in the following scheme.[341]

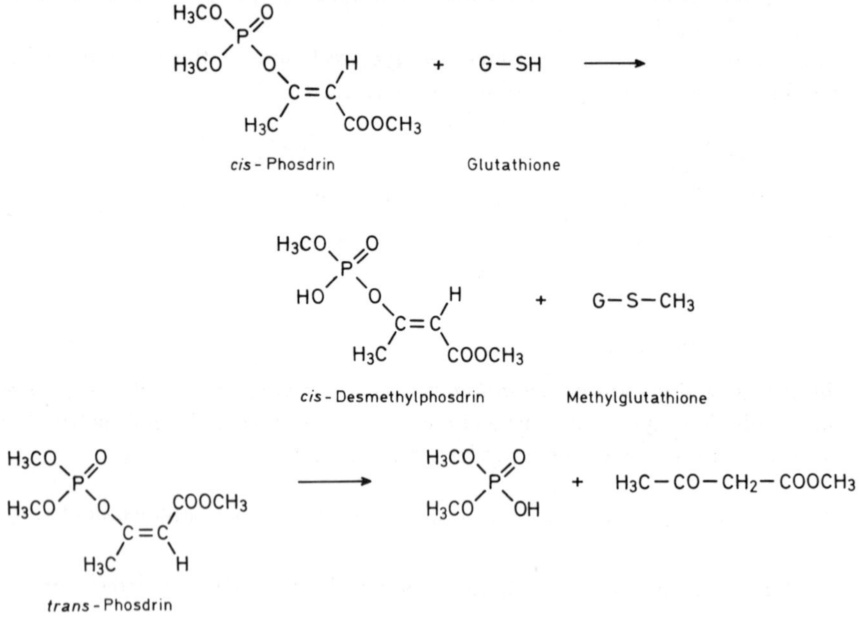

cis - Phosdrin   Glutathione

cis - Desmethylphosdrin   Methylglutathione

trans - Phosdrin

Similar considerations have been discussed in the case of *trichlorfon*. If glutathione plays the decisive role in dealkylation, then the methyl group is transferred to glutathione, forming S-[14]CH_3-glutathione, which is further degraded to S-[14]methyl cysteine. The S-[14]CH_3 is metabolized further to carbon-[14] dioxide. Tests with radiolabeled *trichlorfon* ([14]CH_3) showed that P-O-CH_3 cleavage also occurred, as was demonstrated by the formation of $H^{14}CO_2H$ that originated from the stepwise oxidation of [14]CH_3OH.

The process of methyl transfer to glutathione is also involved as well as the hydrolytic and oxidative pathways in the *in vitro* degradation of *azinphos-methyl* by mouse liver. This reaction is inhibited by methyl iodide thus proving that glutathione *S*-alkyl transferase is responsible. Occasionally another type of transferase is reported, namely, a system that transfers aryl groups. Final proof of the existence of this system is lacking. The possible breakdown of *azinphos-methyl* is shown in the following scheme.[342]

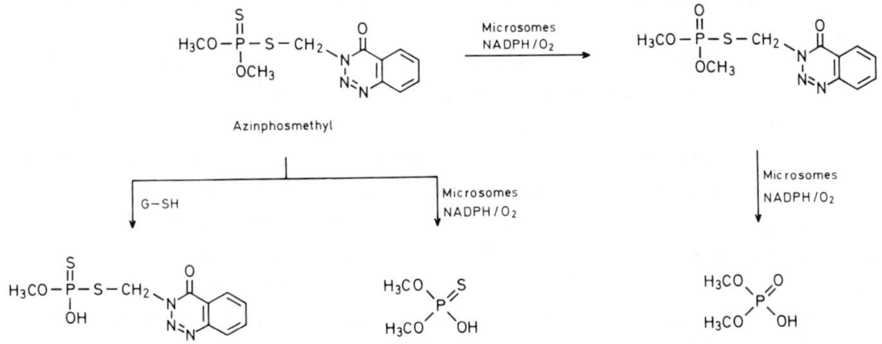

Azinphosmethyl

Glutathione is just as necessary for the degradation of *parathion-methyl* and Sumithion® in rat liver extracts and insect midgut homogenates. The enzyme system is only activated by glutathione; other SH compounds such as 1-cysteine and thioglycolic acid have no activating effect. The principal metabolite in the presence of glutathione (GSH) is demethyl-parathion. Two principal metabolites are found in insect homogenates: demethyl-parathion and phosphoric acid[343] (see scheme).

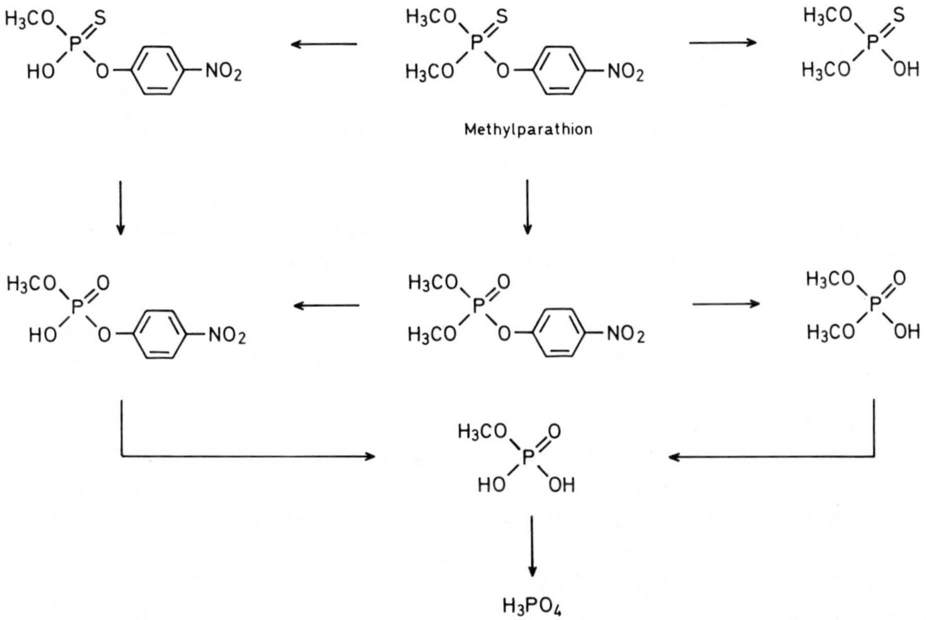

Methylparathion

H_3PO_4

In recent years another aspect of the methylating properties of certain methyl phosphates has come to the forefront of discussion: the question whether such compounds can directly methylate the nucleotide bases in DNA and thus have a mutagenic or carcinogenic activity. As it has not yet proved possible to demonstrate such a relationship in animal experiments, the primitive model

$$\text{methylating activity} \longrightarrow \text{7-methylguanine} \longrightarrow \text{mutagenic activity}$$

should be abandoned. Chemical experiments give no support to such a relationship either.

Chromosomal changes in patients after suicide attempts have been reported in a recent publication.[344] Here it must be taken into account that the normal metabolism, that is, the detoxifying ability of the organism, is naturally severely damaged in such circumstances. The authors found an initially increasing number of chromosomal breaks and aberrations and conclude that the phosphate insecticides used have mutagenic activity. However, after six months no differences with respect to the control groups were to be found. The authors have thus overseen the true significance of their results. Here has been shown for the first time *in vivo* that the organism has an efficient repair system in the cell that was not even damaged by almost lethal doses of the organophosphates so misused. Nor can this study be used as evidence of a correlation between direct methylating action and chromosomal damage, because no comparison was made with the effect of nonmethylating compounds in almost lethal doses.

It would be more correct to elucidate the mechanisms of mutagenesis *in vivo* caused by proven mutagens and then examine the influence of a test compound on these mechanisms and verify it *in vivo*, rather than adjust the theory of mutagenesis to a chemical property like methylating action and thus ignore metabolism *in vivo*.

2.3.4.6. Degradation at the Carboxy Group

Degradation at the carboxy group of an insecticide molecule also belongs to the hydrolytic breakdown reactions and can be a main degradative pathway particularly in mammals. Compounds such as *malathion* and *acethion* are cleaved at the carboxylate group by the enzyme carboxylesterhydrolase. This enzyme is more widespread in animal rather than in insect tissue, thus explaining the selectivity of *malathion. Malathion* is metabolized by this reaction to a nontoxic α-monoacid.

Carboxamides are subject to the same degradation. *Dimethoate* is metabolized by the enzyme carboxyamidase to the corresponding nontoxic carboxylic acid. This enzyme should be distinguished from the carboxylesterhydrolase that attacks *malathion*, and from the enzymes that catalyze the oxidative *N*-dealkylation of *dimethoate* and related compounds. A high content of carboxyamidase is an important precondition for the low mammalian toxicity of *dimethoate*.[340] When discussing the degradation of *dimethoate* one must define precisely the organism under discussion. In sheep liver the influence of the amidase reaction predominates and "dimethoate acid" is formed, which, owing to its hydrophilic character, is only a weak cholinesterase inhibitor. In many insects and plants the phosphatase reaction is more important. One exception, however, is the boll weevil, where the degradation starts with the amidase reaction as in mammalian organisms. The scheme below summarizes the fate of *dimethoate* in various organisms.

In the degradation of *formathion* in bean plants the C–N–C bond is first attacked (amide cleavage) with the formation of equal quantities of *dimethoate* and *O,O*-dimethyl *S*-(carboxymethyl) phosphorodithioate. Oxidation products such as *dimethoxon* and *O,O*-dimethyl phosphorodithioic acid are, of course, also formed. The appearance of bis[*O,O*-dimethyl-thiophosphoryl]disulfan is surprising. The only insecticidally active metabolites are *dimethoate* and *dimethoxon*. *N*-Demethyl metabolites are not found here in the degradation by bean plants.*

The initial activity observed after application of *formothion* is ascribed to *formothion* itself; the long-lasting biological activity to *dimethoate* and *dimeth-oxon*.[345]

In general it may be said that, in plants and insects, the phosphatase activity plays a more important part in the total hydrolysis than the carboxylesterase or amidase activity. It should be stressed again at this point that several degradative mechanisms almost always occur simultaneously and in parallel, the most important

*See scheme on p. 122.

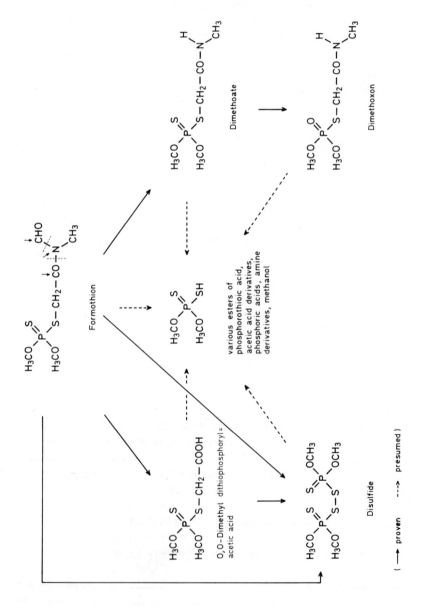

of which are oxidation and hydrolysis. A reminder should also be given here that the same compound has different degradation pathways in different organisms.[346]

2.3.4.7. Conjugation

As has just been shown, the numerous degradative reactions of organophosphates occur simultaneously or consecutively. Occasionally a new substituent is introduced, for example, the hydroxy group, and this is the site for conjugation. The hydroxylated compounds can be toxic, and they are detoxified by conjugation with glucose, glucuronic acid, sulfuric acid, amino acids, and the like. Conjugation is an easy reaction leading to the formation of inactive, hydrophilic derivatives that are excreted by animal organisms or stored by plants. (See Table 12.) A high energy state and a suitable enzyme are always required for a conjugation reaction.

Glucuronic acid reacts with phenols, enols, alcohols, hydroxylamines, carboxy, amino, imino, and sulfhydryl groups. The last reaction step requires uridine-5'-diphosphoglucuronic acid and glucuronyltransferase as enzyme. The mammalian liver and kidneys are the organs responsible for glucuronide synthesis. In general the reaction may be imagined as follows.

Uridine-5'-diphosphoglucuronic acid

UDP
Uridine-diphosphate

TABLE 12. Relationship between Conjugating Agent and Animal or Plant Organism

Conjugating Agent	Found in
Glucuronic acid	Vertebrates
Sulfuric acid	Vertebrates
Glucose	Plants, insects
Glutathione	Animals, plants
Phosphoric acid	Insects
Amino acid (glycine)	Animals, plants
Methyl groups	Animals, plants
Acetyl groups	Animals, plants

The following sequences illustrate specific examples.

This reaction does not seem to take place in insects.

The important detoxification mechanism in plants and insects is the formation of glucosides. A uridine diphosphatoglucose donor and a glucosyltransferase are necessary for the reaction. Other sugars or acids and other nucleosides are hardly ever cited. *Trichlorfon*, the hydroxylated derivatives of *monocrotophos* and *dicrotophos*, the aromatic thioethers, and their sulfoxides and sulfones are all detoxified by direct conjugation.

Sulfate conjugation occurs mainly with phenols and occasionally with aromatic amino acids. 3'-Phosphoadenosine-5'-phosphosulfate is required as energy donor, the necessary catalyst being sulfokinase. This reaction is illustrated for *cyanophos* (see Section 2.3.2.3.3).

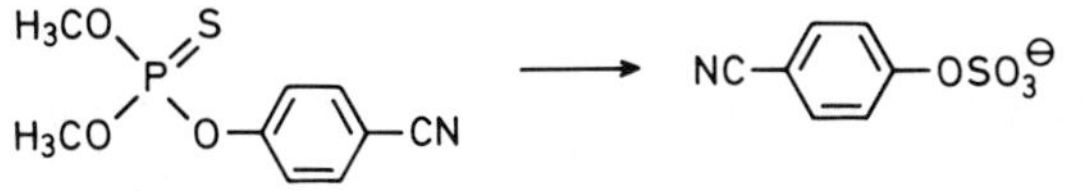

If the conjugating agent is an amino acid, it is usually glycine. The aromatic acceptor is an aromatic carboxylic group and the enzyme aminoacidacylase.

2.4. Insecticidal Carbamates

W. DRABER
Bayer AG, Wuppertal-Elberfeld

2.4.1. INTRODUCTION

The insecticidal carbamates are easily defined as a class, being characterized by the general formula:

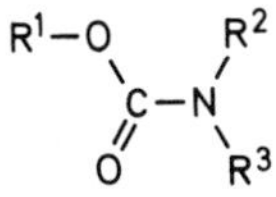

In all insecticidal carbamates of commercial significance, the hydroxy group in the parent compound R^1-OH is weakly acidic, R^2 is a methyl group, and R^3 is usually hydrogen but can also be a residue cleaved away easily either chemically or biologically.

These structural features are found in a natural product isolated as early as 1864, physostigmin (eserin) **(1)**, the structure of which was finally elucidated in 1925:

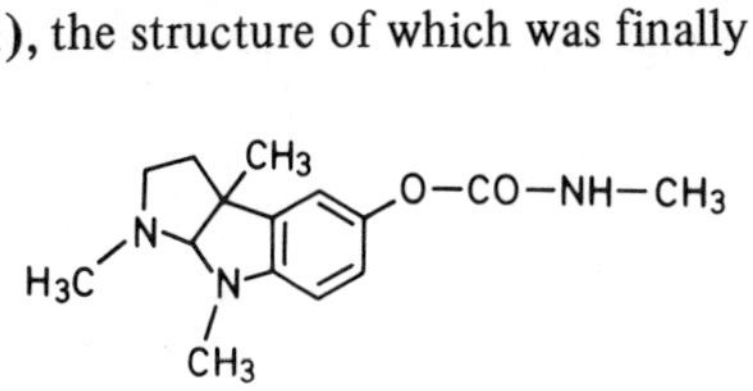

1

In 1926, Stedman[347] showed that the *N*-(methyl)carbamoyl group is essential for the miotic (i.e., pupil-contracting) activity of physostigmin and analogous compounds. Simultaneously, Loewi and Nawratil[348] found that this physiological effect depends on inhibition of cholinesterase. The activity of the insecticidal carbamates also depends on inhibition of this enzyme. Physostigmin (**1**) itself is insecticidally inactive and thus played no part as a model for insecticides, even when it became known from the organophosphates that inhibition of cholinesterase could be a highly relevant mechanism of action in the control of insects. The first carbamate insecticides were developed from 1947 onward at Geigy AG by Gysin et al.[349] from other considerations. *Dimetan* (no. 1 in Table 13), an enolcarbamate of no relevance today, was introduced in 1951.[350] Neither *Pyrolan* (no. 2 in Table 13) nor *isolan* (no. 3 in Table 13), also developed by Geigy, is any longer on the market. In contrast, the first monomethylcarbamate to be developed as an insecticide, *carbaryl* (Sevin®, Union Carbide, 1953), is still of great commercial importance. *Carbaryl* was also discovered without the identity of its mechanism of action with that of physostigmin being recognized immediately,[351] even though reference had been made as early as 1951 in publications from Geigy AG[352] to the relation between dimethylcarbamates and organophosphates.

Patents from E. Merck/Darmstadt, registered in the years 1952–1953, in which the insecticidal activity of phenyl carbamates is claimed,[353] remained without commercial consequence.

Kolbezen, Metcalf, and Fukuto[354] were the first scientists to investigate the quantitative relationship between cholinesterase inhibition and insecticidal activity among phenylcarbamates.

In 1962 it was found in the laboratories of the Boots Co. that *N*-acylation of *N*-monomethylcarbamates to give **2** can lead to a detoxification of the parent compounds without much of the insecticidal activity being sacrificed.[355] Such variations

$$Ar-O-CO-N\begin{smallmatrix}CH_3\\ \\CO-R\end{smallmatrix}$$

2

of the methylcarbamoyl group, also with other acyl residues, are playing an increasing role in the patent literature since the practical use of many carbamates is limited by their high mammalian toxicities. However, none of these "detoxified" carbamates has yet achieved commercial significance.

In contrast, numerous market products originate from the oximecarbamate (**3**) class, which has also been investigated since 1962.

$$\begin{smallmatrix}R^1\\ \\R^2\end{smallmatrix}C=N-O-CO-N\begin{smallmatrix}R^3\\ \\CH_3\end{smallmatrix}$$

3

Further work on the oximecarbamates, among which both highly toxic and highly active insecticides are to be found, is mainly directed to the search for less toxic products.

Considering the applications of carbamates in agriculture and hygiene, there are numerous parallels with the organophosphate insecticides, a class capable of much wider chemical variation, but also some important differences of practical relevance.

2.4.2. ECONOMIC SIGNIFICANCE AND USE

Since the introduction of *carbaryl* in 1958 the carbamates have steadily increased their share of the total tonnage of insecticides produced. In 1965, 23,500 tonnes of carbamates (100% a.i.) was produced in the United States, of which carbaryl accounted for 75-80%. This product owes some of its success to its activity against insect pests resistant to *DDT*.

More recent carbamates such as *carbofuran* (no. 17 in Table 13) are also taking over areas previously the preserve of organochlorine compounds. *Carbofuran* production in the United States in 1972 was ca. ~3000 tonnes (*carbaryl*: 25,000 tonnes).

The main areas of use in agriculture are in cotton, fruit, vegetables, and fodder crops. Certain products have been specially developed against rice pests, mainly against leafhoppers.

Numerous carbamates are systemically active; that is, they are translocated in the transport systems of the plant, mainly in the xylem. It is thus possible to control pests on the shoots and in the roots, which are otherwise difficult to reach. This property is of particular value, enabling their use as soil insecticides and nematicides (e.g., *aldicarb*, *carbofuran*, *oxamyl*, see Table 13).

Products of low toxicity are necessary for use in hygiene and veterinary medicine. Some carbamates are even used as poisons for slugs and snails and as bird repellents (*mercaptodimethur*, no. 8 in Table 13).

Among the carbamates, even an extremely high insecticidal activity is often accompanied only by moderate effectiveness against spider mites. One exception is Tranid®, an experimental product that is, however, no longer on the market. Another product that is primarily recommended as an acaricide, *formetanate*, owes this property to the presence of a formamidine group in the molecule. Some of the more recent carbamates fulfill the desire of integrated pest control for insecticides that are harmless to beneficial insects.

In spite of being closely related chemically to some herbicides, carbamate insecticides are generally well tolerated by plants. A (desirable) thinning-out effect on apples has been reported for *carbaryl*.[356]

2.4.3. COMMERCIAL AND EXPERIMENTAL PRODUCTS

Table 13 gives a summary of 39 carbamates that are either on the market, well in to the development stage, or at least were there for a certain period of time. They have been grouped into three classes.

1. *N,N*-Dimethylcarbamates of enols and hydroxy heterocycles (nos. 1-5).
2. Phenylcarbamates (nos. 6-30).
3. Oximecarbamates (nos. 31-38).

TABLE 13. Commercial and Experimental Carbamates

Common Name, Trade Name, Code	Structure	Manuf., Introduced	Patent	LD$_{50}$ (mg/kg) (Species)	Properties, Use	Ref.
A. N,N-Dimethylcarbamates from Enols and Hydroxy Heterocyclics						
1 *Dimetan*® Dimetan® G 19258		Geigy, 1951	DBP 826133 (1948/51), Geigy, Inv.: Gysin; *C.A.* **46**, 1214b (1952)	150 (rat)	Contact insecticide with partly systemic action; as dusts against aphids, no longer available	349, 350
2 *Pyrolan*® Pyrolan® G 22008		Geigy, 1951	Swiss P 279553 (1949/52), Geigy, Inv.: Gysin; *C.A.* **47**, 10172a (1953)	62 (rat)	Was used in sprays against flies, no longer available	349, 357
3 *Isolan*® Primin® G 23611		Geigy	See *pyrolan* above	.54 (rat), 14–18 (mouse)	Systemic aphicide used in Europe	349
4 *Dimetilan*®, Snip® Dimetilan®, Snip® G22870		Geigy, 1962	DBP 844741 (1949/51), Geigy, Inv.: Gysin et al.; *C.A.* **72**, 30580t (1970)	50 (rat), 60–65 (mouse)	Water-soluble, as a bait formulation against flies, not marketed in USA	349
5 *Pirimicarb*® Primor® PP 062		ICI, 1969	USP 493974 (1966/70), ICI, Inv.: Baranyovits et al.; *C.A.* **71**, 13137r (1969)	147 (rat), 107 (mouse)	Systemic aphicide also effective against organo-phos.-resistant strains	358

B. Phenylcarbamates

No. / Name / Trade names	Structure	Company, year	Patent / Reference	LD_{50} (mg/kg)	Use	Ref.
6 *Carbaryl* Sevin®, Hexavin®, Karbaspray®, Ravyon® UC 7744		Union Carbide, 1956	USP 2903478 (1958/59), Union Carbide, Inv.: Lambrech; C.A. **54**, 2293c (1960)	850 (rat)	Broad spectrum contact insecticide, nonsystemic, used in cotton, fruit, vegetables, forage crops, etc.	359, 360
7 *Mexacarbate* Zectran® DOW C 0129		Dow, 1961	USP 3084098 (1959/63), Dow Chem. Corp., Inv.: Shulgin; C.A. **60**, 457f (1964)	19 (rat), 15–30 (dog)	Insecticide and acaricide; weakly systemic; toxic to bees; used against forest pests	362
8 *Mercaptodimethur*[a] *Methiocarb*[b] Mesurol®, Draza® BAY 37344		Bayer, 1962	DAS 1162352 (1959/64), Bayer, Inv.: Schegk et al.; C.A. **60**, 1644c (1964)	87–135 (rat)	Contact and stomach poison with good residual effect, acaricide (fruit, hops); also used against slugs and snails and as bird repellent; toxic to bees (like most carbamates)	362
9 *BPMC* Baycarb®, Osbac®, Bassa®		Bayer, Sumitomo, Kumiai, 1962	DAS 1159929 (1959/63), Bayer, Inv.: Böcker et al.; C.A. **60**, 7956d (1964)	410 (rat), 340 (mouse)	Contact insecticide against leaf and plant hoppers in rice	363
10 *Aminocarb* Matacil® Bay 44646		Bayer, 1963	DAS 1145162 (1960/63), Bayer, Inv.: Heiss et al.; C.A. **59**, 9885d (1963)	30–50 (rat)	Nonsystemic insecticide and molluscicide; Used in cotton, tomato, tobacco, fruit; effective against forest pests	364
11 *Propoxur* Baygon®, Blattanex®[c], Unden®[d], Suncide®[e]		Bayer, 1964	DAS 1108202 (1959/61), Bayer, Inv.: Böcker et al.; C.A. **56**, 5886i (1962)	90–128 (rat)	Broad-spectrum contact and stomach poison with good knockdown properties; weakly systemic; flushing effect with cockroaches and bugs; used in agriculture and hygiene	364

TABLE 13. (*Continued*)

Common Name, Trade Name, Code	Structure	Manuf., Introduced	Patent	LD$_{50}$ (mg/kg) (Species)	Properties, Use	Ref.
12 *Carbanolate* Banol®, Sok® f	O—CO—NH—CH$_3$ (aromatic ring with CH$_3$, CH$_3$, Cl substituents)	Upjohn, 1965	USP 3131215 (1960/64), Upjohn; *C.A.* **59**, 12708g (1963)	30–44 (rat), 300 (mouse)	Broad-spectrum insecticide used in veterinary sector; no longer on the market	365
13 *Promecarb* Carbamult® Schering 34615	O—CO—NH—CH$_3$ (aromatic ring with CH(CH$_3$)$_2$ and H$_3$C substituents)	Schering AG, 1965	DAS 1156272 (1960/63), Schering, Inv.: Cyzewski et al.; *C.A.* **58**, 7316c (1963)	74–90 (rat)	Nonsystemic contact insecticide; against Lepidoptera and Coleoptera, mainly in forests	366
14 Mobam® MCA 600®	O—CO—NH—CH$_3$ (benzothiophene ring with S)	Mobil, 1966	USP 3288673, 3288808 (1965/66), Mobil, Inv.: Kilsheimer, Kaufmann; *C.A.* **66**, 104900x, 75904u (1967)	234 (rat)	Contact insecticide, mainly against hygiene pests	367
15 *Allyxycarb* g Hydrol® BAY 50282	O—CO—NH—CH$_3$ (aromatic ring with CH$_3$, H$_3$C, N(CH$_2$—CH=CH$_2$)$_2$ substituents)	Bayer, 1967	DAS 1153012 (1961/63), Bayer, Inv.: Heiss et al.: *C.A.* **60**, 2838f (1964)	90–99 (rat), 48–71 (mouse)	Nonsystemic insecticide against biting, sucking, mining insects in fruit, vegetables, citrus, tea, rice; no longer sold.	—
16 *Butacarb*	O—CO—NH—CH$_3$ (aromatic ring with C(CH$_3$)$_3$ and (H$_3$C)$_3$C substituents)	Boots, 1967	British P 987254 (1963/65), Boots, Inv.: Fraser, Harrison; *C.A.* **62**, 16121f (1965)	4000 (rat)	Veterinary insecticide against sheep blowfly; stable for 10–16 weeks in the dip	368

No. / Name	Structure	Company, Year	Reference	Toxicity LD₅₀	Uses	

No. / Name	Structure	Company, Year	Reference	Toxicity (LD₅₀)	Uses	Ref.
17 *Carbofuran* Curaterr®*i*, Furadan®*h* BAY 70142, NIA 10242	O–CO–NH–CH₃ (2,2-dimethyl-2,3-dihydrobenzofuranyl carbamate; H₃C, H₃C)	Bayer, FMC, 1967	DOS 1493646 (1963/69), Bayer, Inv.: C.A. **63**, 583a (1965) USP 3474171 (1964/69), FMC, Inv.: Scharpf; C.A. **64**, 3484e (1966)	8–14 (rat), 19 (dog)	Broad-spectrum insecticide, nematicide, and miticide	369
18 *Formetanate HCl* Carzol® Schering 36056	O–CO–NH–CH₃ ; N=CH–N(CH₃)₂ · HCl	Schering, 1967	DAS 1169194 (1962/64), Schering AG, Inv.: Preissker et al.; C.A. **61**, 3636g (1964)	20 (rat)	Slightly systemic acaricide; combined with *chlordimeform* in Fundal®, used in orchard fruit	370
19 *Meobal*® S-1046	O–CO–NH–CH₃ (dimethylphenyl, CH₃, CH₃)	Sumitomo, 1967	USP 3134712 (1962/64), Sumitomo, Inv.: Bywater, Price; C.A. **61**, 7641c (1964)	290–380 (rat)	Used for control of hoppers on rice and scales on fruit	371
20 *MTMC* Tsumacide®, Metacrate®	O–CO–NH–CH₃ (methylphenyl, CH₃)	Sumitomo, Nihon Nohyaku, 1967		268 (mouse)	Insecticide used against hoppers on rice	
21 *Bufencarb*ᵍ (formerly *metakamate*) Bux® Ortho 5353	O–CO–NH–CH₃ (CH–C₃H₇, CH₃) + O–CO–NH–CH₃ (CH(C₂H₅)₂) 3 : 1	Chevron, 1968	USP 3062864, 3062867 (1959/62), Chevron, Inv.: Openson et al.; C.A. **58**, 8969c (1963)	87–170 (rat)	Soil insecticide against corn and rice pests (Diabrotica); moderate residual effect; used as 10% granulate	

TABLE 13. (*Continued*)

Common Name, Trade Name, Code	Structure	Manuf., Introduced	Patent	LD$_{50}$ (mg/kg) (Species)	Properties, Use	Ref.
22 *Dioxacarb* Elocron[®] [d], Famid[®] [c] C 8353	O–CO–NH–CH$_3$ (with 1,3-dioxolane-substituted benzene)	Ciba, 1968	DOS 1518675 (1964/69), Ciba, Inv.: Nikles et al.: *C.A.* **65**, 7102b (1966)	107–156 (rat)	Contact and stomach poison used against hygiene pests and Colorado beetle, capsid bugs (cacao), ricehoppers, aphids	372
23 Macbal[®] H-69	O–CO–NH–CH$_3$ (dimethyl-substituted benzene, CH$_3$ and H$_3$C)	Hokko, 1968		542–697 (rat)	Against ricehoppers	
24 Landrin[®] SD 8530	O–CO–NH–CH$_3$ (trimethyl-substituted benzene, CH$_3$, CH$_3$, H$_3$C)	Shell, 1969	USP 3130122 (1962/64), Shell, Inv.: Kuderna, Phillips; *C.A.* **61**, 8837d (1964)	208 (rat)	Soil insecticide against corn root worm; effective against cacao capsid bugs resistant to *lindane*	373
25 *CMPO* Hopcide[®], Etrofol[®]	O–CO–NH–CH$_3$ (chloro-substituted benzene, Cl)	Kumiai, Bayer, 1970		648 (rat), 150 (mouse)	Against ricehoppers	
26 *Isoprocarb, MIPC* Etrofolan[®] Bay 105807	O–CO–NH–CH$_3$ (isopropyl-substituted benzene, CH(CH$_3$)$_2$)	Bayer, 1970		403–485 (rat), 487–512 (mouse)	Contact insecticide effective against rice and cacao pests	
27 Saprecon C[®] C-10015	O–CO–NH–CH$_3$ (2,2-dimethyl-1,3-dioxolane-substituted benzene, H$_3$C, H$_3$C)	Ciby-Geigy, 1970	DOS 1518675 (1964/69), Ciba-Geigy, Inv.: Nickles et al.: *C.A.* **65**, 7102b (1966)	110 (rat)	Soil insecticide with good residual action	372

No. / Name	Structure	Company, Year	Patents / References	LD$_{50}$	Uses	Ref.
28 Knockbal® TBPMC	O—CO—NH—CH$_3$ on ring with C(CH$_3$)$_3$	Hokko, 1970		470 (mouse)	Contact insecticide against pests in rice, fruit, tea	374
29 Bendiocarb Ficam® NC 6897	O—CO—NH—CH$_3$ on benzodioxole ring (H$_3$C, H$_3$C)	Fisons, 1971	DOS 1667979 (1967/71), Fisons, Inv.: Gates, Grillon; C.A. 71, 38941 (1969)	35–100 (misc. mammals)	Contact and stomach poison for Coleoptera, bugs, and other hygiene pests	
30 Ethiofencarb Crotenon® HOX 1901	O—CO—NH—CH$_3$; CH$_2$—S—C$_2$H$_5$ on ring	Bayer, 1975	DOS 1910588 (1969/70), Bayer, Inv.: Hoffmann et al.; C.A. 73, 109533j (1970)	411 (rat)	Systemic insecticide highly effective against aphids in fruit, vegetables, ornamentals; less toxic toward beneficials	375

C. Oximecarbamates

No. / Name	Structure	Company, Year	Patents / References	LD$_{50}$	Uses	Ref.
31 Aldicarb Temik®, Ambush® UC 21149	H$_3$C—S—C(CH$_3$)$_2$—CH=N—O—CO—NH—CH$_3$	Union Carbide, 1965	USP 3217037 (1962/65), Union Carbide, Inv.: Payne, Weiden; C.A. 63, 2900a (1965)	0.93 (rat)	Systemic insecticide, acaricide, nematicide for soil use: only available as granules to reduce handling hazards	
32 Methomyl Lannate® j, Nudrin® k DuPont 1179	H$_3$C—S—C(H$_3$C)=N—O—CO—NH—CH$_3$	Dupont, Shell, 1966	USP 3506698 (1967/70), DuPont, Inv.: Jelinek; C.A. 70, 57219c (1969) DOS 1567142 (1965/70), Shell, Inv.: Donninger et al.; C.A. 69, 35430s (1968)	17–24 (rat)	Broad-spectrum systemic insecticide and acaricide; low residual action as soil insecticide; less toxic to bees	376, 377
33 Tranid® UC 20047A	N—O—CO—NH—CH$_3$ oxime on chloro-bicyclic ring, NC substituent	Union Carbide, 1966	USP3231599 (1961/66), Union Carbide, Inv.: Kilsheimer, Manning; C.A. 60, 10568f (1964) USP 3328457 (1964/67), Union Carbide, Inv.: Payne; C.A. 67, 73247d (1967)	17 (rat)	Contact acaricide, also effective against phosphate-resistant strains; not ovicidal; less toxic to beneficials; product no longer on market	378

TABLE 13. (Continued)

Common Name, Trade Name, Code	Structure	Manuf., Introduced	Patent	LD$_{50}$ (mg/kg) (Species)	Properties, Use	Ref.
34 *Oxamyl* Vydate® DPX 1410	H$_3$C–S, C=N–O–CO–NH–CH$_3$, (H$_3$C)$_2$N–C(=O)	DuPont, 1969	USP 3530220 (1967/70), DuPont, Inv.: Buchanan; C.A. 70, 114646r (1969)	5.4 (rat)	Contact insecticide and acaricide; broadly active, systemic nematicide	379
35 *Thiocarboxime* WL-21959	NC–CH$_2$–CH$_2$–S, C=N–O–CO–NH–CH$_3$, H$_3$C	Shell, 1970	DOS 1912294 (1968/69), Shell, Inv.: Davies, Davies; C.A. 72, 21362w (1970)	5 (rat)	Broad-spectrum insecticide for control of fruit and cotton pests; acaricide and molluscicide; no longer sold	380
36 *Butoxicarboxim* Plant pin®	H$_3$C, C=N–O–CO–NH–CH$_3$, H$_3$C–SO$_2$–CH, CH$_3$	Wacker, 1970	DAS 2036491 (1970/72), Consortium f. elektrochem. Ind., Inv.: Müller et al.; C.A. 76, 99148h (1972)	458 (rat)	Water-soluble, systemic insecticide against sucking insects in potted ornamentals (used in form of cardboard sticks)	
37 *Thiofanox* Dacamox®	H$_3$C–S–CH$_2$, C=N–O–CC–NH–CH$_3$, (H$_3$C)$_3$C	Diamond Shamrock, 1973	DOS 2216838 (1971/72), Diamond Shamrock, Inv.: Magee; C.A. 78, 29938e (1973)	8.5 (rat)	Soil systemic insecticide effective against spider mites	381
38 *Aldoxycarb* Standak® Aldicarb sulfone UC 21865	CH$_3$, H$_3$C–SO$_2$–C–CH=N–O–CO–NH–CH$_3$, CH$_3$	Union Carbide, 1978	See *aldicarb* (no. 31)	27 (rat)	Effective against many nematode, insect, and mite pests of tobacco, cotton, peanut	
39 *Cartap* Padan® T 1258	H$_3$C, CH$_2$–S–CO–NH$_2$, N–CH, H$_3$C, CH$_2$–S–CO–NH$_2$	Takeda, 1967	USP 3332943 (1964/67), Takeda, Inv.: Konishi et al.: C.A. 66, 115339x (1967)	380 (rat)	Contact and stomach poison against beetles and caterpillars; not a cholinesterase inhibitor	382

[a] ISO.
[b] BSI.
[c] Hygiene, veterinary medicine.
[d] Agriculture.
[e] Japanese name.
[f] Veterinary medicine.
[g] Proposed common name.
[h] Product of FMC Corp.
[i] Product of Bayer AG.
[j] Product of DuPont.
[k] Product of Shell Chemical Co.

The thiolcarbamate *cartap* is a special case, also with respect to its mechanism of action (see no. 39). Within these three groups the compounds have been arranged according to their year of introduction, which as a rule is not identical with the first year of sales, because most products run for several years under experimental permits. This arrangement means that the most topical products in these three groups are to be found toward the end. Some acute, oral toxicological data, usually obtained on the rat or mouse, have been included to illustrate the wide range in carbamate toxicities.

2.4.4. EXPERIMENTAL PRODUCTS

Experimental products with the corresponding patent and literature references are listed in Table 14. They are given in approximately chronological order within the following three classes.

1. Phenylcarbamates.
2. Oximecarbamates.
3. Carbamates substituted at the carbamoyl *N*-atom by acyl, sulfenyl, or similar groups.

This list is not a complete overview of all carbamate patents. Only classes that can be clearly differentiated chemically have been included to illustrate, on the one hand, the extent of variability of the basic structures and, on the other, the preference shown to certain structural elements (e.g., to *S*-alkyl residues in the oxime-carbamates). Naturally, only a small proportion of these compounds will have the opportunity of becoming commercial products.

2.4.5. CHEMISTRY

Only a few methods come under consideration for the introduction of the carbamoyl unit into compounds bearing hydroxy groups: heterocycles and enols react with dimethylcarbamoyl chloride to give the *N,N*-dimethylcarbamates 4; almost all HO compounds react with methyl (or alkyl) isocyanate to the *N*-monomethyl-carbamates 5, which in the case of phenols may also be obtained via the chloro-formates 6:

TABLE 14. Some Interesting Experimental Products

	General Formula	Experimental Product

A. Phenylcarbamates

1

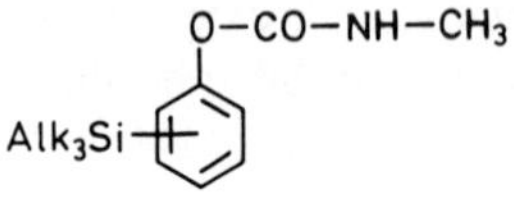

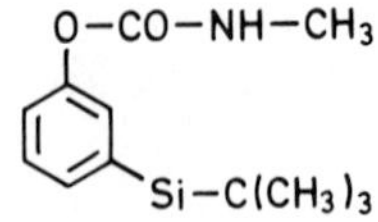

O—CO—NH—CH$_3$

Alk$_3$Si—⟨ring⟩

O—CO—NH—CH$_3$ ⟨ring⟩ Si—C(CH$_3$)$_3$

2

O—CO—N⟨R^1, R^2⟩ on R—⟨bicyclic ring⟩

O—CO—NH—CH$_3$ ⟨tetrahydronaphthalene ring⟩

3

UC 8454

O—CO—NH—CH$_3$

H$_3$C—⟨quinoline ring with N⟩

GS-13798

4

O—CO—NH—CH$_3$
O—CH$_2$—CH=CH$_2$
⟨benzene ring⟩

H-9485

5

O—CO—N⟨R^1, R^2⟩
⟨benzofuran ring with O⟩—X
H$_3$C—C—CH$_3$

O—CO—NH—CH$_3$
⟨benzofuran ring with O⟩
H$_3$C—C—CH$_3$ Cl

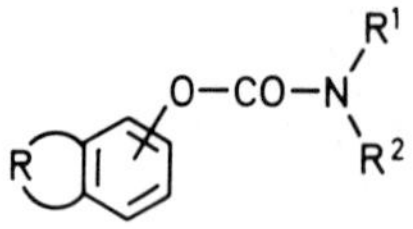

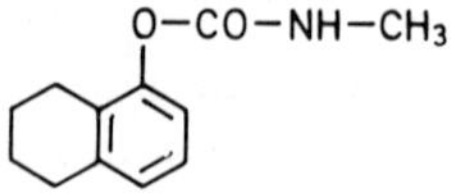

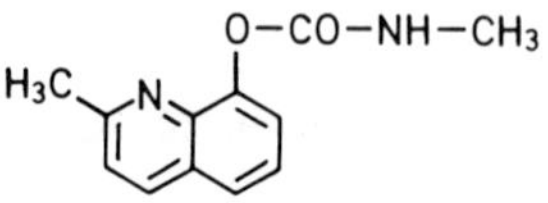

NIA-10559

6

O—CO—NH—CH$_3$
X—⟨benzene ring⟩
NH—COOR

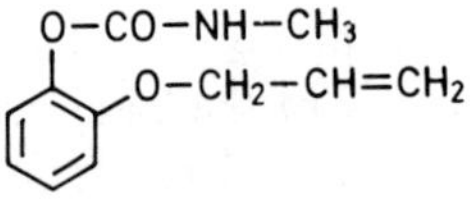

O—CO—NH—CH$_3$
H$_3$C—⟨benzene ring⟩—CH$_3$
NH—COOCH$_3$

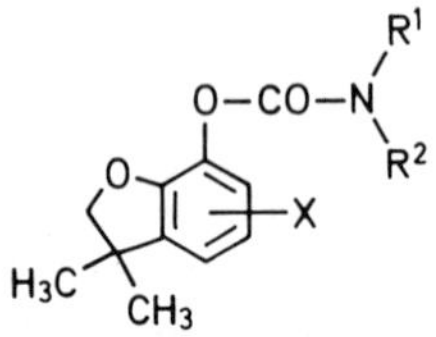

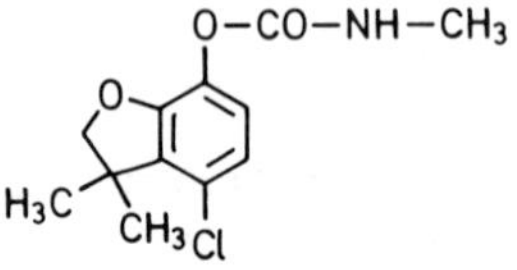

UC-30044

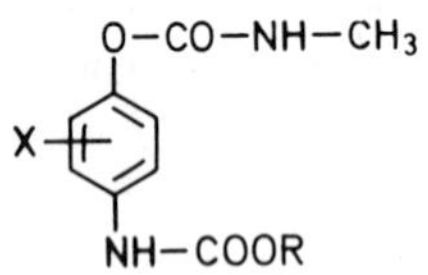

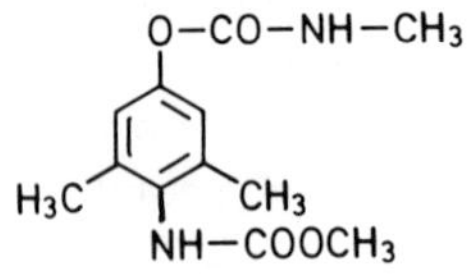

7

O—CO—NH—CH$_3$
(H$_3$C)$_3$C—⟨benzene ring⟩
CH$_3$
N(CH$_3$)$_2$

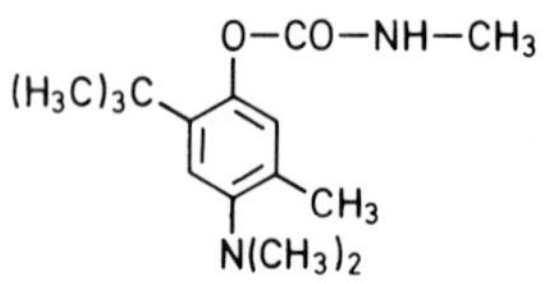

NC 1493

TABLE 14. *(Continued)*

Patent	Biological Properties of the Experimental Product	Refs.
	Highly effective against *Musca domestica* and *Culex pipiens*; synergist necessary; potent ChE inhibitors	383
USP 3084096 (1955/63), Union Carbide, Inv.: Lambrech; *C.A.* **59**, 11381a (1963)	Highly effective against cockroaches resistant to organophosphate compounds, low acute toxicity (LD_{50})	384
USP 3538099 (1965/70), Ciba, Inv.: Rohr et al.; *C.A.* **67**, 21843a (1967)	Insecticide effective against organophosphate-resistant cockroaches	385
USP 3296068 (1960/67), Amer. Cyanamid, Inv.: Adder; *C.A.* **58**, 7867d (1963)	Good activity against hygiene pests, particularly mosquitoes	386
DOS 1493691 (1964/69), FMC, Inv.: Scharpf; *C.A.* **64**, 3484e (1966)	Active against *Melanoplus sanguinipes* (migratory grasshopper)	387
USP 3546343 (1966/70), Union Carbide, Inv.: Payne, Weiden; *C.A.* **71**, 3132w (1969)	Systemic insecticide with high residual action	388
DOS 1207143 (1962/65), Fisons, Inv.: Newbold et al.; *C.A.* **60**, 1658f (1964)	Effective against aphids and spider mites but low residual action	389

TABLE 14. *(Continued)*

	General Formula	Experimental Product

A. Phenylcarbamates

8

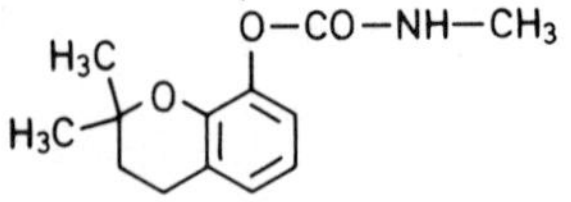

H3C—, H3C—(chroman) O—CO—NH—CH3

NIA-11637

9

O—CO—NH—CH3

NH—CO—R

O—CO—NH—CH3

NH—CO—C$_2$H$_5$

10

$$O-CO-N\begin{smallmatrix}R^2\\CH_3\end{smallmatrix}$$

O—P(OR^1)$_2$ ‖ O(S)

O—CO—N(CH$_3$)$_2$

O—P(OC$_2$H$_5$)$_2$ ‖ O

11

H$_3$C—NH—CO—O

$$N\begin{smallmatrix}CH_3\\CH_2-C\equiv CH\end{smallmatrix}$$

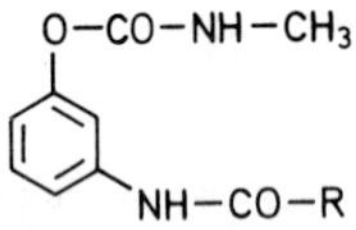

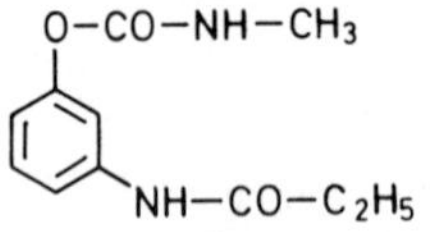

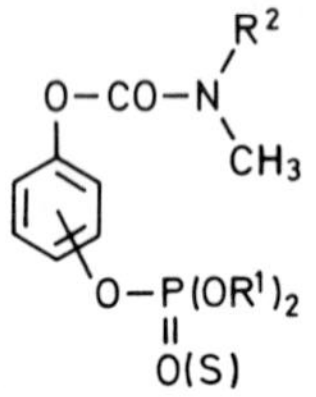

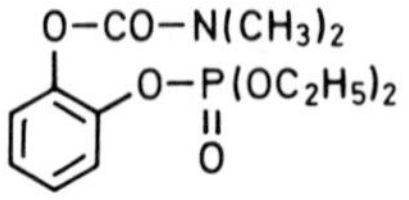

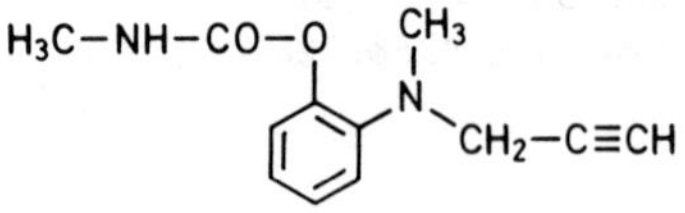

C-17018

12

O—CO—NH—CH$_3$

O—CH—OCH$_3$

CH$_2$—Cl

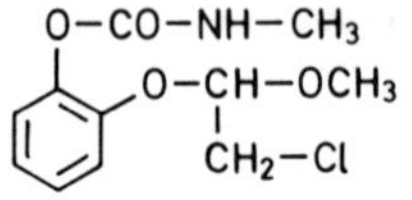

BAS 263 I

B. Oximecarbamates

13

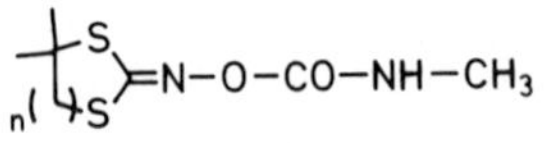

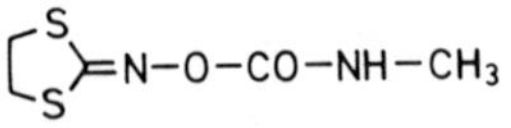

$$\underset{n}{(}S\underset{S}{\overset{S}{)}}=N-O-CO-NH-CH_3$$

$$S\underset{S}{\overset{S}{)}}=N-O-CO-NH-CH_3$$

ACC-38906

14

$$Y-\overset{|}{C}-\overset{|}{C}=N-O-CO-N\begin{smallmatrix}R^1\\R^2\end{smallmatrix}$$

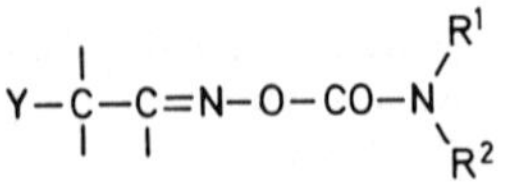

Y = CN, NO$_2$, SCN

H$_3$C

NO$_2$

=N—O—CO—NH—CH$_3$

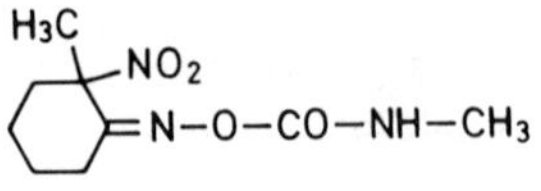

UC-23746

TABLE 14. *(Continued)*

Patent	Biological Properties of the Experimental Product	Refs.
USP 3468913 (1965/69), FMC, Inv.: Scharpf; *C.A.* **67**, 100004f (1967)		390
Monsanto	Good inhibitor of ChE but only moderately effective against *Musca domestica*; not potentiated by synergists	392
Japanese P 1126/62 (1959/62), Takeda, Inv.: Harukawa, Konishi	Good inhibitor of ChE, little effect against *Musca domestica*	392
DOS 1793189 (1967/72), Ciba-Geigy, Inv.: Nikles; *C.A.* **71**, 70300n (1969)	Insecticide, acaricide, nematicide	393
DOS 2231249 (1972/74), BASF, Inv.: Kiehs et al.; *C.A.* **80**, 95574d (1974)	Contact insecticide highly effective against aphids, beetles, caterpillars, etc.; toxic (LD_{50}: 13 mg/kg, rat)	
USP 3193561 (1962/65), Amer. Cyanamid, Inv.: Addor; *C.A.* **63**, 11577b (1965)	Insecticide, acaricide; high mammalian toxicity	394
USP 3400153 (1964/68), Union Carbide, Inv.: Payne, Weiden; *C.A.* **69**, 106001r (1966)		390

TABLE 14. (*Continued*)

	General Formula	Experimental Product

B. Oximecarbamates (contd.)

15

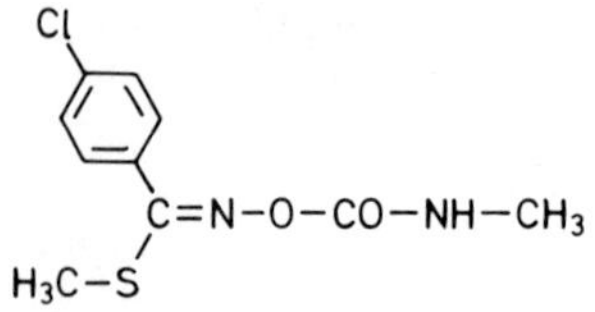

$$\text{Cl}-\bigcirc-\underset{H_3C-S}{\overset{}{C}}=N-O-CO-NH-CH_3$$

Dupont 1335

16

$$R^1-CO-\underset{NC}{\overset{}{C}}=N-O-CO-N\overset{R^2}{\underset{R^3}{\diagup}}$$

R^1= OR,SR,NR$_2$; R^2= Aryl

$$H_3COOC-\underset{NC}{\overset{}{C}}=N-O-CO-NH-\bigcirc\overset{CF_3}{\underset{Cl}{}}$$

17

$$=N-O-CO-N\overset{R^1}{\underset{R^2}{\diagup}}$$

$$=N-O-CO-NH-CH_3$$

PP-156

18

$$R^1-S,\ \underset{R^2-S}{\overset{R^3}{\diagup}}C=N-O-CO-N\overset{R^4}{\underset{R^5}{\diagup}}$$

$$CH=N-O-CO-NH-CH_3$$

19

$$NC-\underset{R^2}{\overset{R^1}{\diagup}}C-\overset{R^3}{\underset{}{C}}=N-O-CO-N\overset{R^4}{\underset{R^5}{\diagup}}$$

$$NC-\underset{CH_3}{\overset{CH_3}{\diagup}}C-CH=N-O-CO-NH-CH_3$$

20

$$R^1-\langle\underset{S}{}\rangle-\underset{R^2}{\overset{}{C}}=N-O-CO-NH-CH_3$$

$$Cl-\langle\underset{S}{}\rangle-\underset{CH_3}{\overset{}{C}}=N-O-CO-NH-CH_3$$

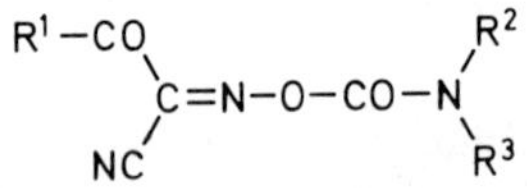

21

$$(H_2C)_n\underset{S-R^1}{\overset{CH=N-O-CO-N\overset{R^2}{\underset{R^3}{}}}{}}$$

$$\underset{S-CH_3}{\overset{CH=N-O-CO-NH-CH_3}{}}$$

22

$$\underset{R^2}{\overset{R^1}{\diagup}}N-CO-\overset{O-R^3}{\underset{}{C}}=N-O-CO-N\overset{R^4}{\underset{R^5}{\diagup}}$$

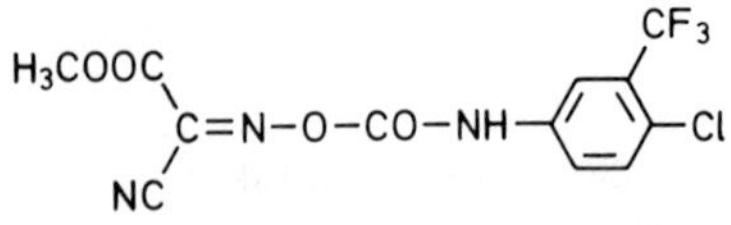

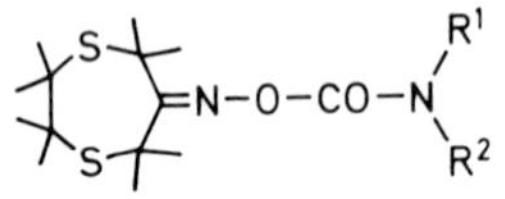

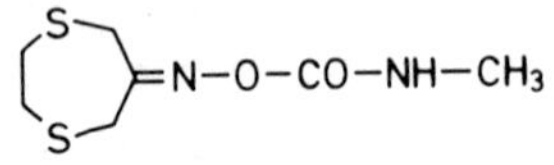

23

$$\underset{R^1-CO}{\overset{R^2-S(O)}{\diagup}}C=N-O-CO-N\overset{R^3}{\underset{R^4}{\diagup}}$$

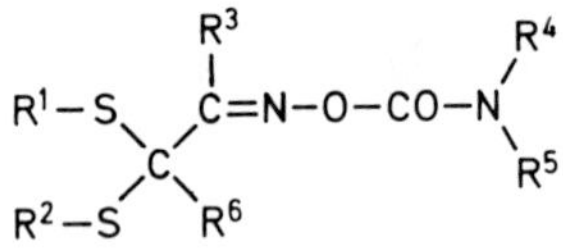

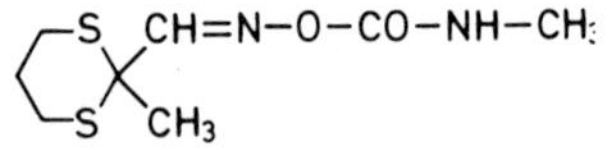

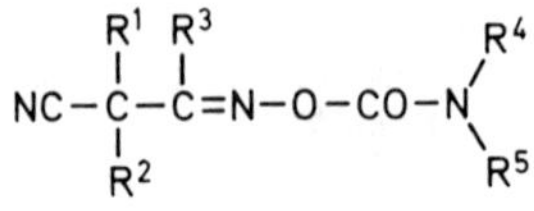

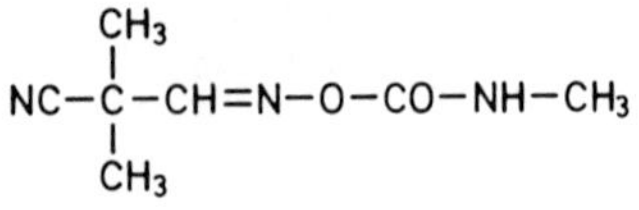

$$\underset{(H_3C)_2CH-CO}{\overset{H_3C-S}{\diagup}}C=N-O-CO-NH-CH_3$$

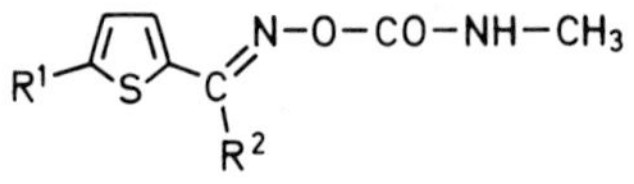

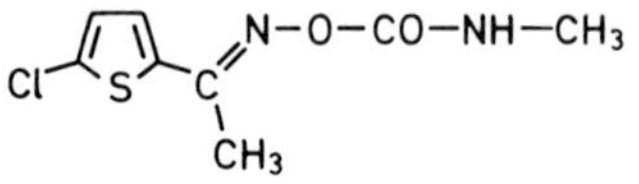

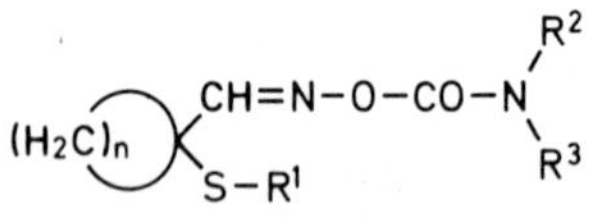

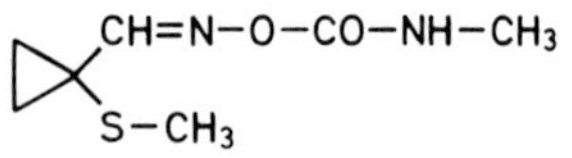

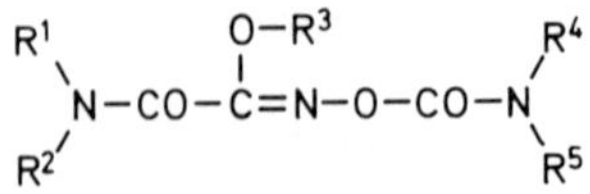

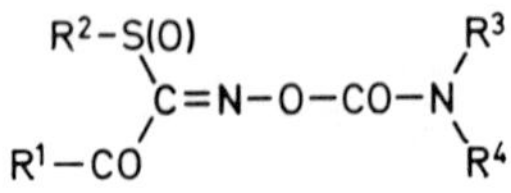

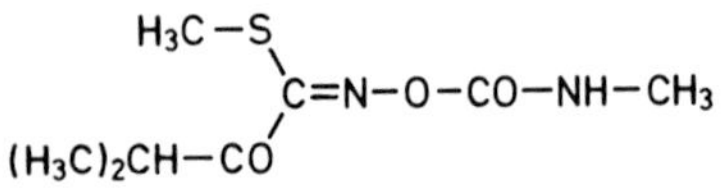

TABLE 14. (*Continued*)

Patent	Biological Properties of the Experimental Product	Refs.
DOS 1543402 (1966/71), DuPont, Inv.: Buchanan, Corty; *C.A.* **75**, 109888r (1971)	Not systemically active against blowflies	395
DOS 1693052 (1966), Ciba, Inv.: Hubele; *C.A.* **70**, 57388g (1969)	Fungicidal and bactericidal activity; insecticidal activity not claimed	
DOS 1812762 (1967/69), ICI, Inv.: Gosh et al.; *C.A.* **71**, 101901r (1969)	Effective against blowflies	396
USP 3681386 (1969/1972), 3M Comp., Inv.: Fridinger, Mutsch; *C.A.* **77**, 140092w (1972)	Good acaricidal and nematicidal activity	397
USP 3466316 (1968/69) [Part of USP 3400153 (1964)], Union Carbide, Inv.: Payne, Weiden; *C.A.* **69**, 106001r (1968)	Insecticide, acaricide, nematicide	
Japanese P 7502733 (1970/75), Mitsubishi; *C.A.* **83**, 75524x (1975)	Effective against spider mites resistant to organophosphates	
USP 3721711 (1970/73), Esso, Inv.: Maravetz; *C.A.* **76**, 72099f (1972)		
DOS 2135251 (1970/72), DuPont, Inv.: Bellina; *C.A.* **76**, 85423g (1972)	Insecticide	
DOS 2111459 (1971/72), C. H. Boehringer, Inv.: Grabinger, Sehring; *C.A.* **77**, 151479d (1972)		

TABLE 14. *(Continued)*

	General Formula	Experimental Product

B. Oximecarbamates (contd.)

24

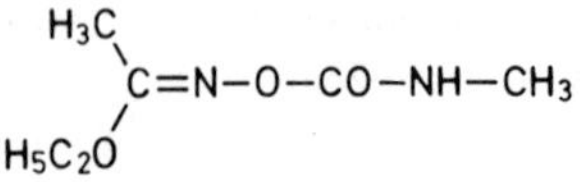

$$H_3C \\ \quad \diagdown C=N-O-CO-NH-CH_3 \\ H_5C_2O \diagup$$

WL 19561

25

$$R^2-\underset{\underset{R^3}{|}}{\overset{\overset{R^1}{|}}{C}}-CH=N-O-CO-N\overset{R^5}{\underset{R^6}{\diagup}}$$

$R^1 = CH_2-O-Alk$

$$H_3CO-CH_2-\underset{\underset{CH_3}{|}}{\overset{\overset{S-CH_3}{|}}{C}}-CH=N-O-CO-NH-CH_3$$

26

$$\overset{R^1}{\underset{R^2}{\diagdown}}C=N-O-CO-NH-CH_3$$

$R^1 = Aryl, Thionyl$

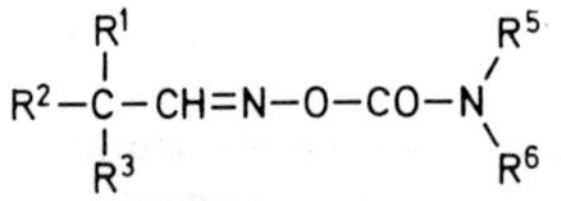

$R^2 = H, \triangleleft, \diamondsuit$

$$\text{(thienyl)(cyclopropyl)} C=N-O-CO-NH-CH_3$$

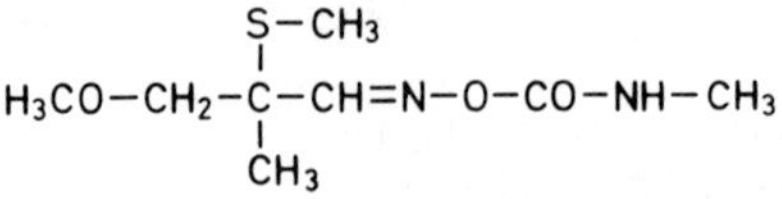

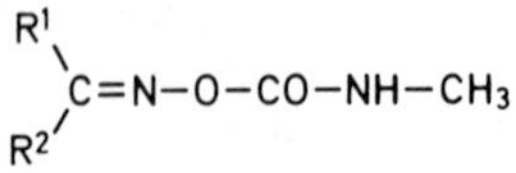

27

$$H_3C-SO_2-O-CH_2-\underset{\underset{CH_3}{|}}{\overset{\overset{CH_3}{|}}{C}}-CH=N-O-CO-NH-CH_3$$

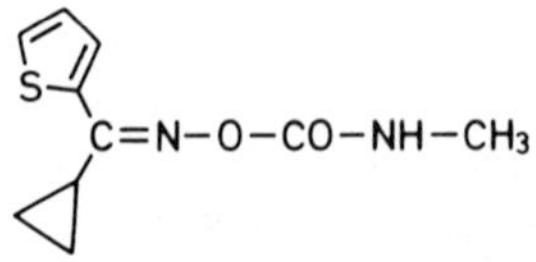

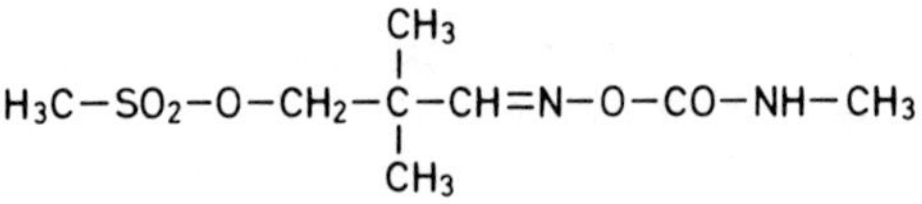

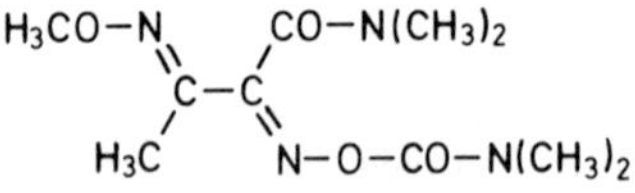

28

$$\underset{H_3C}{\overset{H_3CO-N}{\diagdown}}C=C\underset{N-O-CO-N(CH_3)_2}{\overset{CO-N(CH_3)_2}{\diagup}}$$

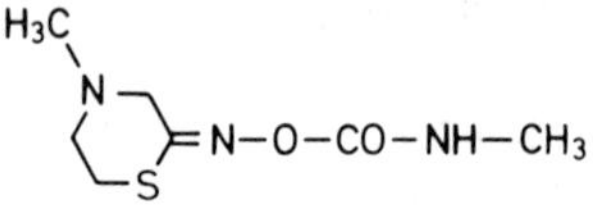

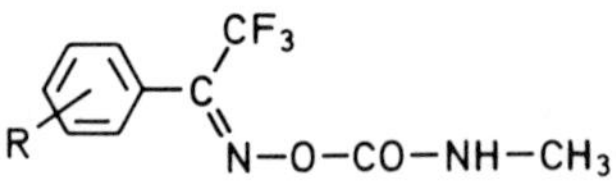

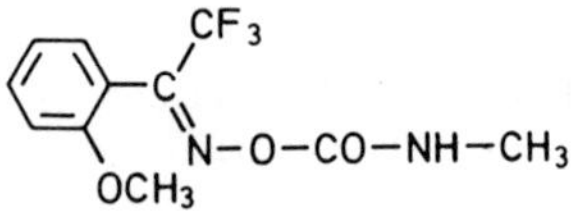

29

$$\underset{H_3C}{\diagdown}N\text{-thiomorpholine} =N-O-CO-NH-CH_3$$

30

$$R-\text{(phenyl)}-\underset{N-O-CO-NH-CH_3}{\overset{CF_3}{|}}C$$

$$\text{(phenyl, OCH}_3)-\underset{N-O-CO-NH-CH_3}{\overset{CF_3}{|}}C$$

C. $R-O-CO-N\overset{X}{\underset{CH_3}{\diagup}}$ *N-Acyl, N-Sulfenyl, and Similar Carbamates*

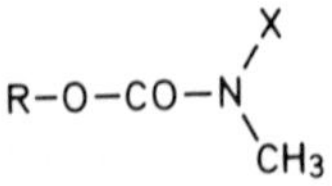

31

$$R^1-O-CO-N\underset{CH_3}{\overset{CO-R^2}{\diagup}}$$

$R^1 = subst. Aryl$

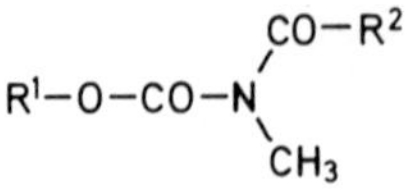

$$H_5C_2-\underset{\underset{CH_3}{|}}{CH}-\text{(phenyl)}-O-CO-N\underset{CH_3}{\overset{CO-CH_2-S-CH_3}{\diagup}}$$

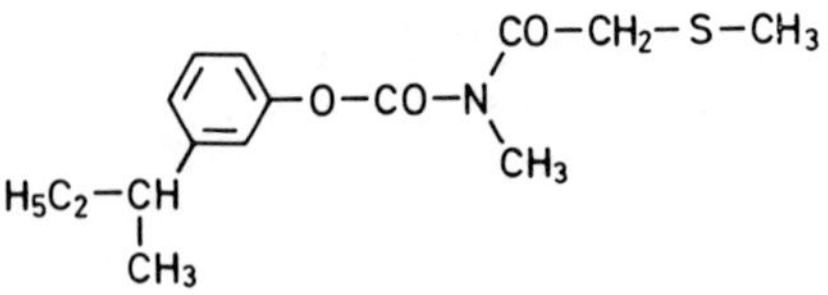

Boots RE 17955

TABLE 14. *(Continued)*

Patent	Biological Properties of the Experimental Product	Refs.
DOS 1567142 (1965/70), Shell, Inv.: Donninger et al.; *C.A.* **69**, 35430s (1968)	Good activity against *Musca domestica*	398
USP 3832400 (1972/74), Ciba-Geigy, Inv.: Meyer, Böhner; *C.A.* **79**, 42000m (1973)	Insecticide, nematicide	
	Only moderately effective against *Musca domestica*; most active ChE inhibitor in the series	399
USP 3835174 (1973/74), Ciba-Geigy, Inv.: Kristiansen; *C.A.* **82**, 97709s (1975)	Insecticide	
DOS 2428070 (1973/74), DuPont, Inv.: Bellina; *C.A.* **82**, 139802z (1975)	Aphicide, systemic; low toxicity toward beneficial insects	
USP 3883510 (1973/75), DuPont, Inv.: Bellina; *C.A.* **83**, 79268p (1975)	Insecticide	
USP 3748361 (1968/73), Esso, Inv.: Rosenfeld et al.; *C.A.* **79**, 78432t (1973)	Moderately toxic to *Musca domestica* and *Culex pipiens*	400
DOS 1693155 (1964/71), Boots, Inv.: Robertson et al.; *C.A.* **69**, 67087b (1968)	Effective against *Culex pipiens*, mosquitoes; low acute toxicity (LD_{50})	401

TABLE 14. (Continued)

	General Formula	Experimental Product

$$C.\quad R-O-CO-N\overset{X}{\underset{CH_3}{<}}$$

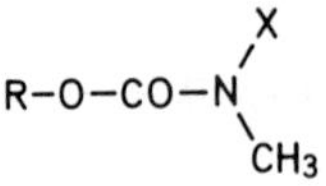

N-Acyl, N-Sulfenyl, and Similar Carbamates (contd.)

32

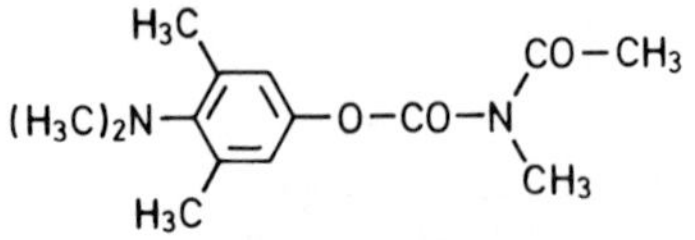

$$(H_3C)_2N-\underset{H_3C}{\overset{H_3C}{\underset{}{\bigcirc}}}-O-CO-N\overset{CO-CH_3}{\underset{CH_3}{<}}$$

33

$$\underset{O-CH(CH_3)_2}{\bigcirc}-O-CO-N\overset{CO-CH_2-OCH_3}{\underset{CH_3}{<}}$$

U-18120

34

$$\underset{(H_3C)_2CH}{\bigcirc}-O-CO-N\overset{CO-CH_2-Cl}{\underset{CH_3}{<}}$$

Hercules 6007

35

$$\underset{(H_3C)_2CH}{\overset{H_3C}{\bigcirc}}-O-CO-N\overset{CO-C_3H_7}{\underset{CH_3}{<}}$$

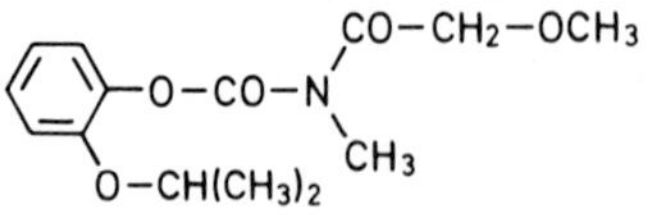

Promacyl

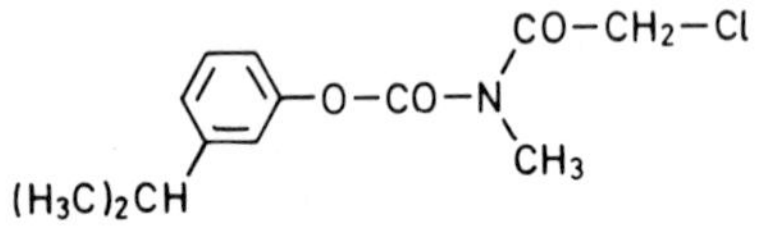

36

$$R^1-O-CO-N\overset{\overset{X}{\overset{\|}{P(OR^2)_2}}}{\underset{CH_3}{<}}$$

X = O,S   R² = CH₃, C₂H₅

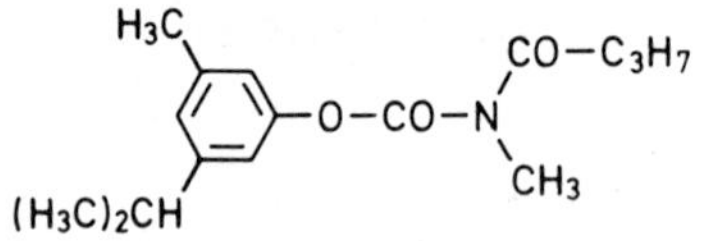

$$\underset{OCH(CH_3)_2}{\bigcirc}-O-CO-N\overset{\overset{S}{\overset{\|}{P(OCH_3)_2}}}{\underset{CH_3}{<}}$$

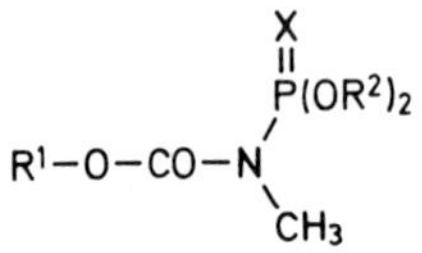

37

$$R^1-O-CO-N\overset{S-R^2}{\underset{CH_3}{<}}$$

R² = Alkyl, Aryl

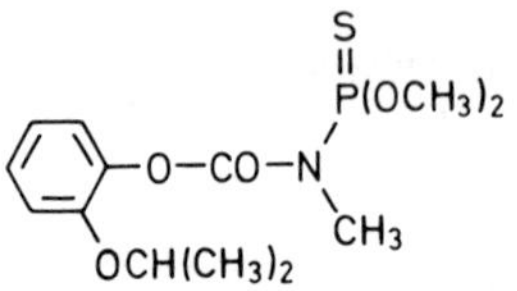

$$\underset{H_5C_2-CH\atop |\atop CH_3}{\bigcirc}-O-CO-N\overset{S-\bigcirc}{\underset{CH_3}{<}}$$

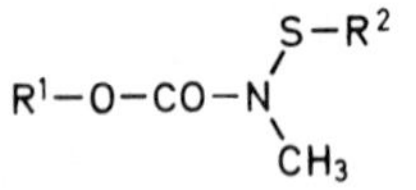

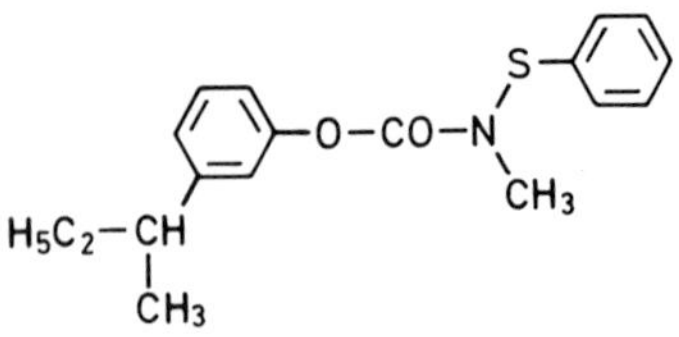

RE-11775

38

$$\underset{(H_3C)_2CH}{\bigcirc}-O-CO-N\overset{CH=C(C_2H_5)_2}{\underset{CH_3}{<}}$$

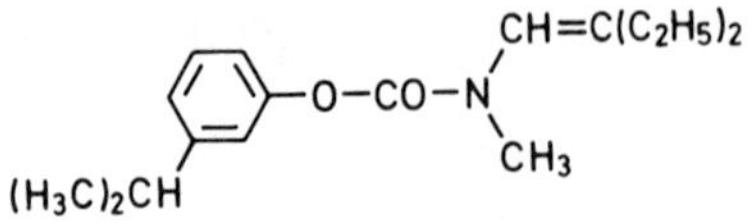

TABLE 14. (*Continued*)

Patent	Biological Properties of the Experimental Product	Refs.
DOS 1542669 (1962/69), Boots, Inv.: Fraser et al.; *C.A.* **62**, 16145b (1965)	Insufficient activity against *Horpales rufipes*	402
DOS 1693155 (1964/71), Boots, Inv.: Robertson et al.; *C.A.* **69**, 67087b (1968)	As effective as *malathion* against adult mosquitoes	403
	Effective against mosquitoes, aphids, caterpillars; low acute toxicity (LD_{50})	355
DOS 2027058 (1969/71), ICI (Australia), Inv.: Baklin, et al.; *C.A.* **75**, 140524x (1971)	Effective against cattle ticks; low toxicity (LD_{50}: 1500 mg/kg, mouse)	404
	Acylated products at least as toxic as the parent carbamates toward flies, but mammalian toxicity (mouse) reduced by factor 3–50	405
USP 3663594 (1968/72), Chevron, Inv.: Brown, Kohn; *C.A.* **77**, 61631t (1972)	Very effective against larval and adult mosquitoes resistant to organophosphates (LD_{50}: 131–275 mg/kg, rat)	406
USP 3764694 (1970/73), Monsanto, Inv.: Kirchner; *C.A.* **80**, 6950h (1974)	Broad-spectrum insecticide and acaricide	

TABLE 14. (*Continued*)

	General Formula	Experimental Product

C. $R-O-CO-N\overset{X}{\underset{CH_3}{}}$ *N-Acyl, N-Sulfenyl, and Similar Carbamates* (*contd.*)

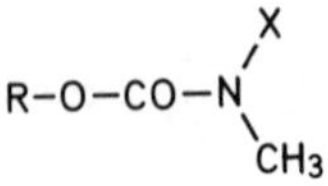

39

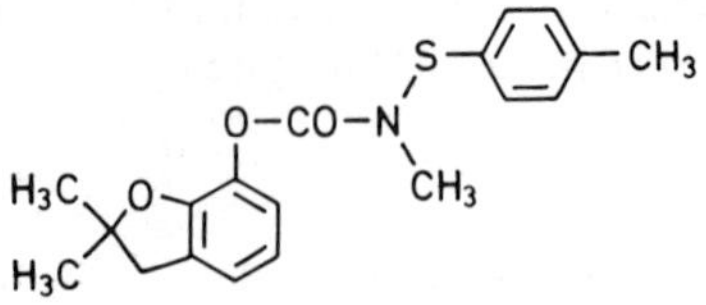

SIT 560

40

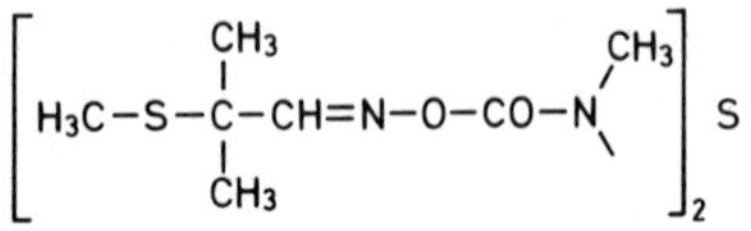

41

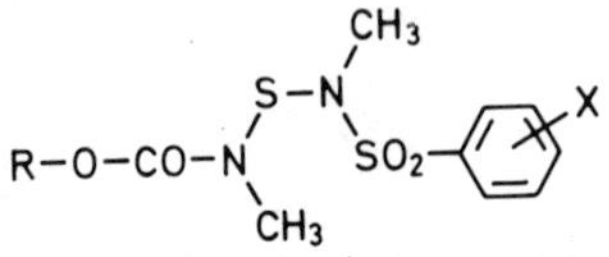

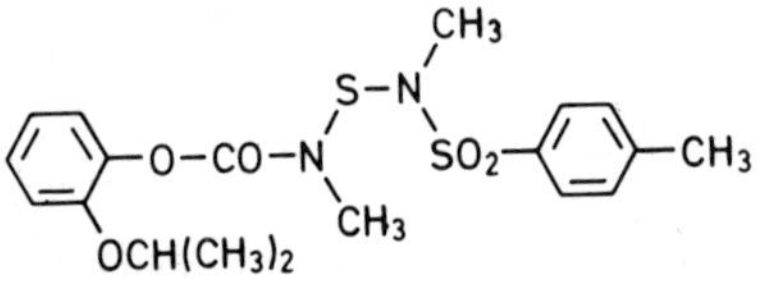

42

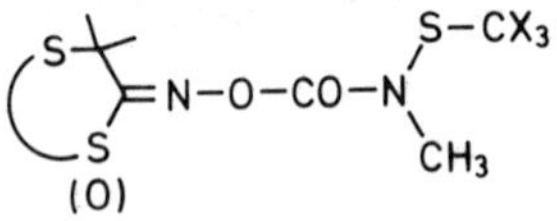

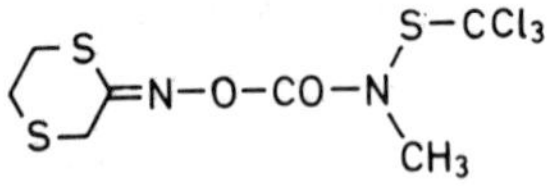

Oximecarbamates are usually made by the isocyanate route since Beckmann rearrangement or loss of water (with aldoximes) frequently occurs with carboxylic chlorides. Tertiary amines (e.g., triethylamine) or organotin compounds [e.g., bis(acetyloxy)dibutylstannane] function as catalysts for the addition of methyl isocyanate.

In the preparation of carbamates acylated or sulfenylated at the carbamate N atom, the corresponding NH compounds can be treated with anhydride-sulfuric acid or halide-base. *N*-Formylcarbamates **7** must be made by another method.[368]

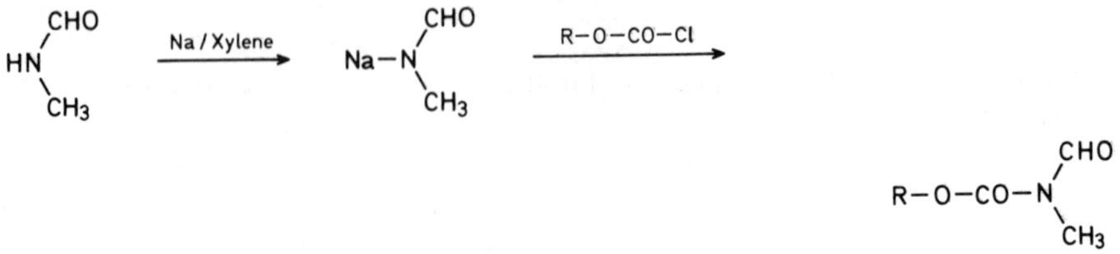

7

TABLE 14. (*Continued*)

Patent	Biological Properties of the Experimental Product	Refs.
DOS 2045441 (1970/72), Bayer, Inv.: Zumach et al.; *C.A.* **77**, 48215r (1972)	Equally or more effective than *carbofuran* against *Musca domestica* and *Culex fatigans* but much less toxic (LD_{50}: 100–125 mg/kg, mouse)	407
	Effective against aphids, spider mites, mosquito larvae; better systemic properties than *aldicarb*	408
DOS 2254354 (1972/74), Bayer, Inv.: Siegle et al.; *C.A.* **81**, 37408n (1974)	Better residual action than *propoxur* against *Musca domestica*	
DOS 2425211 (1973/74), Union Carbide, Inv.: Durden, D'Silva; *C.A.* **83**, 164240t (1975)	Broad-spectrum insecticide	

The following reaction sequence is an alternative for the preparation of *N*-acyl carbamates by direct acylation, and is of particular interest in the case of large acyl residues on oximecarbamates:

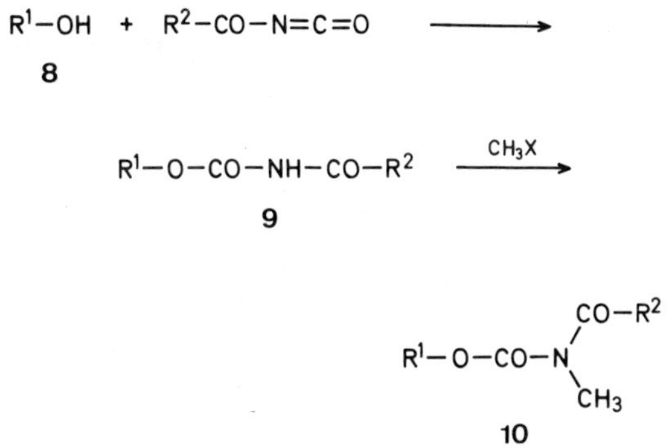

$$R^1-OH \ + \ R^2-CO-N=C=O \longrightarrow$$

8

$$R^1-O-CO-NH-CO-R^2 \xrightarrow{\ CH_3X\ }$$

9

$$R^1-O-CO-N \big\langle \begin{matrix} CO-R^2 \\ CH_3 \end{matrix}$$

10

Other synthetic routes have been developed for the preparation of *N*-sulfenylated carbamates.[409,410]

The oximes used as starting materials are normally prepared by the conventional method from the aldehyde and hydroxylamine. Introduction of alkylthio groups in the α-position is often carried out on the oxime. In a few cases the oximes are prepared using nitrosyl chloride. The syntheses of *aldicarb* **11**,[411] *tranid* **12**,[378] and *oxamyl* **13**[412] are given as examples.

$$H_3COOC-CH_2-CO-CH_3 \xrightarrow{HNO_2} H_3COOC-\overset{\overset{\displaystyle NOH}{\|}}{C}\diagdown_{CO-CH_3} \xrightarrow{Cl_2} H_3COOC-\overset{\overset{\displaystyle NOH}{\|}}{C}\diagdown_{Cl} \xrightarrow{CH_3SH/B}$$

$$H_3COOC-\overset{\overset{\displaystyle NOH}{\|}}{C}\diagdown_{S-CH_3} \xrightarrow{HN(CH_3)_2} (H_3C)_2N-CO-\overset{\overset{\displaystyle NOH}{\|}}{C}\diagdown_{S-CH_3} \xrightarrow{H_3C-N=C=O}$$

$$(H_3C)_2N-CO-\overset{\overset{\displaystyle N-O-CO-NH-CH_3}{\|}}{C}\diagdown_{S-CH_3}$$

13

As esters of carbamic acid, the carbamate insecticides are somewhat labile under basic conditions. Several authors have studied the kinetics and mechanism of the alkaline hydrolysis of phenyl carbamates, mainly with a view toward finding parallels to the inhibition of cholinesterase.[413–415] In the case of *N*-(methyl)phenyl-carbamates, thermal cleavage can occur to give the phenol and methyl isocyanate; among the oximecarbamates the nitrile and carbamic acid (or its salts) are obtained.[412] The oximecarbamates can exist in the *syn* and *anti* geometrically isomeric forms. The question of this isomerism, and a precise structural allocation, is seldom discussed. In many cases only one of the two forms seems to be present,[411,416] apparently energetically favored by steric factors or hydrogen bonding. Where *syn* and *anti* forms have been separated, they have had different biological activities, albeit not markedly so.

2.4.6. MECHANISM OF ACTION

The mechanism of action of the insecticidal carbamates is identical to that of the organophosphates, namely, inhibition of the enzyme cholinesterase or, more precisely, of acetylcholinesterase E. C. no. 3.1.1.7. This enzyme has, both in vertebrates and arthropods, the function of hydrolyzing the postsynaptic effector, acetylcholine, into choline and acetic acid. The function and properties of this enzyme have been described in detail.[417,418]

Inhibition of acetylcholinesterase leads to a buildup of acetylcholine in the postsynaptic membrane and hence to a permanent nerve stimulation with—in extreme cases—lethal results. This stimulation in insects manifests itself in uncontrolled movements and paralysis. Although the differences in anatomy and chemistry of nerve impulse transfer and in enzyme specificity between vertebrates and insects are not inconsiderable, there is no reason to doubt that acetylcholinesterase (AChE) is also the primary site of action of carbamates in insects.[419] According to Wilson[420] and Main,[421] the kinetic scheme of hydrolysis of acetylcholine (AC) by AChE (E) and the inhibition by an inhibitor (I) can be described as follows.

Hydrolysis

$$E + AC \underset{}{\overset{K_A}{\rightleftharpoons}} E-AC \xrightarrow{K_2} E-A + C \xrightarrow{K_3} E + A$$

Inhibition

$$E + I \underset{K_{-1}}{\overset{K_1}{\rightleftharpoons}} E{-}I \xrightarrow{K_2'} E{-}I' \xrightarrow{K_3'} E + I'$$

Here E-I is the enzyme/inhibitor complex that reacts to E-I' under covalent bonding of the carbamoyl moiety (e.g., $-CO-NH-CH_3$) to the serine-OH group in the active center of the enzyme. This scheme is valid both for carbamates and organophosphates.

In comparative studies on AChE inhibitors, often only the I_{50} value is measured.[412,416] This is the concentration at which a 50% inhibition is achieved under constant conditions.

More precise kinetic analyses of the inhibition of AChE isolated from fly brains have given an insight into the mechanism of inhibition and into important differences between carbamates and organophosphates.[422-425] Thus it has been found that K_3', the rate constant for the step that determines the regeneration of the enzyme, is practically equal to zero for organophosphate insecticides, whereas the value of K_3' for carbamates is small but finite. For example, when I' = $CO-NH-CH_3$, $K_3' = 0.01$, and when I' = $CO-N(CH_3)_2$, $K_3' = 0.003$ (min^{-1}, fly-brain AChE, 25°, pH 7.4).[425] For monomethylcarbamates the inhibited enzyme is thus regenerated (detoxified) much more quickly than for dimethylcarbamates. When I' = $PO(OR)_2$, $K_3' = 0$, which has consequences for the medical treatment of poisoning cases.

At low inhibitor concentrations the overall reaction rate, important for the insecticidal effect, is determined by the bimolecular rate constant $K_1' = K_2'/K_I$,[421,424] in which K_I is the binding constant K_1/K_{-1} of the inhibitor-enzyme complex. K_1 is highly dependent on the structure of the residue R in the carbamates $R-O-CO-NR^1R^2$. The distance between the acyl moiety to be transferred (e.g., $CO-NH-CH_3$) and a second grouping in the molecule, which should be isosteric with the trimethylammonium group of acetylcholine, plays an important role. This distance is 5.9 Å in acetylcholine, the natural substrate of AChE. Certain oxime-carbamates fulfil these conditions well, as may be illustrated with *aldicarb*.[411]

$$H_3C \overset{\oplus}{\underset{\underset{CH_3}{|}}{\overset{\overset{CH_3}{|}}{N}}} - CH_2 - CH_2 - O - CO - CH_3$$

anionic   esteratic
Binding site

$$H_3C - S - \underset{\underset{CH_3}{|}}{\overset{\overset{CH_3}{|}}{C}} - CH = N - O - CO - NH - CH_3$$

Aldicarb

Binding to the enzyme is strengthened when the grouping corresponding to trimethylammonium is electron deficient. In actual fact, the sulfoxide of *aldicarb* is a much better AChE inhibitor, which explains the good *in vivo* effect of the latter since oxidation to sulfoxide has been detected in insects.[411] Carbamates with positively charged groups such as the compound:

$$\left[\underset{(H_3C)_3\overset{\oplus}{N}}{\bigcirc}-O-CO-NH-CH_3 \right] X^{\ominus}$$

are good AChE inhibitors (as shown by Metcalf and Fukuto[419] and as already known in principle from Stedman's work) but are incapable of reaching the site of action in the insect, because the nerve fibers are protected by a lipophilic sheath. This simple isosteric model does not explain all the facts. Metcalf and Fukuto themselves pointed out[419] that the difference in activities of enantiomeric carbamates with a chiral substituent in the *ortho* position makes a three-centered binding necessary. Hetnarski and O'Brien found that—at least with arylmethylcarbamates—hydrophobic binding and apparently also charge-transfer complex formation contribute to the binding constant K_1.[427]

The model character of all work on enzymes must be kept in mind when it comes to the interpretation of *in vivo* activities. Furthermore, the possibility of a direct effect of the carbamates—as of the organophosphates—on the acetylcholine receptor at the postsynaptic level cannot be excluded.[428] The thiolcarbamate *cartap* (no. 39 in Table 13) is such an insecticide. It acts as a ganglion blocker, not an AChE inhibitor,[382] and thus sets itself apart, both chemically and biochemically, from the other carbamate insecticides.

2.4.7. STRUCTURE AND BIOLOGICAL ACTIVITY

The first studies on the relationships between structure and biological activity among phenyl carbamates, both *in vitro* and *in vivo*, stem mainly from Metcalf and Fukuto.[419] Gysin[347] had already pointed out in 1954 that a strict parallel between cholinesterase inhibition and toxicity toward insects and mammals was not to be expected for several reasons, one being the differing specificities of the enzymes. This was later demonstrated when AChE isolated from flies was studied.[422] The fly enzyme has a greater affinity toward carbamates than that isolated from bovine erythrocytes, the explanation lying in the different distances between the anionic and esteratic sites.[423]

In trials on insects, differences in penetration and metabolism among compounds of a congeneric series must be taken into account. These can severely distort or mask a structure–activity relationship. Some general conclusions are, however, possible.

The preference enjoyed by the CO—NH—CH$_3$ group has grounds that can be explained by the kinetics of inhibition (see Section 2.4.6). It has a sufficiently high affinity to the esteratic site (i.e., the rate of acylation, governed by K'_2, is high), whereas the regeneration of the enzyme is relatively slow compared to CO—NH$_2$, even though it is somewhat faster compared to CO—N(CH$_3$)$_2$ [K'_3 of CO—NH$_2$; CO—NH—CH$_3$; CO—N(CH$_3$)$_2$ at 25°, pH 7.5, oximecarbamates of the *methomyl* type:[425] 45; 9.8; 2.5 (10^{-3} min^{-1})].

Phenylcarbamates require a substituent in the 2 or 3 position that is as electron rich as possible. Electron-deficient substituents have an unfavorable effect on activity. This observation is compatible with the assumption that this substituent must contribute to the binding to the anionic site of the enzyme. In a somewhat limited series of phenylcarbamates, Hansch and Deutsch[429] found a positive correlation between log I_{50} (AChE inhibition) and π, the parameter representing the lipophilicity or hydrophobic binding. In fact, all commercially interesting phenylcarbamates have lipophilic substituents (alkyl, alkoxy, or alkylthio groups). Hetnarski and O'Brien[427] have analyzed the binding constants K_I and confirmed the positive correlation with π and a further parameter C_r, which was defined as a measure of the tendency to form charge-transfer complexes. However, an inaccurate prediction of the K_I values of the two most active members of the series [*carbaryl* and 3-(1-methylethyl)phenyl methylcarbamate] is made. In all analyses to date it has not proved possible to take account of steric effects, which are obviously of particular importance with *ortho* substituents.

Relatively few studies have been published on the oximecarbamates. Inductive, electron-withdrawing groups in the α or β position to the oxime carbon atom seem to be necessary, as shown by investigations on analogues of *aldicarb*.[411] This stands in apparent contradiction to the fact that a remarkable number of oximecarbamates bear an alkylthio residue, that is, electronically an almost neutral group, on the α or β carbon. It has been shown for *aldicarb*[430] that this product is quite rapidly oxidized to the sulfoxide in the plant, this metabolite being a more effective AChE inhibitor by a factor of 10 and also having a higher water solubility. This explains the high insecticidal and systemic efficacy of *aldicarb*. Since the metabolic conversion of sulfide to sulfoxide (and sulfone) is also a rapid process in mammals, the alkylthio group offers no advantages in terms of selective toxicity (compare nos. 31, 32, 34, 37, and 38 in Table 13).

Several interesting reports have been published concerning the *selective* detoxification of carbamates.[355,368] Mammalian toxicity can be significantly reduced by *N*-acylation of methylcarbamates without any great loss of insecticidal activity.[355,368] Methylcarbamates have thus been acylated at nitrogen with PO(OR)$_2$, PS(OR)$_2$, *S*-aryl, *S*-alkyl, and *S*-[bis(carbamoyl)sulfide] residues[405,407,408] with the purpose of exploiting differences in metabolism between mammals and insects. In the case of the *carbofuran* derivative 14, whose LD$_{50}$ (mouse) is 25–50 times greater than that of the parent carbamate 15, a rapid degradation to 15 has been found in the housefly.[408,431]

The insecticidal activity (against the housefly) of 14 is only a factor 2–3 weaker than that of 15. Sulfides of the 14 type are weak AChE inhibitors. With oximecarbamates such as *methomyl* it has been found that *N*-acylation leads to a drop in

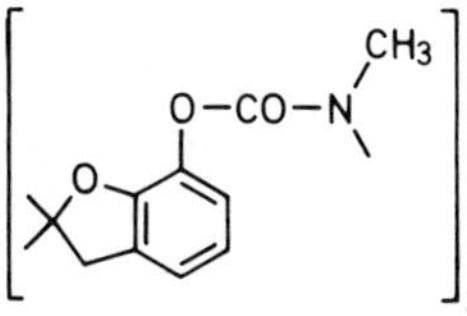

14

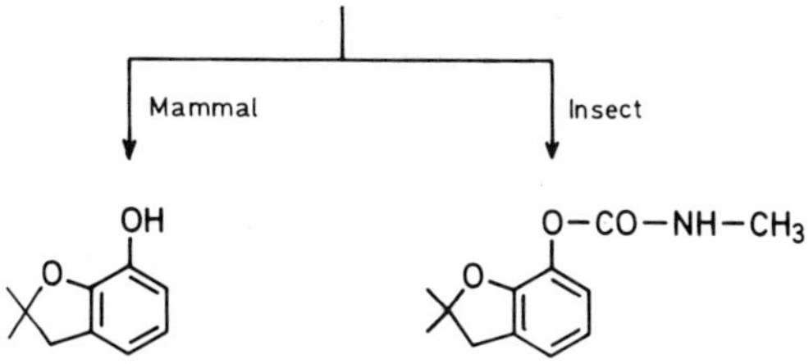

15

mammalian toxicity but also to an almost complete loss of insecticidal activity.[432] Apparently, in the *N*-acetyl carbamates **(16)** in which RO is an oximino group:

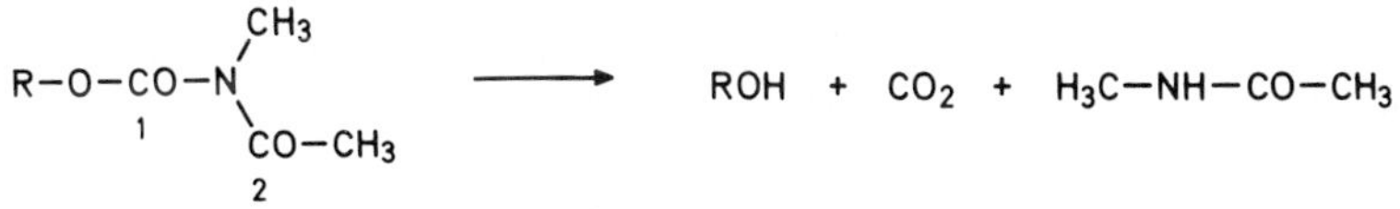

$$R-O-CO-N \overset{CH_3}{\underset{CO-CH_3}{\big|}} \longrightarrow ROH + CO_2 + H_3C-NH-CO-CH_3$$

16

deacylation in insects is favored at the C-1 atom, but when R = phenyl, at the C-2 atom. Numerous other examples of *N*-derivatization of carbamates with the aim of decreasing toxicity are to be found in the patent literature (see Table 14). One experimental product with such a structure is the tickicide *promacyl* (Table 14, no. 35), which is also effective against organophosphate-resistant ticks.[404]

2.4.8. TOXICOLOGY

The mammalian toxicity of the carbamates varies within wide limits (see Table 13). *Carbaryl*, the product with the largest market share at the present, is practically nontoxic (LD_{50}: 850 mg/kg rat, oral, acute). This property is a major contributory factor toward the success of this insecticide.[433] On the other hand, some economically significant carbamates such as *aldicarb* (1 mg/kg) and *carbofuran* (8–14 mg/kg, rat, oral, acute) belong to the most toxic insecticides known (cf. *parathion*, 3.5–8 mg/kg, and *endrin*, 7.5–17.5 mg/kg). In spite of the high acute toxicities of numerous carbamates, particularly the oxime derivatives, their toxicological properties are frequently viewed more favorably than those of the organophosphates even though both groups of insecticides have the same mechanism of action, namely, AChE inhibition.

The explanation lies in the details of mechanism of AChE inhibition (Section

2.4.6). There is a large difference between carbamates and phosphates in the size of K'_3, the rate constant governing the reactivation of the acylated enzyme. For mono- and dimethyl carbamoyl-AChE complexes, the half-lives of spontaneous reactivation are of the order of minutes; for phosphorylated AChE, hours or days.[434] In addition, "ageing" of the phosphorylated enzyme can lead to the inhibition becoming irreversible.[435] This difference means that, in cases of carbamate poisoning, only symptomatic treatment and administration of atropine sulfate are indicated. Administration of the cholinesterase reactivator PAM (pyridine aldoxime methiodide) or toxogonin, often necessary with organophosphate poisoning, is strictly contraindicated with carbamates.[436] Furthermore, the rapid spontaneous hydrolysis of carbamoylated AChE means that the toxic effects can hardly accumulate. This holds true for both insects and mammals. Thus no AChE inhibition can be detected in the blood of animals or humans exposed to carbamates, whereas, with certain organophosphates, even a relatively low concentration over a long period can lead to a slowly increasing AChE inhibition.

2.4.9. METABOLISM, RESISTANCE, SYNERGISM

Studies on metabolism in insects, mammals, and (in the case of systemic agents) plants have been published for numerous commercially important carbamates. Such studies are required by the authorities for the registration of new products.

Generally the carbamoyl grouping is attacked at two positions:

$$R-O-CO-NH-CH_3 \quad \Bigg[\begin{array}{l} \longrightarrow \ ROH \ + \ CO_2 \ + \ NH_2CH_3 \\[2ex] \longrightarrow \ R-O-CO-NH-CH_2-OH \end{array}$$

17

Either *O*-acyl cleavage or a hydroxylation of the methyl group occurs, which leads to the (usually unstable) compounds **17**. Phenylcarbamates are hydroxylated—mainly in the *para* position—by the mixed-function oxidase complex. Alkylthio groups, as are usually to be found in oximecarbamates, are rapidly oxidized to sulfoxide and sulfone. The primary products then enter into numerous further degradation reactions that are very difficult to generalize. In most cases metabolic breakdown leads rapidly to atoxic products. Some exceptions are discussed in Section 2.4.7.

Resistance toward carbamates can be created by generation tests in the laboratory but has not played any significant role to date in practice. Some reports have been made of resistance to *carbaryl*, the carbamate available for the longest time. On the other hand, *carbaryl* is effective against insect species that are resistant toward *DDT* and other chlorohydrocarbons.[360]

Organophosphate-resistant insects can show cross-resistance to carbamates. One reason could be that such insects are better equipped with metabolizing enzymes, particularly hydrolases and oxidases.

Hydroxylation, which already plays an important part in metabolism by nonresistant insects, can be inhibited by *synergists*. Synergists active in the sense that they increase *in vivo* activity against strains of normal sensitivity are mostly benzo-dioxoles of the structure **18**. Other classes such as the propargyl ester **19**[437] or ether **20**[438] are also active.

An inhibition of mixed function oxidase,[439] a microsomal enzyme complex, is generally assumed to be the mechanism of synergist action. Some studies on structure and activity have been published.[440,441] Although the activity of *carbaryl*, for example, can be increased by up to 400-fold against flies of normal susceptibility (synergist: carbamate = 5 : 1),[442,443] carbamate–synergist combinations have played no part in practice to date. The relatively high costs, both of the synergists and of the official registration of a new mixture of insecticide and synergist, are probably responsible.

$$R\text{—}\underset{\text{18}}{\left\langle\!\!\!\!\!\overset{O}{\underset{O}{\text{—}}}\right\rangle} \qquad \underset{\text{19}}{\left\langle\!\!\!\!\!\right\rangle\text{—NH—CO—O—CH}_2\text{—C}\equiv\text{CH}}$$

$$\underset{\text{20}}{\overset{\text{NO}_2}{\left\langle\!\!\!\!\!\right\rangle}\text{—O—CH}_2\text{—C}\equiv\text{CH}}$$

2.5. *Insecticides from Miscellaneous Classes*

W. LUNKENHEIMER
Bayer AG, Wuppertal-Elberfeld

Most compounds discussed in this chapter are of historical interest, being no longer on the market. These so-called first-generation insecticides have made way for the far more active products of the second generation (chlorohydrocarbons, organo-phosphates, carbamates), which themselves are now facing competition from the so-called third generation. This rather diffuse term covers the naturally occurring insecticides, their synthetic analogues (see Section 2.1), and products with new mechanisms of action. The latter are represented in this chapter by the acylureas.

Some readers may miss certain classes such as *DDT* analogues, nitrophenol derivatives, and amidines. Owing to their acaricidal properties these are dealt with in Section 2.7.

Literature references to the older products are to be found in the standard works.[444–448]

2.5.1. INORGANIC INSECTICIDES

At one time arsenic products such as Paris green, sodium arsenite, basic copper arsenate, and lead arsenate were popular insecticides in fruit and vegetable growing. In some countries such products are still marketed today, the most important being calcium arsenate$[Ca_3(AsO_4)_2$, LD_{50}: 35 mg/kg rat, oral, acute]. Formerly this was used against Colorado beetle and other chewing insects. Owing to its high mammalian toxicity and phytotoxicity, however, it was later used only against the boll weevil in cotton. The U.S. Department of Agriculture (USDA) recently canceled this use as well.

Cryolite, a naturally occurring mineral (Na_3AlF_6; Kryocide®), acts as a contact and stomach poison in the form of a suspension and is nonphytotoxic. This and the low mammalian toxicity (LD_{50}: 10,000 mg/kg rat, oral, acute) permit use in fruit, vegetable, cotton, soybean, and sugarcane crops.

The use of Silica Aerogel® is of more recent origin (1956). This mixture of very finely divided silica gel and aluminum fluosilicate finds application for the control of structural pests (roaches, silverfish, ants, etc.) in farm buildings. It acts by contact, removing the protective oil film covering the insect's body that prevents loss of moisture.[447] The purely physical drying-out effect is complemented by the insecticidal properties of the fluosilicate added.

2.5.2. ORGANIC INSECTICIDES

2.5.2.1. Thiocyanates

The first insecticidal thiocyanates were marketed at the start of the 1930s under the name Lethane® (Rohm and Haas). They were 2-(thiocyano)ethyl ethers and 2-(thiocyano)ethyl esters of long-chain fatty acids (C_{10}-C_{18}).

$$H_9C_4-O-CH_2-CH_2-O-CH_2-CH_2-SCN$$

2-(2-Butoxyethoxy)ethyl thiocyanate; Lethane 384®

Made by reaction of the corresponding chloro compound with sodium rhodanide, Lethane 384® is an effective contact poison and, owing to its good knockdown properties, is frequently used in combination with other agents as a spray against mosquitoes, flies, and household pests. Its disadvantages are an unpleasant odor, an irritant effect on skin and mucosa, and a not inconsiderable acute toxicity (LD_{50}: 90 mg/kg rat, oral, acute).

Thanite®, prepared by esterification of chloroacetic acid with isoborneol and subsequent reaction with ammonium thiocyanate, contains ~18% of other, related terpene esters. This is also a contact insecticide with good knockdown effect and is used as a spray against household pests. Owing to its favorable acute toxicity, it finds application in veterinary medicine against ectoparasites.

NCS—CH$_2$—CO—O   CH$_3$

1,7,7-Trimethylbicyclo[2.2.1]hept-2-yl thiocyanatoacetate
(Hercules Inc., 1945); LD_{50}: 1603 mg/kg rat, oral, acute

2.5.2.2. Heterocyclics

Phenothiazine, the first synthetic organic larvicide (1925) and the forerunner of *DDT*, replaced the highly toxic arsenicals in *Anopheles* control. It was used on a large scale. In the United States alone the annual consumption of phenothiazine as insecticide and anthelmintic amounted to ~1700 tonnes in 1944.

Phenothiazine may be prepared by melting diphenylamine with sulfur at $180°$ in the presence of iodide or aluminum chloride as catalyst.

LD_{50}: 5000 mg/kg rat, oral, acute

The susceptibility of phenothiazine to oxidation prevented to its wide application in agriculture (e.g., against codling moth larvae) and in veterinary medicine against ectoparasites. Even the addition of antioxidants did not improve the stability under field conditions.

Poor absorption of phenothiazine in the digestive tract permits use in the control of intestinal worms, mainly in sheep.

Even diphenylamine itself, a precursor in the manufacture of phenothiazine, played a certain role as insecticide and ectoparasiticide in pre-*DDT* times. The American blowfly insecticide Smear 62® contained 35% diphenylamine. Treating open wounds in cattle prevented infection by blowflies.

From the nitrocarbazole class (IG Farbenindustrie), developed from 1938 onward, sprang the first organic insecticide to be produced on a large scale in Europe—Nirosan®, which was joined by Nirosit® (or Holfidal®) after 1945.

Nirosan®

Nirosit® (Hoechst)

Both compounds are mainly stomach poisons, the contact effect being weak. They have low mammalian toxicities and are effective generally against biting, but not sucking, insects. Numerous beneficial insects such as bees are unaffected. The nitrocarbazoles were of particular value in grape crops for the control of larvae of the grape berry moth. In this indication they made the toxic arsenicals superfluous and were the leading vineyard insecticides for more than a decade.

2.5.2.3. Alkyl Aryl Sulfones

Chloromethyl 4-chlorophenylsulfone is effective against certain sucking insects, spiders, woodlice, ants, and the European corn borer, as well as ectoparasites.

$$Cl-CH_2-SO_2-\langle\bigcirc\rangle-Cl$$

Lauseto neu®, CCS (IG Farbenindustrie, Leverkusen; Wolfen since 1941)

Its main practical use was against body lice in the German Wehrmacht. Clothing impregnated with the product protected the wearer from lice for several months. The chloromethylsulfones were displaced by *DDT* and *lindane* after 1945.

2.5.2.4. Indandiones

The toxicity of 1,3-indandione toward flies is increased by acylation in the 2-position, depending on the chain length of the acyl residue, and reaches a maximum at C_5. Two indandiones with C_5-acyl groups were launched by the Kilgore Chemical Company at the beginning of the 1940s: 2-isovaleryl- and 2-pivaloyl-1,3-indandione under the names Valone® and Pival®, respectively.

R = $-CH_2-CH(CH_3)_3$; Valone®

R = $-C(CH_3)_3$; *Pindone*, Pival®, Pivalyn®

These compounds are prepared by condensation of dimethyl 1,2-benzenedicarboxylate with 3-methyl-2-butanone and 3,3-dimethyl-2-butanone, respectively, in the presence of sodium in benzene. They are yellow, crystalline products that form salts easily with aqueous alkali.

Valone® was used with pyrethroids in fly sprays. *Pindone* is also ovicidal toward the nits of body lice. A practical exploitation of this property is prevented by the high toxicity of the compound, dogs being killed by 2.5 mg/kg daily.

Pindone and other indandiones have gained interest as rodenticides as a result of their anticoagulant properties.

2.5.2.5. Xanthogenates

Potassium xanthogenate was used in grape crops in France toward the end of the last century. Some success was achieved in the control of nematodes and insects owing to the slow release of carbon disulfide from the product, which was incorporated into the soil. Dixanthogen, obtained by oxidation of xanthogenates, is an effective insecticide.

$$\underset{H_5C_2O}{\overset{S}{\|}}\!\!\!\!C-S-S-\underset{OC_2H_5}{\overset{S}{\|}}\!\!\!\!C$$

It acts against body lice, fleas, mange mites, and similar parasites. In the form of ointments, powders, and soaps, it was mainly used in Russia for control of lice and thus served to reduce typhus infections.

2.5.2.6. Acylureas

During the search for new derivatives of the herbicide *dichlobenil* (Casoron®) **(1)**, the nonphytotoxic acylurea DU 19111 **(2a)** was synthesized and was found to have very interesting insecticidal properties.[449,450]

1   **2**

2a; X = Y = Cl, n = 1; DU 19111 (Philips-Duphar)

2b; X = Y = Cl, n = 0; PH 60-38

2c; X = F, Y = CL, n = 0; PH 60-40, *Diflubenzuron*, Dimilin®; LD_{50}: 4640 mg/kg rat, oral, acute

2d; X = F, Y = CF_3, n = 0

The more active derivatives **2b–2d** were found as a result of intensive investigations in this class.[451] One method for their preparation involves the reaction of 2,6-disubstituted benzoylisocyanates with suitably substituted anilines.[450]

The products are only effective as stomach poisons. They kill the larvae of various different insects (e.g., cabbage white caterpillar, Colorado beetle, housefly, yellow-fever mosquito). The preimaginal stages of sucking insects (e.g., hoppers, plant bugs, aphids) are not affected, because the agent can penetrate neither the insect cuticle nor the plant epidermis. The formation of the insect cuticle in the larval stage is disrupted owing to blocking of chitin synthesis, and the larvae do not survive the molt.[452,453]

Dimilin® **(2c)** is used as a mosquito larvicide and for control of certain major pests of cotton, soybeans, citrus fruits, and vegetables.

2.5.2.7. Benzyl Esters

The insecticidal properties of certain benzyl esters and their carbinols (1-aryl-trihaloethanols and their *O*-acyl derivatives) have been known for many years[454,455] but were never sufficient for a practical exploitation. A commercial product has recently been developed from this class, however, and is sold in combination with *dichlorvos*, pyrethrum, or *tetramethrin* as Baygon® MEB.[456]

3,4-Dichloro-α-(trichloromethyl)benzenemethanol acetate; *Plifenate* (proposed),
Baygon® MEB (Bayer AG); LD_{50}: >10,000 mg/kg rat, oral, acute

This colorless crystalline compound is prepared by Friedel–Crafts reaction of chloral with 1,2-dichlorobenzene in the presence of aluminum chloride and subsequent acetylation of the product with acetic anhydride. Being an activated ester, it is sensitive to base.

A contact poison with a residual action of several weeks, *plifenate* kills flies, mosquitoes, clothes moths, and hide and carpet beetles. It is also effective against strains that are resistant to chlorohydrocarbons and organophosphates.

As a household spray *plifenate* is used in combination with *dichlorvos*, pyrethrum, or *tetramethrin* (Neo-Pynamin®).

Although *plifenate* has certain similarities to *DDT* (five chlorine atoms in the molecule, chloral chemistry), there is one most important difference. In contrast to *DDT*, *plifenate* is an ester and is rapidly hydrolyzed in the organism, the alcohol formed being excreted as a glucuronide in the urine. In feeding trials with *plifenate* neither the intact product nor the alcohol could be detected in the liver, kidneys, brain, or fatty tissue of the test animals.[457]

2.5.3. FUMIGANTS

Fumigants are gaseous or easily vaporized compounds that frequently possess a nonspecific pesticidal activity. Thus the insecticidal fumigants, which alone are of interest here, are often also nematicidal and rodenticidal and are sometimes fungicidal and bactericidal. Fumigants have various uses such as soil insecticides, stored-product protectants, and for control of household and hygiene pests.

The major requirement for soil insecticides is that they are able to spread in the soil so that the insects present come into contact with the insecticide. This requirement is fulfilled by the fumigants. The requirement of highest possible insecticidal potency, however, which is understandable from economic considerations, is not fulfilled by fumigants. Generally, much higher application rates are necessary than with foliar insecticides. In addition, soil fumigation is very labor intensive. For these reasons the fumigants were replaced by the chlorinated hydrocarbons as soil insecticides.

Nevertheless, fumigants are irreplaceable as stored-product protectants. They protect the millions of tonnes of harvested crops in storage from insects, fungi, bacteria, and rodents. Although effective commodity protection is, strictly speaking, not part of crop protection, it is a logical extension thereof.

Examples of important stored-product insect pests are the grain weevil, rice weevil, and angoumois grain moth. In controlling commodity pests, the insects to be found in houses, mills, warehouses, ships, and railroad cars are also frequently destroyed—for example, ants, termites, roaches, spiders, and wood lice. The major requirement of a stored-product protectant is that it exterminate all development stages of the pests mentioned above. Since there is an excessive food supply in these cases, a mere 95% pest kill is insufficient to prevent a rapid and unhindered multiplication of the remainder.

The odor, taste, and color of the treated commodities must not be affected by the fumigant. Absorption and water solubility of the gas should be low so that the fumigant disappears completely during subsequent ventilation. The insecticidal activity should be of rapid onset; phytotoxicity and mammalian toxicity should be low. As no single ideal fumigant exists that meets all these requirements, combinations are often used. Several important fumigants are listed in the following.

Methyl bromide.   bp: $4.5°C$; LC_{100} (rat): 514 ppm. Fumigation of grain, fruit and vegetables. Debugging of grain elevators, mills, ships.

Ethylene dichloride. bp: 83.5°; LD_{50}: 670 mg/kg rat, oral, acute. Mainly for fumigation of stored grain.

Ethylene dibromide. bp: 131.7°; LD_{50}: 146 mg/kg rat, oral, acute. Mixtures with carbon tetrachloride and ethylene dichloride are used for mill, warehouse, or household fumigation.

Carbon tetrachloride. bp: 77°; LD_{50}: 7460 mg/kg rat, oral, acute. Grain fumigant, often used in mixtures with more potent fumigants to reduce fire hazards.

Carbon disulfide. bp: 46°; LC_{100} (man): 15 g/m^3, inhaled for 3-4 hr. Fumigation of grain during the filling of the silos.

Acrylonitrile, Ventox®, Acritet®. bp: 77°; LD_{50}: 93 mg/kg rat, oral, acute. For fumigation of milling, baking, and food-processing machinery and of packed grain products and stored tobacco. Registration withdrawn for fumigating food commodities.

Ethylene oxide, Cartox®. bp: 10.7°; LC_{50} (rat): 7.2 mg/l, inhaled for 4 hr. Mainly used as a mixture with carbon dioxide for vault fumigation of stored food products.

Hydrogen cyanide, Cyclon®. bp: 26.5°; LD_{100}: 1 mg/kg man, oral, acute. Fumigation of grain silos, warehouses, ships, and mills.

Phosphine, Phostoxin®. bp: −87.8°; LC_{100} (man): ~10 mg/m^3, inhaled for 6 hr. A mixture of aluminum phosphide and ammonium carbamate for grain fumigation. Phosphine is produced on contact with moist air, its flammability being reduced by the carbon dioxide and ammonia generated at the same time.

Chloropicrin. bp: 112.2°; LD_{50}: 250 mg/kg rat, oral, acute. Irritating to skin, eyes, and respiratory tract. For insect and rodent control in grain elevators, bins, and other grain-storage places.

Sulfuryl fluoride, Vikane®. bp: −55.2°; LD_{50}: 100 mg/kg rat, oral, acute. For fumigation of structures and wood products for control of drywood termites and wood-infesting beetles.

2.6. Nematicides

H.-J. RIEBEL
Bayer AG, Wuppertal-Elberfeld

2.6.1. GENERAL BIOLOGY

Chemical products used for control of nematodes are termed nematicides. Nematodes appear in various shapes and forms on practically all culture plants* and can cause such damage that the harvest yields are much impaired both qualitatively and quantitatively. In extreme cases the plants die.[458,459]

*Nematodes pathogenic to animals and humans are also known, apart from the phytopathogenic species discussed here.

The most important groups and species of phytopathogenic nematodes are as follows.

2.6.1.1. Free-Living Root Nematodes[460]

In taxonomical terms the free-living root nematodes are not a unit, but belong to various different families and orders. The most important representatives are *Pratylenchus, Paratylenchus, Rotylenchus, Tylenchorhyncus, Trichodorus, Longidorus,* and *Xiphinema* spp. These thin nematodes, most often only ~1 mm long, attack the roots of plants. Like all phytopathogenic nematodes, these nematodes— also called eelworms owing to their eel-like form—puncture plant cells with the aid of a stylet or hollow spear and draw out the cell contents. The transport systems of the plant are thus constricted, and the transport of water, food, and assimilates is considerably reduced or interrupted. Rot bacteria and fungi often enter the plant through the puncture sites and cause decomposition of the root tissue. Depending on the degree of nematode contamination in the soil, the damage ranges from slight growth depression to stunting or even death of the plant.

The free-living root nematodes appear as pests on numerous broadleaf and woody plants. The most important host plants are grain, beet, potato, vegetables, apple, banana, cotton, tea, and coffee.

2.6.1.2. Root-Gall Nematodes (*Meloidogyne* spp.)[460]

The typical characteristic of this group of nematodes is their ability to induce the formation of galls in plant roots.

The larvae of such nematodes invade the roots from the soil. The surrounding root tissue is induced to extra growth by growth-stimulating secretions excreted by the nematodes. Galls are then formed that, depending on the species of plant, can swell to the size of a walnut or even a fist. The nematodes feed on and multiply in these galls.

Mainly broadleaf crop plants are attacked by root-gall nematodes. They cause the greatest damage in tobacco, cotton, tomato, cucurbits, carrot, carnation, and numerous other vegetables, fodder crops, and ornamentals.

2.6.1.3. Cyst Hematodes (*Heterodera* spp.)[460]

The females of this genus form an egg capsule that is attached via the epidermis to the outer surface of the root. This capsule, which is a lemon- or ball-shaped structure the size of a pinhead and contains 300 to 400 eggs, is the so-called cyst.

Cyst nematodes are mainly pests in the temperate zones. The most important hosts are potato, beet, oat, and clover.

2.6.1.4. Stem Nematodes (*Ditylenchus* spp.)[460]

These nematodes feed and reproduce predominantly in the stem tissue of their host plants. An infection can normally be recognized from stunting of the plant or stem rot.

Stem nematodes normally attack corn, rye, oat, clover, beet, tobacco, and bulbs (tulips, narcissi, and hyacinths).

2.6.1.5. Leaf Nematodes (*Aphelenchoides* spp.)[460]

Leaf nematodes invade the plants from the soil. They climb up on the outside of the leaf stems and enter the interior of the leaf via the stomata. There they inhabit the intercellular space of the mesophyll and feed on cell contents, which they draw out with the aid of a stylet.

Leaf nematodes are mainly pests in temperate and warm climates and attack rice, strawberries, and numerous ornamental plants such as chrysanthemum, lily, begonia, and gloxinia.

2.6.2. GENERAL NEMATODE CONTROL

Of all the phytopathogenic nematodes the group of root-parasitic species (free-living root nematodes, root-gall nematodes, and cyst nematodes) is the most significant in terms of economic losses. On a worldwide basis the stem and leaf nematodes are clearly second in importance to the root parasites. Furthermore, as they only appear on aerial plant parts, their control is that much easier. They are usually killed by insecticides, particularly those with systemic properties.

Under the term nematicide one therefore understands a chemical agent that kills the so-called soil nematodes (i.e., the root-parasitic species). The soil, as biotope for these pests, presents very high demands on the biological and physical properties of nematicidal agents.

1. A nematicide must be able to distribute itself well in the soil. Distribution can take place via the air- or water-capillary systems of the soil. Nematicides are therefore divided into fumigants and water-soluble agents.

2. Since the distribution of an agent in the soil only takes place slowly depending on weather conditions and since the agent must retain its activity until the distribution process is completed, nematicides should persist in active form in the soil for 1 to 4 months according to experience to date. The lower limit is the minimum for the full deployment of activity and the upper limit is the maximum tolerable for the residue behavior of a chemical in soil and plants.

2.6.3. NEMATICIDAL AGENTS

2.6.3.1. Fumigants[461]

The most important and well-known fumigants belong to two classes, the halogenated hydrocarbons[462] and the isothiocyanates.[461] Both classes have an unspecific pesticidal action, soil insects, soil-borne fungi, and weed seeds frequently being killed as well.

One great drawback to the fumigants is their high phytotoxicity. They can only be used when the harvest has been gathered from the fields to be treated. Furthermore, a withholding time of usually several weeks must be observed before the next crop is sown or planted. The most important fumigants in use today are listed in the following.

1. *Methyl bromide*; Meth-O-Gas®.[463]
2. *Ethylene dibromide*; Bromafume®, Nephis®, Dowfume®, N 85®, Soilbrom®.[463]
3. *1,3-Dichloropropene*; Telone®, Vidden D®.[464]
4. A mixture of *1,3-dichloropropene* and 1,2-dichloropropane; D-D®, Dowfume N®, Vidden D®.[462]
5. 1,2-Dibromo-3-choropropane; Fumazone®, Nemagon®.[465]
6. A mixture of *chloropicrin* and methyl isocyanate; Di-Trapex CF®.
7. *Metham-sodium* (sodium methylcarbamodithioate); Vapam®.[466]

$$H_3C-NH-C\underset{S-Na}{\overset{S}{\ll}}$$

Vapam® decomposes slowly in soil by hydrolysis to hydrogen sulfide and methyl isothiocyanate, which is the actual active pesticide.

$$H_3C-NH-C\underset{S-Na}{\overset{S}{\ll}} \quad \xrightarrow[-\,NaOH]{H_2O}$$

$$H_3C-NH-C\underset{SH}{\overset{S}{\ll}} \quad \xrightarrow[-\,H_2S]{} \quad H_3C-N{=}C{=}S$$

8. *Dazomet* (tetrahydro-3,5-dimethyl-2H-1,3,5-thiadiazine-2-thione;[467] Mylone®, DMTT®, Basamid®.

The active constituent is again methyl isothiocyanate.

$$H_3C-N{=}C{=}S \quad + \quad H_3C-NH_2 \quad + \quad 2\,H_2C{=}O \quad + \quad H_2S$$

9. Potassium *N*-hydroxymethyl-*N*-methylcarbamodithioate; Bunema[®].[468]

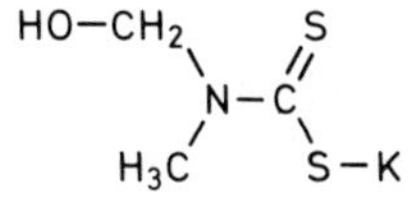

$$\begin{array}{c} HO-CH_2 \qquad S \\ \diagdown \qquad \parallel \\ N-C \\ \diagup \qquad \diagdown \\ H_3C \qquad S-K \end{array}$$

Bunema[®] also releases methyl isothiocyanate in the soil.

10. Further fumigants with exceptional nematicidal properties.

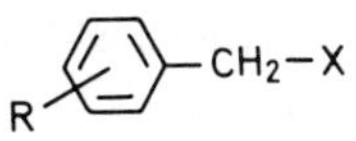

$$R \overbrace{}^{} -CH_2-X$$

R = X, CX$_3$, NO$_2$, NO, CN, SCN
X = Halogen

Benzylhalides[469,470]

$$R-CO-CH_2-X$$

R = H, CH$_3$, CH$_2$-CH$_3$, CH=CH$_2$, CH$_2$-CH=CH$_2$
X = Halogen

α-Haloketones[470]

No compound from either of these classes has yet been developed to the marketing stage.

2.6.3.2. Water-Soluble Agents

Chemical compounds that spread via the water-capillary system of the soil, are taken up by the roots (systemic action), and poison the so-called soil nematodes are termed water-soluble nematicides. The most active agents have been found among the insecticidal organophosphates and carbamates (see Sections 2.3 and 2.4). In contrast to fumigants the toxic action of these compounds is specific. Both the phosphate and carbamate types of nematicide are inhibitors of cholinesterase.

Apart from their specific pesticidal effect the modern water-soluble nematicides are well tolerated by plants and thus, in contrast to fumigants (which can only be used preventatively owing to their phytotoxicity), may be used in crop stands.

The most important organophosphate and carbamate nematicides are listed in the following.

1. *O*-2,4-Dichlorophenyl *O,O*-diethyl phosphorothioate; *Dichlofenthion*, Nemacide[®], Mobilawn[®];[471] LD$_{50}$: 270 mg/kg rat, oral, acute.

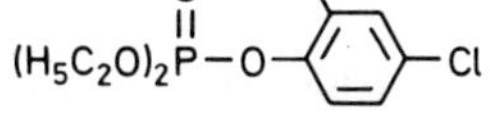

$$(H_5C_2O)_2\overset{\displaystyle S}{\overset{\displaystyle \parallel}{P}}-O-\overbrace{}^{Cl}-Cl$$

This compound was the first organophosphate to be used against soil nematodes. Apart from its nematicidal effect, *dichlofenthion* has a certain insecticidal action.

2. *O,O*-Diethyl *O*-4-(methylsulfinyl)phenyl phosphorothioate; *Fensulfothion*, Terracur-P®;[472] LD_{50}: 2-10 mg/kg rat, oral, acute.

$$(H_5C_2O)_2\overset{\overset{S}{\|}}{P}-O-\!\!\!\langle\!\!\!\rangle\!\!\!-SO-CH_3$$

A nematicide and insecticide with a persistence of 4-6 months.

3. Ethyl 3-methyl-4-(methylthio)phenyl-1-methylethylphosphoramidate; *Fenamiphos*, Nemacur®;[473] LD_{50}: 15-19 mg/kg rat, oral, acute.

$$(H_3C)_2CH-NH-\overset{\overset{O}{\|}}{\underset{\underset{H_5C_2O}{|}}{P}}-O-\!\!\!\langle\!\!\!\rangle\!\!\!\overset{CH_3}{-}S-CH_3$$

A particularly valuable systemic nematicide for control of free-living, cyst-forming, and root-gall nematodes in beet and potato crops.

4. Phenyl *N,N'*-dimethylphosphorodiamidate; *Diamidafos*, Nellite®, Dowco 169®;[474] LD_{50}: 140-200 mg/kg rat, oral, acute.

$$(H_3C-NH)_2\overset{\overset{O}{\|}}{P}-O-\!\!\!\langle\!\!\!\rangle$$

5. *O,O*-Diethyl *O*-2-pyrazinyl phosphorothioate; *Thionazin*, Nemafos®, Zinophos®;[475] LD_{50}: 12 mg/kg rat, oral, acute.

$$(H_5C_2O)_2\overset{\overset{S}{\|}}{P}-O-\!\!\!\langle\!\!\!\rangle_N^{N}$$

A systemic soil insecticide and nematicide effective against plant-parasitic and free-living nematodes including those attacking bulbs, buds, leaves, and roots.

6. *O,O*-Diethyl *O*-(1-phenyl-1H-1,2,4-triazol-3-yl) phosphorothioate; *Triazophos*, Hostathion®;[476] LD_{50}: 82 mg/kg rat, oral, acute.

$$(H_5C_2O)_2\overset{\overset{S}{\|}}{P}-O-\!\!\!\langle\!\!\!\rangle$$

A broad-spectrum pesticide effective against insects, mites, and nematodes.

7. *O*-Ethyl *S,S*-dipropyl phosphorodithioate; *Ethoprophos*, Mocap®;[477] LD_{50}: 61 mg/kg rat, oral, acute.

$$H_5C_2O-\overset{\overset{\displaystyle O}{\|}}{P}(S-C_3H_7)_2$$

8. 2-Chloroethenyl ethyl diethylphosphoramidate; SD 8832;[478] LD_{50}: 76 mg/kg rat, oral, acute.

$$Cl-CH=CH-O-\overset{\overset{\displaystyle O}{\|}}{\underset{\underset{\displaystyle OC_2H_5}{|}}{P}}-N(C_2H_5)_2$$

9. *O,O*-Diethyl *O*-(6-fluoro-2-pyridinyl) phosphorothioate; Dowco 275.[479]

$$(H_5C_2O)_2\overset{\overset{\displaystyle S}{\|}}{P}-O$$

10. *O*-[5-Chloro-1-(1-methylethyl)-1H-1,2,4-triazol-3-yl] *O,O*-diethyl phosphorothioate; *Isazophos*, Miral®;[480] LD_{50}: 60 mg/kg rat, oral, acute.

Soil insecticide and nematicide.

11. 2,3-Dihydro-2,2-dimethyl-7-benzofuranyl methylcarbamate; *Carbofuran*, Furadan®, Curaterr®;[481] LD_{50}: 8–14 mg/kg rat, oral, acute.

Highly active insecticide, acaricide, and nematicide.

12. 2-Methyl-2-(methylthio)propanal *O*-[(methylamino)carbonyl] oxime; *Aldicarb*, Temik®;[482] LD_{50}: 0.93 mg/kg rat, oral, acute.

$$H_3C-S-\overset{\overset{\displaystyle CH_3}{|}}{\underset{\underset{\displaystyle CH_3}{|}}{C}}-CH=N-O-CO-NH-CH_3$$

Systemic insecticide and nematicide.

13. Methyl 2-(dimethylamino)-*N*-{[(methylamino)carbonyl] oxy}-2-oxoethanimidothioate; *Oxamyl*, Vydate L®;[483] LD_{50}: 5.4 mg/kg rat, oral, acute.

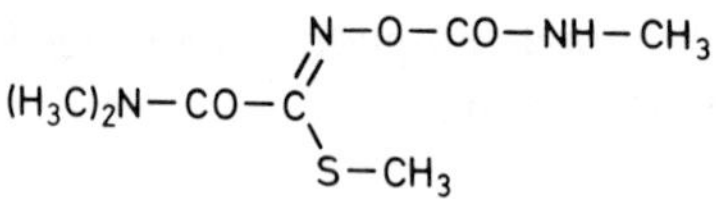

$$(H_3C)_2N-CO-C\overset{\displaystyle N-O-CO-NH-CH_3}{\underset{\displaystyle S-CH_3}{\big\|}}$$

Apart from *aldicarb* and *oxamyl*, some experimental products have been reported from among the oximecarbamates that also have a high nematicidal activity.

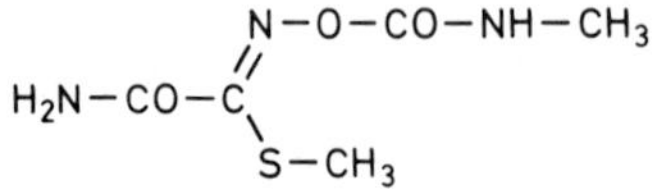

$$H_2N-CO-C\overset{\displaystyle N-O-CO-NH-CH_3}{\underset{\displaystyle S-CH_3}{\big\|}}$$

Methyl 2-amino-*N*-{[(methylamino)carbonyl] oxy}-2-oxo-ethanimidothioate; Du-Pont 1764.[483]

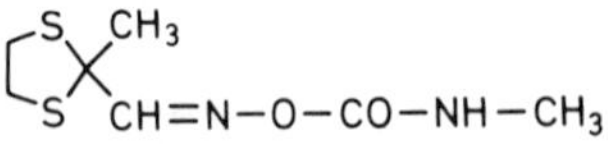

$$\underset{S}{\overset{S}{\big[}}\!\!\!>\!\!\!\underset{CH=N-O-CO-NH-CH_3}{\overset{CH_3}{\big<}}$$

2-Methyl-1,3-dithiolane-2-carboxaldehyde *O*-[(methylamino)carbonyl] oxime; MBR 5667.[484]

2.7.   *Acaricides*

W. LUNKENHEIMER
Bayer AG, Wuppertal-Elberfeld

Acaricides are agents for the control of mites that damage plants. Only commercial products are described in this section, but among them are also products of purely historical interest. Other compounds that had to be excluded here, literature references, and additional information are to be found in the standard works.[485-489] Original literature is, however, cited in this section for the more recent products that are missing from the standard works. The reader is also referred to the chapter on acaricides in Ullmann.[490]

2.7.1.   INTRODUCTION TO THE BIOLOGY AND CONTROL OF PHYTOPARASITIC MITES

Mites (Acarina) belong to the phylum Arthropoda and to the class Arachnida. The most important families and species of phytophagous mites in economic terms are:

1.   Tetranychidae (Spider mites), for example:
  Panonychus ulmi (fruit tree red spider mite).
  Panonychus citri (citrus red mite).

Tetranychus urticae (common red spider mite), on fruit and greenhouse crops, roses and vegetables.

Tetranychus cinnabarinus, on cotton in Egypt.

Tetranychus kanzawai, on tea in Asia.

2. Phytoptipalpidae (false spider mites), for example:

Brevipalpus oudemansi in greenhouse crops.

3. Tarsonemidae (soft-bodied mites), for example:

Hemitarsonemus latus (broad mite), on cotton and tea.

4. Eriophyidae (gall mites), for example:

Phyllocoptruta oleivora (citrus rust mite).

In this section the control of spider mites (Tetranychidae) is mainly discussed. Spider mites are up to 0.5 mm long and inhabit the undersides of leaves, mostly in dense colonies. All development stages are to be found side by side in a colony: eggs, larvae, protonymphs, deutonymphs, and adults. Under favorable circumstances 1-2 weeks suffices for a whole development cycle. During one vegetation period in greenhouses up to 30 generations of *Tetranychus urticae* can be produced. Toward the end of summer, winter eggs and adults are produced that overwinter on bushes, in cracks, or in the earth.

Spider mites destroy the assimilation cells in the foliar tissue in that they suck out the cell contents. Mottled areas of dead tissue are formed. Heavy infestation can lead to whole leaves dying and dropping prematurely. The loss of assimilation area causes a reduction in yield from the plant.

Injury to crop plants by spider mites has increased markedly in the last 30 years. The major reasons are intensively grown monocultures and the use of nonspecific insecticides of insufficient acaricidal potency that eliminate the natural enemies of the mites. An additional factor, favored by the short generation cycle of mites, is the rapid development of resistance toward originally effective acaricides, particularly of the organophosphate type.

Control of spider mites can be directed against the overwintering stages or, more effectively, against the active stages. Only products based on mineral oils have proved of value as winter sprays. The ideal product for control during the vegetation period combines the best possible ovicidal effect with a residual action against the mobile stages. As a rule, resistant spider mites are immune not just to one agent, but also to others with the same mechanism of action (group resistance). Compounds that act by another mechanism are therefore much in demand.

2.7.2. ACARICIDAL AGENTS

Some agents from the early days of chemical crop protection are still in use today. One is sulfur, which in a finely divided form (wettable sulfur), is mainly used in grape crops as a fungicide with a remarkable acaricidal side effect. Mineral oils are used as winter sprays against orchard fruit spider mites owing to their ovicidal properties. Their phytotoxicity prohibits general use on green plants.

2.7.2.1. Nitrophenol Derivatives

The nitrophenols are the longest known synthetic insecticides. As early as 1892 the potassium salt of 2-methyl-4,6-dinitrophenol (*DNOC*) was in use for control of the black-arched tussock caterpillar (*Lymantria monacha*). It also has a certain ovicidal action against spider mite eggs.

2-Cyclohexyl-4,6-dinitrophenol (*DCNP*), available by nitration of 2-cyclohexyl-phenol;

R = CH₃   ; *DNOC*; Antinonnin® (Bayer & Co.)

R = ⟨⟩ ; *DCNP*; Dinex® (Dow)

is more active than *DNOC* and is mainly used in orchard fruit as an ovicide against spider mite eggs. It also has insecticidal properties against, for example, scale insects.

The dinitrophenols have serious drawbacks:

1. Intense color with a high affinity for skin.
2. High mammalian toxicity (e.g., *DNOC*, LD_{50}: 180 mg/kg rat, oral, acute).
3. Phytotoxicity.

Owing to their marked herbicidal properties (see Section 4.3.2.2.1), they can only be used as dormant sprays, for instance, as a mixture with mineral oils.

The amine salts of dinitrophenols are somewhat less phytotoxic, for example, the dicyclohexylamine salt of *DCNP* (DN-111®, Dow Chemical Company). This also has an improved residual action and a more favorable acute toxicity (LD_{50}: 300–600 mg/kg rat, oral, acute).

The phytotoxicity of the dinitrophenols is drastically reduced by esterification with unsaturated carboxylic acids. These esters are highly active acaricides and, in addition, are effective against powdery mildew fungi (see Chapter 3, Section 3.2.5.3). *Dinocap*, a mixture of isomeric 2,4-dinitro-6-isooctylphenyl and 2,6-dinitro-4-isooctylphenyl crotonates, was the first product from this series.

R = H₃C–(CH₂)₅₋ₙ–CH–(CH₂)ₙ–CH₃ ; n = 0, 1 or 2

Dinocap; Karathane® (Rohm & Haas, 1946)

It is manufactured by nitration of the product of condensation between phenol and octanol, and subsequent esterification with crotonyl chloride. The end product is a dark-brown oil.

Dinocap is used as a nonsystemic acaricide and contact fungicide against orchard tree spider mites and powdery mildews. The insecticidal activity is low.

2-(1-Methylpropyl)-4,6-dinitrophenyl 3-methyl-2-butenoate, similar in activity to *Dinocap*:

Binapacryl; Acricid®, Morocide® (Hoechst, 1960);
LD_{50}: 130–165 mg/kg rat, oral, acute

is synthesized by nitration of 2-(1-methylpropyl)phenol and subsequent esterification with dimethylacrylyl chloride.

Binapacryl is nontoxic to bees and has a good residual action. The onset of action is delayed in cooler weather.

Dinitrophenylcarbonates prepared from dinitrophenols and short-chain chloroformates are also effective acaricides and fungicides.

1-Methylethyl 2-(1-methylpropyl)-4,6-dinitrophenyl carbonate;
Dinobuton, Acrex® (Murphy Chem. Ltd., 1963);
LD_{50}: 140 mg/kg rat, oral, acute; 1500 mg/kg mouse, oral, acute

Dinobuton is mainly effective against the mobile stages of spider mites and is used in greenhouses.

2.7.2.2. Azo and Hydrazine Compounds

Azobenzene is one of the oldest acaricides.

Long after its insecticidal properties were discovered, it was also developed in 1945 as an acaricide.

It mainly served as a fumigant or smoke in greenhouses against the eggs of mites and insects. Dogs fed 63 days on a diet containing 600 ppm suffered high mortality and liver damage.

[(4-Chlorophenyl)thio] (2,4,5-trichlorophenyl)diazene and *chlorfenethol* (see Section 2.7.2.5) were components of the combination product Milbex® (Nippon Soda, 1964).

Chlorfensulphide; *CPAS*; LD_{50}: >3000 mg/kg mouse, oral, acute

Chlorfensulphide is prepared by diazotization of 2,4,5-trichlorobenzenamine and subsequent reaction with 4-chlorobenzenethiol. It is an ovicide and larvicide with good residual action. *Chlorfenethol* kills the adult mites. Milbex®, thus active against all stages of spider mites, was used in Japan and Europe in orchard fruit and ornamentals.

Numerous hydrazine derivatives have insecticidal and acaricidal properties. Only one commercial product is known to date, the hydrazone **1** obtained by chlorination of benzaldehyde (2,4,6-trichlorophenyl)hydrazone in carbon tetrachloride.[491]

1; *N*-(2,4,6-Trichlorophenyl)-benzenecarbohydrazonyl chloride; U 27415, Banamite® (Upjohn); LD_{50}: 389 mg/kg rat, oral, acute

Banamite® is highly effective against the fruit tree red spider mite[492] and citrus rust mite.[493] However, it was only used for a short time in the United States on a temporary permit in citrus.[494]

2.7.2.3. Sulfides, Sulfones, and Sulfonates

The common features of the numerous active compounds from the groups described in this section are two substituted phenyl rings joined by a simple sulfur-containing bridge. They are purely contact acaricides with long residual action, low phytotoxicity, and low mammalian toxicity and are mainly effective against egg and larval stages.

1,2,4-Trichloro-5-[4-(chlorophenyl)thio] benzene:

Tetrasul; Animert V-101® (Philips-Duphar, 1957);
LD_{50}: 6810 mg/kg female rat, oral, acute

produced by the reaction of the sodium salt of 4-chlorobenzenethiol with 1,2,4,5-tetrachlorobenzene, is oxidized to the sulfone (*tetradifon*, see below) on prolonged exposure to sunlight. It controls various spider mite species that hibernate in the egg stage, and must be applied during the hatching of these eggs. *Tetrasul* is highly selective, not being a hazard to beneficial insects or to wildlife.

The 1-halo-4-{[(4-chlorophenyl)methyl]thio}benzenes, available by reaction of 4-chlorobenzyl chloride with sodium 4-halobenzenethiolate, have almost disappeared from the market today.

X = Cl; *Chlorbenside*, Chlorparacide®, Chlorocide® (Boots Co., 1953)
Use: against most spider mite species in orchard crops before the blossoming stage
LD_{50}: >10,000 mg/kg rat, oral, acute
X = F; *Fluorbenside*, Fluorparacide®, Fluorsulphacide® (Boots Co.)
Use: better suited for aerosols and smokes because of its 20× higher volatility
LD_{50}: 3000 mg/kg rat, oral, acute

In the diarylsulfone series the chlorine-free diphenylsulfone itself (no longer on the market; Sulfobenzide®, Shell), prepared by heating benzene with concentrated sulfuric acid under continuous removal of the water formed, has a remarkable ovicidal effect. 4-Chloro-diphenylsulfone (Sulfenone®, Stauffer Chemical Company), made by heating benzenesulfonic acid and chlorobenzene at 200–250°, is also no longer marketed. In addition to good ovicidal properties, it showed a marked activity against the mobile stages of spider mites.

The most active compound among the chlorinated diphenylsulfones is 1,2,4-trichloro-5-[(4-chlorophenyl)sulfonyl]-benzene:

Tetradifon; Tedion V-18® (Philips-Duphar, 1954);
LD_{50}: >15 mg/kg rat, oral, acute

made either by Friedel–Crafts reaction between 2,4,5-trichlorobenzenesulfonyl chloride and chlorobenzene or by oxidation of *tetrasul* (see above) with hydrogen peroxide. It kills spider mites in the larval stages of development and causes sterility in female mites contacting or feeding on the deposit. *Tetradifon* has a good residual action and low phytotoxicity in contrast to Sulfobenzide® and Sulfenone®. It is used on many species of fruit, vegetables, cotton, hops, tea, and ornamentals.

Phenyl esters of benzenesulfonic acid must be chlorinated at least in the phenyl residue in order to be acaricidal. 2,4-Dichlorophenyl benzenesulfonate, a waxy solid no longer produced, is ovicidal and larvicidal.

$$\langle\!\!\!\bigcirc\!\!\!\rangle\!-SO_2-O-\langle\!\!\!\bigcirc\!\!\!\rangle\!-Cl$$
Cl

Genite®, Genitol® (Allied Chem. Corp., 1947);
LD_{50}: 1400 ± 420 mg/kg male rat, oral, acute;
1900 ± 240 mg/kg female

4-Chlorophenyl benzenesulfonate, introduced in Britain in 1952 under the name Murvesco®, is made by reaction between benzenesulfonyl chloride and 4-chlorophenol.

$$\langle\!\!\!\bigcirc\!\!\!\rangle\!-SO_2-O-\langle\!\!\!\bigcirc\!\!\!\rangle\!-Cl$$

Fenson; Murvesco® (Murphy Chem. Ltd.);
LD_{50}: 1560–1740 mg/kg rat, oral, acute

Fenson has the usual spectrum of activity but is somewhat phytotoxic to apples and tomatoes.

Chlorfenson, the ester of 4-chlorobenzenesulfonic acid and 4-chlorophenol, is used for control of the spring–summer eggs and young forms (larvae and nymphae) of the red spider mite of fruit trees, citrus, vegetables, and ornamentals; of the yellow spider of vines; Eriophyidae; and other mite species. Today it is one of the most important acaricides.

$$Cl-\langle\!\!\!\bigcirc\!\!\!\rangle\!-SO_2-O-\langle\!\!\!\bigcirc\!\!\!\rangle\!-Cl$$

Chlorfenson, Ovex®, in Japan Sappiran® (Nippon Soda, 1969);
LD_{50}: ~2000 mg/kg rat, oral, acute

In the combination product Neosappiran® (Nippon Soda) the ovicidal activity of *chlorfenson* is complemented by the quick knockdown property of *DCPM*, 1,1'-[methylenebis(oxy)]bis(4-chlorobenzene), which also has a synergistic effect.

DCPM itself is applied as an acaricide to navel orange trees when no fruit is present.

$$Cl-\langle\!\!\!\bigcirc\!\!\!\rangle\!-O-CH_2-O-\langle\!\!\!\bigcirc\!\!\!\rangle\!-Cl$$

DCPM; Neotran® (Dow)

The combination of *DCPM* with *chlorfenethol* (see below)—marketed under the name Mitran® (Nippon Soda)—has a similar effect to that of Neosappiran®.

2.7.2.4. Sulfites

These are nonsystemic contact acaricides that are effective against all stages of numerous spider mite species, but are well tolerated by plants and have low acute mammalian toxicities.

Aramite®, 2-chloroethyl 2-[4-(1,1-dimethylethyl)phenoxy]-1-methylethyl sulfite, was introduced as an acaricide by the U.S. Rubber Company (now Uniroyal, Inc.) in 1948.

$$(H_3C)_3C-\!\!\!\!\!\langle\ \rangle\!\!\!\!-O-CH_2-CH-O-SO-O-CH_2-CH_2-Cl$$
$$| \atop CH_3$$

It is prepared by reaction of 2-[4-(1,1-dimethylethyl)phenoxy]-1-methylethanol, available from 4-tert-butylphenol and methyloxirane (propylene oxide), with 2-chloroethyl chlorosulfite (from 2-chloroethanol and thionyl chloride).

The ester is hydrolyzed by alkali and strong acid and decomposes slowly in sunlight with formation of sulfur dioxide. As a suspected potential carcinogen (possible alkylating effect of the 2-chloroethyl ester moiety), it is only registered for postharvest use in the United States. LD_{50}: 2000 mg/kg mouse, oral, acute.

2-[4-(1,1-Dimethylethyl)phenoxy]cyclohexyl 2-propynyl sulfite, closely related to Aramite®, has no alkylating or carcinogenic properties.

$$(H_3C)_3C-\!\!\!\!\!\langle\ \rangle\!\!\!\!-O \quad O-SO-O-CH_2-C\equiv CH$$

Propargite; Omite® (Uniroyal, Inc.);
LD_{50}: 2200 mg/kg rat, oral, acute

Propargite does not affect bees and is less harmful to predatory mites than any other acaricide. It is prepared by reacting the glycol ether obtained from 4-tert-butylphenol and 1,2-epoxycyclohexane with thionyl chloride to give the chlorosulfinate, which is then reacted with propargyl alcohol.

2.7.2.5. Diphenylcarbinols

The introduction of a hydroxy group at the bridging carbon in the potent insecticide but inactive acaricide *DDT* causes a complete loss of insecticidal activity. A high acaricidal activity appears in its place (e.g., Kelthane®). Such diphenylcarbinols are contact acaricides, with a slow onset of action but long residual effect, mainly used in fruit, grape, hops, and ornamentals.

The first diphenylcarbinol acaricide *chlorfenethol*:

$$Cl-\!\!\!\!\!\langle\ \rangle\!\!\!\!-\overset{\overset{\displaystyle OH}{|}}{\underset{\underset{\displaystyle CH_3}{|}}{C}}-\!\!\!\!\!\langle\ \rangle\!\!\!\!-Cl$$

4-Chloro-α-(4-chlorophenyl)-α(methyl)benzenemethanol; *Chlorfenethol*, Dimite® (Sherwin Williams Co.);
LD_{50}: 500 mg/kg rat, oral, acute

is made by reaction of 4,4'-dichlorobenzophenone with methylmagnesium bromide. It is unstable to heat and concentrated acids.

4-Chloro-α-(4-chlorophenyl)-α-(trichloromethyl)benzenemethanol is more active than *chlorfenethol*.

Dicofol; Kelthane® (Rohm & Haas, 1955);
LD_{50}: 685 mg/kg rat, oral, acute

Dicofol is manufactured by partial hydrolysis of 1,1'-(1,2,2,2-tetrachloroethylidene)bis(4-chlorobenzene) (obtained by chlorination of *DDT*) with 80–90% formic acid in the presence of a sulfonic acid at 100–125°. It is mainly used against spider, soft-bodied, and gall mites in vegetables, fruit, hops, grape, and ornamentals.

Whereas the chlorinated diphenylcarbinols intermediate between *chlorfenethol* and *dicofol* are much weaker acaricides, the benzilates also have high activity.

Ethyl 4-chloro-α-(4-chlorophenyl)-α-hydroxybenzeneacetate; *Chlorobenzilate*, Acaraben®, Akar®, Folbex® (Geigy, 1952)

Chlorobenzilate, prepared by esterification of the corresponding acid with diethyl sulfate, is hydrolyzed by alkali and strong acid. It is used against many species of mite in all stages of development on fruit, ornamentals, and cotton. As it has no insecticidal activity, it can be used as a smoke for control of acarine disease in bees, caused by the mite *Acarapis woodi*, which infests the bee trachea.

Closely related to *chlorobenzilate* is the corresponding isopropyl ester, *chloropropylate*.

Chloropropylate; Acaralate®, Chlormite®, Rospin® (Geigy, 1964);
LD_{50}: >5000 mg/kg rat, oral, acute

This crystalline ester is obtained by esterification of the corresponding acid with isopropanol. No longer marketed by Ciba-Geigy in the United States, *chloropropylate* is mainly used in fruit, tea, cotton, and sugar beet growing.

The corresponding bromo compound, *bromopropylate*:

Bromopropylate; Neoron®, Acarol® (Geigy, 1966);
LD_{50}: 5000 mg/kg rat, oral, acute

differs from *chloropropylate* and *chlorobenzilate* in that it has a longer residual action, especially against organophosphate-resistant spider mites, and is less phytotoxic. It is used in deciduous fruit and citrus (not available in the United States).

2.7.2.6. Chlorohydrocarbons

The chlorohydrocarbons are almost exclusively insecticides. One exception is 1,1'2,2',3,3',4,4',5,5'-decachlorobi-2,4-cyclopentadien-1-yl, a specific acaricide.

Dienochlor; Pentac® (Hooker Chem. Corp., 1960);
LD_{50}: 3160 mg/kg rat, oral, acute

Obtained by catalytic reduction of hexachlorocyclopentadiene, *dienochlor* is a tan crystalline solid that is stable toward acids and bases but that loses activity when exposed to direct sunlight or UV light.

Pentac® is very effective against the two-spotted spider mite (*Tetranychus urticae*). It has a slow onset of activity but long residual action and is mainly used in greenhouses for control of mites on ornamentals.

2.7.2.7. Fluorinated Compounds

Derivatives of 2-fluoroethanol and fluoroacetic acid are of general biological activity (blocking of the citric acid cycle by fluoroacetic acid). They have acaricidal properties as well as being insecticidal.

2-Fluoroethyl [1,1'-biphenyl]-4-acetate, which is no longer made:

Fluenetil, Fluenethyl; Lambrol® (Montecatini Edison)

was sold as an insecticide and acaricide. It was mainly used as a dormant spray in orchard fruit against the overwintering stages of spider mites. *Fluenetil* is phytotoxic to green plants and highly toxic to mammals (LD_{50}: 1.5–2.0 mg/kg dog, oral, acute).

Sodium fluoroacetate and fluoroacetamide are effective insecticides but, owing to their high toxicity, only find application as rodenticides—for example, fluoroacetamide (LD_{50}: 15 mg/kg rat, oral, acute) under the name Fluorakil 100®. *N*-Aryl-substituted fluoroacetamides are also effective insecticides, some also being recommended in Japan for use as acaricides.

2-Fluoro-*N*-methyl-*N*-(1-naphthalenyl)acetamide:

MNFA; Nissol® (Nippon Soda, 1965); LD_{50}: 115 mg/kg rat, oral, acute is a nonsystemic acaricide and aphicide with contact and respiratory toxic action. No longer manufactured today, it was used against spider mites and aphids in citrus, orchard fruit, and grape.

2.7.2.8. Organophosphates

The majority of the organophosphate insecticides on the market also have acaricidal properties (see also Section 2.3). However, spider mite strains resistant to this class develop very rapidly. Apart from the organophosphate itself, this development depends on the species of spider mite, the crop plant, climatic conditions, and the previous control methods used. The sheer diversity of these influences makes it practically impossible to assess the present utility of organophosphates as acaricides.

Basically there is no simple relationship between the acaricidal properties of an organophosphate and its insecticidal potency. Rather a species specificity exists, as is to be found within the insecticidal spectrum as well. Thus it is not surprising that many organophosphates are inactive even against normally susceptible spider mites.

Systemic phosphates of the following general structure have acaricidal activity:

$$RO-\underset{\underset{RO}{|}}{\overset{\overset{X}{\|}}{P}}-Y-CH_2-CH_2-\overset{(O)_n}{\overset{\|}{S}}-C_2H_5$$

4

$R = C_2H_5$	*Demeton*
$X = S, O$	Systox® (Bayer AG, 1951)
$Y = O, S$	
$n = 0$	
$R = CH_3$	
$X = O$	*Demeton-S-methyl*
$Y = S$	Metasystox (i)® (Bayer AG, 1957)
$n = 0$	
$n = 1$	*Demeton-S-methylsulfoxide*
	Metasystox R® (Bayer AG, 1960)
$R = C_2H_5$	*Disulfoton*
$X = Y = S$	Disyston® (Bayer AG, 1956)
$n = 0$	
$n = 1$	*Oxydisulfoton*
	Disyston S® (Bayer AG, 1965)

Development of resistance toward the group of organophosphates characterized by the general structure **5** is remarkably slow. *Carbophenathion* and *phenkapton* are worthy of special mention. Both products were developed as acaricides and have only weak insecticidal properties.

$$RO-\overset{\overset{\textstyle X}{\|}}{\underset{\underset{\textstyle RO}{|}}{P}}-S-CH_2-Z$$

5

$R = C_2H_5$	*Ethion*
$X = S$	Nialate® (FMC Corp., 1956)
$Z = -S-P(S)(OC_2H_5)_2$	
$R = CH_3$	*Omethoate*
$X = O$	Folimat® (Bayer AG, 1965)
$Z = -CO-NH-CH_3$	
$R = CH_3$	*Dimethoate*
$X = S$	Cygon® (American Cyanamid, 1956)
$Z = -CO-NH-CH_3$	
$R = CH_3$	*Formothion*
$X = S$	Anthio®, Aflix® (Sandoz, 1959)
$Z = -CO-N(CH_3)-CHO$	
$R = C_2H_5$	*Prothoate*
$X = S$	Fac® (Montecatini, 1956)
$Z = -CO-NH-CH(CH_3)_2$	

$R = C_2H_5$ *Azinphos-ethyl*

$X = S$ Gusathion A® (Bayer AG, 1953)

$R = CH_3$ *Endothion*

$X = O$ Endocide® (Rhône-Poulenc, 1958)

$R = C_2H_5$ *Dioxathion*

$X = S$ Delnav® (Hercules, Inc., 1954)

R = C$_2$H$_5$

X = S

Z = —S—⟨ ⟩—Cl

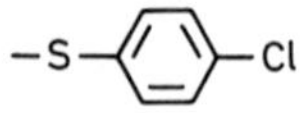

Carbophenothion

Trithion®, Garrathion® (Stauffer Chem. Co., 1955)

R = C$_2$H$_5$

X = S

Z = —S—(structure with Cl, Cl substituents)

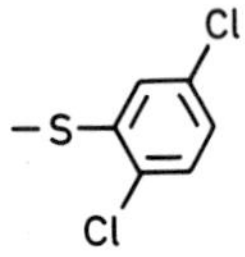

Phenkapton

G 28029 (Ciba-Geigy)

The following insecticides have a remarkable acaricidal effect, *methamidophos*, *EPN*, and *dimefox*:

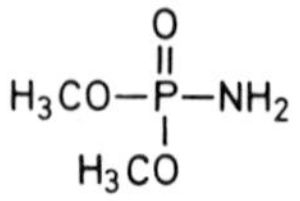

$$H_3CO-\underset{\underset{H_3CO}{|}}{\overset{\overset{O}{||}}{P}}-NH_2$$

Methamidophos; Tamaron®, Monitor® (Bayer AG)

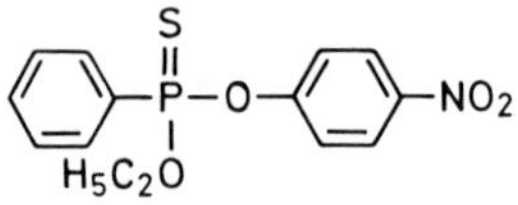

$$\text{⟨ ⟩}-\underset{\underset{H_5C_2O}{|}}{\overset{\overset{S}{||}}{P}}-O-\text{⟨ ⟩}-NO_2$$

EPN

$$(H_5C_2)_2N-\underset{\underset{(H_2C_2)_2N}{|}}{\overset{\overset{O}{||}}{P}}-F$$

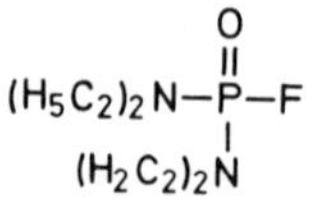

Dimefox

Dimefox is highly toxic and is used for soil treatment of hops to control aphids and spider mites.

2.7.2.9. Carbamates

In contrast to the organophosphates, few of the insecticidal *N*-methylcarbamates (see also Section 2.4) exhibit any acaricidal properties. The exceptions are as follows.

3,5-Dimethyl-4-(dimethylamino)phenyl methylcarbamate, no longer produced today:

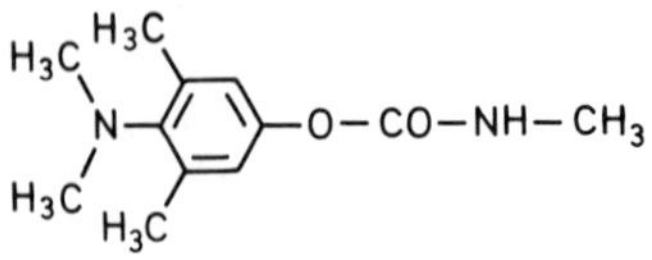

$$\underset{H_3C}{\overset{H_3C}{\underset{|}{N}}}-\text{(aromatic ring with 2 CH}_3\text{)}-O-CO-NH-CH_3$$

Mexacarbate; Zectran® (Dow Chem. Co., 1961);
LD$_{50}$: 15–63 mg/kg rat, oral, acute

had a useful activity against ticks on cattle in addition to its insecticidal properties.

The only acaricide among the carbamates is 3,5-dimethyl-4-(methylthio)phenyl methylcarbamate.

$$\text{H}_3\text{C}-\text{S}-\underset{\underset{\text{H}_3\text{C}}{}}{\overset{\overset{\text{H}_3\text{C}}{}}{\bigcirc}}-\text{O}-\text{CO}-\text{NH}-\text{CH}_3$$

Methiocarb (BSI), *Mercaptodimethur* (ISO); Mesurol® (Bayer AG, 1962);
LD$_{50}$: 130 mg/kg rat, oral, acute

Prepared from 3,5-dimethyl-4-(methylthio)phenol and methyl isocyanate, *mercaptodimethur* is an acaricide with a rapid onset of action and a good residual activity. As a contact and stomach poison it is also an effective insecticide against sucking and biting insects. Furthermore, it has a molluscidal effect and a good repellent action for birds.

2-Methyl-2-(methylthio)propanal *O*-[(methylamino)carbonyl]oxime is a systemic contact poison used as an insecticide, acaricide, and nematicide in cotton, sugar beet, and ornamentals.

$$\text{H}_3\text{C}-\text{S}-\overset{\overset{\text{CH}_3}{|}}{\underset{\underset{\text{CH}_3}{|}}{\text{C}}}-\text{CH}=\text{N}-\text{O}-\text{CO}-\text{NH}-\text{CH}_3$$

Aldicarb; Temik® (Union Carbide, 1965);
LD$_{50}$: 0.93 mg/kg rat, oral, acute

Aldicarb is extremely toxic and is absorbed through the skin. It is therefore only marketed as a granular formulation.

3,3-Dimethyl-1-(methylthio)-2-butanone *O*-[(methylamino)carbonyl]oxime:

$$\text{H}_3\text{C}-\overset{\overset{\text{H}_3\text{C}}{|}}{\underset{\underset{\text{H}_3\text{C}}{|}}{\text{C}}}-\text{C}\overset{\text{N}-\text{O}-\text{CO}-\text{NH}-\text{CH}_3}{\underset{\text{CH}_2-\text{S}-\text{CH}_3}{}}$$

Thiofanox; Dacamox® (Diamond Shamrock Co., 1974);
LD$_{50}$: 8.5 mg/kg rat, oral, acute

is a systemic insecticide and acaricide of exceptionally long residual action that is under investigation in many crops at the present time. Being an oximecarbamate it is highly toxic.

2.7.2.10. Heterocyclics

Apart from their specific activity against powdery mildew fungi, cyclic esters of 2,3-dimercaptoquinoxaline have a remarkable acaricidal potency.

1,3-Dithiolo[4,5-*b*]quinoxaline-2-thione mainly acts as a larvicide.

Thioquinox; Eradex neu[®] (Bayer AG, 1957);
LD_{50}: 3400 mg/kg rat, oral, acute

No longer on the market, *thioquinox* is obtainable by reaction of thiophosgene with 2,3-dimercaptoquinoxaline (available from 2,3-dichloroquinoxaline). It is readily oxidized to sulfoxides without, however, any loss in activity.

6-Methyl-1,3-dithiolo[4,5-*b*]quinoxalin-2-one (made from 2,3-dimercapto-6-methylquinoxaline and phosgene) is more stable toward oxidation and is used in fruit, vegetables, and ornamentals against powdery mildews but also acts as an acaricide, mainly against the larvae and eggs of Tetranychidae.

Oxythioquinox (ESA, in USA), *Chinomethionat* (ISO); Morestan[®]
(Bayer AG, 1962); LD_{50}: 2500 mg/kg male rat, oral, acute

NH-acidic benzimidazoles act as insecticides and acaricides by uncoupling oxidative phosphorylation in mitochondria[495-498] and as herbicides by inhibiting photosynthesis.[497-500]

In phenyl 5,6-dichloro-2-(trifluoromethyl)-1H-benzimidazole-1-carboxylate:

Fenazaflor; Lovosal[®] (Fisons Ltd., 1966);
LD_{50}: 283 mg/kg rat, oral, acute

the phytotoxicity has been reduced by *N*-acylation without loss of the acaricidal properties.[501] *Fenazaflor*, no longer produced today, acts as a contact poison and was used for control of spider mites, particularly of resistant strains, in orchard fruit.

2.7.2.11. Formamidines

The members of this class are characterized by a pronounced acaricidal and ovicidal activity, even against organophosphate-resistant strains. Noteworthy is their good loco-systemic action. They are basic and form water-soluble salts with strong acids.

N'-(4-Chloro-2-methylphenyl)-*N,N*-dimethylmethanimidamide is obtained by condensation of 4-chloro-2-methylbenzenamine and dimethylformamide with phosphorus oxychloride. It is formulated either as free base or hydrochloride.

Chlordimeform; Fundal® (Schering, 1966), Galecron® (Ciba-Geigy, 1966);
LD$_{50}$: 340 mg/kg rat, oral, acute

Chlordimeform is also effective against the eggs and early instars of Lepidoptera (e.g., codling moth, rice stem borer). Harmless to bees, it is phytotoxic only to certain ornamental plants. It is used for control of bollworm and tobacco budworm in cotton and can be tank-mixed with cotton insecticides.

In plants, animals, and soil, *chlordimeform* is demethylated to N'-(4-chloro-2-methylphenyl)-N-methylmethanimidamide, also active as an insecticide and acaricide, and then hydrolyzed further.

N,N-Dimethyl-N'-{3-[(methylamino)carbonyl]oxy}phenylmethanimidamide:

Formetanate hydrochloride; Dicarzol®, Carzol® (Schering);
LD$_{50}$: 21 mg/kg rat, oral, acute

at first marketed as a combination with *chlordimeform* in Fundal forte 750® (1967), has been sold on its own since 1970. It is formulated as the hydrochloride, the free base being somewhat unstable.

Formetanate hydrochloride is effective against the mobile stages of spider mites and against some insects.

The symmetrical formamidine derivative, N'-(2,4-dimethylphenyl)-N-{[2,4-dimethylphenyl)imino]methyl}-N-methylmethaniminamide:

Amitraz; Tactic®, Mitac® (Boots Co., 1973);
LD$_{50}$: 600 mg/kg rat, oral, acute

is effective against all stages of numerous phytophagous mites, as well as against ticks and some insects.

Amitraz is a colorless, crystalline compound made by condensation of 2,4-

dimethylbenzenamine, triethyl orthoformate, and methanamine. It is unstable at a pH below 7.

2.7.2.12. Organotin Compounds

The acaricidal tricyclohexylstannane derivatives *cyhexatin* and *azocyclotin* bear a formal resemblance to the fungicide *fentin hydroxide* (hydroxytriphenylstannane).

X = OH: *Cyhexatin*; Plictran® (Dow Chem. Co., 1968);
LD_{50}: 540 mg/kg rat, oral, acute

Azocyclotin; Peropal® (Bayer AG, 1976);
LD_{50}: 99 mg/kg rat, oral, acute

Cyhexatin and *azocyclotin* have a long residual, nonsystemic action against mobile stages of spider mites, even against resistant strains, and are well tolerated by plants. *Azocyclotin*, with a particularly wide spectrum of activity, contains relatively little tin owing to its high molecular weight.

Hexakis(2-methyl-2-phenylpropyl)distannoxane is prepared by a method analogous to that used for Plictran®.

Fenbutatin oxide (proposed); Vendex®, Torque® (Shell, 1974);
LD_{50}: 2630 mg/kg rat, oral, acute

(2-Methyl-2-phenylpropyl)magnesium chloride is first reacted with tin(IV) chloride to give tris(2-methyl-2-phenylpropyl)tin chloride, which forms the oxide (distannoxane) on treatment with sodium hydroxide.[502,503] Water converts the oxide to the corresponding hydroxide, which reverts the parent compound slowly at room temperature and rapidly at 98°.

Fenbutatin oxide is a nonsystemic acaricide with a good residual action against all mobile stages of a wide range of phytophagous mites[504] on orchard fruit, citrus, vines, vegetables, and ornamentals.

2.8.   Other Methods of Pest Control

W. SIRRENBERG
Bayer AG, Wuppertal-Elberfeld

2.8.1. HORMONES AND HORMONE MIMICS*

2.8.1.1. General

The intensive crop protection methods practiced worldwide in the past few decades can solve most of the pest control problems that confront the farmer today. During this time crop protection has come much nearer to the target of controlling pests by "natural" means. New developments within the field of insect hormones and their analogues and the development of chemosterilants, attractants, repellents, and the like have contributed to this and also given new, powerful impulses to the old field of biological control methods. A common feature of these control methods is their increased specificity, which destroys the pests but preserves the harmless and beneficial insects in many cases. The development of resistant insects is thus greatly delayed and reduced to a minimum.

In 1956 Williams[505] mooted the possibility of whether the insect hormones involved in molting and metamorphosis could form the basis for a new generation of pesticides. Since that time the insect hormones and compounds that are able to imitate their activity have enjoyed a growing interest as potential insecticides.

2.8.1.2. Insect Hormones

Invertebrates have hormones that, by virtue of their regulation and coordination functions, induce the various organs of a polycellular organism to act in a uniform and balanced manner. These hormonal controls in invertebrates cover a large number of physiological processes. The regulation of growth, metamorphosis, and reproduction in arthropods by hormones has been studied in great depth. In many cases, however, we must still be satisfied today with the outward phenomena of hormone effects without knowing anything about the actual mechanism of action.

Three different hormones or groups of hormones are responsible for the regulation of growth, metamorphosis, and reproduction in insects.[506-510]

1. Brain hormones (BH, activation hormone) in the neurosecretory cells of the brain,
2. Molting hormones (MH, ecdysones, prothorax hormones) in the prothorax glands,
3. Juvenile hormones (JH, neotenine) in the *Corpora allata*.

*Review literature, ref. 505A.

The brain hormones activate the prothorax glands, which then release the molting hormones, the ecdysones. The molting hormones initiate molting in the larvae or pupae. The juvenile hormones from the *Corpora allata* determine only the type of cuticle to be newly formed by the epidermis at each molting stage. If a sufficiently high level of juvenile hormones is present in addition to the ecdysones in the insect, then the larva always molts to another larva. Otherwise an imaginal molt occurs among the Hemimetabola, and a pupal molt among the Holometabola followed by the imaginal molt at the end of the pupal stage.

During the development of the insect the hormones are present in a precisely regulated ratio appropriate to the momentary stage of development. A disturbance to this ratio caused by the loss or artificial supply of these hormones or their analogues causes numerous, mainly morphogenetic disruptions that normally end with the death of the insect. This effect is the basis for the use of these insect hormones for pest control.

During the time before the isolation and structural elucidation of the insect hormones, this knowledge was gained by studies involving application of ligatures, extirpation, transplantation, and parabiosis tests.

2.8.1.3. Molting

Insects, crustaceans, and other arthropods possess a hard cuticle that serves as an exoskeleton for support of the body and protection of the internal organs and muscle. The exoskeleton does not change its dimensions during an insect stage, and the growth of insects is thus characterized by molts that end with metamorphosis to the imago. Molting[511–513] is the central process in the postembryonic development of insects. (See Figure 3.) It is a function of the epidermis and begins with *apolysis*, that is, detachment of the epidermis from the cuticle. A major proportion of the old cuticle is enzymatically degraded and replaced by a new one, which remains soft and elastic for a certain time, giving the insect the opportunity to expand its body—to grow, before it hardens (sclerotizes). Discarding of the old remaining cuticle follows the expansion of the insect's body. This event is termed *ecdysis* (molting).

The new larval cuticle is similar to the old. During metamorphosis one or two special molting processes take place in which a new cuticle of different character is

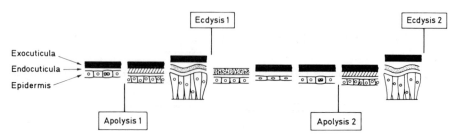

FIGURE 3. Stages of molting in the insect. For an explanation, see the text in this section.

formed. The histolysis of superfluous organs and formation of new organs for the adult stage occur simultaneously.

The number of molts varies from species to species of insect, frequently being 4 to 6, sometimes 10 to 14 or even more.

2.8.1.4. Brain Hormones

Kopec[506] first recognized the significance of the brain as a source of hormones and the essential factor in the regulation of insect molting and metamorphosis.

It has not yet proved possible to isolate and identify the brain hormones. The results of experiments carried out to this end are not in agreement. This is caused partly by the absence of a good test and partly by the fact that many other compounds are able to activate the prothorax glands.

The currently held view is that the brain hormones are proteins or polypeptides with a molecular weight of 20,000–40,000 daltons.[514–516]

As proteins or polypeptides the natural brain hormones are unsuitable for use as insecticides because they would be too unstable under field conditions. They cannot function as contact poisons, being unable to pass through the insect cuticle. As stomach poisons they would be degraded and inactivated too rapidly in the insect gut, and, finally, their economic production on a large scale is unthinkable.[517] Compounds with comparable activity, however, would be interesting insecticides. With such compounds it would be possible to regulate the diapause of numerous insects and shift their development into an extremely unfavorable time of year (winter, period of drought, etc.). Pests would thus be induced to commit "ecological suicide" without causing any damage.

2.8.1.5. Molting Hormones

Insects utilize α- and β-ecdysone and other ecdysones as molting hormones.[518,519] α-Ecdysone is converted into β-ecdysone in the fat body of the insect. Only β-ecdysone can induce molting. Oberlander[520,521] was the first to point out that the individual ecdysones fulfill different functions in the insect. Ecdysones cause molting in all arthropods.[522,523] Molting is initiated by 3–50 μg β-ecdysone per gram insect. As a general rule, the higher the dose the earlier the molt and the higher the percentage of the insects that molt.[524]

Of practical interest in crop protection is the observation that most insects die after injection or topical application of a large dose amounting to 10–100X the quantity the insect contains itself (1 ppm of body weight). Here α-ecdysone is less active than β-ecdysone. Possibly the insect is able to regulate the conversion of α- into the more active β-ecdysone.

2.8.1.5.1. Zooecdysones. In 1954 Butenandt and Karlson[525] isolated the first pure crystalline arthropod hormone with molting-hormonal activity from dried silkworm pupae (*Bombyx mori*). The structure was elucidated by Huber and Hoppe.[526]

Ecdysone; α-Ecdysone, (2β,3β,5β,22R)-2,3,14,22,25-pentahydroxy(cholest)-7-en-6-one

The structure was confirmed by synthesis almost simultaneously by two research groups working independently.[527,528]

α-Ecdysone causes the same effects as are brought about by the molting hormones secreted by the prothorax glands.[529]

Butenandt and Karlson detected a second molting hormone in smaller quantities, which they designated *β-ecdysone*, giving the name *α-ecdysone* to ecdysone itself. The name *ecdysone* is used today as a collective term for steroids with molting-hormonal activity.

β-Ecdysone has also been isolated from crustaceans, from various different insects, and from plants.[530–533]

β-Ecdysone; Crustecdysone, Ecdysterone, 20-Hydroxyecdysone, Polypodin A

The structure allocated, (2β,3β,5β,22R)-2,3,14,20,22,25-hexahydroxy(cholest)-7-en-6-one, has been confirmed by several syntheses.[534,535]

β-Ecdysone is regarded today as the true molting hormone of most insects. α-Ecdysone is the precursor[536] of β-ecdysone in various arthropods and dominates only in certain insects.[533,537,538]

Since the isolation of the first arthropod hormone (α-ecdysone) other zooecdysones have been isolated and their structures have been elucidated[539] (see also ref. 540).

2.8.1.5.2. Phytoecdysones.

Shortly after the isolation and structural elucidation of α- and β-ecdysone, compounds with activity as molting hormones were isolated almost simultaneously from plants by various research groups. Their ecdysone structures were recognized from a comparison of their NMR spectra with those of α- and β-ecdysone.[541]

Nakanishi et al.[542] isolated the first phytoecdysone from *Podocarpus nakaii*, found the molting-hormonal activity in the *Calliphora* test, and were able to identify the structure:

Ponasterone A

During their studies of *Podocarpus elatus*, Galbraith and Horn[543] found considerable quantities of a compound with high molting-hormonal activity in this plant. From comparative chromatography it was recognized as β-ecdysone. Extensive studies have shown that ecdysones are widespread in plants.[544,545] According to quantity and frequency of occurrence, β-ecdysone heads the list, whereas α-ecdysone is only found in small quantities if at all. To date approximately forty phytoecdysones have been isolated, and the structures of most have been identified. Extensive literature on phytoecdysones is available.[546]

Owing to the relatively high concentrations of ecdysones in plants (up to 0.5% of the dry weight), larger quantities of these compounds were available for the first time. It was thus possible to assess ecdysones as crop protection agents on the firm basis of laboratory tests.

2.8.1.5.3. Extraction of the Ecdysones. The ecdysones are isolated by extraction of large quantities of animal or plant material and subsequent concentration and purification. The basis of this method was developed with the isolation of α-ecdysone from silkworm pupae.[547-549] The flow chart of isolation and purification steps is given in ref. 550.

Isolation from plant material is simpler than isolation from animal material owing to the higher content of active material. However, the phytoextracts are normally composed of a mixture of several ecdysones, the separation of which is often very difficult. As a rule this is carried out by chromatographic methods (see ref. 551).

2.8.1.5.4. Tests for Molting-Hormone Activity. Numerous bioassays[552] are available for the assessment of molting hormone activity, most of which involve the study of morphogenetic effects on the insect cuticle. In the *Calliphora* test the solution under investigation is injected into the ligatured abdomen of *Calliphora* larvae of a definite stage of development. Active extracts cause the white larval cuticle to turn brown (sclerotize). This test may be arranged to give quantitative and reproducible results.[533]

The *Chilo* dipping test is simpler and quicker but less sensitive and thus better suited for large test series.[554] Here the ligatured abdomen of the last larval instar of the rice stem borer, *Chilo suppressalis*, is dipped for 5 sec into a methanolic extract of the material under test. A molting-hormone activity is indicated by pupation of the abdomen within 24–48 hr.

2.8.1.5.5. Application Forms. A widely variable application form is a desirable property in a pest control agent. The ecdysones are inactive when they are applied topically as a solution in a small amount of highly volatile solvent (methanol, acetone).[555,556] If the insect is dipped for 5 sec into the methanolic solution of an ecdysone, the ecdysone can penetrate into the insect (*Chilo* dipping test).

For topical application of ecdysones in nonvolatile solvents (undecyclic acid, α-tocopherol, caprylic acid), the dose must be increased by the factor 10 before an effect is observed.[557] No effect is found with α- and β-ecdysone when they are taken up by the insect with the food. Various phytoecdysones (e.g., ponasterone A) and synthetic ecdysone analogues cause disruptions to growth and metamorphosis when administered in the food at a concentration of 150 ppm.[558]

2.8.1.5.6. Effect of Ecdysones on Insects. The ecdysones act on the epidermis cells and initiate molting. It is accepted as proven that they act as molting hormones for all arthropods,[559-562] even ticks[563] and mites.[564] This molting-hormonal activity does not seem to be limited to arthropods but also plays a role among nematodes.[565]

Remarkably, the epidermis cells are sensitive to ecdysones at all stages apart from the adult. Thus the administration of appropriate dosages can cause an insect to molt at any stage.

High doses of ecdysones are lethal to larvae and pupae.[566,567] Depending on the time of application the following can occur: accelerated molting, the generation of extralarva, extrapupa, or nonviable intermediate larva–pupa or pupa–adult forms.

The action of a large ecdysone dose to larvae or pupae before metamorphosis leads to effects similar to those of juvenile hormones. The reason is that the excessive ecdysone dose forces the epidermis cells to form a further larval or pupal cuticle rapidly before a DNA replication is possible, which must precede the formation of a pupal or adult cuticle. High ecdysone concentrations stimulate the epidermis cells to form the previous cuticle once more because they have no time available to reprogram themselves for the synthesis of the new tissue.[568-570]

The injection of 3000 μg α-ecdysone per animal rarely causes molting in adult insects[568] but leads to a reduction in fertility among some.[571] This observation extends the action of ecdysones to include the adults. Ecdysones thus act on all insect stages from the newly hatched larva to the adult.

2.8.1.5.7. Effect of Ecdysones on Mammals. The widespread occurrence of ecdysones in plants is no proof that these are innocuous compounds. With regard to their well-known stability, the use of ecdysones as pest control agents must be judged with caution and no differently to that of conventional pesticides. The effect of ecdysones on mammals is of particular interest.[572,573] *In vitro* effects of ecdysone extracts have been observed in growing mammalian cells.[574] Feeding ecdysones to young mice over a long period promoted their growth. Indications of increased metabolism and cell regeneration were found in the liver tissue.[575] Some ecdysones stimulate protein synthesis in mouse liver.[576] No molting-hormonal activity has been detected in human tissue.[577] Ecdysones are effective against neuralgia, 0.1 g ecdysone relieving the pain for a long time without any side effects.[578] No other noteworthy effects have been found from the screening of ecdysones for pharmacological activity.[579] The ecdysones have very low acute toxicities. The LD$_{50}$ for β-ecdysone is 6400 mg/kg (IP) in the mouse.[580]

2.8.1.5.8. Biosynthesis and Metabolism of Ecdysones. Insects are unable to construct their essential steroids by biosynthesis from simple, basic precursors. Instead, they take them up with their plant food or obtain them from their gut symbionts. They are able to alkylate, hydroxylate, dealkylate, dehydroxylate, and so on and thus form their essential ecdysones from steroidal precursors.[581–584] Cholesterol seems to play a major role here in many insect species. Whether the ring system is first synthesized and then the side chain or vice versa is a matter of dispute and possibly varies from one insect species to another.

The inactivation of the ecdysones takes place mainly in the fat body of insects by enzymatic means.[585] One conceivable degradation pathway is the oxidative cleavage possible for all 20,22-diols that converts ecdysterone into poststerone and 4-hydroxy-4-methylpentanoic acid.[586,587] Although poststerone has not yet been isolated from insects, it has been demonstrated that poststerone is converted to rubrosterone in plants.[588]

Ecdysterone   Poststerone   Rubrosterone

2.8.1.5.9. Superecdysones. Phytoecdysones have been found that have a much greater molting-hormonal activity than α- or β-ecdysone. They are active at lower concentrations. Cyasterone is 20 times more active than α-ecdysone.[589]

A 50% inactivation of α-ecdysone occurs within 7 hr in silkworm pupae; that of cysterone within 33 hr. Apparently some ecdysone structures are more rapidly degraded by insects than others. This is the foundation of the hope that even more active compounds remain to be found.

Cyasterone

2.8.1.5.10. Antiecdysones. With ajugalactone:

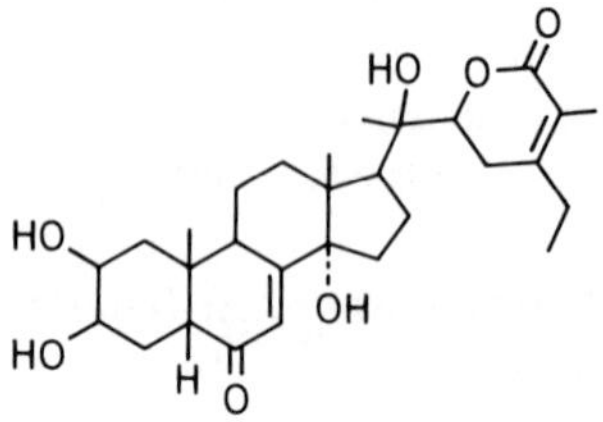

Ajugalactone

a compound has been found[591] that inhibits the molting-hormonal activity of ponasterone A in the *Chilo* dipping test. β-Ecdysone is not inactivated.

The following synthetic compounds also have a high antiecdysone activity.

(2β,3β,5β)-2,3,14-trihydroxy(cholest)-7-en-6-one **(1)**, the corresponding 2-methoxy- **(2)**, 3-methoxy- **(3)**, and the 25-desmethyl derivatives **(4)**:

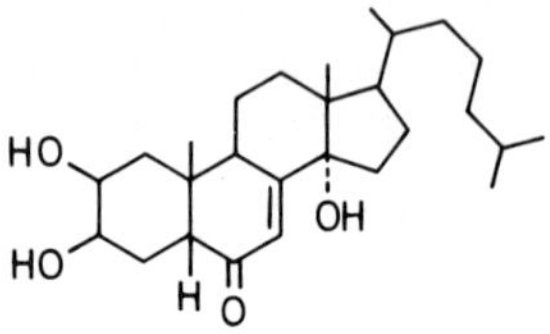

1

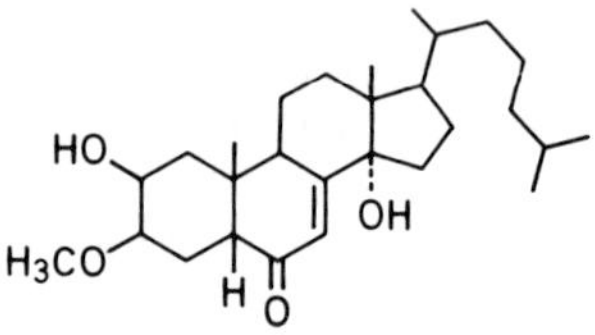

3

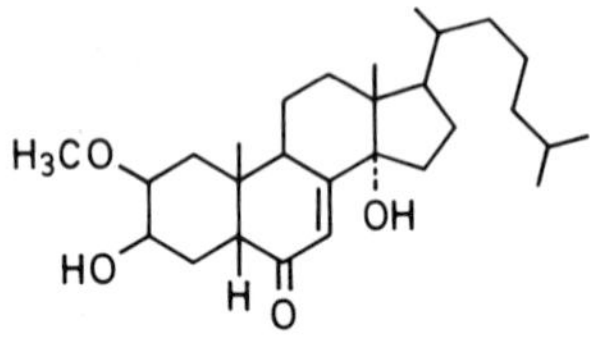

2

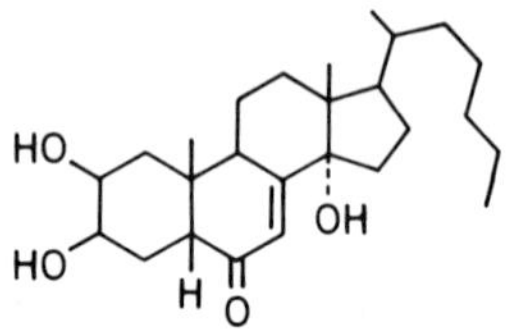

4

which, as stomach poisons, inhibit larval growth and development in numerous insect species.[592]

Compounds of the type (3β,5α)-3-hydroxy-cholestan-6-one **(5)**:

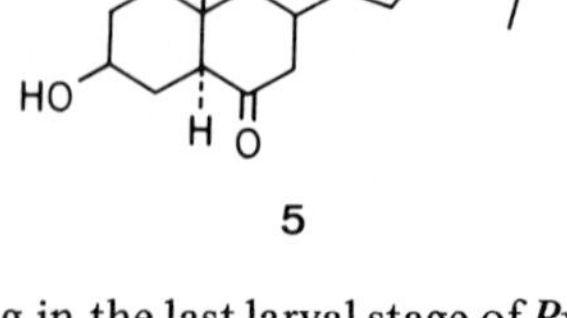

5

cause a disruption to molting in the last larval stage of *Pyrrhocoris apterus*. The insect cuticle remains soft and does not sclerotize. The insect dies after a few hours.[593]

2.8.1.5.11. Structure-Activity Relationships. The relative significance of the various structural elements in the ecdysone molecule for biological activity has been roughly elucidated by studies on the relationship between constitution and activity. The individual results have been reviewed in numerous publications.[594]

The following elements are essential for high molting-hormonal activity among the ecdysones:

1. *Cis* combination of the rings A and B.
2. A β-hydroxy group in the 3-position.
3. An oxo group in the 6-position in conjugation with a double bond at Δ^7.
4. A side chain at C-17 with a hydroxy group in the *R* configuration at C-22.

2.8.1.5.12. Conclusion. No pesticide based on ecdysones is known to date. Reports on field trials have not yet been published. The utility of ecdysones and their analogues in crop protection can only be assessed from the results of laboratory tests. Pest control with ecdysones along the lines of the "high-dose" method is regarded by Karlson as unrealistic[595] because insects have enzymes with the aid of which they are able to degrade rapidly an excess of ecdysones.[596-598] Many plants contain unexpectedly high quantities of ecdysones and are still used as food by insects without the latter coming to any harm.[599-601]

Furthermore, there is still no possibility today of producing the compounds at a reasonable price. It has been possible to replace the unproductive isolation from insects by a more productive one from plants, but even the productivity from this source is low.

Syntheses of ecdysones have been reported in many papers. The complex chemical structure of such compounds requires much synthetic effort with low overall yields. The use of readily available compounds such as ergosterol as precursors does not alter this hard fact.

At the present time the question of practical application of ecdysones is completely open.

2.8.1.6. Juvenile Hormones, Juvenoids and Precocenes*

2.8.1.6.1. General. The juvenile hormones, secreted by the *Corpora allata*, are essential for the growth and development of the larval stages. They mainly influence the nature of the cuticle produced from the epidermis at each molt. As long

---

*Review literature.[601A]

as juvenile hormones are present, insect larvae molt into larvae. If the secretion of juvenile hormones is stopped by some internal mechanism then the Hemimetabola undergo an imaginal molt; the Holometabola undergo a pupal molt followed by the imaginal molt.

The removal or premature failure of the *Corpora allata* leads to a premature metamorphosis that has the formation of correspondingly small pupae and adults as consequence.

The implantation of an active *Corpora allata* into a larva in its last larval stage causes the formation of an extralarva (superlarva). The administration of juvenile hormones to pupae in the form of transplanted active *Corpora allata* causes the formation of a second pupal stage, that is, the formation of adults is prevented.

Artificial changes in juvenile hormone (JH) concentration in the last larval stage or during metamorphosis cause considerable irreversible disruptions to the insect's development.

2.8.1.6.2. Historical Development. In the *cecropia* silkworm (*Hyalophora cecropia*), Williams[602,603] found an insect that contains large quantities of juvenile hormones. He was able to extract these juvenile hormones from the abdomens of male *cecropia* silkworms with ether and, by applying this extract to pupae, to cause the same reactions as are observed on implantation of active *Corpora allata*. Injection or topical application of the extract brought about the same reaction. He also observed the occurrence of nonviable intermediate forms having pupal and adult characteristics.

The disruption to metamorphosis that Williams' topical application of the JH extract occasioned was the first indication that juvenile hormones could be of value as insecticides. It was supposed that insects could hardly develop resistance to their own hormones.[604]

The first hints as to the chemical structure of the juvenile hormones originate from work by Schmialek.[605] In 1961 he isolated farnesol from the feces of the mealworm, *Tenebrio molitor*, and demonstrated its JH activity.

Farnesol

Later he reported the higher activity of farnesolmethyl ether and farnesyldiethylamine.[606]

Bowers et al.[607] came to the conclusion that *trans,trans*-10,11-epoxy-farnesic acid methyl ester was the structure of the natural *cecropia* juvenile hormone.

Methyl *trans,trans*-10,11-epoxy-farnesate

The synthetic compound had high biological activity but turned out not to be the major component of the *cecropia* silkworm JH extract. In 1967 Röller et al.[608] reported their isolation and structural elucidation of this major component, methyl 7-ethyl-9-(3-ethyl-3-methyloxiranyl)-3-methyl-2,6-nonadienoate (JH I).

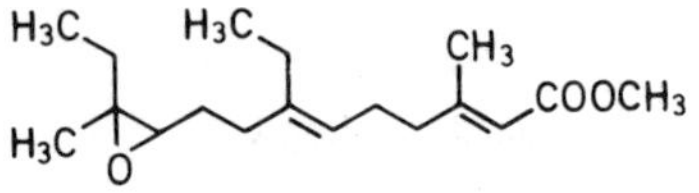

JH 1

They showed that JH I causes all the physiological reactions in insects that are also brought about by the *Corpora allata*. Meyer et al.[609] later confirmed the structure of JH I and found a second compound with JH activity in the *cecropia* extract. It has the same structure as JH I except for a methyl group at C-7 instead of ethyl and was designated as JH II:

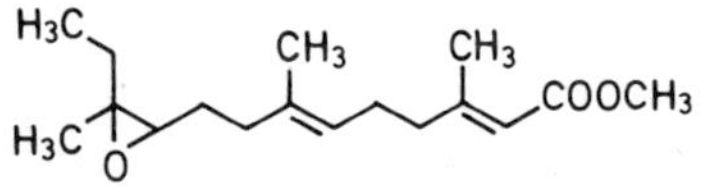

JH II; methyl 7-methyl-9-(3-ethyl-3-methyloxiranyl)-3-methyl-2,6-nonadienoate

JH III was isolated from culture media of *Corpora allata*.[610] It is identical with the compound synthesized by Bowers et al.[607]

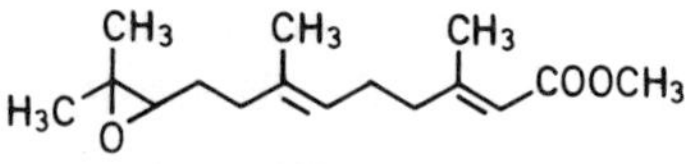

JH III

Slama and Williams[611] also discovered a compound with JH activity in American pine timber, which is normally used for papermaking. This compound prevents larvae of *Pyrrocoris apterus* in their fifth (last) larval instar from starting metamorphosis and forming the sexually mature adult form. Instead a sixth larval stage occurs, sometimes even a seventh, in which the insects grow to an enormous size. They die without ever undergoing a complete metamorphosis.

This compound acts selectively only against insects of the Pyrrhocoridae family and became known under the designation *Paper Factor*. It is of significance because here, for the first time, a specifically acting compound with JH activity was found.

Juvabione, the methyl ester of todomatuic acid, was isolated and identified by Bowers et al.[612] as the active compound in the Paper Factor. Cerny[613] found the more active dehydrojuvabione in Slovakian pine.

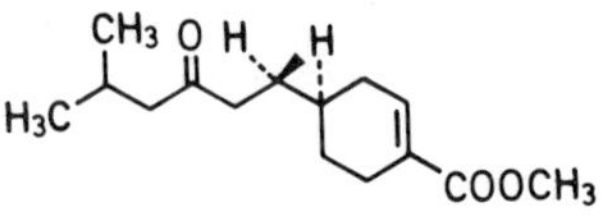

Juvabione

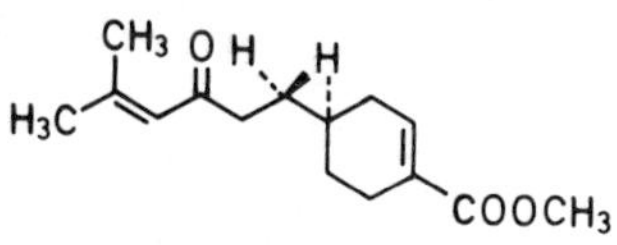

Dehydrojuvabione

During the screening of known synergists for use to enhance the activity of compounds with JH properties, Bowers[614] found that piperonyl butoxide, sesamex, and others had a remarkable intrinsic JH activity.

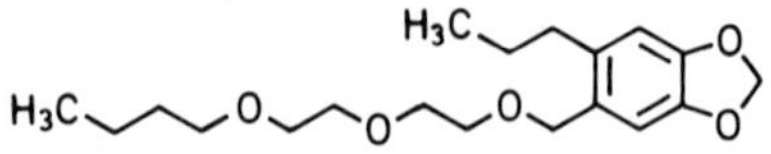

Piperonyl butoxide

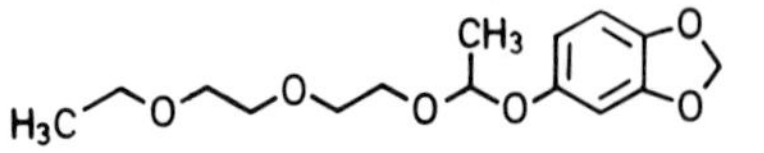

Sesamex

He then synthesized compounds in which he systematically combined the structural elements of known synergists with those of juvenile hormones. With these new "hybrid compounds," for example:

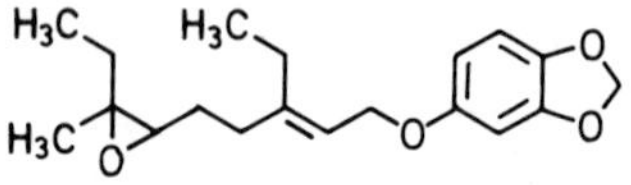

he attained levels of JH activity far in excess of those of the natural juvenile hormones.[615] The question remains open whether these compounds act as juvenoids or their mechanism of action corresponds to that of a synergist.

Since Schmialek's[605] discovery of the JH activity of farnesol, many research groups have worked intensively in the search for other compounds with such activity. All compounds with JH activity are referred to today as juvenoids. Compounds with either a close structural relationship to the natural JH or having partial structures regarded as essential features have always been in the foreground of synthetic work. In 1971 Slama[616] divided the juvenoids known up until then into eight types.

Ia 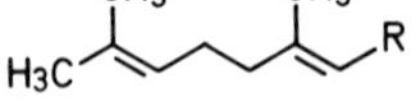
$R = COOR^1, CH_2-OR^1,$
$\quad CH_2-OH$
$R^1 = Alkyl, Aryl$

Ib 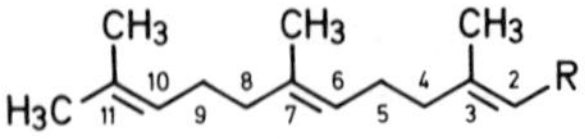
$R = CH_2-OH, COOCH_3$

IIa
$R = CH_2-CH-COOCH_3,$
$\quad\quad\quad CH_3$
CH_2
$\quad\quad\quad\backslash C=CH-COOCH_3$
H_3C

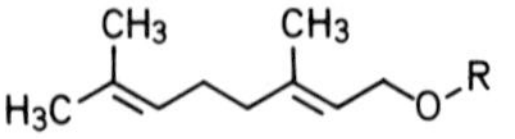

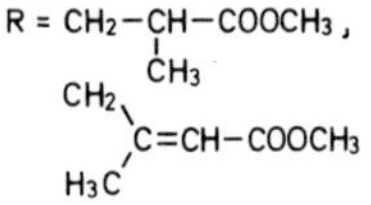

IIb

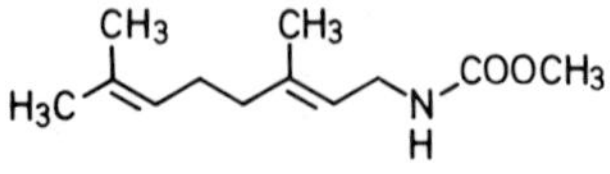

III

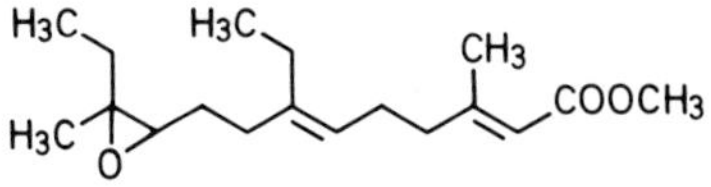

$$\text{H}_3\text{C}, \text{H}_3\text{C}, \text{CH}_3, \text{COOCH}_3$$

IVa

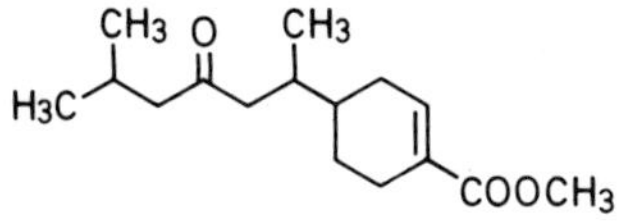

$$\text{CH}_3 \quad \text{O} \quad \text{CH}_3 \quad \text{H}_3\text{C} \quad \text{COOCH}_3$$

IVb

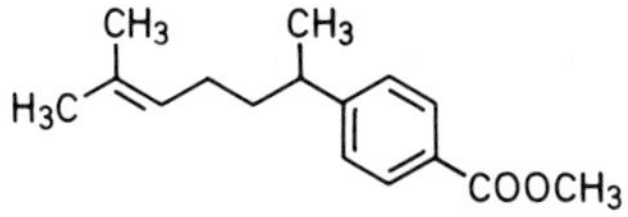

$$\text{CH}_3 \quad \text{CH}_3 \quad \text{H}_3\text{C} \quad \text{COOCH}_3$$

V

$$\text{CH}_3 \quad \text{CH}_3 \quad \text{H}_3\text{C} \quad \text{X} \quad \text{R} \qquad \text{X} = \text{O, S, NH}$$

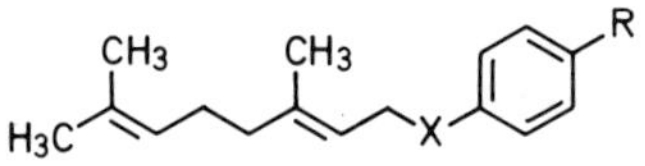

$$\text{R} = \text{COOR}^1, \text{COOCH}_3, \text{CO}-\text{N(CH}_3)_2, \text{OCH}_3, \text{CO}-\text{CH}_3, \text{NO}_2, \text{Cl, Br,}$$

$$\text{SO}_2-\text{NH}_2, \text{SO}_2-\text{N(CH}_3)_2, \overset{\text{O}}{\underset{\text{O}}{\text{CH}_2}} \text{ (in 3,4-position)}$$

VI

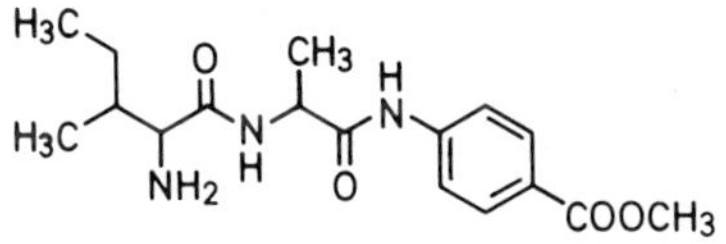

$$\text{H}_3\text{C}, \text{H}_3\text{C}, \text{O} \quad \text{CH}_3 \quad \text{H} \quad \text{N} \quad \text{N} \quad \text{COOCH}_3 \quad \text{NH}_2 \quad \text{H} \quad \text{O}$$

VII

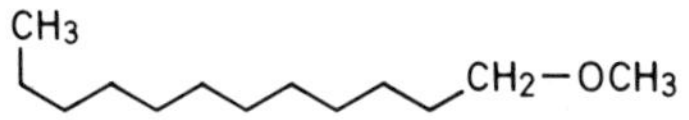

$$\text{CH}_3 \quad \text{CH}_2-\text{OCH}_3$$

VIII

$$\text{CH}_3 \quad \text{O} \quad \text{O} \quad \text{O} \quad \text{CH}_3 \quad \text{O} \quad \text{O} \quad \text{O}$$

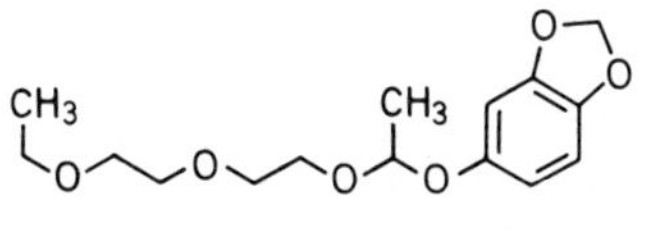

These have been widely varied by many research groups (see review ref. 601A).

2.8.1.6.3. Tests for Juvenile Hormone Activity. The tests for JH activity require the presence of ecdysones, which stimulate the cuticular secretion and ecdysis. Pupae are used as a rule as test insects, no juvenile hormones being present in the pupal stage.

Morphogenetic effects that are easily observed are favored as test criteria. Local application causes a local effect.

Juvenile hormones and juvenoids act directly on the target cells without the intermediacy of other, specialized organs. These compounds are able to penetrate the insect cuticle rapidly[619] and are also rapidly deactivated inside the insect. This inactivation is hindered by application of the test compound as a dilute solution in oil. This oil solution has a depot action, the juvenoid being released slowly to the insect.[620] The oldest test involves injection. The test compounds or extracts are injected into young moth pupae. Any JH activity present leads to a surplus pupal molt.[621]

The "*Galleria* wax test" is highly sensitive.[622,623] A small piece of cuticle complete with epidermis is removed from young *Galleria* or *Polyphemus* pupae and the wound is closed with a piece of hard wax in which the test compound has previously been mixed. After the subsequent imaginal molt any JH activity is recognizable from the pupal character of this cuticular area. The test lends itself to quantitative assessment.

In the "*Tenebrio* test" the (potential) juvenoid is injected into *Tenebrio* pupae. An active juvenoid is indicated by retention of the pupal cuticle or a second pupal molt. This test is very sensitive and can be used for crude JH extracts with low activity.[624,625]

2.8.1.6.4. Structure-Activity Relationships.

Most juvenoids are only very slightly soluble in water. Their lipophilic character enables them to penetrate through the insect cuticle and to be taken up easily by the hemolymph. Their JH activity is lost on injection in the form of an aqueous emulsion.[626]

The activity of juvenoids seems to be determined by the distance between certain active centers in the molecule rather than by the molecular size.[627,628]

The branching at the end of the carbon chain is important for activity among acylic juvenoids. The *trans* double bond in the 2,3-position, frequently conjugated with a carbonyl group, is essential for JH activity. In alicyclic juvenoids the ring contains a double bond, mostly in conjugation with a carbonyl group.

Benzene derivatives also belong to the active juvenoid structures[629] and are to be found in some promising experimental products from industrial laboratories.

2.8.1.6.5. Action of Juvenoids on Insects.

Juvenoids are only effective during a limited period of embryonic and post-embryonic development. The susceptible phases are the egg stage and the last larval or pupal stage. Younger insect larvae are not killed by juvenoids. The juvenile hormones and probably some of the juvenoids act directly on the epidermis cells and control the sloughing of the larval cuticle. They only act on those cells that have not ceased their DNA synthesis.[630] On the other hand, they activate the prothorax glands via a feedback mechanism and imitate the activation hormones.[631] The diapause of insects can thus be ended with juvenoids (ecological suicide).[632,633]

When females about to oviposit are treated with juvenoids either egg hatching is prevented or the emerged larvae are not viable.[634] Juvenoids are also ovicidal if brought into direct contact with the eggs.[635,636]

The males of *Pyrrhocoris apterus* can be sterilized by the application of methyl 7,11-dichlorofarnesate[637] and are able then to pass over the factor, causing sterility to their female partners during copulation.[638,639]

Methyl 7,11-dichlorofarnesate

Application of juvenoids to larvae or pupae causes the formation of surplus larval or pupal stages and the formation of intermediate forms (larva–pupa, larva–adult, pupa–adult).

Application of the right concentration at the right time is essential for juvenoid activity. Both are difficult to judge when controlling nonhomogeneous (in terms of development) natural populations.

2.8.1.6.6. *Juvenoids as Enzyme Inhibitors.*

Juvenoids have also been discovered that bear little or no structural resemblance to the natural juvenile hormones but are also highly active. These compounds probably do not act upon the same specific hormone receptors as the natural JH but function as inhibitors of the enzyme systems that degrade JH—mostly esterases and epoxide hydrases.[640] They help to stabilize the insect's own juvenile hormones and thus merely simulate an intrinsic JH activity.

The major degradation pathways for *Cecropia* juvenile hormone in insects[640] are as follows.

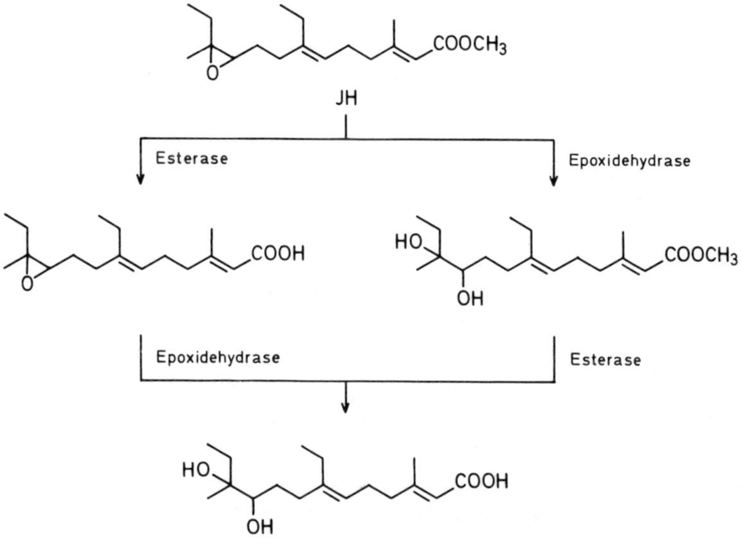

The following compounds have been found to act as enzyme inhibitors.

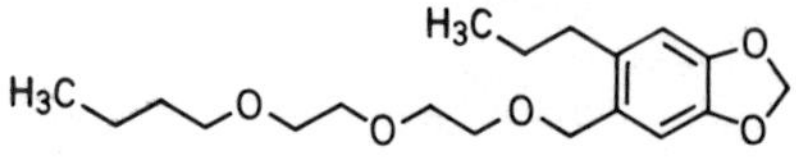

1 Piperonyl butoxide

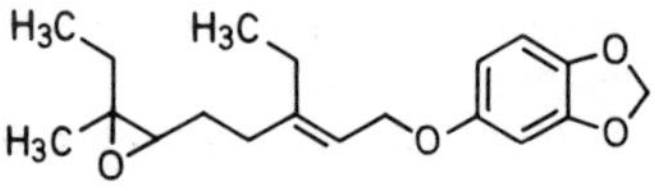

2 RO7-9767

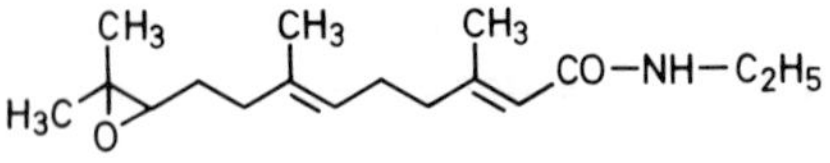

3 NIA 23509

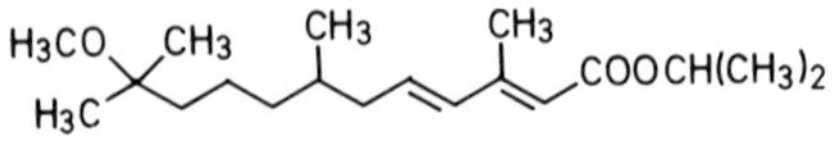

4 ZR-515

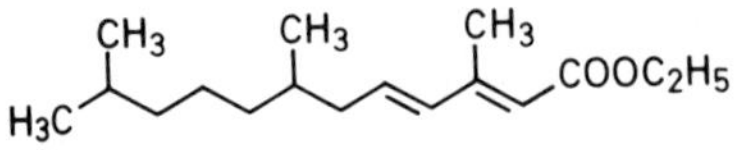

5 ZR-512

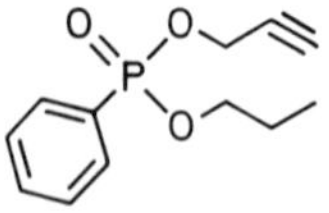

6 NIA 16388

Compounds **1–3** are supposed to act mainly as inhibitors of epoxide hydrase; **4–6** of esterase.

These findings may have far-reaching consequences for further research and synthesis work on juvenoids, for their toxicology, and for considerations of resistance.

2.8.1.6.7. Resistance. According to Williams[604] insects should not be able to develop resistance to their own hormones. However, Crow[641] had earlier drawn attention to the inevitable formation of resistance during the use of pesticides. Schneiderman[642] regards the development of resistance to juvenoids as a possibility as, in the course of time, the resistant individuals present in any insect population increase because of selection. Remarkably, Scolytidae and Curculionidae show no response to applications of juvenoids.[643]

In several instances development of resistance among *Musca domestica*[644] and *Tribolium castaneum*[645] has been confirmed in laboratory tests.

2.8.1.6.8. Commercial and Experimental Products. Zoecon, Ciba-Geigy, Stauffer, and Hoffmann-La Roche are some of the leading companies involved in the development of insect growth regulators (IGRs) based on juvenoids. Zoecon has developed efficient syntheses that permit the preparation of compounds with greater stability and more specific activity that are closely related to the natural juvenile hormones.

Publications by Staal[646,647] and by Henrick, Staal, and Siddall[648] give an interesting insight into the history, concepts, and experience of the Zoecon Corporation in their work on the development of IGRs.

Zoecon introduced the first market product to be registered that was based on the juvenile hormone structures (Altosid®).

$$H_3CO \quad CH_3 \qquad CH_3 \qquad CH_3$$
$$H_3C \qquad \qquad \qquad \qquad COOCH(CH_3)_2$$

Metroprene, Altosid®, ZR-515; 1-Methylethyl (*E,E*)-11-methoxy-3,7,11-trimethyl-2,4-dodec-adienoate;
LD_{50}: 34 g/kg rat, oral, acute

The structure bears a close resemblance to that of the natural juvenile hormones. A methoxy group has replaced the labile epoxide ring. Both double bonds are conjugated to the carbonyl group, thus lending a greater stability to the whole molecule. The methyl ester, which is susceptible to hydrolysis, has been replaced by the more stable isopropyl ester.

The technical product is an amber liquid soluble in organic solvents. Solubility in water is 1.39 ppm, and the vapor pressure at 25°C is 2.37×10^{-5} mmHg. It causes no irritation to the rabbit skin or eye.

Methoprene, particularly well-suited for mosquito control, prevents the formation of adult mosquitoes. It is used against the 2nd, 3rd, and 4th instar larvae. The larvae develop and pupate normally. The fresh pupae appear to be normal but die before they reach the imaginal stage. A 100% control success is attained with 75–300 g a.i./ha. As the long-term administration of lower doses is more effective than a single large dose, the product has also been formulated in a slow-release micro-granulate form. Apart from mosquito control, *methoprene* also finds application for hornfly control in cattle and chicken feces. The compound is administered with the food or supplements to cattle and chickens, a major proportion being excreted unchanged in the feces, where it develops its activity (feed-through technique).

Tripene has high activity against certain Lepidoptera and Diptera species.

$$H_3CO \quad CH_3 \qquad CH_3 \qquad CH_3$$
$$H_3C \qquad \qquad \qquad \qquad CO-S-C_2H_5$$

Tripene, ZR-619 (Zoecon); *S*-Ethyl (*E,E*)-11-methoxy-3,7,11-trimethyl-2,4-dodecadienethioate;
LD_{50}: 10 g/kg rat, oral, acute

Hydroprene is effective against the economically important Homoptera, Lepidoptera, and Coleoptera pests.

$$CH_3 \qquad CH_3 \qquad CH_3$$
$$H_3C \qquad \qquad \qquad COOC_2H_5$$

Hydroprene, Altozar®, ZR-512 (Zoecon); Ethyl (*E,E*)-3,7,11-trimethyl-2,4-dodecadienoate;
LD_{50}: 34 g/kg rat, oral, acute; Water solubility 0.54 ppm, soluble in organic solvents

Kinoprene, in contrast, is practically only active against Homopteran species.

$$CH_3 \qquad CH_3 \qquad CH_3$$
$$H_3C \qquad \qquad \qquad COO-CH_2-C \equiv CH$$

Kinoprene, Enstar®, ZR-777 (Zoecon); 2-Propynyl (*E,E*)-3,7,11-trimethyl-2,4-dodecadienoate;
LD_{50}: 3.5 g/kg rat, oral, acute

The 2,4-diene structure conjugated to the carbonyl group is the common feature of these four Zoecon products. Synthesis involves Wittig–Horner reaction of suitable aldehydes (e.g., dihydrocitronellal) with the appropriate phosphonates.[649–652]

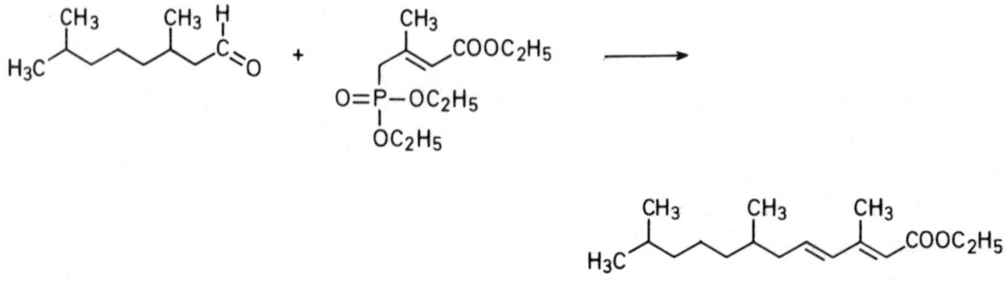

A Ciba-Geigy concept for the development of IGRs is illustrated in the following.[653] Starting with an alternative way of drawing the structure of JH I:

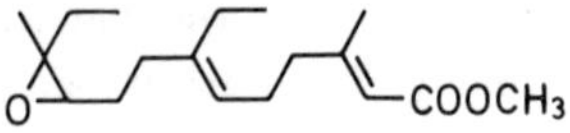

JH I

one can extrapolate to the following four structural possibilities for compounds with potential JH activity:

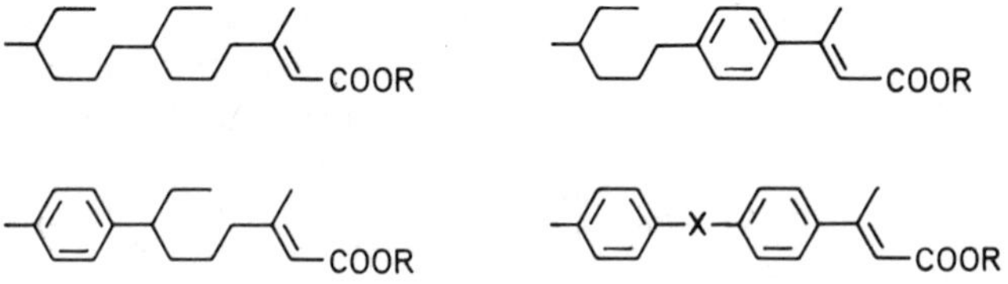

X = O, S, CH$_2$

Of the variations on these structures synthesized, some had high JH activity.

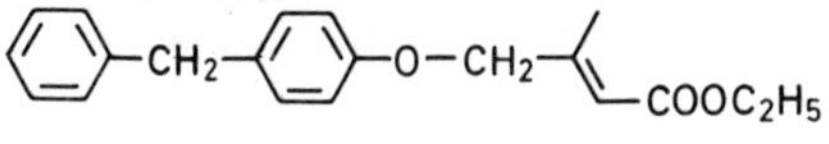

CGA 13353

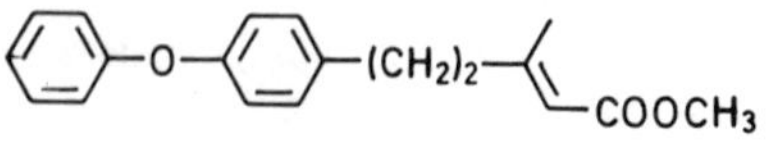

CGA 34300

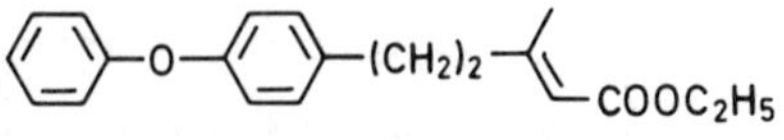

CGA 34301

These compounds act as contact and stomach poisons towards Coleoptera, Lepidoptera, Hemiptera, and Diptera. They are formulated as emulsifiable concentrates with a 40% a.i. content. A disruption to fertility among *Plutella xylostella* and *Porthetria dispar* needs further confirmatory studies. These products have no effect on the development of parasites. A grave disadvantage in their practical use is that the lethal effect occurs at the end of the larval period. Extralarval stages are observed. The period of damage is even extended when compared to untreated controls, thus limiting the utility of these compounds for pest control.

The value of these compounds in forest crops is assessed more favorably since, in contrast to field crops, the damage caused by one generation of pests can sometimes be tolerated.[654]

The experimental product (*E*)-3-[5-(4-ethylphenoxy)-3-pentenyl]-2,2-dimethyl-oxirane (R 20458) is synthesized by epoxidization of geranyl bromide and subsequent reaction with 4-ethylphenol.[615,655]

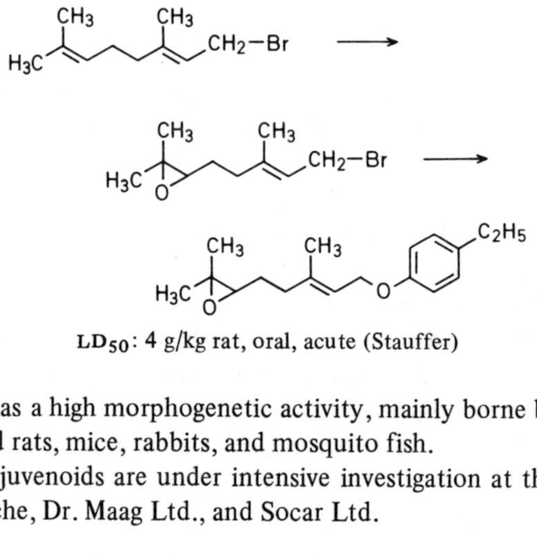

LD_{50}: 4 g/kg rat, oral, acute (Stauffer)

This compound has a high morphogenetic activity, mainly borne by the *trans* form. It is atoxic toward rats, mice, rabbits, and mosquito fish.

The following juvenoids are under intensive investigation at the present time at Hoffmann-La Roche, Dr. Maag Ltd., and Socar Ltd.

1. Bowers' compound[615] (a mixture of 4 isomers).

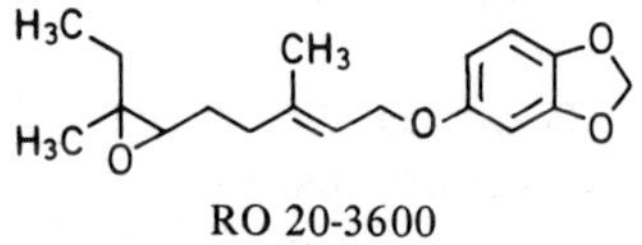

RO 20-3600

2. Epiphenonane.

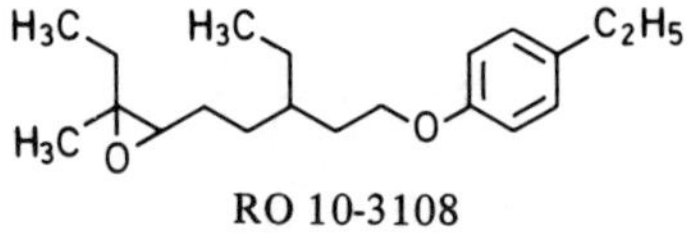

RO 10-3108

See ref. 656 for details of comparative stability trials on a JH I isomer mixture and RO 20-3600.

2.8.1.6.9. Field Trials. See ref. 657 for the results of field trials on the "Bowers' compound" in comparison with "Röller's compound" and "Romanuk's compound."

1. Röller's compound.

Methyl 7-ethyl-9-(3-ethyl-3-methyloxiranyl)-3-methyl-2,6-nonadienoate (mixture of 8 isomers)

2. Romanuk's compound.

Ethyl (*E*)-7,11-dichloro-3,7,11-trimethyl-2-dodecadienoate

Of the three compounds, Bowers' is the most active. In most instances high concentrations of these juvenoids had to be used to bring about juvenilizing effects. The effects observed were often not sufficient to provide satisfactory protection from crop damage.

Further field trials yielded the following results.[658-661]

1. Relatively high concentrations of juvenoids are necessary for control of pests in the field (presumably owing to the instability of the compounds).
2. In the use of juvenoids against caterpillars all possible intermediate forms are produced at metamorphosis.
3. Too high a dose leads to the appearance of surplus caterpillar stages, which, as giant caterpillars, are unusually long lived and cause a corresponding increase in crop damage.

The molting of a few giant caterpillars into viable pupae was observed in the field.[661] Parasitic Hymenoptera and Diptera were not affected by certain juvenoids.

2.8.1.6.10. Precocenes. Bowers et al.[661a] have isolated two compounds from an extract of *Ageratum houstonianum* that cause premature metamorphosis when administered to immature insects and sterility among female adults. In some insects, they prevent the development of the ovaries. These compounds, called precocenes:

R = H   Precocene I
R = OCH$_3$ Precocene II

cause effects in *Oncopeltus fasciatus* that can be reversed by administration of juvenile hormones. The precocenes are thus also referred to as antijuvenile hormones.

Precocenes also act as antifeedants and, in higher concentrations, as general poisons.[661b]

2.8.1.6.11. Conclusion. The natural *cecropia* JH is unsuited for pest control owing to its lack of species specificity and of stability under field conditions. Juvenoids with specific action and greater stability are being sought.

The research leads of Ciba-Geigy, Hoffmann-La Roche, and Stauffer have shown that further chemical development among the juvenoids need not necessarily be related to the terpenoid structures of the natural JH.

Owing to the inhomogeneous state of development of a field insect population, juvenoids must have a residual action of at least 3–4 weeks duration so that all individuals come into contact with the agent during their susceptible phase. In practical terms the action of juvenoids is limited to the first half of the last larval stage.

For most insects it is the larval stages that cause the actual damage to the plants. Since the juvenoids mainly act at the last larval stage crop, damage cannot then be prevented. The use of juvenoids for control of larvae is thus less attractive.

Use of juvenoids is most suitable where damage is caused by the adult stages, for instance, mosquitoes, hornflies, and bean beetles. Excess larval stages, among Lepidoptera, for example, have been observed where juvenoids have been applied in overdoses, thus leading to additional crop damage.

Juvenoids have very low acute toxicities. In spite of this they are chemicals and thus should be judged as any conventional insecticide.

It remains to be seen what significance juvenoids will attain in the future.

2.8.1.7. Acylated Ureas

2.8.1.7.1. General. Recently, acylated ureas have been discovered[662] that act as stomach poisons to insects and disrupt formation of the endocuticle. Typical of this disruption is that complications occur during molting that lead to the death of the insect. All results obtained to date indicate clearly that acylated ureas have no JH activity.[662] Injury occurs when the compounds are taken up orally after ecdysis or at the most 48 hr beforehand.[663,664] Histological examinations have revealed that no proper connection is then formed between exo- and endocuticle. The soft, newly formed cuticle is not strong enough to withstand the muscular strain and turgor during the molting phase and bursts.

Two mechanisms of action have been proposed. Either chitin biosynthesis is affected because of blocking of the chitin synthetase enzyme[665] or the degradation of chitin is accelerated by enhancement of chitinase levels.[666]

2.8.1.7.2. Diflubenzuron. *Diflubenzuron* is not systemic.[667]

N-{[(4-Chlorophenyl)amino]carbonyl}-2,6-difluorobenzamide; Dimilin®, PH 60-40, TH 6040 (Philips-Duphar)

It is effective against all homometabolous insects, for example, Lepidoptera, Hymenoptera, Coleoptera, and Diptera. The only Hemimetabola to be affected are those that bite or lick.

Particle size has a great influence on activity. In the commerical product Dimilin® it is set at 2-5 μm. *Diflubenzuron* is very stable and is degraded only slowly by the action of UV radiation. Degradation in soil is rapid, mainly by a microbiological process.[668,669] The acute oral toxicity (LD_{50}) for male and female rats is above 4640 mg/kg.

A concentration of 0.01-0.025% a.i. is sufficient in practice. An application rate of 75 g *diflubenzuron*/ha gave sustained and effective control of *Lymantria dispar* and *L. monacha* in forests.[670]

An initial damage to the crop is unavoidable because a molt is necessary before the insect is killed. Damage can be reduced by application before egg eclosion has occurred.[671]

Diflubenzuron controls a wide range of leaf-feeding and other insects in forests, woody ornamentals, fruit, vegetables, and field crops. The utility in soybeans and cotton as well as for control of the larvae of flies, mosquitoes, and gnats is under investigation. The compatibility of *diflubenzuron* with most conventional pesticides permits use in integrated spray programs.

2.8.2. CHEMOSTERILANTS*

2.8.2.1. General

Mutation and sterility can be caused by certain chemicals called chemosterilants. These chemosterilants are able to reduce or completely suppress the reproductive ability of insects.

The chromosomes of treated insects are altered by chemosterilants in a multitude of ways. Loss of chromosomal sections can occur, which causes the death of the organism. Other chromosomal changes become apparent only in later generations (translocations and inversions).

Changes in the genetic code can be transmitted as chromosomal errors to subsequent generations. They are nearly always lethal to the organism concerned and result in a reduction in offspring or even the dying out of a population.

Knipling[672] first proposed that treated insects be used to control their own kind. Before chemosterilants were used, male insects were sterilized by high-energy radiation and set free in great numbers. They then competed with the normal males of a natural population, preventing the generation of viable offspring by sterile copulation with the females.

Knipling's "sterile male technique"[673] was tried out in 1955 on the island of Curacao under particularly favorable conditions.[674] Here the use of this self-destruction technique led to the complete elimination of the screwworm fly, *Cochliomyia hominivorax*.

Experience gained with this control method fired the search for other methods that offered more favorable conditions for practical pest control (no radiation source, no mass breeding, no transport problems into the infested area, no uncer-

*Review ref. 671A.

tainty with regard to the radiation dose, etc.). Model calculations showed that a more rapid reduction of a natural population can be achieved when the natural population itself is genetically damaged instead of swamping it with sterile insects. This is not possible with radiation techniques, it being impossible to reach, say, 90% of a natural population spread over a wide area. For this reason chemosterilants were developed.

2.8.2.2. Important Chemosterilants and Their Action

In 1958 the U.S. Department of Agriculture began a systematic search for chemosterilants.[675] The sterilants listed here belong to all different classes of chemicals. Some are only active against the males, some only against females, and others against both sexual partners. The damage caused can be of a temporary or permanent nature. Generally, the most active compounds have little specificity. Some less active compounds do have a certain species specificity.

The compounds investigated most thoroughly play a role in genetic and cancer research apart from their role as chemosterilants. They interefere in ovogenetic and spermatogenetic processes, destroy mature egg cells and sperm or cause dominant lethal mutations in insect reproductive cells.

Chemosterilants are generally divided into three groups:

1. Alkylating agents (cytostatic).
2. Antimetabolites (interfering in metabolic processes).
3. Miscellaneous.

Alkylating agents form the largest group. They normally attack the genetic system directly. The majority contain one or more aziridinyl groups in the molecule. Phosphorus derivatives of aziridine are among the most active.

Typical compounds are:

1. 2,2,4,4,6,6-Hexakis(1-aziridinyl)-2,2,4,4,6,6-hexahydro-1,3,5,2,4,6-triazatriphosphorine.

Apholate[676]

2. 1,1′,1″-Phosphinylidynetrisaziridine.

Tepa, Aphoxide®, APO

3. 2,4,6-Tris(1-aziridinyl)-1,3,5-triazine.

Tretamine, TEM, triethylenemelamine

The second group of chemosterilants is that of the antimetabolites. These are mostly analogues of purines, pyrimidines, and folic acid. They interfere in metabolism and lead to damage and so on by disruption of nucleic acid synthesis, for example, which may be recognized inter alia from the effect on insect fertility.

The third, miscellaneous group of chemosterilants includes such compounds as triphenyl tin derivatives,[677] cyclic ureas, hormones, and antibiotics. Compared with the first group they are of little significance.

The proven advantages of the chemosterilants in insect pest control are countered by serious toxicological problems.[678,679] Teratogenic, carcinogenic, mutagenic, and sterilizing effects have been observed in mammals. In addition, resistance phenomena have been reported among insects.

Of the various opportunities for the use of chemosterilants, the combined attractant and sterilization technique seems to hold promise for the future.

By means of pheromones, large proportions of the insect population are attracted to certain sites where they are exposed to the necessary doses of chemosterilant, either by consumption of treated food or by contact with objects coated with the chemical. Up to now chemosterilants have not been able to replace conventional insecticides and will certainly remain only one of many possible methods of insect control in the future.

2.8.3. ATTRACTANTS (PHEROMONES)*

2.8.3.1. General

Pheromones are compounds released as signals by organisms of a species that, after reception or uptake by other members of the same species, provoke certain reactions in the latter.[681]

Pheromones serve as:

1. Sex attractants (normally released by the females).
2. Stimulants and sedatives (aphrodisiacs).
3. Congregation scents.
4. Alarm signals.
5. Tracking and marking signs.
6. Social signals (for colony differentiation, swarm formation etc.).

*Review ref. 679A.

At the present time only the sex attractants and congregation pheromones have any significance in insect control. The phenomenon of insect pheromones was recognized long ago. With the aid of new or improved laboratory techniques and analytical methods it has now proved possible in some cases to concentrate the substances from animal material, isolate them, elucidate their structures, and then synthesize them. A specific electrophysiological technique—the recording of an electroantennogram (EAG)[682-684] for the determination of pheromone action and concentration, comparison of pheromones, and the like—was specially developed for pheromone research. This technique depends on the fact that electrical potential changes caused in the insect antennae as a result of pheromone activity at the receptor can be measured and visualized.

2.8.3.2. Use

The era of synthetic sex pheromones began with the isolation of the sex attractant "bombykol" [(*Z,E*)-10,12-hexadecadien-1-ol] from the silkworm *Bombyx mori* by Butenandt et al.[685-687]

The specific effect of minute quantities on the target organism is characteristic of this class of compounds. Pheromones have low toxicities and seldom cause any side effects. As naturally occurring compounds they are innocuous to beneficial insects.

Even before they had been isolated and their structures had been elucidated, pheromones were used in the form of extracts or virgin females in traps for early recognition, supervision, and control of insect pests.[688]

Early recognition of insect pests is frequently important for the application of effective control measures. With the aid of traps it is possible to monitor the density and state of development of an insect population. The appearance of pests in crops not yet affected can also be determined by this technique.

Various methods are available for influencing and controlling a pest insect population. A sex attractant can be used to attract the insects to a site where they are then destroyed by treatment with insecticides, chemosterilants, or specific pathogens (e.g., bacteria and viruses). Furthermore, the protracted exposure to the sex pheromone can send the insects into a hyperactive phase that is followed by an inactive phase, due to exhaustion, in which they show no tendency to copulate. With an excessive supply of attractant sources in the form of pieces of paper or cork particles soaked in pheromone it is possible to provoke insects into numerous unproductive copulations, thus reducing the probability of a productive one. However, only when almost 95% of the males in an insect population have been rendered ineffective does the infestation level drop by half in the following year.[689] Measures that do not render such a high percentage of sex partners ineffective thus provide little control success in the long term.

The activity of a pheromone in field trials depends to a great extent on its purity and stability.[690] Adequate purity is essential for accurate assessment of activity, particularly in the early trial stage. It has been found that the presence even of small quantities of geometric isomers can greatly reduce or eliminate the attractive effect of pheromones.[691] Impurities can form in unstable pheromones under field trial conditions and impair the attractive qualities.

One problem hardly solved in the field is the adjustment of the pheromone concentration. Only certain concentrations attract the interest of the insects; excessive concentrations can even have a repellent effect. This problem becomes even more critical when the pheromone is a mixture of several compounds, as is the case with the cotton boll weevil, *Anthonomus grandis*.[629] The conditions of a field test must be so arranged that the pheromone source exerts the same physiological attraction on the insect to be controlled as would a virgin female of the same species. This target is difficult to meet when one considers that the actual attractive effect is also made up of components contributed by the host plant and that nonolfactory factors, for example, optical attraction, also play a part.

Apart from the sex attractants, the aggregation pheromones have a certain practical significance.[693] These bring about the congregation of both sexes at one place. If the wood engraver beetle, for example, finds a suitable tree on which to feed, it emits a pheromone after the first feed that attracts other members of the species, whereupon copulation and egg deposition take place. A control technique has evolved from this behavioral process. So-called trap trees are smeared with the pheromone and, after beetle infestation, cut down and burnt. Attempts have also been made to treat trap trees with an insecticide that would kill the beetles and their eggs.

Aggregation pheromones play an important part in the control of the stored-grain pest *Trogoderma granartum*.[694] They are also used in cotton crops for control of boll weevil, *Anthonomus grandis*. This aggregation pheromone, made up of four components, also acts as a sex attractant and is produced and marketed as *grandlure*.[695]

The use of pheromones enables pests to be controlled specifically. The minute application quantities avoid contamination of food and fodder as well as of the environment. The development of resistance or other such phenomena has not yet been observed. Suitable combinations of pheromones with conventional insecticides seem likely to be of increasing importance in the future.

The precise chemical constitution of many pheromones has not been established beyond all doubt. Many failures under field conditions could thus be explained.[696]

2.8.3.3. Important Attractants

Some important attractants are given in the following list.

1. (Z)-9-Tricosene.[697]

$$H_3C-(CH_2)_{12}\diagdown \quad \diagup (CH_2)_7-CH_3$$
$$C=C$$
$$H \qquad H$$

Muscalure, Muscamone®

2. *cis*-2-Decyl-3-(5-methylhexyl)-oxirane.[698]

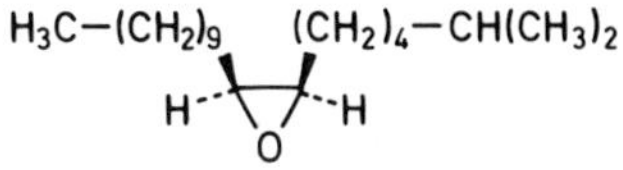

$$H_3C-(CH_2)_9 \quad (CH_2)_4-CH(CH_3)_2$$

$$H \cdots \overset{\triangle}{\underset{O}{}} \cdots H$$

Disparlure, Disparmone®

3. (*Z*)-7-Hexadecen-1-ol, acetate.[699]

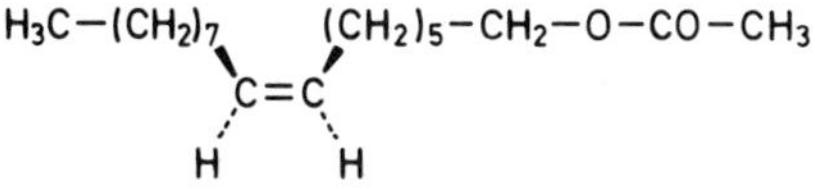

$$H_3C-(CH_2)_7 \quad (CH_2)_5-CH_2-O-CO-CH_3$$

$$C=C$$

$$H \qquad H$$

Hexalure, Hexamone®

4. Ethyl 2,2-dimethyl-3-(2-methyl-1-propenyl)cyclopropanecarboxylate.[700]

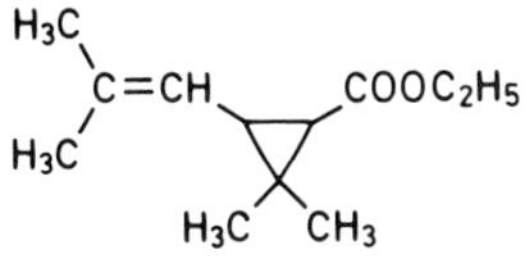

$$\begin{array}{c} H_3C \\ \quad C=CH \qquad COOC_2H_5 \\ H_3C \\ \qquad H_3C \quad CH_3 \end{array}$$

Ethyl chrysanthemate, *Rhinolure*

5. 1,1-Dimethylethyl 4(or 5)-chloro-2-methyl(cyclohexane)carboxylate.[701]

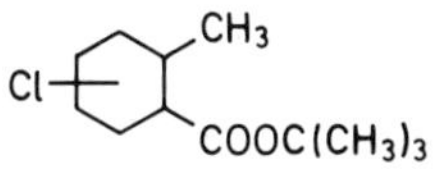

$$Cl - \overset{CH_3}{\underset{COOC(CH_3)_3}{}}$$

Trimedlure

6. Eugenol.[702]

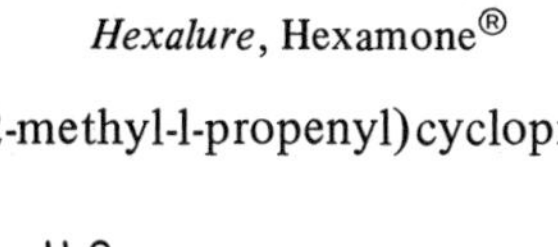

$$\begin{array}{c} HO \\ H_3CO \qquad CH_2-CH=CH_2 \end{array}$$

7. A mixture of two terpene alcohols and two terpene aldehydes.

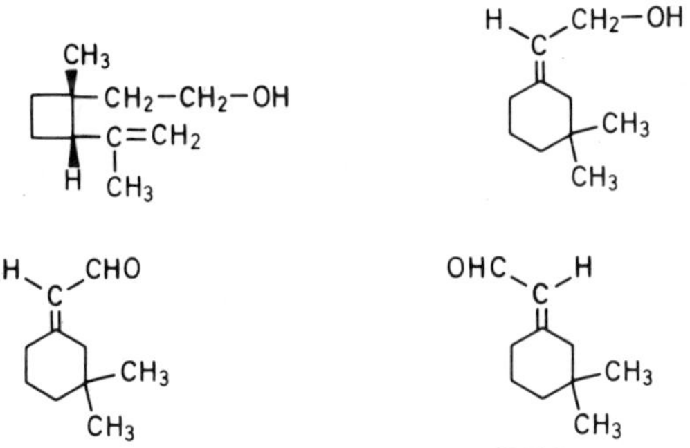

Grandlure, Grandamone® [703]

2.8.4. INSECT REPELLENTS*

2.8.4.1. General

Repellents drive troublesome insects away rather than kill them. The insects are discouraged from settling on surfaces treated with repellents. The mechanism of action is not yet understood.

The main application of these compounds is in the human and veterinary sectors for the control of troublesome insects that are frequently vectors of infectious diseases and thus a public health problem.

Repellents act in the vapor phase surrounding the object treated—frequently parts of the body or clothing. Possibly specific body odors or perspiration components that are normally attractive to the insects are masked or suppressed.

In the past, plant extracts, essential oils (e.g., oil of citronella), smoke, and the presence in the immediate vicinity of certain plants or plant parts have been used to repel insects. Now more effective, mostly synthetic repellents with more favorable cosmetic properties are in use.

2.8.4.2. Synthetic Repellents

Owing to their wider spectrum of activity, repellents are mostly used in the form of combinations for practical purposes. Two such mixtures are very well known.[704]

1. M-250 or 622 mixture; 60% dimethyl phthalate, 20% 2-ethyl-1,3-hexanediol; 20% *butopyronoxyl* (see no. 3 below).
2. M 2020 mixture; 40% dimethyl phthalate, 30% 2-ethyl-1,3-hexanediol; 30% *dimethylcarbate* (see no. 5 below).

The following repellents are of economic importance.

1. *N,N*-Diethyl-3-methylbenzamide[705,706]

Deet, Delphone®, Detamide®, Autan®

Use: repellent for various insects, particularly mosquitoes.

2. 2-Ethyl-1,3-hexanediol.[707]

$$H_3C-CH_2-CH_2-CH-CH-CH_2-OH$$
$$\overset{|}{OH} \quad \overset{|}{C_2H_5}$$

Ethyl hexanediol, Rutgers 612

Use: against flies and mosquitoes, frequently in combination with Indalone® or dimethyl phthalate.

*Review ref. 703A.

3. Butyl 3,4-dihydro-2,2-dimethyl-4-oxo-2H-pyran-6-carboxylate.[708]

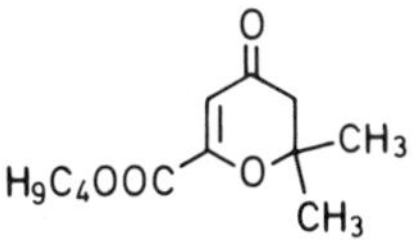

Butopyronoxyl, Indalone®

Use: highly repellent toward biting insects.

4. Dimethyl phthalate.[709]

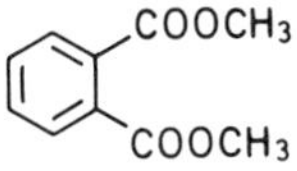

Use: effective mosquito repellent in man. Disadvantages: not stable to sweat, causes irritation to the eyes and mucous membranes.

5. Dimethyl *cis*-bicyclo[2.2.1]hept-5-ene-2,3-dicarboxylate.[710]

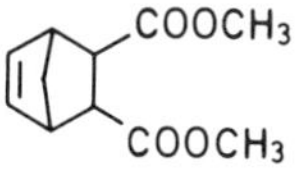

Dimethylcarbate, Nisy®, Dimelone®

Use: repellent for mosquitoes, particularly in combination with dimethyl phthalate (see no. 4 above) or Indalone® (see no. 3 above).

6. 1,5a,6,9,9a,9b-Hexahydro-4a(4H)-dibenzofurancarboxaldehyde.[711]

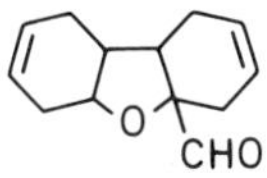

MGK Repellent 11

Use: as a repellent for cockroaches.

7. Di-*n*-propyl 2,5-pyridinedicarboxylate.

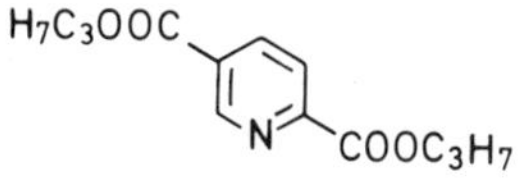

MGK Repellent 326

Use: effective housefly repellent.

2.8.5. BIOLOGICAL METHODS*

2.8.5.1. General

In biological pest control, other organisms are used to limit the population density of certain harmful animals or plants. In some cases the intent is not the complete

*Review ref. 712A.

elimination of a pest but, rather, a reduction of a pest population to a level at which the damage caused by the pest remains below the economic threshold.[713] Sweetman[713] and Franz and Krieg[714] have written surveys of the whole area of biological pest control. Refer to ref. 715 for a more specialized summary.

2.8.5.2. Control with the Aid of Beneficial Organisms

In this context beneficials are the natural enemies of the particular pest.

1. *Predators*, which require the pest as prey.
2. *Parasites*, which develop at the expense of the host—the pest—and thereby cause lasting injury to the host or even death.
3. *Pathogens*, which cause an often lethal infection of the host.
4. *Pests* themselves, provided that they differ from other members of their own species in terms of reproductive ability or other essential functions. The difference can be genetic, either naturally present or artificially induced, as, for example, sterility caused by chemicals or high-energy radiation. The difference can be a plasma incompatibility between animals of the same species but of different geographical origins.

The pests themselves thus become natural enemies of their own species and introduce the sometimes hereditary differences as a lethal factor into the natural population.

2.8.5.2.1. Introduction of Beneficial Insects. The most well-known and frequently practiced form of biological pest control is the introduction of new beneficial insects, mainly arthropods from a foreign fauna region, to control an unwittingly introduced pest. The assumption is that a pest introduced unwittingly is able to multiply so quickly because of the absence of natural enemies in the new environment that would otherwise limit the population.

Some of the entomophages successfully introduced and released in new areas are:

1. *Novius cardinalis* (lady bug) for control of *Icerya purchasi* (Australian cottony-cushion scale).
2. *Aphelinus mali* (sawfly) for control of *Eriosoma lanigerum* (woolly apple aphid).
3. *Prospaltella berlesii* for control of *Pseudaulacaspis pentagona* (mulberry scale).
4. *Prospaltella perniciosi* for control of *Quadraspidiotus perniciosus* (San-José scale).
5. *Eretmocerus serius* for control of *Aleurocantus woglumi* (citrus blackfly).[716]

One tries to find the most specific natural enemies possible for the control of pest arthropods introduced earlier by accident. This then gives a certain guarantee that the beneficial does not become a pest itself in time.

It has been assumed that this method of biological pest control may lead to formation of resistance that originates in a certain genetic uniformity of the beneficial arthropod introduced.

No spontaneous effect is observed following the introduction of beneficials for biological pest control. The effect of beneficials introduced in warmer climates is seen earlier as a consequence of the shorter succession of generations.

2.8.5.2.2. Maintenance and Promotion of Beneficial Insects. Relatively monotonous, intensively farmed landscapes usually offer poor accommodation for the maintenance of beneficials either already present or newly introduced. Improvements can usually be made at little cost, for example, by retention of banks and ridges at the sides of fields and pathways, of hedges and copses, and of trees and bushes. The stock of entomophage can be maintained or promoted by provision of food, nesting places, hides, and so on, frequently also by releasing the pests themselves as food source or host animals in times when nature itself cannot provide the pests in sufficient numbers. Beneficials can also be protected by the use of selective insecticides.

2.8.5.2.3. Periodic Release of Beneficials. Periodical release of beneficials is recommended when their permanent naturalization fails. Facilities for mass rearing of the entomophage are, however, a precondition. Apart from entomophagic arthropods, insect pathogens such as bacteria and viruses are also used for control purposes. For control measures to be carried out at short notice, these pathogens must be produced beforehand and stockpiled.

Genetically altered or sterilized beneficials have a great advantage as mobile organisms in that they seek out the pests themselves.

2.8.5.3. Self-Destruction Techniques

2.8.5.3.1. General. Introduction of beneficial insects for the control of previously imported pests is the classical technique of biological pest control.

In the course of the last 30 years, new knowledge has enabled further control techniques to be developed that differ from the classical in that the pest is used as its own enemy instead of specific natural enemy species. Various methods are available for the conversion of a pest to a beneficial, which then acts in a different way toward its own kind. These methods can be steered to give a precise, desired effect.

2.8.5.3.2. Sterilization. The influence of high-energy radiation or chemical agents can cause mutations that, at adequate dosage, lead to sterility. One must make sure that the other life functions, viability, and mating capacity are maintained.

As the more active and mobile sexual partners, male insects are normally chosen for sterilization. The first large-scale field trial was carried out on the island of Curacao, where the introduction of sterilized insects for the control of the screwworm fly (*Cochliomyia hominivorax*, affecting cattle and goats) led to complete eradication of this pest.[717]

This success was made possible by particularly favorable circumstances: the isolation (an island), low pest population density, easy mass rearing of the beneficials, and the mating behavior of the screwworm fly (the female mates only once).

2.8.5.3.3. Translocations.
Chromosomes can be broken apart by high-energy radiation or treatment with certain chemicals. If the break heals then everything remains normal, provided that the genetic codes at the breakpoint have not been damaged. A so-called translocation occurs when several breaks occur simultaneously and the segments attach themselves in the wrong place on another nonhomologous chromosome. The consequence of translocation is a structural change at the gene level. When such genetically damaged individuals are crossed with normal insects, it is frequently observed that a large proportion of the offspring does not develop. This reduction of fertility is called semisterility.

A translocation that has once appeared is handed down to the offspring. For example, in the case of the common gnat, *Culex pipiens*, it has proved possible to select races with different semisterility. On being inbred these strains behave normally and produce a full and viable progeny. However, if they are crossed with normal insects or a natural population the offspring are again heterozygotic with respect to the translocation and hence are semisterile.

Theoretically, insect control with semisterile insects should prove more effective than that to be expected from the release of many, fully sterile insects. The reason is that sterilized insects produce no offspring, whereas the effectiveness of semisterile insects among a natural population lies in the production of offspring that pass on the lethal genetic coding to the next generation.[718]

Development of resistance among organisms with translocations is regarded as impossible owing to the genetic fixing of such properties. The effect of a semisterile system is absolutely selective and only hits the organism against which it was developed.

2.8.5.3.4. Plasmatic Incompatibility.
An unusual hereditary mode has been observed in some mosquito species (*Aedes, Culex*). If one crosses insects of the same species but different geographical origins, one frequently finds that no offspring are produced. The female lays eggs, but they do not hatch. Detailed investigations have shown that copulation and fertilization occur normally. The sperm penetrate into the eggs, where, as foreign sperm, they are then prevented from unification with the nucleus by a factor present in the egg plasma. It has been assumed that this spermatic blockade depends on a qualitative difference of the egg plasma and that the genes in the chromosomes have no influence on it. The use of spermatic blockade for the control of pests is an attractive possibility. The sterilizing principle is already present in nature. The effect of this principle can be increased by selection of more efficient, incompatible males that copulate more frequently and more successfully with the females of the other strain than the latter's own males do.

Field trials in small isolated areas have been carried out successfully. They required the mass rearing of incompatible males of another strain.[719]

There is as yet no definite general picture concerning the distribution of plasmatic incompatibility in nature.

2.8.5.4. Microbiological Pest Control

2.8.5.4.1. General. Microbiological pest control works with mechanisms of limitation that depend on knowledge of insect pathology.

Only fungi, bacteria, and viruses are treated here as insect pathogens. These pathogens all have one factor in common: they cause disease in other foreign organisms. When isolated from one diseased organism, they are able to cause the same disease in other organisms (Koch's postulate).

2.8.5.4.2. Fungi. The first practical attempts at control of insects with fungi were made at the end of the 19th century. The effect of fungi pathogenic to insects is not just limited to larvae but embraces all stages of insect development. Infection can take place orally, through the chitinous shell, or by both means simultaneously.

Beauveria and *Metarrhizium* spp., which are not very host specific and have a relatively polyphagic behavior, have reached practical use. Biotrol BB®, which is based on *Beaveria bassiana*, is marketed in the United States.

Host-specific fungi pathogenic to insects are found among the Entomophthoracea. Wide practical use has been hindered by as-yet-unsolved problems in the large-scale production of these fungi, as multiplication normally only takes place in the corresponding host.

2.8.5.4.3. Bacteria. The most promising insect pathogens of today are found among the bacteria. In the genus *Bacillus*, the species *Bacillus popilliae* and *Bacillus thuringiensis* are of importance.

Bacillus popilliae is only effective against the grubs of Scarabaeidae (Coleoptera). Its practical use is mainly directed against the grubs of the Japanese beetle (*Popilliae japonica*), a serious economic pest in the United States.

Multiplication of *B. popilliae* takes place in the Japanese beetle itself with a yield of $\sim10^9$ spores per larval carcass. The dead larvae are ground and formulated as a dustable powder with the usual carriers and additives.

These products containing *B. popilliae* spores are marketed in the United States under the names Japidemic® and Doom®. They have a storage life of several years and are remarkably stable toward heat.

Bacillus thuringiensis is the insect pathogen used to the greatest extent in microbiological pest control. Its activity is limited specifically to Lepidoptera larvae. There is no evidence that this bacterium is either toxic or infectious to mammals.

During the development cycle of *B. thuringiensis* a crystalline toxin is formed in the sporulation phase, which is released with the spores when the sporangium discharges. The crucial microbiological effect of *B. Thuringiensis* depends on this crystalline toxin. In the abdominal tract of sensitive larvae it acts as a specific poison, causing gut paralysis or destruction of the gut epithelium. The insects die as a result of the subsequent infection (septicemia), which, among other things, can be caused by the spores of *B. thuringiensis*. *Bacillus thuringiensis*, contained in such products as Bactur®, Biotrol BTB®, and Thuricide HP®, is produced on an industrial scale by fermentation on an artificial medium.

Certain Lepidoptera species are resistant to *B. thuringiensis*.

Known enemies of *B. thuringiensis* have no effect on its activity provided that the crystalline toxin remains intact. Where necessary, non-Lepidoptera must be controlled by additional methods.

Approximately 150 Lepidoptera species are known, the larvae of which are sensitive to *B. thuringiensis*. Control measures have been successful against about 30 of them.

The development of commercial products based on *B. thuringiensis* has given a significant impulse to microbiological pest control. Many favorable circumstances have contributed to this success.

1. Easy fermentative production on artificial media.
2. Good shelf life of the formulated products (>1 yr).
3. Specific activity against Lepidoptera larvae.
4. Application with the usual equipment according to standard techniques.
5. No danger, either toxic or infectious, to mammals.

2.8.5.4.4. Viruses.[720]   The use of viruses in microbiological pest control is mainly directed against lepidopterous pests, to a lesser extent against Hymenoptera, Diptera, and Coleoptera. The significance of viruses in microbiological control has dropped considerably since the development of effective products based on *Bacillus thuringiensis*.

Apart from a regional significance in areas where silkworms are reared (polyhedral viruses are more host specific than *B. thuringiensis*), virus-based products will probably gain a certain significance in cotton against cotton bollworm (*Heliothis zea*).

Multiplication of insect viruses is host specific. Insect viruses, the virions of which are embedded in polyhedral inclusion bodies of protein, are called polyhedral viruses. They have a particularly high host specificity in comparison to viruses without inclusion bodies. Such polyhedral viruses seem to be generally more host specific than bacteria.

In the United States an industrially prepared virus product, based on *Heliothis* nuclear polyhedrosis virus, is on the market under the name Viron/H®.[720] The product is ingested by the insects during feeding. The virions are released by the action of the gut juices and paralyze the abdominal tract. Young larvae die within 2 days, older ones within 6 days.

The question whether such host-specific organisms as the polyhedral viruses can change hosts cannot be answered directly. It is concluded from the fact that polyhedral viruses have lived side by side with their host for millions of years that the host–virus relationship has evolved to a state of equilibrium. The great advantage of absolute host specificity is countered by the disadvantage of delayed onset of effect. An initial damage of 2 to 6 days must be tolerated before the virus takes its toll.

2.8.5.5. Conclusion.   Biological pest control has been employed successfully many times in the past, mainly where a certain initial damage is acceptable, for example, in forest crops. However, the main tasks of modern pest control lie in the optimization of crop yields and eradication of storage and hygiene pests. Here pesticides are required with immediate onset of effect. This immediate effect cannot be achieved with the use of biological control methods. In spite of this, biological pest control has the right to its place in integrated systems of pest management.

2.9. Molluscicides[721]

W. LUNKENHEIMER
Bayer AG, Wuppertal-Elberfeld

2.9.1. MOLLUSCS AS CROP PESTS AND DISEASE CARRIERS

Of all animal pests the terrestrial molluscs (slugs and snails) generally play a relatively insignificant role as chewing pests in agricultural crops, gardens, and ornamental plant nurseries. However, the slug problem is by no means solved. Occasionally they appear in such large numbers that catastrophic damage results. A few years ago in Florida a veritable snail plague broke out, caused by the agate snail (*Achatina fulica*), three specimens of which had been brought back from Hawaii to Miami by a small boy.[721a]

Mainly it is the slugs that cause the most damage to crops. An example is the gray field slug, each of which can lay 200 to 400 eggs between August and November. Other slug pests are the earth slug, which lives in the soil and feeds on subterranean plant parts, and the cellar slug, which feeds on stored products in warehouses. The large garden snail (*Helix pomatia*) can eat up to 200 cm^2 of lettuce in a single night.

Molluscs also serve as intermediate hosts for certain development stages of a whole series of worms parasitic on humans and animals. Freshwater snails are intermediate hosts for blood flukes (*Schistosoma*), the common liver fluke (*Fasciola hepatica*), the large intestinal fluke (*Fasciolopsis buski*), and the lung fluke (*Paragonimus westermanni*), as well as the lungworm of sheep and goats, to name a few examples. Control of such mollusc hosts has only succeeded in the case of the blood fluke, which causes schistosomiasis, a human parasitic disease widespread in tropical localities.

2.9.2. AGENTS FOR THE CONTROL OF MOLLUSCS

The number of compounds that find use as molluscicides is very small. Some inorganic compounds have a little significance: quicklime (calcium oxide) can be strewn on insensitive plants such as cabbage on dry days and causes a drying-out of the slugs affected. Calcium cyanamide is also used in this way. Copper(II) sulfate pentahydrate was used for control of freshwater snails. The necessary concentration of 20 ppm is tolerated by fish and is nontoxic to humans and domestic animals.

The adjustment of the molluscicidal concentration in lakes, ponds, streams, and so on is difficult because of precipitation and inactivation by suspended matter and organic material.

The only molluscicide that has found general application is *metaldehyde*.

$$\text{H}_3\text{C} \diagdown \text{O} \diagup \text{CH}_3$$

LD$_{50}$: 600–1000 mg/kg dog, oral, acute

Metaldehyde is prepared by acid-catalyzed tetramerization of acetaldehyde in ethanol and has been used since 1934 in various products, usually in the form of a bran-based bait.

It is a contact and stomach poison, causing mucous secretion, immobilization, and death. However, species- and environment-specific differences in the susceptibility of various slugs and snails are frequently observed. The monomer, acetaldehyde, and trimer (paraldehyde) have no molluscicidal properties.

The insecticide and acaricide 3,5-dimethyl-4-(methylthio)phenyl methylcarbamate (see also Section 2.7.2.9) is also an effective contact and stomach poison against slugs and snails.

Mercaptodimethur; Mesurol® (4% bait formulation, Bayer AG)

Sodium pentachlorophenolate is used for control of the intermediate snail host of schistosomiasis living in static or slowly flowing waters.

LD$_{50}$: 210 mg/kg rat, oral, acute

It is more toxic to fish than copper sulfate at the necessary concentration (10–15 ppm) and is easily inactivated by suspended matter.

Bilharziasis (schistosomiasis) was first controlled successfully with 5-chloro-*N*-(2-chloro-4-nitrophenyl)-2-hydroxybenzamide.

Niclosamide; as the ethanolamine salt: Bayluscid® (Bayer AG, 1959);
LD$_{50}$: >5000 mg/kg rat, oral, acute

A concentration of 0.3 ppm suffices for a total kill of the freshwater snail *Australorbis glabratus*.

Niclosamide is nontoxic to humans and domestic animals but harmful to the water fauna in concentrations not far in excess of the recommended application rate. It is produced by the condensation of 5-chloro-2-hydroxybenzoic acid with 2-chloro-4-nitrobenzenamine. The ethanolamine salt is a yellow solid with a water solubility of 230 ± 50 ppm at room temperature.

No longer manufactured, 4-(triphenylmethyl)morpholine was also used for control of the freshwater snail host of bilharziasis.

Trifenmorph; Frescon® (Shell, 1966)
LD_{50}: 1400 mg/kg rat, oral, acute

Owing to its low water solubility (0.02 ppm at 20°), *trifenmorph* was formulated as an emulsifiable concentrate. It was recommended for application to irrigation and other moving-water systems by a drip-feed technique at 0.03–0.10 ppm and for static water at 1.0–2.0 ppm. Above 0.03 ppm it can be harmful to fish. The microflora and fauna are unaffected at 0.02 ppm. The mammalian toxicity is low.

2.10. Rodenticides[721a]

W. LUNKENHEIMER
Bayer AG, Wuppertal-Elberfeld

2.10.1. RODENT PESTS AND THEIR CONTROL

The most important pests among the rodents (Rodentia) are the brown or Norway rat (*Rattus norvegicus*), the black rat (*Rattus rattus*), and the house mouse (*Mus musculus*).

The above have associated themselves particularly closely with Man (commensal life-style). The black rat prefers living near water, often walks great distances for food, and can swim for long periods and dive and thus penetrate into inhabited areas via the sewer system. The less common black rat prefers a dry environment and lives almost without exception within a building, usually in the attic. It always settles near the source of food. Apart from its dietary preferences, the house mouse behaves completely differently from the rats, It is strictly nonnomadic. Usually one male lives with several females in a small territory often only a few meters in cross section. The house mouse investigates strange objects immediately, in contrast to the neophobic rat.

Rats and mice are pests in three respects: they consume or destroy vast quantities of food and act as vectors for disease.

Far greater losses are caused by contamination than by direct consumption. Urine, feces, hair, and even carcasses spoil stored products. Mouse feces are difficult to separate from grain owing to their similarity in size. In most cases the germs of various diseases distributed by rodents (e.g., typhus, murine typhus, and amebic dysentery) are transferred to foodstuffs by such contaminants or by direct contact with paws and fur.

Rodenticides are mostly used as bait formulations that are placed along or in the paths used by the rodent pests. It is most important that the bait has no repellent taste or odor. The effect should not be too rapid so that the rodents do not develop bait shyness and, finally, the a.i. should have the lowest possible toxicity for pets, domestic animals, and humans while retaining the highest possible toxicity for the pests. These requirements are best met by the anticoagulants, which have a chronic effect and are used almost exclusively in the industrialized countries for control of commensal rodents.

2.10.2. CHRONIC POISONS

The chronically acting, or multiple-dose, rodenticides are inhibitors of blood platelet coagulation that reach their full effect only after the consumption of several sublethal doses. As competitive antagonists of vitamin K_1, even small quantities of anticoagulant displace the former from its role in the formation of prothrombin, which is necessary for blood coagulation. Prothrombinopenia is intensified by repeated doses, and eventually the blood can no longer coagulate. As the anticoagulants also have a toxic capillary side effect, internal bleeding occurs in organs and tissue, leading to an inconspicuous death. The rodent has no perception of the incipient poisoning and thus does not develop bait shyness.

Any danger to humans and domestic animals is mainly determined by the acute toxicity. The low dosages that are oriented on the much higher chronic toxicity give adequate safety margins in the handling of the product formulations. Apart from this a reliable antidote is available in the form of vitamin K_1.

The anticoagulants are coumarin and indandione derivatives that have a certain similarity to vitamin K_1. They differ from one another in terms of acute toxicity, rapidity of action, and acceptance by the rodent. The most important commercial products are:

R = H: *Warfarin*
R = Cl: *Coumachlor*; Tomorin®, Ratilan® (Geigy, 1953)

Pyranocoumarin; Actosin-Fertigköder® (Schering AG, 1963)

Coumafuryl; Fumarin®, Cumarax® (C. F. Spiess & Sohn, Norddeutsche Affinerie, 1954)

R = H: *Coumatetralyl*; Racumin® (Bayer AG, 1956/7)

R = Biphenyl-4-yl: *Difenacoum*;[722,723] Neosorexa®, Ratak® (Sorex Ltd., 1974)

R = 4'-Bromo(1,1'-biphenyl)-4-yl: *Brodifacoum*; Talon®, Ratak Plus® (ICI/Sorex, 1978)

Pindone; Pival® (Kilgore Chem. Co., 1952)

R = H: *Diphacinone*; Diphacin® (Upjohn Co., 1953)

R = Cl: *Chlorophacinone*; Caid®, Liphadione® (Lipha SA, 1961)

Large-scale rodent control campaigns were first successful with these compounds to such an extent that, with the breakthrough of the coumarin rodenticides after 1948, the rat problem seemed to be solved. However, so-called superrats resistant to anticoagulants first appeared in Great Britain at the end of the 1950s, then in the Northwest of continental Europe, and finally in the United States and Australia. This resistance does not embrace *brodifacoum* and *difenacoum*[722,723] and is weak toward *coumatetralyl*. A 4:1 mixture of *warfarin* and *calciferol* (vitamin D_2), marketed under the name Sorexa $CR^®$,[724,725] is also effective against resistant rats and mice. *Calciferol* is also harmful to humans in high doses. It disturbs calcium metabolism and causes kidney damage, among other things. There is no known antidote.

Acute poisons are not affected by the development of resistance to anticoagulants. After more than 20 years of obscurity this group has again moved into the field of general interest.

2.10.3. ACUTE POISONS

Acute or single-dose rodenticides are fully effective even when only consumed once. In some cases cumulation of multiple sublethal doses causes an additive increase in action but never a potentiation as with the anticoagulants. The rapid onset of poisoning symptoms causes bait shyness in rats, whereas mice, which do not live in social packs, can be controlled effectively.

Thallium(I) sulfate, the a.i. of $Zelio^®$-Giftkörner and $Zelio^®$-Giftpaste (Bayer AG), is accepted well because of its long latency period (36 hr–6 days). LD_{50} (oral, acute): 15 mg/kg brown rat, 76 mg/kg black rat.

The effect of zinc phosphide is due to the phosphine it develops in contact with dilute acid (gastric juice). LD_{50} (oral, acute): 40–50 mg/kg mouse and brown rat; latency period: 2–4 hr (mouse).

The powdered bulbs of red squill (*Urginea maritima*), or an extract of the same, have been used for centuries for rodent control. The active principle is scilliroside.

LD_{50}: 0.7 mg/kg male rat, 0.43 mg/kg female rat, oral, acute

Scilliroside is highly toxic to the brown rat, but the black rat tolerates much higher doses. Humans and domestic animals are not endangered, because red squill induces violent vomiting. For anatomical reasons, rodents are incapable of vomiting, and the toxic effect can take its course.

2-Chloro-N,N-6-trimethyl-4-pyrimidinamine has been marketed since 1937 for control of house and field mice.

Crimidine; Castrix$^®$ (Bayer AG);
LD_{50}: 1.25 mg/kg rat, oral, acute

Although also highly toxic to rats, *crimidine* cannot be used against them because of the short latency period (15–45 min). Since *crimidine* is rapidly metabolized, poisoned mice are nontoxic to predators.

The danger of such secondary poisonings is very high in the case of fluoroacetic acid derivatives. Such compounds are generally toxic to mammals because they block the citric acid cycle (see Section 2.7.2.7).

Rats have been controlled with sodium fluoroacetate since the early 1940s.

$$F-CH_2-C\overset{O}{\underset{O^{\ominus}}{\diagup}} \quad Na^{\oplus}$$

"1080"; LD$_{50}$: 0.22 mg/kg rat, oral, acute

Owing to the rapidity of effect, a total success against rats is not possible.

1-Naphthalenylthiourea is a rodenticide specific for adult brown rats.

$$\overset{S}{NH-C\diagdown NH_2}$$

Antu (since 1945); LD$_{50}$: 6.4–7.4 mg/kg rat, oral, acute

The toxicity is insufficient for control of mice and black rats. Pronounced bait shyness develops rapidly among rat populations even though the toxic action (increased capillary permeability) requires 16–30 hr to take effect. In addition, consumption of sublethal doses leads to development of tolerance.

Norbormide (mixture of isomers) is a highly selective toxicant for the brown rat.

$$\text{(structure)}$$

Norbormide; Shoxin®, Raticate® (McNeil Lab. Inc., 1964);
LD$_{50}$: 11.5 mg/kg rat, oral, acute

It is not harmful to pets, domestic animals, or even mice. In practice, *norbormide*, which is relatively expensive, has not been a great commercial success. Baits with a concentration of a.i. as low as 0.5% are somewhat repellent, and the short latency period (0.5–2 hr) leads to bait shyness.

The narcotic *chloralose*, used for control of bird pests, can also be used against mice. It causes deep sleep in which small mammals such as mice, with their large body surface area relative to weight, die from hypothermia. All larger animals, in-

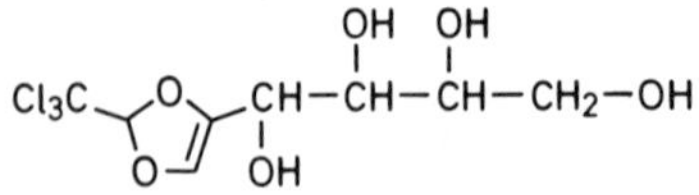

$$Cl_3C \diagdown O \diagdown CH-CH-CH-CH_2-OH$$

with substituent groups OH, OH, OH shown above the chain

cluding rats, are not endangered. At temperatures above 12° the effect against mice is uncertain.

The Japanese invention 2,2'-methylenebis(hydrazinecarbothioamide)[726] is one of the recent developments.

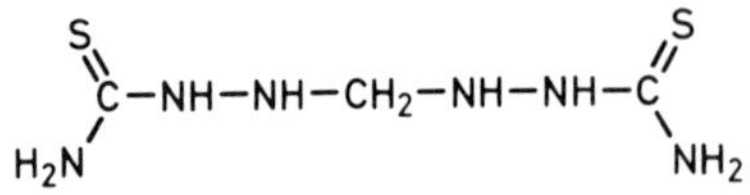

$$H_2N \diagdown \overset{S}{\underset{\|}{C}}-NH-NH-CH_2-NH-NH-\overset{S}{\underset{\|}{C}} \diagdown NH_2$$

Kayanex® (Nippon Kayaku, 1971);
LD_{50}: 25–32 mg/kg brown rat, oral, acute

Kayamex® is effective against both brown and black rats. It is better accepted in baits than *antu*, *norbormide*, and red squill.

1-(4-Chlorophenyl)-2,8,9-trioxa-5-aza-1-silabicyclo[3.3.3]undecane, otherwise known as 5-(4-chlorophenyl)silatrane, a highly toxic rodenticide developed in the United States:

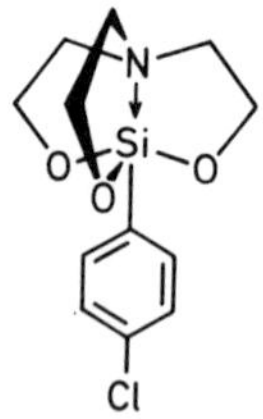

RS 150 (M & T Chem. Inc.);
LD_{50}: 1–4 mg/kg brown rat, oral, acute

is characterized by a rapid, hydrolytic self-detoxification (~3 days).[727,728] There is thus no danger of secondary poisonings.

A promising alternative to the anticoagulants has been found within the class of *N*-phenyl-*N'*-pyridylmethylureas, for example:[729,730]

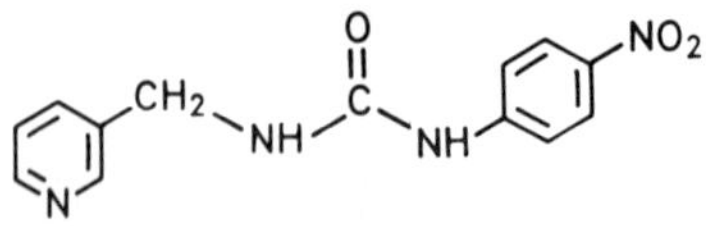

Pyriminil; Vacor® (Rohm & Haas, 1975);
LD_{50}: 4.75 mg/kg rat, oral, acute

Vacor® kills rats within a few hours through disruption of vitamin B metabolism. It is also effective against most other rodent pests but relatively nontoxic to domestic animals. Exhaustive tests are claimed to show that *pyriminil* does not cause bait shyness, but this remains to be confirmed in practice.

Chapter Three

Fungicides and Bactericides

W. KRÄMER
Bayer AG, Wuppertal-Elberfeld

3.1. Plant Diseases

3.1.1. CAUSES, EXTENT, CONSEQUENCES

"Since man came down from the trees and settled on the land," as Horsfall expressed it,[731] humans have been confronted with enemies to their crops. Although the causative relationship of animal pests to plant damage was, by its very nature, recognized very early on, the causes of plant diseases were first perceived correctly only by scientists in the 19th century. For example, de Bary[732] illustrated the disease symptoms of potato blight and studied the development cycle of the causal fungus.

Today viruses, bacteria, and fungi are all recognized as causal organisms of plant diseases, known correspondingly as viroses, bacterioses, and mycoses.[733,734]

Plant diseases are caused by pathogenic organisms and express themselves in changing natural plant processes owing to the interference of the pathogen. As a consequence, partial or total damage occurs after the invasion of the plant by the parasite.

The graduations in the demands a parasite makes on the host permit classification of the former: *Obligate parasites* are completely dependent on living cells for their nutrition and multiplication. Viruses and some fungal species (rusts, powdery mildews) come into this category. *Facultative parasites* (most fungi and bacteria) feed on living tissue in their parasitic phase but can live on dead plant material in a nonparasitic (saprophytic) phase. It is thus possible to cultivate them on artificial nutrient media. The *perthophytes* are a special class that first have to kill the host in order to be able to utilize it.

The parasite is often highly specialized, that is, dependent on a particular host, on certain stages of growth or maturity, and on certain parts of the host.

According to their colonization characteristics, one distinguishes between ecto- and endoparasites. Ectoparasites, such as the powdery mildew fungi, grow and fructify on the outer surface of the host and penetrate the host's cells with special organs (haustoria). Endoparasites, in contrast, penetrate into the plant and spread themselves in the tissue (viruses, bacteria, most fungi).

The *pathogenesis* is characterized by various stages that are reached before the disease breaks out. The prerequisite for pathogenesis is a contact between parasite and host plant under conditions favoring disease (development stage, temperature, light, humidity, pH, etc.). This inoculation is a consequence of the distribution of the pathogen. Although the flagella on various bacteria and fungi (zoospores) make active motion possible, distribution takes place passively as a rule from plant to plant (direct) or via wind, water, and vectors (indirect).

The *infection* is the process by which the preconditions for a stable parasitic relationship with the host plant are fulfilled. In the cases of viruses this occurs, without exception, passively from plant to plant or through wounds whereby vectors (cell puncture by sucking insects) can be involved. Bacteria can also penetrate via wounds or enzymatically through noncutinized areas (stigmas, root hairs) into the host tissue. Such openings as hydathodes and stomata are possible additional points of entry.

As well as the possibilities named above, the fungi are also able to enter the host by active penetration of the cutinized outer surface of the plant. The inoculum is usually composed of fungal spores and so, in general, germination and formation of special attachment (appressoria) and penetration organs are required. Enzymatic and mechanical processes play a role in the penetration of the cuticle and epidermis.

The *incubation* is the period between infection and formation of the first symptoms. During this phase the parasite disseminates itself intercellularly (bacteria, fungi) or intracellularly (viruses, fungi) throughout the host plant. The multiplication of viruses takes place by reduplication, of phytopathogenic bacteria by division and of fungi by formation of mycelia, which, during the course of the disease, produce spores and spread the parasite further.

This phase is characterized by the pathogenic properties of the parasite, which go beyond simple damage to plant tissue. The formation of enzymes, toxins, and other injurious metabolic products by fungi and bacteria and the stimulation of plant cells by viruses to synthesize injurious substances are just as necessary for a successful pathogenesis as a weak or absent defense reaction by the plant. In general, the enzymes from the pathogen destroy cell structure, degrade natural products in the cell, or attack the protoplast directly and interfere in its function. Toxins affect the protoplast and disrupt permeability and function. Growth regulators, as excreted in the host by fungi and bacteria, disturb the hormone balance in the cell and thus its ability to divide and grow.

According to which of these principles plays the dominant role in pathogenesis, the plant disease manifests itself in typical symptoms that can be used in the description and diagnosis of the disease. The most frequent symptoms are stunting, malformation, discolorations, yellowing, spotting, wilting, withering, and rotting as well as exudations. Plants are attacked by pathogens in all stages of development and in all parts.

Thus, at the seedling stage, the most frequent and typical diseases are those of damping-off induced by rots, themselves caused mainly by bacteria of the genera *Pseudomonas* and *Xanthomonas* and fungi of the genera *Pythium*, *Fusarium*, *Phoma*, and *Rhizoctonia*. The plants can be attacked and damaged prior to and post emergence.

In the root area of the developed plant, damage appears from rots, necroses, and growth disturbances. Rots cause the dying-off of larger tissue complexes. Necroses are sites of tissue destruction that remain localized. Apart from bacteria, typical root parasites are to be found in fungi of all classes.

The above-ground parts of the plant exhibit signs of injury (senescence) as a result of rots. Leaf and stem rots are caused by bacteria (*Erwinia*) and by fungi (*Phytophthora*, *Sclerotinia*, *Botrytis*, *Cerosporella*-eyespot), bark and trunk rots by bacteria of the genus *Erwinia* and by fungi, for example, of the genera *Phytophthora*, *Nectria*, *Monilia*, *Fomes*, and *Polyporus*. Furthermore, as a result of infection by viruses, bacteria, and fungi, symptoms such as leaf spot disease, wilting, and discolorations (e.g., yellow diseases) occur, as well as changes in form (e.g., curling, crinkling, stunting, dwarfing, and distortions of the stalks and leaves). Buds and flowers infected by a pathogen can die from rot and necrosis. For example, bacterial blight of pome fruit (*Erwinia amylovora*) leads to a brown-to-black discoloration of the blossom. The pathogen attacks the ovary, prevents the development of the fruit, and then attacks young twigs and stronger branches. Certain smuts attack the cereal ovary and form their sporangia in the ears, which as a result turn black or bluish grey. Fruits are also attacked by bacteria, fungi, and viruses, resulting in tissue destruction. Rots are found mainly on fleshy fruits; usually fungi, occasionally bacteria, but never viruses are involved.

The economically most significant plant pathogens in central Europe are fungal in nature. Of the major diseases of central European crops 83% are caused by fungi.[733] H. Cramer[735] has estimated the world harvest losses in 1965 due to plant diseases to be $24.5 billion, 11.8% of the potential world harvest. Within the total world harvest losses (35% of the potential harvest) due to pests, diseases, and weeds, the losses from disease amount to approximately one third, of which about two-thirds occur before harvesting and one-third after. The harvest loss, divided into pathogen groups, varies widely from crop to crop, as may be seen from Table 15. The figures given are only approximate.

Plant diseases have led to significant sociological upheavals. *Phytophthora infestans* was the direct cause of the famine in Ireland of 1840, which resulted in a mass Irish emigration to North America.[736] The discovery of agents active against plant disease—for example, of Bordeaux mixture by Millardet[737]—and the development of organic crop protection chemicals by industry, coupled with knowledge of the factors favoring infection and the early-warning schemes of the crop protection authorities, have made such catastrophic epidemics avoidable.

3.1.2. PHYTOPATHOGENIC FUNGI

In contrast to the higher plants (Cormophyta), the fungi are eucaryotic thallophytes devoid of chlorophyll. Unlike bacteria they have genuine cell nuclei—sur-

TABLE 15.   Losses Due to Plant Diseases[735]

Crop	Percentage of the Potential Harvest (%)	Value (Billion $)	Total Damage (%) Caused by		
			Fungi	Bacteria	Viruses
Wheat	9.1	2.2	~80		~10
Rice	8.9	3.2	~80	~10	~10
Maize/corn	9.4	1.6	53	4	
Potatoes	21.8	3.4	48	13	32
Sugar beet	10.4	0.5	~37		~37
Tomatoes	14.7		43	10	24
Apples	6.4		72	14	2
Pears	14.4		8	16	72
Plums	5.5		79	13	
Cherries	21.4		42		45

rounded by nuclear membranes each containing several chromosomes. As they are not capable of photosynthesis, unlike green plants (seed plants, ferns, mosses, and algae) and feed on organically bound carbon, a precondition for fungal growth in nature is the simultaneous or prior growth of other organisms. Fungi therefore live in a symbiotic or parasitic relationship with other forms of life. They are ubiquitous and flourish in all climatic conditions on Earth.

Within the phytopathogenic fungi two important groups can be differentiated.

1. Myxomycota (slime fungi).
2. Eumycota (true fungi) with the subdivisions lower fungi (according to Gäu-mann, Phycomycetes to which the Oomycetes are also assigned) and higher fungi (Ascomycetes, Basidiomycetes, and Deuteromycetes).[734,738] The higher fungi include 98% of all known fungal species.[733]

Within the Myxomycota the most important phytopathogens are found in the class of Plasmodiophoromycetes (parasitic slime fungi).

1. *Plasmodiophora brassicae*, responsible for cabbage clubroot, and
2. *Spongospora subterranea*, responsible for potato powdery scab,

are pathogens of crops from within this class. The class is characterized by amoeboid plasmodia with naked thalli (i.e., no cell wall). The infection by parasitic members of the slime fungi starts with zoospores. For example, *Plasmodiophora brassicae* penetrates into the root hair cells of cabbage with flagellated zoospores. After the germination of the persistent spores these move actively in the soil. Within the host cells each individual zoospore grows into a multinuclear plasmodium, which can divide and wander from host cell to host cell. Further infections are caused by flagellated zoospores released from summer sporangia. The plasmodia cause hypertrophia of the host cells and excessive tissue proliferation, which can lead to the death of the plant.

The Eumycota comprise the fungi with thalli surrounded by a cell wall. The subdivision Mastigomycotina—to which belong fungi with a mobile reproductive phase (zoospores) and an aseptate thallus—includes the classes Chytridiomycetes and Oomycetes.

To the class Chytridiomycetes, in which sexual reproduction occurs uniformly via zoosporangia, belong such plant parasites as *Synchytrium endobioticum* (potato wart) and *Olpidium brassicae* (cabbage seedling disease).

Within the class Oomycetes, having over 70 genera, is found the order Perenosporales, which, in the genera *Pythium* and *Phytophthora*, contains economically important species of pathogens. *Phytophthora infestans*, which causes late blight of potatoes, infects the potato leaves in humid weather in June–July with the germination of zoospores, the hyphae of which penetrate the host tissue and form an abundant, intramatricular mycelium. Hyphal conglomerates develop within 6–12 days from this mycelium, migrate outside and form sporangia on the tips. The sporangia fall off, germinate with flagellated zoospores or with germ tubes, and thus lead to a very rapid, epidemiclike infection. The loss of the infected leaves as assimilation organs causes heavy harvest losses compounded by the infection and rotting of the tubers.

The fungus survives the winter in tubers or aerial plant parts left on the field. During sexual reproduction a germ hypha grows out of the oospores that forms a zoosprangium at its tip from which the new infection radiates.

The nearest relatives of *Phytophthora infestans*, which belong to the genus *Pythium* (e.g., *Pythium debaryanum*, responsible for damping-off of seedlings), vegetate in the soil on organic residues (saprophytic) and also damage host plants with their toxins. The so-called damping-off diseases lead to high economic losses, especially in seedbeds.

Plasmopara viticola (downy mildew of grapevine), *Pseudoperonospora humili* (downy mildew of hops), and *Peronospora tabacina* (blue mold of tobacco) germinate predominantly with germ tubes in contrast to *Phytophthora infestans* and thus are more closely related to the higher fungi.

The Zygomycetes form unflagellated spores. Within the individual orders, transitions are to be observed from sporangia to conidia formation in the asexual phase. One member, *Rhizopus stolonifer* (Mucuraceae), causes fruit rot in fruit and vegetables.

The Ascomycotina, the largest subdivision (described as the class of Ascomycetes by Gäumann), embraces about 30% of all known fungal species. The characteristic of these fungi is the ascus, a sporangium in which ascospores are formed endogenically and sexually.

Apart from this perfect reproductive form the Ascomycotina also have imperfect reproductive forms in which conidia are formed exogenically and asexually on conidiophores. These are responsible for the propagation of the fungus during the vegetative phase.

The powdery mildews, of great economic significance as phytopathogens, belong to the family Erysiphaceae (order Erysiphales). This family, very rich in terms of number of species, has a very wide spectrum of potential hosts and can lead to marked plant infections, particularly in climates with low precipitation, as the spore germination and infection—in contrast to most other fungi—occur easiest on dry surfaces. As ectoparasites these fungi colonize the surface of the plant and penetrate into the epidermis cells with haustoria. The white, powdery mycelial coating has given these fungi their name. The following are representative members of the family.

1. The phytopathogens belonging to the genus *Erysiphe*:
   Erysiphe graminis (powdery mildew of cereals).
   Erysiphe cichoracearum (powdery mildew of cucurbits).
2. The pathogen of the genus *Sphaerotheca*:
   Sphaerotheca pannosa (powdery mildew of roses).
3. Pathogens such as:
   Podosphaera leucotricha (powdery mildew of apple).
   Uncinula necator (powdery mildew of grapevine).

Venturia inaequalis, as the causal organism of apple scab the economically most important pome fruit pathogen on a worldwide basis, belongs to the class Loculoascomycetes (order Pleosporales). *Venturia* species attack leaves, shoots, and fruit; grow beneath the cuticle; and break through to the surface with conidiophores. The conidia spread the fungus whereby further infection occurs only during periods when the leaves are damp.

Further important phytopathogens within the Ascomycotina are:

1. *Taphrina* species (*Taphrina pruni*, responsible for plum pocket) from the order Hemiascomycetes.
2. *Ophiobolus graminis* (take-all of cereals).
3. *Helminthosporium gramineum* (leaf stripe of barley).
4. *Claviceps purpurea* (ergot) from the class Pyrenomycetes.

Nectria galligena (nectria canker of pome fruit), *Calonectria nivalis* (*Fusarium nivale*, snow mold of cereals), and *Gibberella fujikuroi* (*Fusarium moniliforme*, bakanae disease of rice) from the order Hypocreales are facultative parasites that are found in their imperfect reproductive form (conidia stage) on diseased plant parts.

Sclerotina species (Discomycetes, Helotiales) cause blossom, fruit, branch, stem, and root rots. *Sclerotina fuckeliana* attacks strawberries and grapes in its imperfect reproductive form, *Botrytis cinerea*, causing a gray mold coating particularly on blossom and fruit in typical cases. Infection takes place at the blossoming time and leads to rotting of the fruit and stalks. Apart from the *Sclerotina* species, such foliar parasites as

1. *Diplocarpon rosae* (black spot of roses), and
2. *Pseudopeziza tracheiphila* (red fire of grapevines)

also belong to the order Helotiales.

Numerous phytopathogens belong to the genus *Mycosphaerella. Mycosphaerella musicola* (imperfect reproductive form: *Cercospora musae*) causes banana leaf spot (sigatoka disease); *Mycosphaerella berkeleyi* (*Cercospora personata*) peanut leaf spot.

In the Basidiomycotina, as a contrast to the Ascomycotina, spore formation occurs exogeneously on the basidia after meiosis. (The basidia correspond in their function to the asci.) In evolutionary terms the Basidiomycotina, which embrace 30% of known fungi, are the most highly developed within the fungal kingdom. They include the orders

1. Uredinales,
2. Ustilaginales,

to which belong obligate phytoparasites of great economic significance.

The Uredinales are characterized by the fact that they can form five consecutive spore stages. In the case of *Puccinia graminis* (black stem rust of cereals), for example, the foliage of the barberry (the intermediate host) is infected by basidiospores. Spermagonia develop from which aecidiospores form after sexual development processes. The aecidiospores then infect the main host, cereals. The fungus spreads throughout the cereal crop after the formation of uredospores (the color of which gives the fungus its name). In the ripening cereal the almost black winter spores (teleutospores) develop and cause considerable harvest losses. The teleutospores winter in the stubble and germinate into basidia in the spring, on which new basidiospores form. *Puccinia graminis* is composed of numerous physiological strains that occasionally even specialize in individual wheat varieties.

Numerous other economically important phytoparasites are found among the rusts. Cereals are attacked by many of these rusts, for example.

1. *Puccinia glumarum* (= *striformis*), yellow rust of wheat.
2. *Puccinia triticina*, brown rust of wheat.
3. *Puccinia dispersa*, brown rust of rye.
4. *Puccinia sorghi*, common maize rust.

The above have a similar biological behavior to *Puccinia graminis*.

The smut fungi (Ustilaginales) also cause significant diseases among cereal crops. *Ustilago tritici* (loose smut of wheat) and *Ustilago nuda hordei* (loose smut of barley) in particular cause high economic losses. These obligate parasitic fungi infect the ovaries and form their clusters in the grain as does *Tilletia caries* (stinking smut of wheat). In stinking smut, or bunt, these clusters contain 2 to 4 million chlamydospores and smell of trimethylamine (hence stink). Even an infection rate of a few percent renders the wheat unfit for human consumption. Since the spores may remain infective on the seed for years, the economic losses to breeders and seed producers are also substantial. The level of loose smut in the stand need only be as low as 0.02% for certification of the seed to be refused.

A further important group, the Deuteromycetes (Fungi imperfecti), embraces, like the Ascomycetes and Basidiomycetes, 30% of all known fungi. This group comprises those fungi for which the perfect reproductive form is absent or has not yet been found. Most of the Deuteromycetes, however, have aspects in common with various groups of the Ascomycotina. This is shown in the correlation of hypha and septa types, cell-wall biochemistry, and the combination of the DNA bases. Often the reproductive forms of the Fungi imperfecti are identical to imperfect reproductive forms of Ascomycotina. For example, insofar as the perfect reproductive forms of fungi of the genus *Fusarium* are known, they all belong to the order Hypocreales within the class Pyrenomycetes. *Fusarium* species cause rots and wilts (e.g., *Fusarium oxysporum* f. sp. *lycopersici* on tomatoes, *Fusarium solani* in numerous forms on peas, beans, and tomatoes).

The following are of economic significance as imperfect reproductive forms of Ascomycetes.

1. *Botrytis cinerea*, a collective species embracing many types (fruit, leaf, blossom, and stem rots).
2. *Penicillium digitatum* and *P. italicum* (green and blue molds of citrus fruits).
3. *Monilia laxa* and *Monilia fructigena* (*Sclerotinia laxa*, blossom blight, brown rot, spur, and twig canker of pome and stone fruits).
4. *Cladosporium fulvum* (*Fulvia fulva*, tomato leaf mold).
5. *Fusicladium* species (imperfect forms of *Venturia* species).
6. *Drechslera* spp. (earlier *Helminthosporium*).
7. *Alternaria* spp.
8. *Cercospora* spp.

Soil fungi from the group Deuteromycotina are:

1. *Rhizoctonia* spp. (e.g., *Rhizoctonia solani*, foot rot, and black speck disease of potatoes).
2. *Sclerotium* spp. and *Ozonium* spp. (foot rots).
3. *Verticillium* spp. (wilts of cotton, lucerne, tomatoes, strawberries, hops, and numerous ornamentals).
4. *Pyricularia oryzae* (rice blast).
5. *Cercosporella herpotrichoides* (stem break in cereals).

Colletotrichum spp. appear as causal agents of leaf spot, stem rots, and fruit rots in dicotyledons such as bananas, beans, and clover.

A survey of important phytopathogenic fungi is contained in Table 16.

TABLE 16. The Most Important Phytopathogenic Fungi[734]

Division	Subdivision	Class	Genus	Species	Disease
Myxomycota (Archimycetes)		Plasmodiophoromycetes	Plasmodiophora	P. brassicae	Clubroot, cabbage
			Spongospora	Sp. subterranea	Powdery scab, potato
			Polymyxa	P. betae	Root diseases of sugar beet
				P. graminis	and cereals
Eumycota (Phycomycetes)	Mastigomycotina	Chytridiomycetes	Olpidium	O. brassicae	Damping-off, cabbage
			Synchytrium	S. endobioticum	Potato wart
		Oomycetes	Aphanomyces	A. euteiches	Root rot, pea
				A. cochlioides	Root rot, sugar beet
				A. laevis	Black leg, beet
				A. raphani	Black rot, radish
Lower fungi			Pythium	P. debaryanum	Damping-off diseases of
				P. ultimum	vegetables, potato, straw-
				P. aphanidermatum	berry, beet, gramineae, etc.
				P. irregulare	
				P. arrhenomanes	
			Phytophthora	P. infestans	Late blight, potato
				P. cactorum	Damping-off, root stem and
				P. parasitica	fruit rots
				P. citrophthora	
			Plasmopara	P. viticola	Downy mildew, grapevine
			Bremia	B. lactucae	Downy mildew, lettuce
			Pseudoperonospora	Ps. cubensis	Downy mildew, cucurbits
				Ps. humuli	Downy mildew, hop
			Peronospora	P. parasitica	Downy mildew, crucifers
				P. tabacina	Downy mildew, tobacco
				P. pisi	Downy mildew, pea
			Albugo	A. candida	White rust, crucifers
	Zygomycotina	Zygomycetes	Mucor		Rots and molds of fruit and
			Rhizopus		vegetables

TABLE 16. (*Continued*)

Division	Subdivision	Class	Genus	Species	Disease
Eumycota	Ascomycotina	Hemiascomycetes	Taphrina	*T. deformans*	Leaf curl, peach, and almond
		Plectomycetes	*Eurotium* (conidial form *Aspergillus*)	*Aspergillus niger*	Black mold, vegetables
				A. flavus	Boll rot, cotton
				Penicillium expansum	Blue mold, apple
Higher fungi			*Podosphaera*	*P. leucotricha*	Powdery mildew, apple
				P. tridactyla	Powdery mildew, apricot, peach, plum
			Sphaerotheca	*S. pannosa*	Powdery mildew, rose, peach
				S. mors-uvae	Powdery mildew, gooseberry, currant
			Erysiphe	*S. macularis*	Powdery mildew, hop
				E. graminis	Powdery mildew, cereal
				E. cichoracearum	Powdery mildew, cucurbits
				E. polyphaga	Powdery mildew, cucurbits
			Uncinula	*U. necator*	Powdery mildew, grapevine
			Leveillula	*L. taurica*	Powdery mildew, pepper, artichoke, tomato, eggplant, lucerne
		Pyrenomycetes	*Nectria*	*N. galligena* (*Cylindrocarpon mali*)	Apple canker
			Calonectria	*C. rigidiuscula* (*Fusarium decemcellulare*)	Cushion gall, dieback of cacao
			Gibberella	*C. nivalis* (*F. nivale*)	Snow mold, cereals
				G. fujikuroi (*F. moniliforme*)	Bakanae disease, rice
			Claviceps (conidial *Sphacelia*)	*C. purpurea*	Ergot, cereals
				Ophiobolus graminis	Take-all, cereals
				Helminthosporium salvinii	Stem rot, rice

Class	Genus/species	Disease, host
	H. gramineum	Leaf stripe, barley
	H. maydis	Leaf spot, corn
	Cochliobolus miyabeanus (*Helminthosporium* = *Drechslera oryzae*)	Brown spot, rice
	Glomerella cingulata	Fruit rot of apple, mango, avocado
Discomycetes	*S. laxa* (*Monilia laxa*)	Blossom blight, brown fruit rot, twig canker of stone fruit
Sclerotinia	*S. sclerotiorum*	Root and stem rot, vegetables and ornamentals
	Diplocarpon rosae	Black spot, rose
	S. fuckeliana (*Botrytis cinerea*)	Gray mold of blossom and fruit
Loculoascomycetes		
Elsinoe	*E. fawcetti*	Citrus scab
Venturia	*V. inaequalis*	Apple scab
	V. pirina	Pear scab
	V. cerasi	Cherry scab
Mycosphaerella	*M. musicola* (*Cercospora musae*)	Leaf spot, banana (Sigatoka disease)
	M. berkeleyi (*C. personata*)	Leaf spot, groundnut
Didymella	*D. lycopersici*	Fruit and stem rot, tomato
Basidiomycotina Teliomycetes		
Uromyces	*U. appendiculatus*	Rust, bean
	U. pisi-sativi	Rust, pea
Tranzschelia	*T. pruni-spinosae*	Rust, stone fruit
Gymnosporangium	*G. sabinae*	Rust, pear
Puccinia	*P. hordei*	Dwarf leaf rust, barley
	P. graminis	Black rust, cereals
	P. recondita	Brown leaf rust, rye
	P. recondita f. sp. *tritici*	Orange leaf rust, wheat
	P. striiformis	Yellow rust, wheat

TABLE 16. (*Continued*)

Division	Subdivision	Class	Genus	Species	Disease
Higher fungi (*Cont.*)	Basidiomycotina (*Cont.*)	Teliomycetes (*Cont.*)	*Hemileia*	*H. vastatrix*	Leaf rust, coffee
			Ustilago	*U. nuda*	Loose smut, wheat and barley
				U. avenae f. sp. *nigra*	Black smut, barley
				U. hordei	Covered smut, barley
				U. avenae	Loose smut, oat
				U. maydis	Boil smut, corn
			Sphacelotheca	*S. reiliana*	Head smut, corn
			Tilletia	*T. caries*	Stinking smut or bunt, wheat
				T. contraversa	Dwarf bunt, wheat
			Urocystis	*U. occulta*	Stripe smut, rye
		Hymenomycetes		*Thanatephorus cucumeris (Rhizoctonia solani)*	Damping-off
				Corticium sasakii	Stem blight, rice
				Corticium solani (Rhizoctonia solani)	Black speck, potato
				Typhula incarnata	Typhula blight, cereals
	Deuteromycotina	Hyphomycetes	*Sclerotium*	*S. rolfsii*	Stem and root rot
			Ozonium	*O. omnivorum*	Root rot, e.g., cotton
			Verticillium	*V. albo-atrum*	Verticillium wilt, potato
				Pyricularia oryzae	Rice blast
				Cercosporella herpotrichoides	Stem break, cereals
			Cladosporium	*C. fulvum*	Leaf mold, tomato and vegetables
			Thielaviopsis	*T. basicola*	Root rot; tobacco, ornamentals, etc.

Alternaria	A. solani	Early blight, potato and tomato
Cercospora	C. beticola	Leaf spot, beet
Fusarium	F. solani	Dry rot, potato
	F. avenaceum	Seedling blight, cereals
	F. moniliforme	Root rot; legumes, rice, maize
	F. oxysporum	Wilt; tomato, cotton, soybean
Coelomycetes	Septoria	S. nodorum	Leaf spot, wheat
	Phoma	P. betae	Seedling root rot, sugar beet
	Colletotrichum	C. gloeosporioides	Fruit rot, twig blight, leaf spot

239

3.1.3. PHYTOPATHOGENIC BACTERIA

Of the 1600 bacterial species known, 180 can cause bacterioses on plants. They are capable of living saprophytically and may be cultivated on artificial media. Apart from the *Streptomyces* (which form conidiospores by septation of the mycelial filaments), they have a rod form and reproduce by cell division. The phytopathogenic bacteria are gram-negative with the exception of *Streptomyces* and *Corynebacterium* spp. In the host tissue they occupy the intercellular space and cause wilts (tracheobacterioses due to disturbances in the phloems), hypertrophy (especially gall formation), rots, and leaf spot diseases. The important phytopathogenic orders are listed in Table 17.[739]

Of great economic significance are the *Pseudomonas* and *Xanthomonas* species belonging to the family Pseudomonaceae. Examples are *P. phaseolicola*, which winters on seeds in the soil and causes grease spot of beans; *P. tabaci* (wildfire of tobacco); *P. lachrymans* (bacterial leaf spot of cucumber); *X. oryzae* (bacterial leaf blight of rice); and *X. vesicatoria* (leaf spot of tomato and red pepper). *Xanthomonas campestris* causes black rot of brassicas and *X. citri* citrus canker.

In the family Rhizobiaceae, the causal organism of crown gall in orchard fruit (*Agrabacterium tumifaciens*) is responsible for root hypertrophy and harvest losses in apples, peaches and apricots.

The family Enterobacteriaceae includes the important *Erwinia* bacterioses: *Erwinia amylovora* (fire blight of pome fruit), which, according to the U.S. Department of Agriculture (USDA), was responsible for 14% of the U.S. apple harvest losses in 1965 and is gaining ground in Europe; and *Erwinia carotovora*, which causes bacterial soft rot in potatoes. Pollinating insects play a major role in the spreading of *E. amylovora*.

The bacterial ring rot of potatoes (*Corynebacterium sepedonicum*, a nonmotile, nonflagellated rod bacterium) was responsible for 2.6% of the total U.S. potato harvest losses in 1965, according to USDA estimates.[735] *Corynebacterium michiganese* induces vascular wilt in tomatoes.

The family Streptomycetaceae harbors the important phytopathogen *Streptomyces* (*Actinomyces*) *scabies*, which causes common scab of potatoes, a source of significant economic losses particularly in the United States.

TABLE 17. Classification of Phytopathogenic Bacteria

Order	Family	Species	Phytopathogenic Species
Pseudomonadales	Pseudomonadaceae	*Pseudomonas*	90
		Xanthomonas	60
Eubacterides	Rhizobiaceae	*Agrobacterium*	7
	Enterobacteriaceae	*Erwinia*	17
	Corynebacteriaceae	*Corynebacterium*	11
Actinomycetales	Streptomycetaceae	*Streptomyces*	2

3.1.3.1. Phytopathogenic Viruses

Viruses—nucleoproteins lacking an intrinsic metabolism—were described as being responsible for plant diseases as early as 1886 by Mayer. The first advances in our knowledge of the nature of viruses were made on the tobacco mosaic virus (TMV), which induces typical light-green/dark-green mottling on the leaves of infected plants. In 1935 Stanley managed to isolate TMV and prepare it in a paracrystalline form.[740] Infection of tobacco plants with TMV occurs on rubbing together of infected and noninfected plants. The viruses, themselves nonmotile, thus enter through sites of injury on the plant (even broken plant hairs suffice). They are then translocated throughout the plant with the sap flow in the phloem and cause damage to the chlorophyll apparatus, deformation of newly growing leaves, and stunting of the plant.

Apart from the purely mechanical spread of a virus such as TMV, the transmission of viruses by vectors, such as insects, nematodes, fungi, and phytoparasites that sink their haustoria consecutively into diseased and healthy plants, is also known.

Two types of virus are to be distinguished in the transmission by insect vectors.

1. Persistent viruses, for example, the leaf roll virus of potato. (This multiplies to a certain extent within the insect after being removed in the sap from a diseased plant during the sucking activity. It migrates from the gut to the salivary glands from where it is injected into a healthy plant during feeding.)

2. Nonpersistent viruses, for example, virus Y (which is only infectious for a short time, ~24 hr, after the sucking activity of the vector, and is transmitted immediately[741]).

The virus species important with regard to potatoes also have a rod-type structure similar to TMV.

1. Virus X transmitted in the United States by a grasshopper and by *Synchytrium* zoospores.

2. Virus Y (leaf drop streak of potato), which is spread by numerous insects and causes necrotic, inkspotlike lesions and wilting of the leaves and leaf drop.

3. The leaf roll virus, which leads to an inward curling of the leaves, discoloration of the foliage, and phloem necrosis with consequent stunting. Control is just as problematic as that of the beet mosaic virus, which causes the yellow disease of beet. Generally a spread of the virus can only be prevented by control of the insect vectors, by breeding resistant cultivars (e.g., by meristem culture), and by heat treatment.[742]

Phytopathogenic viruses having spherical or polyhedral shapes are tobacco necrosis, which attacks tobacco, beans, and tulips; tobacco ringspot, which infects potatoes and tobacco; and bushy stunt, which infects tomatoes and tobacco and in the United States also attacks legumes.

In the diseases of orchard crops that of pear decline takes first place. This disease, a virosis transmitted by insects, is responsible for three-quarters of all losses among pears in the states of Oregon, Washington, and California.[743] Viroses of cherries, peaches, apricots, and strawberries also cause substantial damage. Apart from insect transmission, the possibility of transmission by grafting must also be taken into account with viroses affecting fruit trees.

3.1.4. CROP PROTECTION MEASURES

As in medicine, the purpose of phytomedicine (plant pathology) is the maintenance or restoration of health in plants.[744] In many cases this can be attained by the elimination of pathogens (fungi, bacteria, viruses), but at the same time the susceptibility of plants to disease is also dependent on the resistance of the plant. Natural resistance as a result of genetic factors is neither static nor absolute but is also determined by such factors as nutrition, temperature, humidity, light, and the stage of development of the plant and of the plant tissue.[745] For example, only growing apple leaves are susceptible to mildew (*Venturia inaequalis*), the mature leaves being naturally resistant. Resistance to disease is made up of two principal factors: the innate, passive resistance and the defense response, the active resistance.
Passive resistance factors are:

1. Presence of barriers on the plant surface.
2. Presence of barriers within the plant tissue that prevent the invasion by the pathogen.
3. Existence of insufficient nutritive conditions for the pathogen.
4. Presence of antitoxins.
5. Presence of antagonists to the fungal enzymes.

Active resistance depends on the formation of post-infection barriers or on chemical defense reactions. Control measures directed against plant diseases are designed to promote or develop the natural resistance and improve the factors of which it is composed.

This can be achieved both by cultivation and chemical treatment methods. Whereas crop protection measures today are directed at improvement of the hereditary resistance on the biological side and mainly at killing off pathogens on the chemical side, the research intentions of phytomedicine are also aimed at improving the resistance of plants with chemicals (conferred resistance[746]).

Studies[747] illustrating this aspect have been carried out on poplars and vines where treatment with copper oxychloride and *folpet* strengthens the cuticle or epidermis of the leaves. Natural barriers to the penetration of fungi are built up. These fungicides do not just act fungicidally but also raise the passive resistance of the host plant. Growth regulators such as *chlormequat* induce a passive resistance toward fungal infections by reducing the number of stomata on the leaf surface[748] and strengthening the plant tissue.[749]

Some chemicals alter the distribution of sugar and amino acids within the plant by influencing biochemical processes. The conditions of nutrition for the pathogen are changed and resistance to some pathogens results. Heitefuss and Fuchs[750] were able to prevent *Pernospora brassicae* infections by treatment of the plant with thiosemicarbazide, which leads to lower sugar content in the plant. Systemic fungicides, which are often used in transport forms that enter the plant and are converted to the active agent by enzymatic action,[752] confer resistance on the plant throughout the whole growth period. This is in analogy to natural fungitoxic substances such as capsidiol,[752] which is formed by resistant plants as a natural fungicide. Plant resistance can also be improved by defense products acting against the enzymes and toxins formed by the pathogen that cause such disease symptoms as black spot and wilt. Examples are provided by Horsfall and Zentmyer,[753] who attributed the chemotherapeutic effect of 8-hydroxyquinoline derivatives to a partial inactivation of pathogenic toxins, and Grossmann,[754] who was able to reduce the symptoms of *Fusarium oxysporum* infection in tomatoes by use of a pectinase inhibitor. Phenolic compounds and their oxidation products are formed by diseased plants as active resistance factors suppressing the infection. These (phytoalexins) build up around the infection site. Matta[755] was able to stimulate the phenoloxidase production in the stems and roots of tomatoes by use of naphthylacetic acid. The biosynthesis of phenol oxidation products is thus increased, and resistance to *Fusarium oxysporum* is induced.

3.1.4.1. Cultivation Measures

3.1.4.1.1. Breeding for Resistance. After Biffen[756] demonstrated that the susceptibility of wheat plants to yellow rust, *Puccinia glumarum*, was a property obeying Mendel's law, increased attention was paid to the breeding of more resistant varieties by selection and crossing, remarkable successes being achieved.[757,758] Grafting is a further method for creation of resistant varieties that has led to success especially with slow-growing plants such as fruit trees and vines. A successful control of fire blight (*Erwinia amylovora*) was achieved in the eastern United States by grafting the desired variety onto a resistant stock.[759] The drawback of these methods is that, after the formation of physiologically new pathogenic strains, the resistance level of resistant varieties created biologically is insufficient to suppress infection by these strains.[760]

3.1.4.1.2. Plant Nutrition. The level of resistance of a plant toward pathogens depends not only on the nature of the parasite but also on the physiological conditions of plant growth. In the early 19th century Liebig[761] noticed that the same potato variety grown in two differently fertilized fields showed a different resistance to disease.

Excessive fertilizing with available nitrogen is particularly responsible for increased susceptibility to disease.[762] Gassner and Hassebrauk showed a definite connection between susceptibility to rusts and inorganic nutrients.[763] The calcium content of the plant is a dominant factor determining the resistance level. Plants with a high fraction of water-soluble pectin, which is produced increasingly at lower calcium levels, are particularly susceptible to various fungi because the fungal pectinases are able to hydrolyze these pectin fractions very rapidly.[764]

The available potassium and phosphate also exert a marked influence on the level of resistance. Bewley and Paine[765] have shown that control of tomato streak disease is possible by complete fertilizing, special attention being paid to the amount of available potassium.

3.1.4.1.3. Cultivation and Climate.

Soil conditions influence the resistance level of plants not only indirectly by fertilizer action but also via the growth of the plant and parasites. Growth is dependent on soil type, aeration, humidity, and temperature. Müller and Molz[766] have studied the susceptibility of wheat to yellow rust (*Puccinia glumarum*) as a function of soil type. Tisdale[767] found that cabbages normally resistant to *Fusarium conglutinans* are attacked under conditions of water shortage.

The influence of soil temperature was shown by Jones et al.,[768] for example, who showed that *Fusarium conglutinans* did not develop at soil temperatures below 17°C. Understanding of soil conditions, temperature, and air humidity makes it possible to reduce the activity of a parasite in greenhouses without impairing plant growth.[769,770] An optimum sowing time is also decisive in the reduction of plant disease as McKinney[771] and Reed[772] have shown in studies on *Helminthosporium* and bunt (or stinking smut, *Tilletia caries*) outbreaks in winter wheat.

3.1.4.2. Chemical Protection Measures

According to the site of disease and application, one distinguishes between foliar fungicides, soil fungicides, and dressings whereby the same chemical can be used for all three. Foliar fungicides are applied as dusts or sprays to the aerial, green parts of the plant. The compound must, of course, be applied in a formulation that ensures an optimal contact of agent and plant. An additional specification requirement on the formulation is that an even distribution of the agent on the plant be ensured, which makes dilution with a carrier necessary. Sprays are formulated as emulsifiable or suspension concentrates containing wetters and stickers and are diluted, usually with water, to give the required concentration for application to the plant. In contrast, dusts are purchased by the farmer as ready-to-use formulations whereby, as a rule, quartz, talc, or pyrophyllite functions as carriers.

Soil fungicides are used against soil-borne fungi such as *Plasmodiophora brassicae*, *Pythium* spp., *Olpidium brassicae*, *Fusarium* spp. and *Verticillium* spp. These are products that, according to the nature of the active ingredient, are applied to the soil as liquids, dry powders, or granulates and act either through the vapor phase or by systemic properties. However, the fungicidal soil sterilization techniques known to date and the corresponding fungicides are not completely satisfactory. Soil treatment requires high application rates, and the techniques are complicated.

The killing of pathogens on or in the seed and protection from emergence diseases caused by soil-borne fungi are achieved by dressing of seed, tubers, corms, and the like. Various dressing methods are used. The seed is dipped or wetted in liquid formulations (wet dressing), coated with powder (dry dressing) or aqueous suspensions (slurry dressing) in mixers, or sprayed with solutions of fungicide (moist dressing).

Fungicides are described as protective, curative, or eradicative according to their mode of action. Protective fungicides prevent the infection by spores either by a sporicidal activity or by changing the physiological conditions at the leaf surface. Either the spores do not germinate, the fungal hyphae are killed as they penetrate the leaf, or their penetration is prevented. Rumm[773] assumed that copper compounds (particularly Bordeaux mixture) stimulate the resistance of vines and thus prevent *Plasmopara* infections. Swingle[774] put forward the hypothesis that copper ions, absorbed by plant tissue in the form of soluble copper salts, prevent infection, first, by killing the hyphae; second, by impeding penetration owing the changes to the leaf surface; and, third, by interfering with enzymatic processes during penetration. (Support for the hypothesis that fungal growth can be prevented by interfering in the plant physiology comes from the studies of Erwin,[775] who demonstrated prevention of *Verticillium* infections by the application of growth regulators in very low concentrations.) Protective fungicides, as their name implies, must be applied to the host plant before any contact takes place between plant and spores of phytopathogenic fungi. This requires a precise knowledge of the infective pressure exerted by phytopathogenic fungi as a function of temperature and climatic conditions and the early-warning schemes of the crop protection authorities, who give farmers notice of imminent danger from fungal infection. If the fungal hyphae have already penetrated the plant tissue and the mycelium, for example, of *Venturia inaequalis* (apple scab), is growing between cuticle and epidermis, then reproduction and sporulation have not yet occurred in the fungal development cycle. During this incubation time (2–12 days) curative fungicides can still prevent further infection. These fungicides, such as organomercury compounds and salts of dodecylguanidine (*dodine*), penetrate the cuticle and kill off the young fungal mycelium growing in the epidermis or prevent further development of fungal growth.

Agents that make control of the fungus possible even after the symptoms have become visible (frequently after sporulation) and that kill off both newly developed spores and the mycelium are called eradicative fungicides. Successful eradication can be achieved of fungal species that, like powdery mildews, grow on the leaf surface and only form haustoria in the plant tissue. Curative and eradicative agents for pathogens such as *Phytophthora infestans* and *Plasmopara viticola* (Phycomycetes) that grow deep within the leaf tissue (mesophyll) have not yet been introduced into crop protection practice.

Agents such as derivatives of benzimidazole (Benlate®), pyrimidine (*dimethirimol*, PP 675), oxathiin (Vitavax®), morpholine (*tridemorph*), and triazole (*triadimefon*) are systematic fungicides.[776] They are translocated actively in the phloem system, either from root to the leaf tip or vice versa, according to the particular compound. The great advantage of these systemic agents (or genuine therapeutics) is that they protect newly growing plant parts from infection. The problems lie in the demands made on the therapeutic index (i.e., the ratio of fungicidally active concentration to the maximum concentration still tolerated by the host plant) and on the stability in soil.

3.2. Fungicidal Agents

3.2.1. INORGANIC FUNGICIDES

"Fungi and all other organisms require very small quantities of various inorganic components which must be contained in the culture medium. The most important of these trace elements are heavy metal ions. A deficiency of trace elements inhibits growth or activity. High concentrations of the same are often toxic. The threshold values for mercury and copper (and also for sulfur not bound in the form of sulfate ion or in amino acids) are relatively low."[777] As early as 1803 a mixture of to-bacco, sulfur, chalk, and elder buds was recommended by Forsyth[778] for control of mildew disease on fruit. In 1807 Prevost[779] recognized the fungicidal action of copper sulfate against wheat bunt. Today inorganic fungicides still belong to the economically important agents for control of plant disease even though, as a result of toxic and cumulative effects, the use of mercury compounds in particular is limited for ecological reasons.

3.2.1.1. Sulfur and Sulfur Compounds

Sulfur is mainly used as a protective fungicide against powdery mildews. Ducharte recommended sulfur as long ago as 1848 for control of *Oidium* in vines. Apart from its activity against powdery mildews, sulfur is also acaricidal. The activity of sulfur is highly dependent on the particle size, which should lie between 1 and 10 μm. It can be maintained at this size by addition of wetting agents (wettable sulfur) or kaolin, lead arsenate,[780] or bentonite sulfur.[781]

Sulfur is used in the form of powders (vineyard sulfur, Kolodust®), sprays (e.g., flowable sulfur Bayer in 0.2–0.7% concentration in an aqueous suspension), or as a fumigant (Rupprecht technique).[782] The temperature-dependent phytotoxicity of sulfur is especially marked during warm and humid weather periods and leads to scorching or premature defoliation and fruit drop.[783]

Calcium polysulfide (CaS_x) (recommended in 1852 by Grison as Eau Grison[784]) was used mainly against apple scab (*Venturia inaequalis*) and powdery mildew of vine but was also active against San José scale. The disadvantage of lime sulfur was its lack of stability, particularly in solution, and the consequent phytotoxicity.

This disadvantage is not shared by barium polysulfide ($BaS_4 \cdot H_2O$, Solbar®).[785] It also has acaricidal activity and is still used today against fruit scab and mildew. Barium polysulfide is water soluble and can be made by dissolving sulfur in a barium hydroxide suspension under pressure in the absence of air.

According to Parker-Rhodes,[786] the activity of sulfur and polysulfides is due to the formation of the hydrodisulfide ion ($HS_2^{\ominus}$). Horsfall[787] has suggested that sulfur influences the normal hydrogenation and dehydrogenation reactions within the cell.

3.2.1.2. Metal Compounds

Metal salts of the 6th, 7th, and 8th subgroups and particularly of the 1st and 2nd subgroups of the periodic table are in use as fungicides. Wöber has reported studies[788] on the fungicidal properties of metal salts as a function of their positions in the periodic table.

3.2.1.2.1. Copper Compounds. Inorganic copper compounds are used on a large scale today in fruit, grape, potato, tomato, and banana growing and in other tropical plantation crops. They have a wide fungicidal spectrum and are employed especially against *Plasmopara viticola* and *Phytophthora infestans* as protective foliar fungicides and against damping-off diseases as seed dressings and soil fungicides. As early as 1885 Millardet[789] described the preparation of a mixture of copper(II) sulfate and hydrated lime for use against *Plasmospora viticola* (Bordeaux mixture)

$$[Cu(OH)_2]_x \cdot CaSO_4$$

According to the ratio of lime to copper sulfate, double salts of various compositions are obtained.

Today basic copper salts in solid form have been introduced by industry, for example:

$$3\,Cu(OH)_2 \cdot CuCl_2 \cdot xH_2O$$

Basic copper chloride (Cupravit®, Bayer AG) is used, like basic copper sulfate, against *Peronospora* in vines and hops, late blight (*Phytophthora infestans*) of potatoes and tomatoes, blister blight of tea (*Exobasisium vexans*), and coffee rust (*Hemileia coffeicola*).

Malachite [$Cu(OH)_2 \cdot CuCO_3$] at one time served as a seed dressing in cereals and as a spray against foliar diseases but is now only used outside of Europe. It is manufactured from copper sulfate and soda:

$$2CuSO_4 + 2Na_2CO_3 + H_2O \longrightarrow Cu(OH)_2 \cdot CuCO_3 + 2Na_2SO_4 + CO_2$$

Malachite was introduced by Masson[790] in 1887 as Burgundy mixture. Solutions of basic copper carbonate in ammonium hydroxide or carbonate were long used in the United States under the name "Cupram." Cheshunt compound[791] (a mixture of 2 parts by weight of copper sulfate to 11 parts ammonium carbonate) is used against damping-off diseases.

Copper(II) dihydrazinium sulfate [$Cu(N_2H_5)_2(SO_4)_2$] (Omazine®, Olin Mathieson Chemical Corporation) is unstable in aqueous solution and decomposes, under oxidation of the hydrazine, to copper(I) oxide and copper without, however, any loss of fungicidal activity.[792]

Copper(I) oxide (Cu_2O) is available as a dispersable powder with a copper content of 50% or more (Yellow Cuprocide®, Rohm and Haas; Perenox®, ICI Plant Protection Ltd.). It is nontoxic and is used as a seed dressing and foliar fungicide (*Phytophthora*).[793]

Copper-zinc-ammonium silicate $[Cu(NH_4)_x \cdot Zn(SiO_4)_y]$, Coposil®, is used to a limited extent as a protective fungicide.[794]

Copper-zinc-chromate ($15\,CuO \cdot 10\,ZnO \cdot 6\,CrO_3 \cdot 24\,H_2O$), known as "Crag Fungicide 658" (Union Carbide) and "Miller 658" (Miller Chemical and Fertilizer Corporation), is used against *Phytophthora* in tomatoes, potatoes, cucurbits, and peanuts.[795]

Copper compounds serve as protective fungicides. They either are sporicidal or prevent the germination of spores[796] by accumulation of soluble copper salts[797,798] or of lipid-soluble complexes with amino-, hydroxy-, and dicarboxylic acids.[799,800] These complexes are formed with exudates from the host surface or spores. They can penetrate to the interior of the spore cell nucleus and lead to poisoning.

In addition to positive physiological effects on the plant,[801] soluble copper compounds in higher concentrations cause foliar damage, splitting, and malformation.[802]

3.2.1.2.2. Other Metal Salts.

Basic zinc sulfate salts (Zinc-Bordeaux mixture),[803] cadmium salts, particularly cadmium(II) chloride (Caddy®), and cadmium-calcium-copper-zinc-sulfate-chromate (Crag Turf Fungicide, Fungicide 531, Union Carbide) are known as fungicides, mainly as mixtures and for use on turf.

Owing to their high toxicity[805] (LD_{50}: 37 mg/kg, rat, oral, acute) and ecological problems, mercury salts such as mercury(II) chloride (sublimate) (Fusariol®, Chem. Werke Marktredwitz; a seed dressing against *Fusarium nivale*[804]) and mercury(I) chloride (calomel) for soil sterilization are little used today.

3.2.2. ORGANOMETALLIC FUNGICIDES

In contrast to the inorganic metal salts, organometallic fungicides are mostly volatile and thus act via the gas phase. The organomercurials, in particular, have an exceedingly wide and high fungicidal activity and are mostly used for seed dressing. Owing to their wide activity and very low application rates (also against bacterial diseases), the organomercurials are still used today but only in certain indications where, in spite of great residue problems, they cannot be replaced by individual metal-free organic compounds or their combinations.

Apart from the organomercurials, organotin compounds and organoarsenicals are also broadly active fungicides and bactericides.

3.2.2.1. Organomercury Compounds

Alkylmercury compounds. The reaction of alkylmagnesium chlorides with mercury(II) chloride in anhydrous tetrahydrofuran:[806]

$$R-Mg-Cl \ + \ HgCl_2 \ \xrightarrow[-MgCl_2]{} \ R-Hg-Cl$$

or of the salts of alkylsulfinic acids with mercury(II) salts:[807]

$$R-SO_2Na \ + \ HgX_2 \ \xrightarrow[-NaX]{} \ R-SO_2-Hg-X \ \xrightarrow[-SO_2]{} \ R-Hg-X$$

yields alkylmercury halides as crystalline compounds, poorly soluble in water, which are highly volatile in spite of their relatively high melting points and very toxic. Treatment with potassium or sodium hydroxide leads to the akylmercury hydroxides,[808] which are reacted with the appropriate acids to give the commercial alkylmercury salts such as methylmercury nitrile (CH_3—Hg—CN) (Chipcote®, Chipman Chemical Corporation), which has an exceptionally high activity and, being water soluble, is used as an aqueous solution. Very strict precautions are to be observed in use on account of its high toxicity.[809] Ethylmercury chloride (C_2H_5—Hg—Cl) has been sold by many companies (e.g., Du Pont) in seed dressings (®Granosan 2%, ®Ceresan 2%). However, it is increasingly being replaced by the lowly volatile and thus less toxic bis(ethylmercury)phosphate [(C_2H_5—Hg)$_2$HPO$_4$] (Du Pont: Lignasan®, EMP and New Improved Ceresan). Bis(methylmercury) sulfate is also less volatile and was used against seed-borne cereal diseases and for dressing beet seeds, potatoes, and bulbs. Their lower volatility makes the alkylmercury salts of amides suitable for use in various seed treatments, for example (cyanoguanidinato-N') methylmercury, which is the active ingredient:

$$H_3C-Hg-OH \quad + \quad H_2N-C\begin{smallmatrix}NH-CN\\ \\ NH\end{smallmatrix} \quad \xrightarrow[-H_2O]{} \quad H_3C-Hg-NH-C\begin{smallmatrix}NH-CN\\ \\ NH\end{smallmatrix}$$

of the liquid seed dressing Panogen® (Casco, Sweden). Ethyl(4-methyl-N-phenyl-benzenesulfonamidato-N)-mercury is highly active against seed-borne fungi[810] and may be made from ethylmercury acetate:

$$H_5C_2-Hg-O-CO-CH_3 \quad +$$

$$\xrightarrow[\substack{-H_3C-COONa\\-H_2O}]{NaOH}$$

It is contained in the seed dressings ®Ceresan M, ®Ceresan M2X, ®Granosan M, and ®Granosan M2X (Du Pont).

(4,5,6,7,8,8-Hexachloro-1,3,3a,4,7,7a-hexahydro-1,3-dioxo-4,7-methano-2H-isoindol-2-yl)alkyl-mercury

R = CH_3, C_2H_5

Memmi®, Emmi® (Velsicol Chem. Corp.)

was developed as a protective and eradicative fungicide[811] for treatment of wheat and cotton seed and against powdery mildews and scab.

Alkylmercury salts of phenols and thiophenols are easily available from the corresponding alkylmercury hydroxides, for example, the well-known seed protectants:

$$O-Hg-CH_3$$

Methyl(8-quinolinolato-*O*)mercury; Ortho LM® (California Chem. Co. Ortho Div.)

$$COOH(Na)$$
$$S-Hg-C_2H_5$$

Sodium ethyl[2-mercaptobenzoato(2-)-*O,S*]mercurate; ®Elcide 73, Merthiolate® (Eli Lilly Co.), Thimerosal®

Alkoxyalkylmercury compounds. (Acetato-*O*) (2-methoxyethyl)mercury[812] is formed on passing ethylene into a suspension of mercury(II) acetate in methanol:

$$CH_3OH \;+\; H_2C{=}CH_2 \;+\; Hg(OCOCH_3)_2 \xrightarrow[-\,H_3C-COOH]{}$$

$$H_3CO-CH_2-CH_2-Hg-O-CO-CH_3$$

and can be used as a water-soluble intermediate for the production of other 2-methoxyethylmercury salts.

The acetate is itself also used as a seed dressing, for example, in Panogen Metox®, Tayssato®, and Mercuran® (Delmar Chemical Corporation).

The less water-soluble (up to 5%) and thus less skin-irritating chloro(2-methoxyethyl)mercury is manufactured from the corresponding acetate:

$$H_3CO-CH_2-CH_2-Hg-O-CO-CH_3 \;+\; NaCl \xrightarrow[-\,H_3C-COONa]{}$$

$$H_3CO-CH_2-CH_2-Hg-Cl$$

Ceresan Universal Wet® (Bayer AG), Agallol® (Bayer AG), Aretan®; LD_{50}: 40 mg/kg rat, oral, acute

It was introduced as a seed dressing in 1930 and is still used today (®Ceresan Universal Wet), owing to its high fungicidal activity, against leaf stripe of barley, stinking smut of wheat, snow mold of rye, against seedling diseases (damping-off) in beet and legumes, and for dressing of "seed" potatoes, bulbs, and tubers.

(Silicato-*O*) (2-methoxyethyl)mercury ($H_3CO-CH_2-CH_2-Hg-O-SiO_2H$) ($LD_{50}$: 75 mg/kg rat, oral, acute) is, with hexachlorobenzene, an ingredient of Ceresan Universal Dry® (Bayer AG) and is contained in Ceresan-Morkit®, Ceresan Gamma M®, and Ceresan Special®.

(2-Ethoxyethyl)hydroxymercury:[814-816]

$$H_5C_2O-CH_2-CH_2-Hg-OH$$

Tillex® (Sandoz AG, 1967);
LD_{50}: 25.4 mg/kg rat, oral, acute

is also used widely against most seed diseases at a rate of 200–600 g per 100 kg seed. Cautious handling is advisable because of its skin-irritant properties.

Another alkoxymercury seed dressing on the market is ethoxyethylmercury-methylmercury citrate.[817]

$$CH_2-COO-Hg-CH_2-CH_2-OC_2H_5$$
$$HO-\underset{|}{\overset{|}{C}}-COO-Hg-CH_3$$
$$CH_2-COOH$$

Fertix®, Seed Dressing 6334, 6335

In contrast to the alkylmercury salts, the metal–carbon bond in alkoxyalkyl-mercury salts is much less stable. For example, treatment of the alkoxyalkyl deriva-tives with concentrated hydrochloric acid releases the olefin, thus reversing the synthesis.

Arylmercury compounds. (2-Chlorophenol)mercury sulfate:[818]

Uspulun®

prepared from *o*-chlorophenol and mercury(II) sulfate, the oldest organometallic product, was known under the name Uspulun® [819] in many countries. It replaced the much too phytotoxic mercury(II) chloride[820] but is of no significance today.

(Acetato-*O*)phenylmercury:

$$H_5C_6-Hg-O-CO-CH_3$$

PMA, PMAC, PMAS, "Ceresan Trockenbeize," Uspulun secco, Tag HL 331 (Chevron Ltd.),
Ceresan Slaked Lime
Water solubility: 0.438 g/l;
LD_{50}: 40 mg/kg rat, oral, acute[821]

is employed throughout the world as a foliar fungicide and seed dressing, although its phytotoxicity prevents use in some indications (crab apples). It is manufactured by the mercuration process[822] and is used as a curative and eradicative fungicide against apple scab as a pre-blossoming spray, for control of rice diseases (*Pyricularia oryzae*, *Helminthosporium oryzae*) and citrus diseases and as a dressing.

Compared with (acetato-*O*)phenylmercury the chloro derivative has a much lower water solubility, the consequent lower phytotoxicity permitting application as a foliar spray against *Venturia inaequalis* (apple scab).[823,824]

$$[H_5C_6-\overset{\oplus}{N}\equiv N]Cl^{\ominus} \xrightarrow{\text{Hg Cl}_2} [H_5C_6-\overset{\oplus}{N}\equiv N]Cl^{\ominus}\cdot HgCl_2 \xrightarrow[\substack{-2\,\text{Cu Cl} \\ -\,N_2}]{2\ \text{Cu}} H_5C_6-HgCl$$

PMC, 40% (F. W. Berk & Co.), Ceredon® Special (in combination with *benquinox*, Bayer AG); LD_{50}: 40 mg/kg rat, oral, acute

Like almost all arylmercury salts it is highly toxic.
    Other arylmercury salts are:

1. Dihydrogen [orthoborato(3-)-*O*] phenylmercurate (2-).

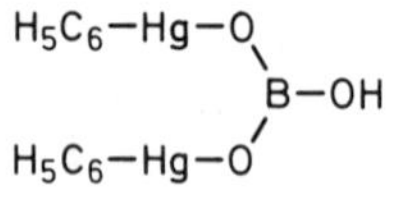

$$\begin{array}{c} H_5C_6-Hg-O \\ \diagdown \\ B-OH \\ \diagup \\ H_5C_6-Hg-O \end{array}$$

Merfen TM

2. (2-Aminoethanolato-*N*)phenylmercury 2-hydroxypropanoate.

$$[H_5C_6-\overset{\oplus}{Hg}-\overset{\oplus}{N}H_2-CH_2-CH_2-OH]\ \overset{\ominus}{O}OC-\underset{\underset{OH}{|}}{CH}-CH_3$$

Puratized Apple Spray (Gallowhur)

3. 2-Hydroxy-*N*,*N*-bis(2-hydroxyethyl)-*N*-(phenylmercurio)ethanaminium   2-hydroxypropanoate.

$$[H_5C_6-Hg-\overset{\oplus}{N}(CH_2-CH_2-OH)_3]\ \overset{\ominus}{O}OC-\underset{\underset{OH}{|}}{CH}-CH_3$$

Puratized Agr. Spray (Gallowhur)

4. Phenyl(propanoato-*O*)mercury.

$$H_5C_6-Hg-O-CO-CH_2-CH_3$$

®Metasol P-G

5. (4,5,6,7,8,8-Hexachloro-1,3,3*a*,4,7,7*a*-hexahydro-1,3-dioxo-4,7-methano-2H-isoindol-2-yl)phenylmercury.

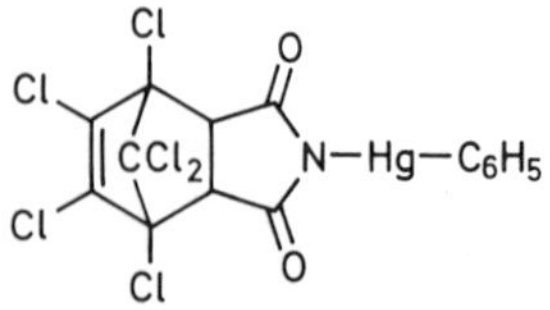

Phimm® (Velsicol Co.)

6. Phenyl(ureato-*N*)mercury.[826]

$$C_6H_5-Hg-NH-CO-NH_2$$

Leytosan® (F. W. Berk & Co.)

In his explanation of the mechanism of action of organomercury compounds, Daines[827] assumes that metallic mercury is formed in the soil by reduction of the mercurates and that this—just like mercury ions—interferes with enzymes containing mercapto groups. Booer[828] confirmed in his studies that organomercury salts decompose rapidly into metallic mercury in soil by reduction and disproportionation reactions via mercury(I) chloride.

3.2.2.2. Organocopper Compounds

Organocopper salts, like the inorganic copper salts, are broad-spectrum fungicides mainly used as foliar sprays. In general, they are less toxic and less phytotoxic than the inorganic copper salts.

A mixture of copper salts of fatty acids is well known under the tradename ®Citcop-4E (Copoloid, copper linoleate) as a foliar fungicide (marketed in the United States since 1964 by Cities Services Co. and Kerr McGee Co.).

It is composed of 20-25% copper abietate, 8-12% copper linoleate, and 10-15% copper oleate and is employed against *Cercospora* spp., powdery mildews, *Phytophthora* spp., and bacterial diseases. Copper oleate:

$$[H_3C-(CH_2)_7-CH=CH-(CH_2)_7-COO-]_2 Cu$$

Chemical Formulators Inc. and Witco Chemical Corp.

may be used as a fungicide in place of Bordeaux mixture and, owing to its low phytotoxicity, may be applied widely throughout the whole growth period. It also serves as a wood preservative as does copper cyclopentanecarboxylate[830] **(1)**, a compound used widely as a preservative and also active against termites, wood parasites, and bacteria.

1; Copper naphthenate; Naptox, ®Wittox C (Witco Chem. Corp.)

Bis(8-quinolinolato-[1]N,[8]O) copper is used, particularly in Japan, against apple scab, *Alternaria* spp., *Phytophthora infestans*, *Peronospora* spp., and *Fusarium* spp. in tomatoes, curcurbits, apples, citrus, peaches, and tea.[831]

Oxine-copper, Milmer® (Ashland Chem. Co., Kanesho Co. Ltd.)

3.2.2.3. Organotin Compounds

Organotin compounds of the type

$$R_2SnX_2 \quad \text{and} \quad R_3SnX$$

have wide biological activity and are used as fungicides, particularly against *Alternaria* spp., *Cercospora* spp., and *Phytophthora* spp. Most have high bactericidal activity and are also used as preservatives for technical materials. They are synthesized by reaction of tin(IV) chloride with organomagnesium, -lithium, -sodium, or -aluminum compounds and subsequent disproportionation reactions.[832,833]

$$4\ R-MgCl \ + \ SnCl_4 \ \longrightarrow \ R_4Sn \ + \ 4\ MgCl_2$$

$$R_4Sn \ + \ SnCl_4 \ \longrightarrow \ 2\ R_2SnCl_2$$

$$4\ R_4Sn \ + \ SnCl_4 \ \longrightarrow \ 4\ R_3Sn-Cl$$

The chlorine may be replaced by numerous other nucleophiles. For example hexabutyldistannoxane is obtained by reaction of chlorotributylstannane with concentrated sodium hydroxide in benzene, the water formed being separated azeotropically by the benzene.[834]

$$2\ (H_9C_4)_3Sn-Cl \ \xrightarrow[\substack{-2\ NaCl \\ -H_2O}]{2\ NaOH} \ [(H_9C_4)_3Sn]_2O$$

Butinox® (Osmose), TBTO®, C-Sn-9®, Biomet®, TMBTO® (Metal and Thermit Corp.)
Practically insoluble in water, highly soluble in organic solvents
LD_{50}: 200 mg/kg rat, oral, acute

It is marketed as a fungicide and bactericide, especially for protection and preservation of wood and technical materials. Hydroxytriphenylstannane:[835]

$$(H_5C_6)_3Sn-OH$$

Fentin hydroxide, Du-Ter® (N. V. Philips-Duphar, 1963), DPTH (Thompson-Hayward Chem. Co.), ENT 28009;
LD_{50}: 100 mg/kg rat, oral, acute

is less phytotoxic and was introduced onto the market as a protective and curative foliar fungicide. Highly active against *Pyricularia oryzae* in rice, *Cercospora* spp. in tropical crops, and *Phytophthora infestans* in potatoes, it has the distinction of being the first organotin fungicide to be used in crop protection. Apart from a long duration of activity, it also exhibits antifeeding properties for surface-feeding insects.

Of greater significance in the European market is (acetyloxy)triphenylstannane.[836]

$$(H_5C_6)_3Sn-O-CO-CH_3$$

Fentin acetate, Brestan® (Hoechst AG), TPHTA, Hoe 2871, GC 6936, TPTA, Batasan®, Suzu® (Hokko Chem. Co.);
LD_{50}: 125 mg/kg rat, oral, acute

It is mainly used for control of *Cercospora* spp. in sugar beet, *Septoria* spp. in celery, and *Alternaria* spp. and *Phytophthora* spp. in potatoes, but is widely active generally in orchard and vegetable crops. Phytotoxicity has been reported in grapes, hops, ornamentals, and in greenhouse use.[837]

Chlorotriphenylstannane (Tenestan®, Tinmate®, Nihon Nohyaku) is used like *fentin acetate*. The acetate is marketed by Fisons as a combination with Maneb® for use in potato growing under the name ®Fennite A.

Complexes such as chlorotriphenyl[sulfinylbis(methane)-*O*] tin:[838]

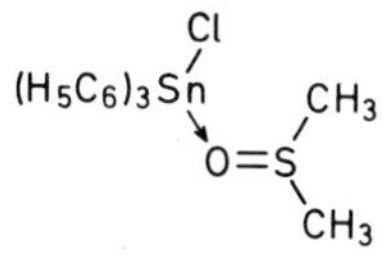

and chlorotriphenyl(quinoline 1-oxide)tin:[839]

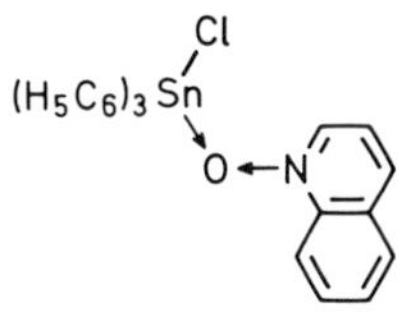

are reported to be superior to the noncomplexes chlorotriphenylstannane and hydroxytriphenylstannane in their activity against *Phytophthora infestans, Septoria appii*, and *Cercospora beticola*.[840]

Decyltriphenylphosphonium bromochlorotriphenylstannate(1-) is used for curative and eradicative control of *Alternaria, Cercospora, Phytophthora, Fusarium, Helminthosporium*, and *Septoria* spp. in potatoes, beet, rice, and celery.[841]

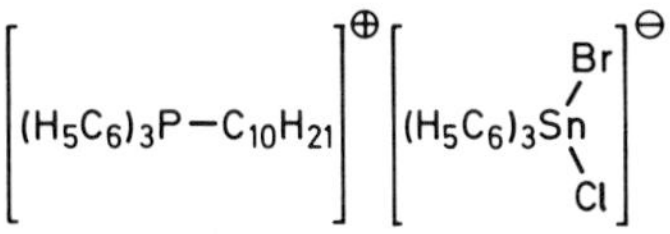

Decafentin, Stannoram®, Stannoplus® (Cela GmbH 1968)

The following compounds are known to be under development: arylthiotriphenyl-stannanes[841] of the type

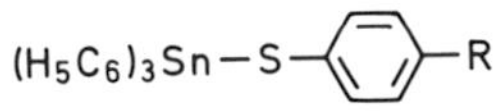

(R = H, Cl)

which act like *fentin acetate*, and 2-[(tributylstannyl)thio]pyridine

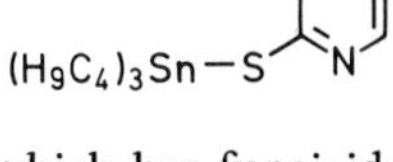

(Fisons Pest Control Ltd.),[843] which has fungicidal, insecticidal, and molluscicidal properties.

3.2.2.4. Organoarsenic Compounds

Although the organoarsenicals in general are not such potent fungicides as the organomercurials and organotin compounds they are still used in some indications. As with the organomercurials, their use in agriculture is limited by the residue problem.

Methylarsonates (1) are the starting materials for all the organoarsenic fungicides and are obtained by reaction of methyl halides with salts of arsonic acid.[844]

$$\text{Na}_3\text{AsO}_3 \quad + \quad \text{H}_3\text{CJ} \quad \xrightarrow[-\text{NaJ}]{} \quad \underset{\substack{\text{H}_3\text{C} \quad \text{ONa}}}{\overset{\substack{\text{O} \quad \text{ONa}}}{\text{As}}}$$

1

The complex of iron(III) monomethylarsonate with ammonium hydroxide (2) finds use as a rice fungicide against *Pellicularia sasakii*.

$$\left[\underset{\substack{\text{H}_3\text{C} \quad \text{O}}}{\overset{\substack{\text{O} \quad \text{O}}}{\text{As}}}\right]^{2\ominus} \text{Fe}^{3\oplus}(\text{NH}_4^{\oplus})_3(\text{OH}^{\ominus})_4$$

2; *Neo-asozin* (Kumiai Chem. Ind. Co.)

Methylthioxoarsine is obtained by reduction of disodium methylarsonate with sulfur dioxide and subsequent reaction with hydrogen sulfide[845] of the methyloxoarsine so formed:

$$\underset{\substack{\text{H}_3\text{C} \quad \text{ONa}}}{\overset{\substack{\text{O} \quad \text{ONa}}}{\text{As}}} \quad \xrightarrow[-\text{Na}_2\text{SO}_4]{\text{SO}_2} \quad \text{H}_3\text{C}-\text{As}=\text{O} \quad \xrightarrow[-\text{H}_2\text{O}]{\text{H}_2\text{S}} \quad \text{H}_3\text{C}-\text{As}=\text{S}$$

MAS®, Urbasulf®;
LD$_{50}$: 100 mg/kg rat, oral, acute

and, owing to its high activity against *Rhizoctonia* spp.,[846] was the active ingredient of the seed dressing Rhizoctol® (Bayer AG) and Rhizoctol-Kombi (together with *benquinox*) used mainly for seed treatment in cotton and rice (350–400 g per 100 kg seed).

Bis(dimethylthiocarbamoylthio)methylarsine[847]

$$\text{H}_3\text{C}-\text{As}=\text{O} \quad \xrightarrow[-\text{H}_2\text{O}]{2\,\text{HCl}} \quad \underset{\substack{\text{Cl}}}{\overset{\substack{\text{Cl}}}{\text{H}_3\text{C}-\text{As}}} \quad \xrightarrow[-2\,\text{NaCl}]{2\,\text{Na}-\text{S}-\overset{\text{S}}{\underset{}{\text{C}}}-\text{N(CH}_3)_2}$$

$$\underset{\substack{\text{H}_3\text{C}-\text{As}}}{\overset{}{}} \begin{array}{l} \text{S}-\overset{\text{S}}{\underset{}{\text{C}}}-\text{N(CH}_3)_2 \\ \text{S}-\underset{\text{S}}{\overset{}{\text{C}}}-\text{N(CH}_3)_2 \end{array}$$

LD$_{50}$: 100 mg/kg rat, oral, acute

found application in the combination product Tuzet® (Bayer AG), together with *thiram* and *ziram*, against fruit scab (*Venturia inaequalis* and *V. pirinia*),[848] *Pellicularia sasakii* in rice, and diseases of coffee.

3.2.3. ORGANOPHOSPHORUS FUNGICIDES

Organophosphates, as insecticides, form the most important class of compounds in chemical crop protection.[849] However, some members have even achieved significance as fungicides.[850] Essentially, phosphates and phosphonates are of interest as protective and curative fungicides for rice diseases (*Pyricularia oryzae, Pellicularia sasakii*), but some have been introduced for control of mildews and soil-borne fungi.

3.2.3.1. Phosphates

Bis[*O*-(2,4-dichlorophenyl)] *O*-ethyl phosphate (**1**)[851] is used in combination with the antibiotic *kasugamycin* for control of rice blast as a 25% wettable powder.

1; Kasumiron® (Hokko Chem. Co.), HOO 34[852]

S-(4-Chlorophenyl) *O*-cyclohexyl *O*-methyl phosphorothioate (**2**):

2; Ceresin® (Bayer AG,[853] Nitokuno/Japan);
LD$_{50}$: 160 mg/kg rat, oral, acute

has protective and curative activity against *Pyricularia oryzae* and is also insecticidal toward rice leafhoppers and planthoppers.

O,O-Diisopropyl *S*-benzyl phosphorothioate (**3**):[854]

3; *IBP*, Kitazin®, Kitazin-P (Kumiai Chem. Ind. Co., Japan);
LD$_{50}$: 490 mg/kg rat, oral, acute

and *O*-butyl *S*-benzyl *S*-ethyl phosphorodithioate (**4**):[855]

4; Conen® (Sumitomo Chem. Corp., Japan);
LD$_{50}$: 120 mg/kg rat, oral, acute

are also used as protective and curative fungicides in rice.

Kitazin® is also insecticidal toward the rice stem borer and rice leafhoppers and planthoppers. It was the first organophosphate fungicide to be introduced onto the market and is transported actively through the roots into the leaves,[856] a property that is exploited by adding a granular formulation to the irrigation water.

O-Ethyl *S,S*-diphenyl phosphorodithioate (5):[857]

5; *Edifenphos*, Hinosan® (Bayer AG, Nitokuno/Japan);
LD$_{50}$: 218 mg/kg rat, oral, acute

has been on the market since 1966 as a curative and protective fungicide for control of *Pyricularia oryzae* and *Pellicularia sasakii*. It is manufactured by reaction of *O*-ethyl phosphorodichloridate with thiophenol.

A heterocyclic phosphate, *O,O*-diethyl *O*-(6-ethoxycarbonyl-5-methylpyrazolo-[1,5-*a*]pyrimidin-2-yl) phosphorothioate (6),[858] has high locosystemic activity against powdery mildews coupled with a low phytotoxicity in more than 30 plant species.[859]

6; *Pyrazophos*, Afugan® (Hoechst AG);
LD$_{50}$: 140 mg/kg rat, oral, acute

A heterocyclic phosphonodiamidate, *P*-(5-amino-3-phenyl-1H-1,2,4-triazol-1-yl) *N,N,N',N'*-tetramethyl phosphonodiamidate (7),[860]

7; *Triamiphos*, Wepsyn® (Thompson-Hayward);
LD$_{50}$: 20 mg/kg rat, oral, acute

controls powdery mildews on apples and ornamentals.

(1,3-Dihydro-1,3-dioxo-2H-isoindol-2-yl)-*O,O*-diethyl phosphonothioate (8):

8; *Ditalimfos*, Plondrel® (Dow Chemical Co.),[861]
LD$_{50}$: 500 mg/kg rat, oral, acute

is used for control of powdery mildews on fruit, vegetables, and ornamentals. There are problems of phytotoxicity in certain sensitive varieties of apple.

According to Tolkmith the activity of the fungicidal organophosphates is, unlike that of the insecticides, not due to inhibition of cholinesterase but to acylation of enzymes containing mercapto groups.[862]

3.2.3.2. Phosphonates

S-Benzyl *O*-methyl phenylphosphonothioate[863] has a curative and protective activity against *Pyricularia oryzae*.[864]

$$H_5C_6-\overset{\overset{O}{\|}}{\underset{\underset{S-CH_2-C_6H_5}{|}}{P}}-OCH_3$$

Inegin® and Inezin® (Nissan Kagaku K. K.);
LD$_{50}$: 720 mg/kg mouse, oral, acute

A systemic fungicide with protective and curative properties and active against Phycomycete fungi was found in this group at the Rhône-Poulenc laboratories[864a]— aluminum tris[*O*-ethyl phosphonate].

$$\left[H-\overset{\overset{O}{\|}}{\underset{\underset{OC_2H_5}{|}}{P}}-O- \right]_3 Al$$

LS 74783, Aliette®;
LD$_{50}$: 5800 mg/kg rat, oral, acute

Aliette®, mainly marketed in combination with a protective fungicide, for instance, Mikal® (50% Aliette, 25% *folpet*), is active against downy mildews in tropical crops (e.g., pineapple, citrus, and avocado) and temperate crops (e.g., lettuce, vegetables, wine grapes, and ornamentals).

3.2.4. ANTIBIOTICS AND PHYTOFUNGICIDES

In 1926 Sanford[865] recognized the inhibitive action of microorganisms on *Streptomyces scabies*, the causal fungus of potato common scab. Millard and Taylor[866] showed that Actinomycetes inhibited the growth of *Streptomyces scabies*, and they ascribed this effect to a competitive inhibition as also reported by Machacek[867] in his studies on the storage rots of fruit and vegetables. In the course of further studies on the interaction of fungi, Weindling[868] found that *Trichoderma viride* reduces the activity of various soil fungi pathogenic to citrus seedlings. He was able to isolate[869] an antifungal toxin, gliotoxin, from culture filtrates. Detailed reports on the fungistatic properties of gliotoxin were given by Brian and Hemming,[870] the structure being elucidated by Bell et al.[871]

Following the discovery of penicillin by Fleming[872] and its chemotherapeutic use, studies on fungal toxins were intensified on a broad front and led to the discovery and technical exploitation of such antibiotic fungicides as *griseofulvin*[873] and *cycloheximide*.[874]

Today such compounds find application as less toxic, selective fungicides against mildews, rusts, *Botrytis*, *Pyricularia oryzae*, *Sclerotinia* spp., and *Alternaria* spp., even though the utility of individual compounds is limited by considerations of stability, phytotoxicity, and price.

During the course of a fungal infection, plants can synthesize defense compounds, phytoalexins,[875] which possess a certain scientific interest as fungistatics and could serve as the basis for synthetic programs in the search for organic fungicides.

At certain stages of development of plants, compounds are produced that, as natural fungicides, are responsible for resistance. These natural resistance factors are the subject of wide-ranging studies by various research groups.[876,877]

3.2.4.1. Fungicidal Antibiotics

4-[2-(3,5-Dimethyl-2-oxocyclohexyl)-2-hydroxyethyl]-2,6-piperidinedione:[878,879]

Cycloheximide, Actidione® (Upjohn Co.);
LD_{50}: 2 mg/kg rat, oral, acute

is isolated from the culture filtrate of *Streptomyces griseus* and is used as a foliar fungicide (also eradicative) against powdery mildews, cherry leaf spot, and rusts. Owing to its high mammalian toxicity, application in the United States has been restricted to treatment of grasses and ornamentals.[880]

Cycloheximide is present in other commercial products in combination with iron(II) sulfate, *thiuram*, and pentachloronitrobenzene. It interferes in protein biosynthesis in the fungal cells.[881]

Griseofulvin,[882]

Griseofulvin (developed by the British Ministry of Agriculture in 1939; marketed by Merck & Co. Inc., Murphy Chem. Co.)

a systemic foliar fungicide, is isolated from *Penicillium griseofulvum*[883,884] and used for control of *Botrytis*, *Alternaria solani*, and powdery mildews.

Griseofulvin acts on the morphogenesis of many fungi and causes a curling of the hyphae at concentrations of 10–1000 ppm. Only those fungi are affected that have chitin as a cell-wall component.[885,886] *Griseofulvin* also interferes in the chitin biosynthesis of insects.[887]

Kasugamycin was isolated from a culture of *Streptomyces kasugaensis*[888] and was introduced as a rice fungicide in 1965.

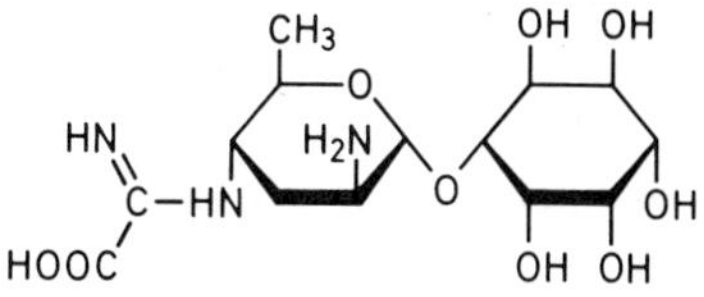

Kasugamycin, Kasumin® (Hokko Chem. Co., Japan);
LD_{50}: 40,000 mg/kg rat, oral, acute

It is used for control of *Pyricularia oryzae*, *Septoria* spp., *Botrytis*, *Fusarium* spp., and *Venturia* spp., either as a foliar fungicide or a seed dressing.

Blasticidin:[889,890]

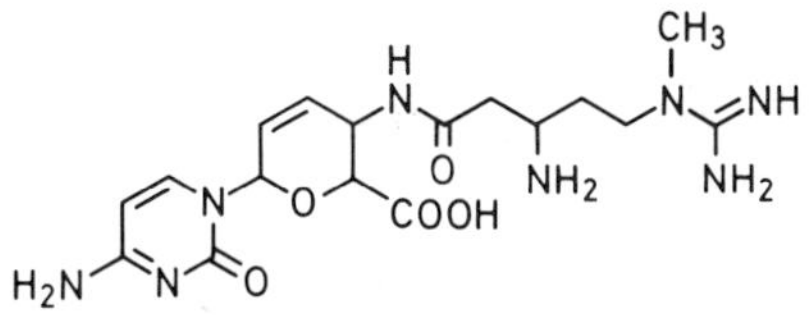

as benzylaminobenzenesulfonate (Nihon Nohyaku Co. Ltd., Kaken Chem. Ltd. of Japan; LD_{50}: 53 mg/kg rat, oral, acute)

is isolated from cultures of *Streptomyces griseochromogenes*. Produced on a large scale since 1958[891,892] and marketed as a rice fungicide against *Pyricularia oryzae*, *blasticidin* has both protective and curative properties and is one of the most important replacement products for mercury compounds on the Japanese market.

Other members of the group of *Streptomyces* antibiotics are the nucleoside antibiotics *polyoxin* A, B, C, D, E, F, G, H, I, J, K, and L,[893] isolated from *Streptomyces cacaoi*.

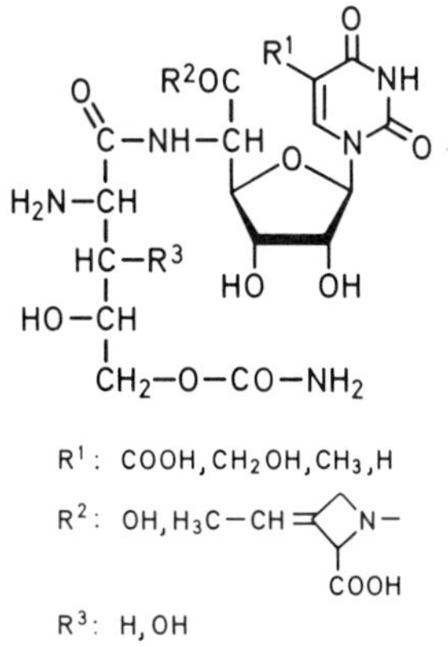

R^1: COOH, CH$_2$OH, CH$_3$, H

R^2: OH, H$_3$C—CH=

R^3: H, OH

Kaken Chem. Co., Japan

They were introduced onto the market in 1965 as a mixture for the control of *Alternaria*, *Botrytis*, *Cercospora*, powdery mildews, *Pellicularia*, and *Sclerotinia* spp. on apples, pears, gooseberries, tomatoes, curcurbits, carrots, and rice. They are protective and curative, being transported actively in the plant.

Piomycin:[894]

$$C_{14}H_{23}N_4O_{11}$$ (structure unknown)
(Hokko Chem. Ind., Japan)

has been in use since 1969 for control of *Alternaria, Cercospora, Cladosporium*, and powdery mildew species.

Prumycin,[895,896] structurally related to the *polyoxins*, is also isolated from *Streptomyces kagawaensis* and has high activity against *Botrytis cinerea, Sclerotina cinerea, Sclerotina sclerotiorum*, and *Sphaerotheca fuliginea*. (It could thus be of interest in wine growing.)

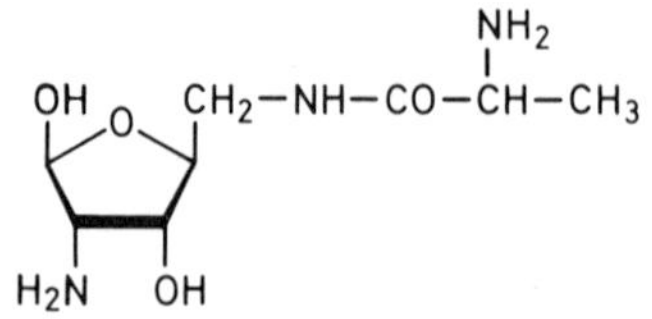

Kumiai Chem. Co.

Validamycin antibiotics with various components, for example, *Validamycin A*[897]

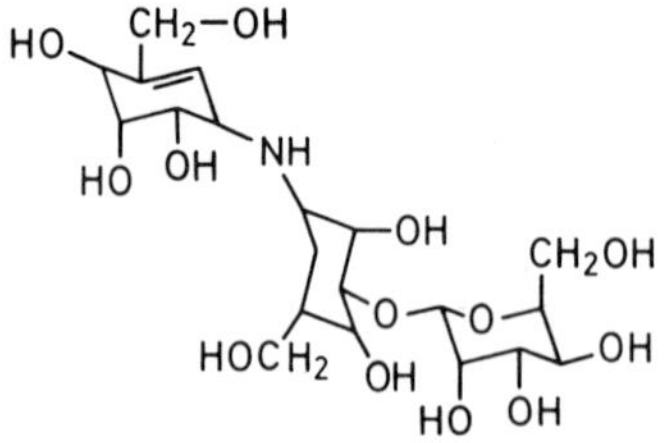

(Takeda Chem. Ind., Japan)

have been registered for control of seed diseases and sheath and stem diseases of rice and are also used in vegetables, citrus, coffee, and cacao against damping-off caused by *Rhizoctonia solani*.

3.2.4.2. Natural Fungicides and Phytoalexins

Natural fungicides are plant products that are responsible for the natural resistance of the plant toward pathogens. They are formed as metabolic products by the plant as a result of genetic coding. Their concentration is independent of the stage of maturity of the plant but varies as a result of external factors such as nutrition and temperature so that natural resistance does not always prevent fungal attack.

3.2.4.2.1. Fungicides

3.2.4.2.1.1. Acetylene Derivatives. 1-Phenyl-2,4-hexadiyn-1-one[898,899] has been isolated as the fungicidal principle of the oil from *Artemisia capillaris*.

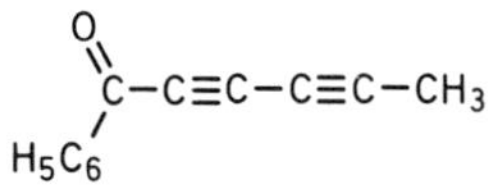

$$\underset{H_5C_6}{\overset{\overset{\textstyle O}{\|}}{C}}-C\equiv C-C\equiv C-CH_3$$

Capillin (Imai, 1956)

Capillin is active against *Trichophyton purpureum* at a concentration of 0.25 µg/ml, and *Penicillium italicum* at 4 µg/ml.

3.2.4.2.1.2. Ketones and Aldehydes. Ripe hops contain up to 20% of a soft resin that is composed of humulone **(1)** and lupulone **(2)**.[900]

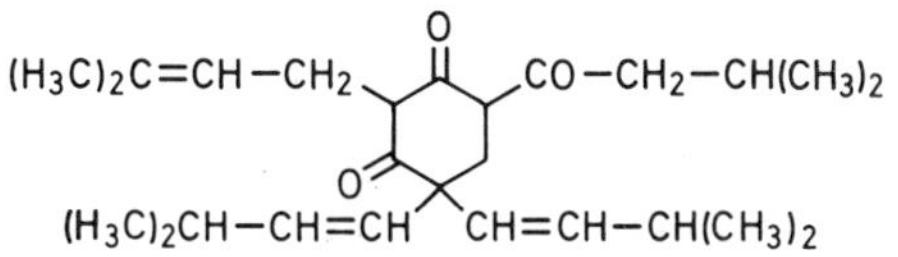

$$(H_3C)_2C=CH-CH_2 \quad CO-CH_2-CH(CH_3)_2$$
$$(H_3C)_2CH-CH=CH \quad CH=CH-CH(CH_3)_2$$

1

$$(H_3C)_2C=CH-CH_2 \quad CO-CH_2-CH(CH_3)_2$$
$$HO \quad CH=CH-CH(CH_3)_2$$

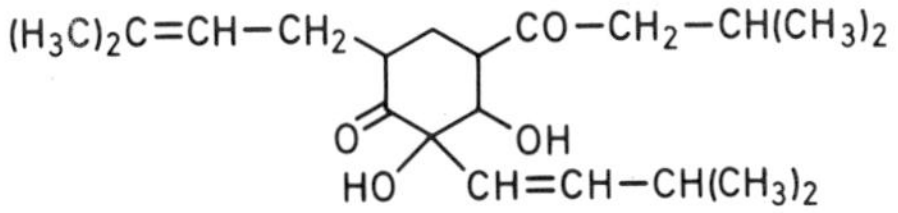

2

Both compounds inhibit growth of various fungi at concentrations of 20–2000 µg/ml.

3.2.4.2.1.3. Carboxylic Acids. Chamic acid **(3)** and chaminic acid **(4)** are responsible for the natural resistance of the yellow or Alaska cedar (*Chamaecaparis nootkatensis*) toward wood-destroying fungi.[901] They are active against these and against *Fusarium* spp. at 100–200 µg/ml.

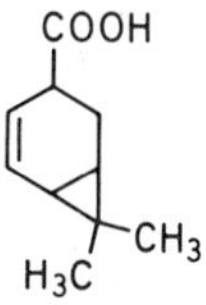

COOH

H₃C   CH₃

3

COOH

H₃C   CH₃

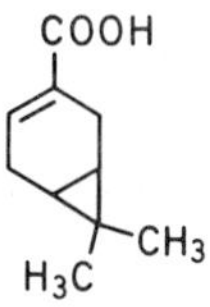

4

3.2.4.2.1.4. Lactones. 5,6-Dihydro-6-methyl-2H-pyran-2-one **(5)** has a fungistatic effect on *Fusarium, Phycomycetes,* and *Nematospora* and is the resistance factor in the berries of the European mountain ash or rowan tree (*Sorbus aucuparia*).[902]

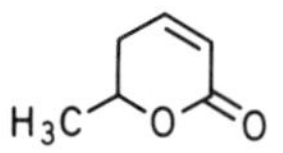

$$H_3C \quad O \quad O$$

5; δ-Hexenolactone, Parasorbic acid

5-Methylene-2(5H)-furanone **(6)** has a remarkable activity against bacteria, fungi, and protozoa (e.g., *Candida albicans*, *Trichophyton* spp., and *Saccharomyces cerevisiae* are inhibited at 3–20 μg/ml). Found in Ranunculaceae (e.g., buttercup, anemones), it polymerizes readily to the less active anemonin.[903]

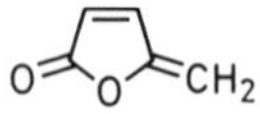

6; Protoanemonin

3.2.4.2.1.5. Phenols. In 1929 Newton and Anderson[904] showed that the wheat variety Khapli, resistant to stem rust, had a higher phenol content than the sensitive variety Little Club. Kargapolova[905] (1936) demonstrated that phenols having hydroxy groups in the *ortho* and *para* positions were more fungitoxic than those with *meta*-hydroxy groups. Furthermore, the concentration of these phenols at the site of infection is crucial for combating the pathogen.[906] 1,2-Benzenediol (catechol, **1**) and 3,4-dihydroxybenzoic acid (protocatechuic acid, **2**) found in the outer skins of onions, are responsible for the resistance of the latter to *Colletotrichum circinans*.[907] Sakai et al. showed that the resistance of potato varieties toward *Phytophthora infestans* is dependent on the *ortho*-diphenol content.[908]

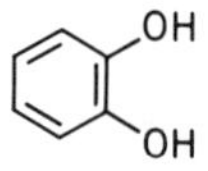

1 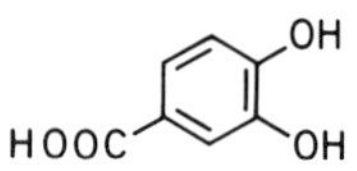   2

Pyrogallol (1,2,3-benzenetriol), isolated from chestnuts, inhibits the growth of *Endothia parasitica* at 40 μg/ml.[909]

The stilbene 5-(2-phenylethenyl)-1,3-benzenediol **(3)** and its monomethyl derivative have been isolated from the Scotch pine (*Pinus silvestris*)[910] and recognized as the resistance factors against wood-destroying fungi, for example, *Merulius lacrymans* (inhibited at 20 μg/ml), *Lentinus lepideus* (at 100 μg/ml), *Polyporus annosus*, and *Coniophora puteana* (at 200 μg/ml).[911]

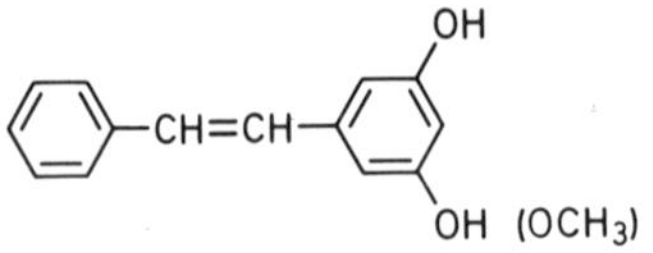

3; Pynosilvin

3.2.4.2.1.6. Quinones. Juglone (5-hydroxy-1,4-naphthalenedione):

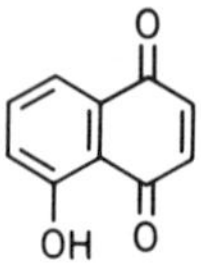

one of the most active fungistatic plant products, has been isolated from the leaves of the walnut.[912] Juglone inhibits the growth of *Botrytis cinerea*, *Uromyces fabae*, and *Aspergillus niger* by 50% at 10 $\mu g/ml$. *Glomerella cingulata* and *Uromyces phaseoli* growth is inhibited by 50% at 5 $\mu g/ml$. The 2-methyl derivative (5-hydroxy-2-methyl-1,4-naphthalenedione) is also widely active against human-pathogenic fungi at 25 $\mu g/ml$.[913]

3.2.4.2.1.7. Tropolones. The tropolone derivatives **1–3** have been isolated from the wood of *Thuja plicata* (canoe cedar).[914] They account for the high resistance of these species toward wood-destroying fungi.[915]

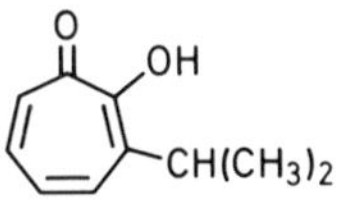

1; α-Thujaplicin
3-Isopropyltropolone

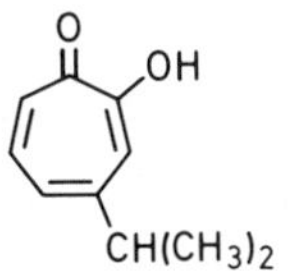

2; β-Thujaplicin
4-Isopropyltropolone

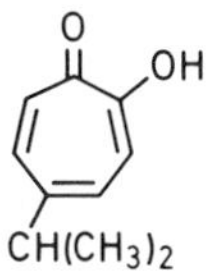

3; γ-Thujaplicin
5-Isopropyltropolone

(Tropolone = 2-hydroxy-2,4,6-cycloheptatrien-1-one.)

3.2.4.2.1.8. Aminoacids. L-Canavanine occurs in concentrations up to 3% in the seeds of many legumes.[916] It inhibits *Puccinia recondita* at 10 $\mu g/ml$ and *Peronospora tabacina* at 25 $\mu g/ml$.[917]

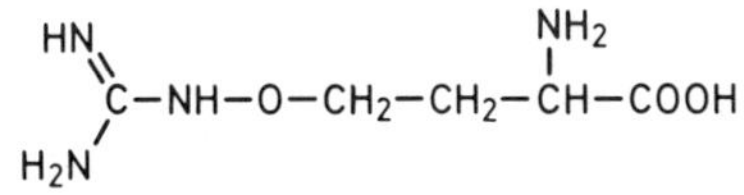

3.2.4.2.1.9. Benzoxazolones. 2(3H)-Benzoxazolones and benzoxazinones, which occur in the form of their glucosides, are responsible for the resistance of rye, corn, and wheat to *Fusarium nivale*.[918] The compounds **1–3**, which may be isolated from seedlings in concentrations up to 100 $\mu g/g$,[919] are fungitoxic *in vitro* to *Fusarium nivale* on artificial culture media at about 500 $\mu g/ml$.

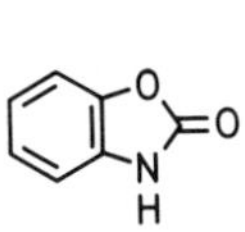

1 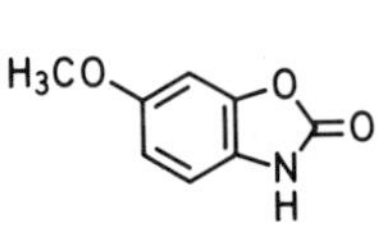   2 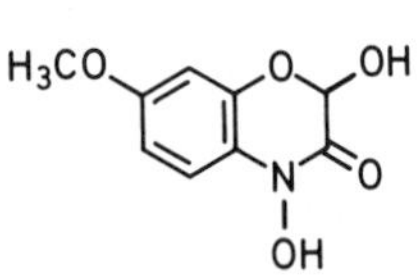   3

3.2.4.2.1.10. Compounds Containing Sulfur. 3-(2-Propenylsulfinyl)-L-alanine (**1**; alliin),[920] which is converted enzymatically into *S*-2-propenyl-2-propene-1-sulfinothioic acid (**2**; allicin):

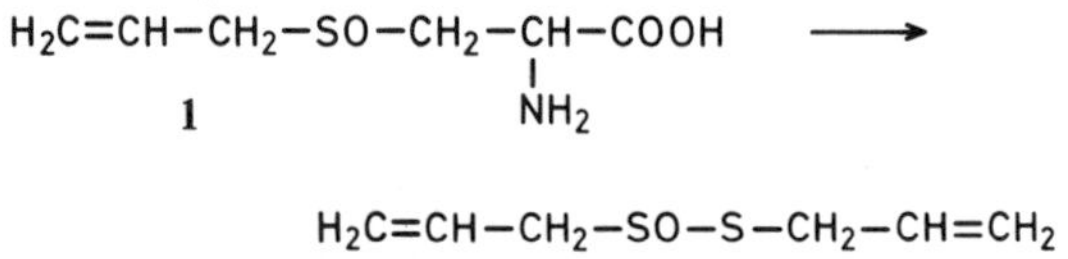

$$H_2C=CH-CH_2-SO-CH_2-CH-COOH \longrightarrow$$

1   NH_2

$$H_2C=CH-CH_2-SO-S-CH_2-CH=CH_2$$

2

is a natural fungicide occurring in garlic at concentrations up to 0.4%.[921]

Isothiocyanates are widespread in cruciferous plants in the form of their gluco-sides.[922] 3-Isothiocyanato-1-propene[923] and 1-isothiocyanato-2-phenylethane,[924] in particular, are found in cabbage roots and account for the resistance of the same toward fungal infection.

3.2.4.2.2. Phytoalexins. Phytoalexins[925] are plant products that are formed as a result of increased enzymatic activity.[926] Factors causing this increase are attack by fungi, bacteria, or viruses; nutrition; temperature; injury; or treatment with certain chemicals (conferred resistance).[927,928] Owing to the nature of the bio-chemical pathways occurring in plants, the phytoalexins come from the following classes of compounds: furanoterpenes, steroids, isocoumarins and coumarins, shikimic acid derivatives, and phenols.

3.2.4.2.2.1. Furans. Wyeronic acid (1) has been isolated from broad bean (*Vicia faba* L.) infected with *Botrytis cinerea* and *Botrytis faber*.[929]

$$H_3C-CH_2-CH=CH-C\equiv C-CO-O-CH=CH-COOH$$

1

The corresponding methyl ester has been found in seedlings of broad bean[930] (see refs. 931, 932 for structure and fungicidal properties).

Ipomeamaron (2) has been known since 1943 and was isolated from sweet potatoes infected with *Ceratostomella fimbrata*[933] (black rot) and *Helicobasidium mompa*[934] (violet root rot). Its production in the stems of sweet potatoes is stim-ulated by treatment with mercury chloride solution.[935]

$$CH_2-CO-CH_2-CH(CH_3)_2$$

2

3.2.4.2.2.2. Phenanthrenes. Bernard's observation[936] in 1909 that orchids in-fected with fungi are protected from further attack was confirmed in 1959 by Gäu-mann and Kern.[937] The latter isolated orchinol (3)[938] from orchids infected with *Rhizoctonia repens*. Orchinol production is also stimulated by injury to the plant and by the influence of temperature.

3; 9,10-Dihydro-5,7-dimethoxy-2-phenanthrenol

See ref. 939 for information on hircinol, also a fungistatic.

Hircinol, 9,10-dihydro-4-methoxy-2,5-phenanthrenediol

3.2.4.2.2.3. Coumarin and Isocoumarin Derivatives. The most widely occurring phytoalexins are derivatives of coumarin and isocoumarin.

Flemichapparin B and C:[940,941]

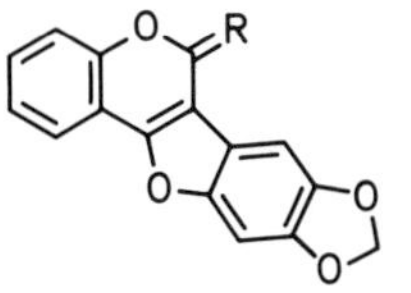

$R = H_2, O$

Pterocarpin:

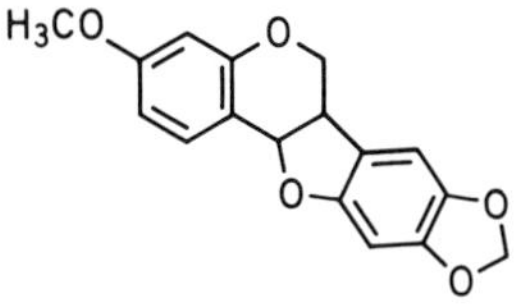

Pisatin:[942]

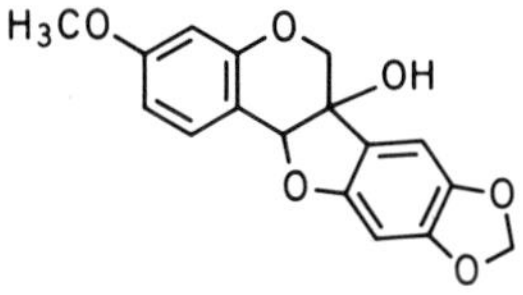

Luteone:[943]

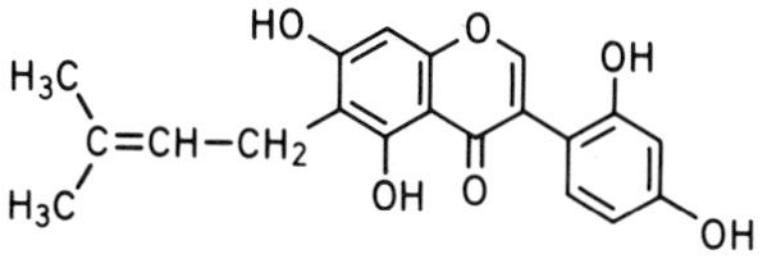

2'-Methoxyphaseollinisoflavan:[944]

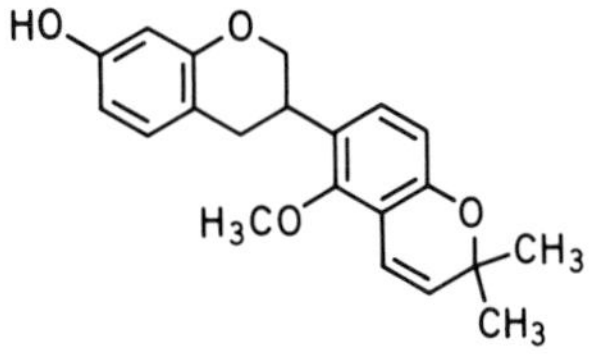

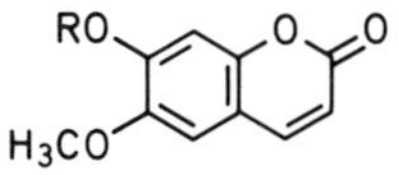

1; R = H, Scopoletin[945]
2; R = $C_6H_{11}O_5$, Scopelin[946]

Umbelliferone:[947]

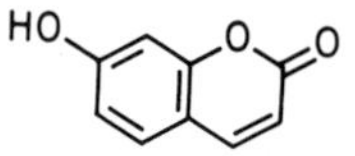

8-Hydroxy-6-methoxy-3-methyl-3,4-dihydro-1H-2-benzopyran-1-one:[948]

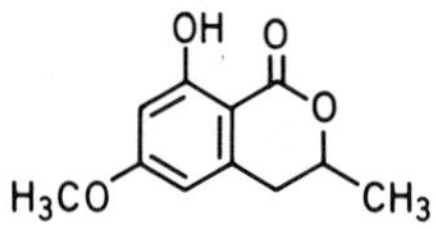

Flemichapparin B and C and Pterocarpin inhibit the growth of *Helminthosporium oryzae*, *Alternaria solani*, and *Curvularia lunata* by 50% at 70 µg/ml. Luteone inhibits the conidial germination of *Cochliobolus miyabeanus* by 50% at 8 µg/ml. 2'-Methoxyphaseollinisoflavan inhibits *Fusarium solani* (Mart.) Appel at 12 µg/ml. Scopoletin is formed in tobacco and tomato plants infected with *Peronospora tabacina*, *Phytophthora infestans*, or viroses.

3.2.4.2.2.4. Phenolcarboxylic Acids. Caffeic acid [1; 3-(3,4-dihydroxyphenyl)-2-propenoic acid] and the ester (2; chlorogenic acid) are widespread in nature. They are held responsible for natural resistance, but it has also been proved that the concentration of these acids in the plants is increased considerably by infection.[949]

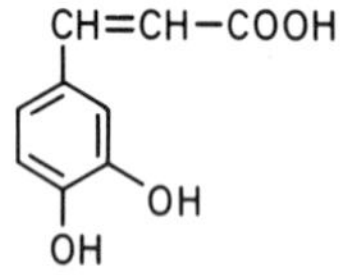

1 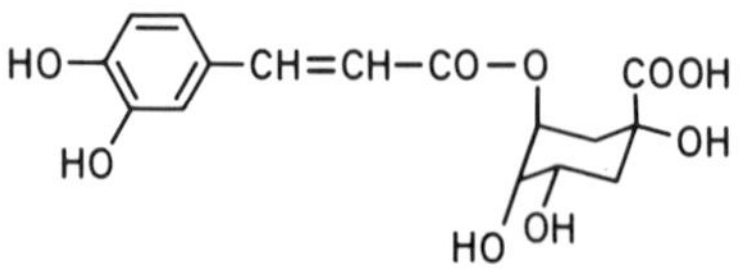   2

In particular, caffeic acid is supposed to induce resistance to *Phytophthora*.[950] Chlorogenic acid forms complexes with aminoacids that are more fungitoxic than the free acid.[951]

3.2.4.2.2.5. Sesquiterpenes. Two phytoalexins are known among the sesquiterpenes: Capsidol (3):[952]

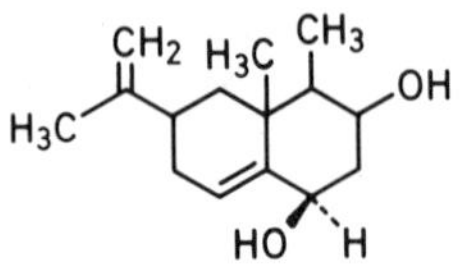

3

which was isolated from sweet pepper and is responsible for the natural resistance of the plant toward fungi, and glutionosone (4),[953] which was isolated from the leaves of *Nicotinia glutinosa* after infection with tobacco mosaic virus.

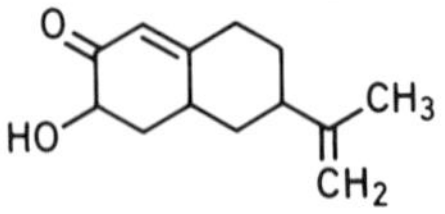

4

3.2.5. SYNTHETIC ORGANIC FUNGICIDES

The history of organic fungicides started in the year 1931 as Tisdale and Williams (Du Pont) registered a patent on disinfectants for the control of fungi and microbes. This patent claimed for this purpose the use of dithiocarbamates of the following structure that were already known as vulcanization promoters.

$$\begin{array}{c} X \\ \diagdown \\ \diagup \\ Y \end{array} N-C \begin{array}{c} S \\ \diagup\diagup \\ \diagdown \\ S-Z \end{array}$$

X = H, Alkyl, Y = H, Alkyl, Aryl

Z = Metal

It took another seven years until Harrington (1941),[955] Anderson (1942),[956] Kincaid (1942),[957] Dunegan (1943),[958] Goldsworthy et al. (1943),[959] and Hamilton and Palmiter (1943)[960] published work on the use and effect of the dithiocarbamates and pointed the way to a general application as fungicides. In England, Martin (1934)[961] had started his studies on sulfur compounds at the same time and discovered the fungicidal properties of the dithiocarbamates independently. Günzler, Heckmanns, and Urbschat (I. G. Farbenindustrie)[962] registered a patent in 1934 that claimed the use of tetramethylthiuram disulfide for control of fruit scab (TMTD was introduced in 1937 for this indication).

Simultaneously, the fungicidal properties of chloranil (tetrachloro-1,4-benzo-quinone), also used as a vulcanization promoter, were discovered and exploited in seed dressings. The quinones had been discovered.[963]

In the 1950s the first fungicidal heterocyclic compounds, the imidazolines (glyodins), became known through the studies of Wellman and McCallan;[964] Kittleson[965] found the perhalogenalkylmercaptoamides (*captan*), and the guanidine derivatives (dodines) were introduced by American Cyanamid.[966]

The organic fungicides are researched intensively in industry today, particularly as a result of the residue situation with regard to inorganic and organometallic fungicides. The targets are less toxic, nonpersistent, highly active fungicides that, above all, permit curative and eradicative control. The systemic fungicides are a promising new development that makes a genuine chemotherapy of plant diseases possible.

3.2.5.1. Sulfur-Containing Organic Compounds

Although countless sulfur compounds have been tested as fungicides[967-971] (such as mercaptans, thiophenols, thioethers, sulfones, thioamides, thiosulfonates), only the dithiocarbamates and perhalogenalkylmercaptides have found worldwide application as fungicides.

3.2.5.1.1. Mercaptans and Thioethers. Mercaptans react with disulfides by a formal transalkylation:[972]

$$R^1-SH \ + \ R^2-S-S-R^2 \ \rightleftharpoons$$

$$R^1-S-S-R^2 \ + \ R^1-SH$$

and with other mercaptans to form disulfides:

$$R^1-SH \ + \ R^2-SH \ + \ Electron \quad acceptor$$

$$\rightleftharpoons \ R^2-S-S-R^1 \ + \ Acceptor-H_2$$

This reactivity explains the wide biological activity of these compounds. In biological systems disulfide enzymes and coenzymes can be inhibited. Complex reactions with electron-transferring, metal-containing enzymes can cause inhibition of cell growth and cell respiration.[973]

The high volatility of the low-molecular-weight mercaptans, thioethers, and disulfides; their intensive, unpleasant odors; and their toxic properties have prevented the use of such compounds as fungicides in spite of numerous attempts.[974]

In contrast, the heterocyclic mercaptans are highly fungicidal but lack these negative properties. 1-Hydroxy-2(1H)-pyridinethione is used as a fungicide in soaps and shampoos but also has a wide spectrum of activity and systemic properties against numerous foliar diseases such as apple scab and peach leaf curl (*Taphrina deformans*).[975]

Omadine® (Olin Mathieson Chemical Corp.)

Omadine® is prepared by nucleophilic substitution of bromine in 2-bromopyridine-N-oxide with sodium hydrosulfide or thiourea.

(2-Hydroxyethyl)ammonium 2-benzothiazolethiolide (1)[976,977] and zinc bis(5-chloro-2-benzothiazolethiolide) (2):

1; Vancide 20-S® (Vanderbilt)

2; LD_{50}: 3000 mg/kg rat, oral, acute

are also vulcanization promoters. Thiazoles and benzothiazoles are reported to be in equilibrium with the corresponding vinyl isothiocyanates:[978]

and so their fungicidal properties may be explained on the basis of an isothiocyanate structure.

2,2'-Thiobis(4-chlorophenol) (3) is a fungistatic with good activity against apple scab. However it is reported to cause a rough fruit finish.[979,980]

3

The corresponding tetrachloro derivative, 2,2'-thiobis(4,6-dichlorophenol) (**4**), has been registered as a rice fungicide for control of *Pyricularia oryzae*.[981]

4

It may be prepared by reaction of the phenol with sulfur dichloride in the presence of aluminum(III) chloride.[982]

1-(Phenylthio)-methanesulfonamide (**5**) is a systemic, protective, and curative fungicide active against *Phytophthora*, *Peronospora*, and *Venturia* spp. It may be obtained by reaction of thiophenol with 1-chloromethanesulfonamide in aqueous sodium hydroxide.[983]

$$H_5C_6-S-CH_2-SO_2-NH_2$$

5; LD$_{50}$: 800 mg/kg rat, oral, acute

3.2.5.1.2. Sulfoxides, Sulfones, and Sulfonates. Sulfur compounds in which the sulfur has a higher oxidation state are mainly known as bactericides (sulfonamides). A few sulfoxides, sulfones, sulfonates, and thiolsulfonates have fungicidal activity. Thiolsulfinates and thiolsulfonates have remarkable activity as fungicides (particularly against *Fusarium culmorum*).[984] The activity is supposed to depend on reaction with cysteine and other thiols present in the cell.[985]

1,1-Sulfinylbis(1,2,2-trichloroethane) (**6**):

$$Cl_2CH-\underset{\underset{Cl}{|}}{CH}-SO-\underset{\underset{Cl}{|}}{CH}-CHCl_2$$

6; Chemagro 4497 (Chemagro Corp./Mobay Ag. Chem. Div.);
LD$_{50}$: 325 mg/kg rat, oral, acute

is active as a seed dressing and soil fungicide for control of *Pythium*, *Rhizoctonia*, *Helminthosporium*, and *Ustilago* spp.[986]

According to a technical bulletin from Lamirsa (Laboratories Miret S.A.), sulfonylbis(tribromomethane) (**7**) and sulfonylbis(trichloromethane) (**8**) have a broad spectrum of activity as bactericides, fungicides, acaricides, algicides, insecticides, and nematicides. They are also used to prevent formation of slimes in paper mills and cooling systems and as additives for plastics and dyes. As experimental products in crop protection they are used in concentrations of 5–50 ppm.

$$Br_3C-SO_2-CBr_3 \qquad\qquad Cl_3C-SO_2-CCl_3$$

7; LD$_{50}$: 3.6 g/kg rat, oral, acute   **8**; LD$_{50}$: 3.8 g/kg rat, oral, acute

The following products are also known.

1. 1,1'-Sulfonylbisethane.[988]

$$H_2C=CH-SO_2-CH=CH_2$$

Fungicide and bactericide (Stauffer Chemical Corp.)

2. 1,2-Dichloro-1-(methylsulfonyl)-ethene.[989]

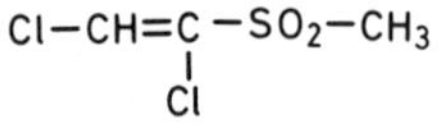

$$Cl-CH=C-SO_2-CH_3$$
$$|$$
$$Cl$$

Fungicide D-113 (Chemagro Corp./Mobay Ag. Chem. Div.)

3. 1,1'-[1,2-Ethenediylbis(sulfonyl)]bispropane.[990]

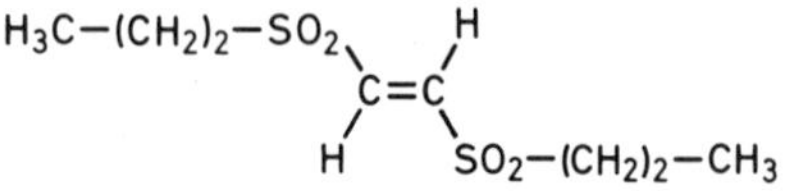

Chemagro 1843 (Chemagro Corp./Mobay)

Seed dressing and soil fungicide for cereals against *Fusarium*, *Pythium*, *Ustilago*, *Tilletia* spp., and other seed diseases.

4. 1,2,3,4,7,7-Hexachloro-5-ethenylsulfonyl-bicyclo[2.2.1]hept-2-ene[991]

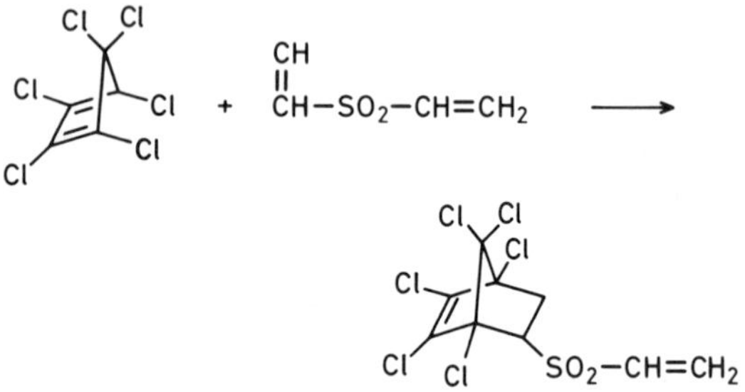

active against *Phytophthora infestans* and *Sclerotium rolfsii*.

5. 2-Chloro-3-[(4-methylphenyl)sulfonyl]propanenitrile.

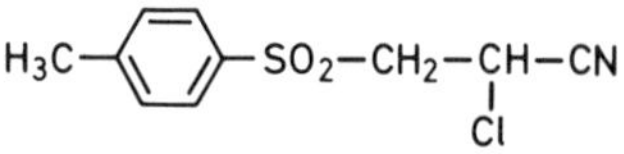

Monsanto CP 30249; Fungicide and bacteriostatic

6. 4-Chlorophenyl ethenesulfonate.[992]

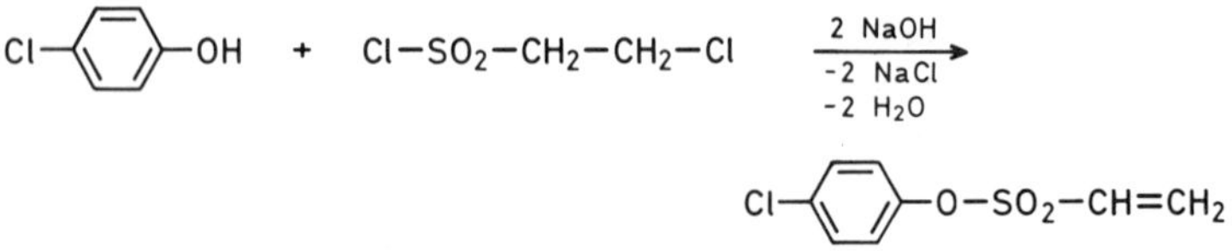

Used in combination with *aldrin* in the ant killer Basformid®. Kills the fungi that the leaf-cutting ants breed as a food supply.

7. Trichloromethyl sulfonothioates.[984]

$$R-SO_2-S-CCl_3$$

$$R = 4-Cl-, 2/4-NO_2- C_6H_4 ; H_5C_6-O-CH_2-CH_2$$

High activity against *Fusarium culmorum*.

S-(2-Hydroxypropyl) methanesulfonothioate **(1)** is a liquid, water-soluble bactericide and fungicide active against bacterial wilts of beans, peas, tomatoes, and seed potatoes[993] and against fungal diseases in mushroom growing.[994]

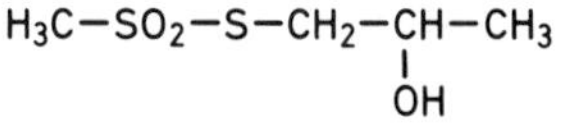

$$H_3C-SO_2-S-CH_2-\underset{\underset{\displaystyle OH}{|}}{CH}-CH_3$$

1; HPTMTS (Buckman Labs. Inc.)

3.2.5.1.3. Thiocarbonate Derivatives. Sulfur analogues of carbonate derivatives are among the most important fungicides. Carbon disulfide, the anhydride of carbonodithioic acid, was used as early as 1872 as an insecticide for control of the vine louse and also demonstrated a fungicidal effect.[995] Owing to its very low flash point, volatility, and toxicity, it did not find wide application.

Starting from the products of reaction between carbon disulfide and alcohols, sulfides and amines, fungicidal thiocarbonate derivatives were developed, such as potassium carbonotrithionate, which decomposes in the presence of water and carbon dioxide to release carbon disulfide.

$$K_2S \ + \ CS_2 \ \longrightarrow \ K_2CS_3$$

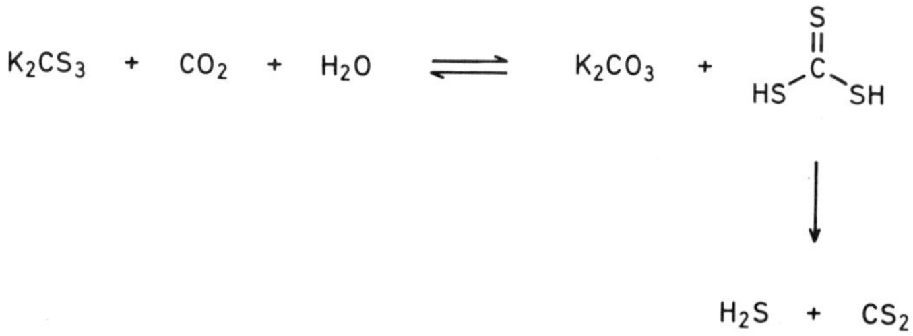

$$K_2CS_3 \ + \ CO_2 \ + \ H_2O \ \rightleftharpoons \ K_2CO_3 \ + \ \underset{HS}{\overset{\displaystyle S}{\underset{\displaystyle \|}{C}}}\!\!-\!SH$$

$$\downarrow$$

$$H_2S \ + \ CS_2$$

Further derivatives are: thiolocarbamates **(1)**, thionocarbamates **(2)**, dithiocarbamates (carbamodithioates) **(3)**, xanthogenates (carbonodithioates) **(4)**, and bis[dithiocarbamates] **(5)**.

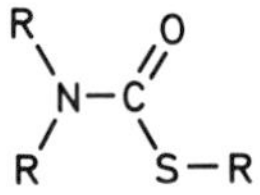

1 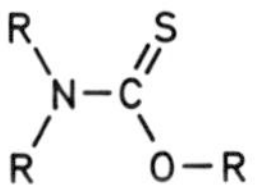 **2** 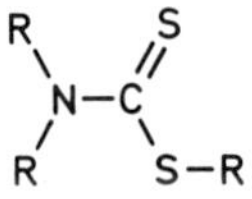 **3**

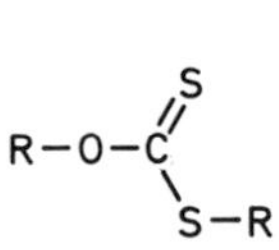

4 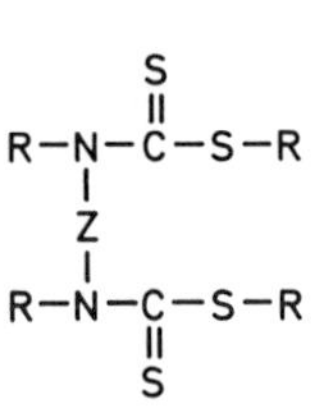 **5**

TABLE 18. Fungicidal Activity of Thiocarbonate Derivatives

Compound	ED_{100} (γ)					Reference
	Bot.	Pen.	Asp.	Rhiz.	Clad	
$(H_3C)_2N-C{\overset{O}{\underset{S-CH_2-COOH}{}}}$	500	500	500	500	500	996
$(H_3C)_2N-C{\overset{S}{\underset{O-CH_2-COOH}{}}}$	500	500	500	500	500	996
$(H_3C)_2N-C{\overset{S}{\underset{S-CH_2-COOH}{}}}$	500	500	500	500	500	996
$H_3CO-C{\overset{S}{\underset{SK}{}}}$	10	10	10	100	—	997
$H_3C-NH-C{\overset{S}{\underset{SNa}{}}}$	10	10	50	200	—	997
$(H_3C)_2N-C{\overset{S}{\underset{SNa}{}}}$	0.2	0.5	20	2	—	997
$\left(-CH_2-NH-C{\overset{S}{\underset{SNa}{}}}\right)_2$	0.1	0.1	0.5	20	—	997

Whereas the thiolocarbamates, thionocarbamates, and xanthates are only weakly fungicidal, the dithiocarbamates and bisdithiocarbamates are, depending on the substitution, highly active fungicides, as may be seen from Table 18.

The fungicidal properties of thioureas and thiocarbazone derivatives have been investigated in depth.[998,999] They have only a moderate fungicidal activity, which is reported to stem from complex formation with either trace elements or the metals of enzymes.[1000,1001]

3.2.5.1.3.1. Dithiocarbamates and Thiuramsulfides. The properties of the dithiocarbamates are changed by replacement of the sulfur hydrogen with metals or other substituents. Extending the alkyl chain leads to a loss of activity. The optimum is reached with dithiocarbamates having two methyl groups on the nitrogen. Chain extension beyond two carbons causes a 20–200-fold drop in activity. Monomethyl derivatives are less active, as are derivatives in which the nitrogen is replaced by alkyl, oxygen, or sulfur (see Table 19). Depending on the metal cation, dimethyl-dithiocarbamate salts are more or less unstable at pH values below 6,[1002] decomposing back into their corresponding starting materials.[1003] Like the mercaptans,

TABLE 19. Dithiocarbamates, Structure–Activity Relationships as a Function of Substitution[997]

Compound	ED_{100} (γ)			
	Bot.	Pen.	Asp.	Rhiz.
$H_3C-C(=S)SNa$	5	5	20	50
$H_3C-NH-C(=S)SNa$	10	10	50	200
$H_3C-O-C(=S)SK$	10	10	10	100
$(H_3C)_2N-C(=S)SNa$	0.2	0.5	20	2
$H_5C_2O-C(=S)SK$	10	10	20	200
$H_5C_2S-C(=S)SK$	500	500	2000	2000
$(H_5C_2)_2N-C(=S)SNa$	1	2	10	5
$(H_7C_3)_2N-C(=S)SNa$	200	200	200	1000
$(H_9C_4)_2N-C(=S)SNa$	1000	2000	2000	1000

they are easily oxidized to the corresponding disulfides. The redox potential dialkyldithiocarbamate $\rightleftharpoons$ tetraalkylthiuramdisulfide decreases with increasing length of the alkyl chain.[1004,1005] The heavy metal salts react with mercaptans by exchange.[1006]

$$(H_3C)_2N-C \overset{S}{\underset{S-Zn-S}{\Big/}} C-N(CH_3)_2$$

$\Big\updownarrow$ HS—R

$$(H_3C)_2N-C \overset{S}{\underset{S-Zn-S-R}{\Big/}} \quad + \quad \overset{S}{\underset{HS}{\Big/}} C-N(CH_3)_2$$

$\Big\updownarrow$ HS—R

$$2\ (H_3C)_2N-C \overset{S}{\underset{SH}{\Big/}} \quad + \quad R-S-Zn-S-R$$

This reaction is fundamental to the mechanism of action of the dithiocarbamates. Thus the dithiocarbamates are, like *captan, folpet*, (acetato-*O*)phenylmercury, and (acetoxy)triphenylstannane, inhibitors of energy production, that is, of ATP production in the cell.[1007] They react with the mercapto groups of the coenzyme lipoic acid dehydrogenase and interfere in the ADP–ATP cycle via inhibition of pyruvate oxidation. Dithiocarbamates also complex with metal-containing enzymes and thus act as inhibitors of countless enzymes such as glutamic acid dehydrogenase, aminoxidase, and polyphenoloxidase.[1008] The alkylenebis[dithiocarbamates] (*maneb, zineb*, etc.) interact similarly with these enzymes. In addition, they can form diisothiocyanates that react with amino and mercapto groups and lead to formation of polymers.[1009]

Sodium methylcarbamodithioate[1010] (**1**) is used as a soil fungicide and nematicide. A herbicidal activity at higher concentration also permits use as a short-term herbicide.[1011] *Metam* is manufactured from methylamine, carbon disulfide, and sodium hydroxide. In dilute aqueous solution, particularly in the presence of acid or heavy metal ions, it decomposes to methyl isothiocyanate.[1012]

$$H_3C-NH_2 \quad + \quad CS_2 \quad + \quad NaOH \quad + \quad H_2O \quad \longrightarrow$$

$$H_3C-NH-C \overset{S}{\underset{SNa}{\Big/}} \quad \cdot \ 2\ H_2O$$

1; *Metam*; LD_{50}: 820 mg/kg rat, oral, acute

$$H_3C-NH-C\underset{SH}{\overset{S}{\big<}} \quad \rightleftharpoons \quad H_3C-N=C=S \ + \ H_2S$$

Sodium dimethylcarbamodithioate (2) is made by the analogous procedure from dimethylamine, carbon disulfide, and sodium hydroxide in aqueous solution. Owing to their water solubility, both the sodium salt and the ammonium salt (Diram A) are only suitable for seed-dressing and soil-treatment purposes.

$$(H_3C)_2N-C\underset{SNa}{\overset{S}{\big<}}$$

2; P 666, Na DMDT, Diram® (in Vancide 51 in combination with sodium 2-benzothiazolethi-olide);
LD_{50}: 2500 mg/kg rat, oral, acute

Bis(dimethylcarbamodithioato-*S,S'*)zinc (3) precipitates as an insoluble salt when an aqueous solution of sodium dimethylcarbamodithioate is mixed with a water-soluble zinc salt. This product has been used worldwide as an important fungicide since the 1930s. *Ziram* finds use as a protective fungicide in fruit, wine grape, and vegetable growing.

$$\left[(H_3C)_2N-C\underset{S}{\overset{S}{\big<}}\right]_2 Zn$$

3; *Ziram*; LD_{50}: 1400 mg/kg rat, oral, acute

Tris(dimethylcarbamodithioato-*S,S'*)iron (4) is used in fruit, vegetables, citrus, and tobacco, particularly in the United States, and, like *ziram*, is active against scab, rust, *Botrytis*, and damping-off diseases.

$$\left[(H_3C)_2N-C\underset{S}{\overset{S}{\big<}}\right]_3 Fe$$

4; *Ferbam* (Du Pont 1931);
LD_{50}: 1000 mg/kg rat, oral, acute

Tris(dimethylcarbamodithioato-*S,S'*)-arsenic, a development product of Ishara Agricultural Chem. Co. Ltd.,[1013] controls downy mildew on strawberries and cur-curbits at an application rate of 2.5 kg (40% a.i.)/ha.[1013]

$$\left[(H_3C)_2N-C\underset{S}{\overset{S}{\big<}}\right]_3 As$$

Bis(dimethylcarbamodithioato-*S,S'*)nickel (5) is a bactericide highly active against *Xanthomonas oryzae*.

$$\left[(H_3C)_2N-C\underset{S}{\overset{S}{\diagup}} \right]_2 Ni$$

5: Sankel® (Sankyo K. K./Mikaso, Japan);
LD$_{50}$: 5200 mg/kg rat, oral, acute

See ref. 1014 for information on the fungicidal activity of 4-morpholinylmethyl dimethylcarbamodithioate (*carbamorph*; Murphy Chem. Co. Ltd.).

$$(H_3C)_2N-C\underset{S-CH_2-N\diagdown O}{\overset{S}{\diagup}}$$

Replacement of a methyl group by hydroxymethyl in dimethylcarbamodithioate leads to potassium (*N*-hydroxymethyl-*N*-methyl) carbamodithioate.[1015]

$$H_3C-NH-C\underset{SK}{\overset{S}{\diagup}} \quad + \quad CH_2O \quad + \quad H_3C-NH_2 \quad \longrightarrow \quad \underset{HO-CH_2}{\overset{H_3C}{\diagdown}}N-C\underset{SK}{\overset{S}{\diagup}}$$

Bunema® (Buckman Labs. Inc.)[1016]

Bunema® is sold as a foliar and soil fungicide, bactericide, and nematicide.

Thiuramdisulfides are available by mild oxidation of an alkali metal salt of dimethylcarbamodithioic acid with oxygen, chlorine, or air.

Probably the most important fungicide of this class, tetramethylthiuramdisulfide (*Chem. Abstr.*: tetramethyl-thioperoxydicarbonic diamide) (6), is the first metal-free organic compound to be described as a foliar fungicide.[1017] Its fungicidal and bactericidal properties were discovered by Tisdale, Flenner, and Williams.[1018,1019]

$$(H_3C)_2N-C\underset{S-S}{\overset{S\quad S}{\diagup \quad \diagdown}}C-N(CH_3)_2$$

6; *TMTD*, *Thiram*, Arasan®, Tersan® (Du Pont), Nomersam® (ICI Plant Protection Ltd.), Pomarsol® forte (Bayer AG);
LD$_{50}$: 750 mg/kg rat, oral, acute

Thiram is a widely active seed dressing and soil treatment for control of damping-off diseases. As a foliar fungicide it gives protective control of scab, shot-hole, and *Botrytis cinerea*. Thiram is also used in combination products in tropical crops, rice, and coffee plantations. In combination with *benquinox* it is marketed as ®Ceredon T, a seed dressing for cotton, beet, peas, and beans.

In contrast to carbamodithioic acids derived from primary amines (*metam*), those derived from diamines (the bis[carbamodithioic acids]) are more stable. The product of reaction between ethylene diamine and carbon disulfide in the presence of sodium hydroxide solution,[1020] disodium 1,2-ethanediylbis(carbamodithioate) (7), finds application particularly as a soil fungicide owing to its water-solubility and stability.

7; *Nabam*, Dithiane® D 14 (Rohm and Haas), Parzate® (Du Pont), DSE;
LD_{50}: 395 mg/kg rat, oral, acute

The corresponding zinc $(8)^{1021}$ and manganese $(9)^{1022,1023}$ salts, only slightly soluble in water, are important foliar fungicides.

8; M = Zn. *Zineb*, Dithiane Z 78® (Rohm and Haas), Parzate® Zineb (Du Pont), Lonacol® (Bayer AG);
LD_{50}: >5200 mg/kg rat, oral, acute
9; M = Mn. *Maneb*, Manzate® (Du Pont), Dithane M-22® (Rohm and Haas), Maneb-Spritz-pulver (Bayer AG);
LD_{50}: 6750 mg/kg rat, oral, acute

Both salts are obtained by addition of zinc or manganese(II) sulfate to *nabam*. Zineb is used against *Phytophthora infestans* and other downy mildews, *Alternaria* spp., rusts, and scabs (it is one of the most widely used fungicides of this group); *maneb* is used on a large scale as a foliar fungicide for control of *Phytophthora infestans*, *Peronospora* of hops and vine, *Pseudopeziza tracheiphila* (red fire disease of grapevine), scab, rust, and downy mildew species in fruit, vegetables, maize, rice, cereals, tobacco, grapevine, and ornamentals, as well as as a seed dressing and soil treatment. Combinations of *maneb* and *zineb* (*mancozeb*, Dithane M-45®) have an even better stability and activity. *Mancozeb* in combination with 2,4-dinitro-phenyl esters (Dikar®, Rohm and Haas) is marketed as a foliar fungicide with an additional acaricidal action.

Zineb and *maneb* form stable addition compounds with ammonia and amines.[1024–1027] *Zineb*-ammonia adducts (Polyram®; BASF and FMC Corp., Niagara Chem. Div.) having a relatively long residual effect are available on the market. *N,N'*-(1,1-Methanediyl)bis({[1,2-ethanediylbis(carbamodithioato)](2-)}-manganese/zinc) (1), the product of reaction between *maneb* or *zineb* and for-maldehyde, is used in hops and grape,[1028] whereas the product from reaction of 1,2-diaminopropane with carbon disulfide–sodium hydroxide and subsequent pre-cipitation with zinc sulfate, {[(1-methyl-1,2-ethanediyl)bis(carbamodithioato)]-(2-)}zinc (2), is also used in orchard fruit and vegetables.[1029] *Propineb*, which also finds application in tropical crops, has a higher fungicidal activity than *maneb* against *Phytophthora infestans*, *Plasmopara viticola*, *Psdudoperonospora humili*, *Helminthosporium oryzae*, and *Peronospora tabacina*. Combinations of the zinc

1; Trizinoc®, Trimanoc®;
LD$_{50}$: >10,000 mg/kg rat, oral, acute

2; *Propineb*, Antracol® (Bayer AG);
LD$_{50}$: 8500 mg/kg rat, oral, acute

and manganese salts of 1,2-propylene[bis(carbamodithioic acid)], prepared by coprecipitation, are also more active than the corresponding salt combinations based on 1,2-diaminoethane.[1030]

Oxidation of *zineb, maneb,* or *propineb* leads to the thiuramsulfides and disulfides. The thiurammonosulfides of monoalkylcarbamodithioic acids are more active than the disulfides, in contrast to the thiuram derivatives of dialkylcarbamodithioic acids.[1031,1032]

Polyethylenethiuramsulfides and disulfides are known under the tradename Thioneb® (Du Pont) and as components of the combination products Polyram® (BASF) and Thiumet® (BASF).

Tetrahydro-1,3,6-thiadiazepine-2,7-dithione **(3)** has high fungicidal activity.[1033] *Tecoram* **(4)** was introduced onto the market as a foliar fungicide for orchard fruit.[1034]

3; Ethylene thiuram monosulfide, Vegelta®

4; *Tecoram*, (N.V. Aagrunol, Neth., 1960);
LD$_{50}$: 300 mg/kg rat, oral, acute

3.2.5.1.3.2. Tetrahydrothiadiazinethiones. Tetrahydro-2H-1,3,5-thiadiazine-2-thiones decompose in the soil to methyl isothiocyanate, formaldehyde, and methylamine. They are prepared by reaction of the ammonium salts of carbamodithioic acids with aldehydes. Thus tetrahydro-3,5-dimethyl-2H-1,3,5-thiadiazine-2-thione is obtained by reaction of the methylammonium salt of methylcarbamodithioic acid with formaldehyde.[1035] *Dazomet* is a soil treatment with a wide spectrum of activity that includes *Pythium* and *Rhizoctonia* spp. and also has nematicidal properties.

Dazomet, DMTT, Mylone® (Stauffer Chem. Co.), Basamid® (BASF);
LD$_{50}$: 650 mg/kg rat, oral, acute

The sodium salt of tetrahydro-3-methyl-5-carboxymethyl-2H-1,3,5-thiadiazine-2-thione (thiodiazinthione) has a nematicidal effect and high activity against soil-borne fungi. It is prepared[1036] from sodium dimethylcarbamodithioate, glycine, and formaldehyde in an analogous synthesis to that of *dazomet*.

Thiodiazinthione (Bayer AG, 1959);
LD_{50}: 1000 mg/kg rat, oral, acute

3,3'-(1,2-Ethanediyl)bis[tetrahydro-4,6-dimethyl-2H-1,3,5-thiadiazine-2-thione] is marketed as a fungicide for control of apple scab, blue mold of tobacco, and late blight of potato.[1037]

Milneb, Banlate®, Du Pont 328;
LD_{50}: 5000 mg/kg rat, oral, acute

3.2.5.1.3.3. Other Thiocarbonate Derivatives. Other thiocarbonates that have been reported recently are:

1. *O*-Benzyl *O*-(4-acetyl-phenyl) thiocarbonate[1038]

highly active against *Pyricularia oryzae* and *Pellicularia sasakii*.

2. *S*-Ethyl [3-(dimethylamino)propyl]carbamothioate hydrochloride[1039]

specifically active against Phycomycetes (*Phythium* and *Phytophthora* spp.)— can be applied as a foliar or soil treatment.

3. Butyl [4-(1,1-dimethylethyl)phenyl]methyl 3-pyridinylcarbonimidodithioate[1040]

$$S-C_4H_9$$

Buthiobate, Denmert® (Sumitomo Chem. Co.)

has high activity against powdery mildews on fruit and vegetables.

4. 1,4,5,6-Tetrahydro-2-(octylthio)pyrimidine hydrobromide[1041]

has high systemic activity against mildews and rusts. See Table 26, Section 3.2.5.6.3 for derivatives of thiourea [*o*-phenylene(bis)thiophanates].

3.2.5.1.3.4. Thiocyanates and Isothiocyanates. Of all the derivatives of thiocyanic acid only the aromatic thiocyanates have any importance as fungicides. For example:

1. 2,4-Dinitrophenylthiocyanate[1042]

Nirit® (Hoechst, 1945)

active against *Plasmopara viticola* on grapevines and scab and rust on orchard fruit (LD_{50}: 3100 mg/kg rat, oral, acute).

2. Ethyl (4-thiocyanatophenyl)carbonate[1043,1044]

$$R = H, 3-CH_3, 2-CH(CH_3)_2$$

Geigy; broad fungicidal activity

3. *x*-(2-Nitro-1-propenyl)phenyl thiocyanate[1045]

Suchirosaido® (Nihon, Kayaku K.K./Japan)

used against powdery mildews on ornamentals.

4. (3,4-Dichlorophenyl)methyl thiocyanate[1046]

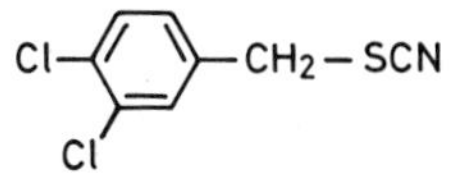

$$Cl-\langle \rangle-CH_2-SCN$$
$$Cl$$

Ishara Noyaku K.K./Japan

5. (Chloro-*p*-phenylene)dimethylene bisthiocyanate[1047]

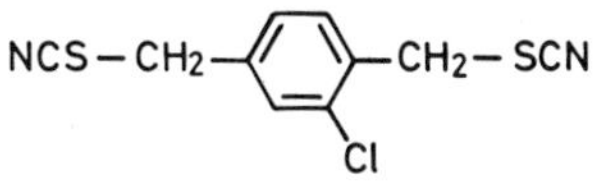

$$NCS-CH_2-\langle \rangle-CH_2-SCN$$
$$Cl$$

DAC 1200 (Diamond Alkali Co.)

(2-Benzothiazolylthio)methyl thiocyanate **(1)**[1048] is a broad-spectrum hetero-cyclic thiocyanate introduced as a soil fungicide and seed dressing for control of root diseases, *Helminthosporium*, *Rhizoctonia*, *Pythium* spp., and other pathogens. It also has bactericidal and insecticidal properties.

The corresponding sulfoxide, (2-benzothiazolylsulfinyl)methyl thiocyanate **(2)**,[1049] is recommended for control of *Gloeotinia temulenta* (blind seed disease of grasses).[1050]

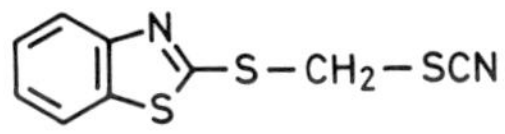

$$\langle \rangle S-CH_2-SCN \qquad \langle \rangle SO-CH_2-SCN$$

1; TCMTB, Busan®-72 (Buckman Labs. Inc., 1970); LD_{50}: 1590 mg/kg rat, oral, acute

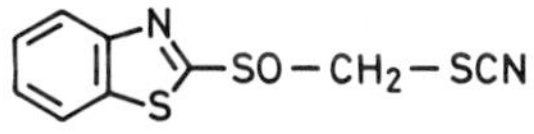

2; TCMTOB

After studies on the mechanisms of action of *N*-monoalkylcarbamodithio-ates[1009,1012] had shown that isothiocyanates are formed as intermediate metab-olites, the latter class was investigated thoroughly.[1051,1052]

Methylisothiocyanate,

$$H_3C-N=C=S$$

Vorlex® (NOR-AM Agricultural Products Inc.), Trapex® (Schering AG); LD_{50}: 97 mg/kg rat, oral, acute

as a 20% formulation in a chlorinated hydrocarbon, is recommended as a soil treat-ment against fungi, nematodes, and—above a certain application rate—against soil pests and weeds.

3.2.5.1.4. Perhaloalkylmercaptan Derivatives. Trichloromethanesulfenyl chloride, prepared by reaction of carbon disulfide with chlorine in the presence of iodine,[1053]

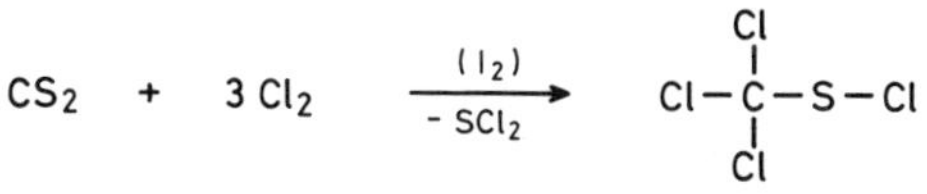

$$CS_2 + 3Cl_2 \xrightarrow[-SCl_2]{(I_2)} Cl-\underset{\underset{Cl}{|}}{\overset{\overset{Cl}{|}}{C}}-S-Cl$$

reacts with such nucleophiles as amines,[1054] amides,[1055,1056] imides,[1055,1057] alco-hols and phenols,[1054] thiols and thiocarbonates,[1058] sulfones,[1059] and thiosulfonates to give trichloromethylmercapto derivatives that have wide fungicidal activity for a broad range of nucleophiles.

The synthesis of fluorodichloromethanesulfenyl chloride, also a starting material for many important fungicides, was achieved by treatment of the trichloro derivative with anhydrous hydrofluoric acid.[1060]

$$Cl-\overset{\underset{\displaystyle Cl}{|}}{\underset{\underset{\displaystyle Cl}{|}}{C}}-S-Cl \quad \xrightarrow[- HCl]{HF} \quad F-\overset{\underset{\displaystyle Cl}{|}}{\underset{\underset{\displaystyle Cl}{|}}{C}}-S-Cl$$

Perhaloalkylmercapto compounds react at the perhaloalkyl group with nucleophiles, particularly with thiols, to give dihalocarbonic acid derivatives that split out carbon disulfide or thiophosgene.[1061,1062]

$$R^1-S-\overset{\underset{\displaystyle Cl}{|}}{\underset{\underset{\displaystyle Cl}{|}}{C}}-Cl \quad \xrightarrow[- HCl]{R^2-SH} \quad R^1-S-\overset{\underset{\displaystyle Cl}{|}}{\underset{\underset{\displaystyle Cl}{|}}{C}}-S-R^2 \quad \xrightarrow[\substack{- CS_2 \\ - CSCl_2}]{R^3-SH}$$

$$R^1H \quad + \quad R^2-S-R^3 \quad + \quad R^3-S-S-R^3 \quad + \quad R^2-S-S-R^3$$

This reaction is the basis of the mechanism of action of these perhaloalkylmercapto derivatives. *Captan* reacts in this manner with the thiols in the cell. The addition of thiols to *captan*-treated suspensions of fungal spores detoxifies *captan* before the cells die off.[1062] However, apart from this nonspecific reaction, the mechanism of action of *captan* probably involves a specific inhibition of energy production in the cell.[1007]

In addition to the reactivity of the perhaloalkylmercapto group,[1063] a sufficient degree of lipophilicity of the compound is necessary for fungicidal activity.[1064]

3.2.5.1.4.1. Perhaloalkylmercaptoimides. Economically important fungicides that play a particularly significant role in fruit growing were discovered through the work of Kettleson[1055] and Daines[1065] in the class of *N*-haloalkylmercaptoimides. For example, 3*a*,4,7,7*a*-tetrahydro-2-[(trichloromethyl)thio]-1H-isoindole-1,3(2H)-dione (1)[1066] has been developed as a foliar fungicide for control of *Venturia* spp. in fruit and as a seed dressing and soil treatment against *Pythium* spp.

1; *Captan*, Orthocide® (Chevron Chem. Co., Ortho Div.);
LD$_{50}$: 9000 mg/kg rat, oral, acute

Captan is used widely today in fruit, vegetables, cereal, rice, grape, citrus, and ornamentals.

The synthesis of *captan*, starting with 1,3-butadiene and maleic anhydride, involves reaction of tetrahydrophthalimide with trichloromethanesulfenyl chloride in aqueous medium.

2-[(Trichloromethyl)thio]-1H-isoindole-1,3(2H)-dione is used as a late spray on orchard fruit because it prevents *Gloeosporium* storage rots of apple. Infections in

Folpet, Phaltan® (Chevron Chemical Co.);
LD_{50}: >10,000 mg/kg rat, oral, acute

grapes caused by *Plasmopara viticola* (downy mildew), *Pseudopeziza tracheiphila* (red fire), and *Botrytis cinerea* (gray mold) are controlled successfully. *Folpet* is also used in vegetables and ornamentals and, like *captan*, serves as a technical fungicide.

3a,4,7,7a-Tetrahydro-2-[(1,1,2,2-tetrachloroethyl)thio]-1H-isoindole-1,3(2H)-dione, a widely active protective and eradicative foliar fungicide,[1067] is synthesized starting from trichloroethylene and sulfur dichloride.[1068]

Captafol, Difolatan® (Chevron Chem. Co., 1965);
LD_{50}: 6200 mg/kg rat, oral, acute

The following fungicides are also known from this group:

1. 2-Phenyl-4-[(trichloromethyl)thio]-1,3,4-oxadiazole-5(4H)-one.[1069]

Trichlazone (Rhône Poulenc), against apple and pear scab and storage rots of pome fruit.

2. 4-(Trichloromethyl)thio-1,2,4-triazolidine-3,5-dione **(1)** and 4-(trichloromethyl)thio-1,2,4-triazine-3,5(2H,4H)-dione **(2)**.[1070]

1   2

Comparable activity to *captan* against *Alternaria tenuis* and *Botrytis allii*.

3.2.5.1.4.2. Perhaloalkylmercaptosulfonamides and Sulfamides. Highly active fungicides have also been found in the group of perhaloalkylmercaptosulfonamides, particularly among the products of reaction of haloalkanesulfenyl chlorides with *N,N*-dimethyl-*N'*-phenyl-sulfamides. However, the fluorodichloromethylmercapto derivatives show an optimum of activity both in the sulfonamides and in the sulfamides.

Whereas *N*-(4-chlorophenyl)-*N*-(trichloromethyl)thio methanesulfonamide **(1)** is used for control of apple scab,[1056] 1,1-dichloro-*N*-[(dimethylamino)sulfonyl]-1-fluoro-*N*-phenylmethanesulfenamide **(2)** has a broad spectrum of activity. Manu-

1; Mesulfan® (Ciba-Geigy AG)

factured by the following procedure,[1071,1072] *dichlofluanid* is used for control of *Peronospora*; red fire disease; and *Botrytis* infections in grape, strawberry, fruit,

Dichlofluanid, Euparen®, Elvaron® (Bayer AG)
LD$_{50}$: >2500 mg/kg rat, oral, acute

vegetable, hop, rose, and ornamental plant growing. The *Botrytis* control is particularly successful. This product is also effective against fruit scab and *Gloeosporium*, *Alternaria*, and downy and powdery mildews, particularly on roses (in higher concentrations).

The corresponding trichloromethyl derivative of **2** is also highly active,[1073] as is the 4-tolyl analogue of **2** (*tolylfluanid*, Euparen M®; Bayer AG), used against scab and powdery and downy mildews and having a miticidal side effect on spider mites.[1073a]

3.2.5.1.5. Heterocyclic Sulfur Compounds. Within the class of sulfur heterocycles, two basic groups are to be differentiated:

   a. Compounds with one or more sulfur atoms in the ring.
   b. Compounds with sulfur and nitrogen atoms in the ring.

Representative products of both groups are listed in the following.

   Group a.
   1. 3,3,4,4-Tetrachlorotetrahydrothiopene 1,1-dioxide.

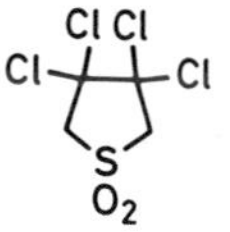

LD$_{50}$: 112 mg/kg rat, oral, acute

As a mixture with *quintozene* (see Section 3.2.5.2) as DAC 649 for seed dressing and soil treatment.

   2. Bis(1-methylethyl) 1,3-dithiolan-2-ylidene-propanedioate.[1074]

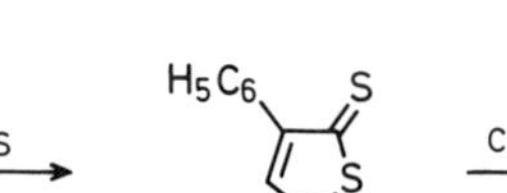

NNF-109, Fujione® (Nihon Noyaku K.K.)

Systemic fungicide against *Pyricularia oryzae*.

   3. 5-Chloro-4-phenyl-3H-1,2-dithiol-3-one.[1075]

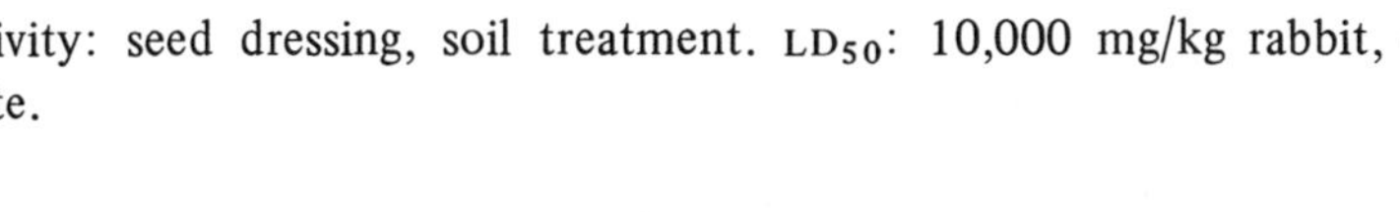

Hercules 3944

Activity: seed dressing, soil treatment. LD$_{50}$: 10,000 mg/kg rabbit, oral, acute.

4. 3,4,6,7-Tetracyano-1,2,5-trithiepin.[1076]

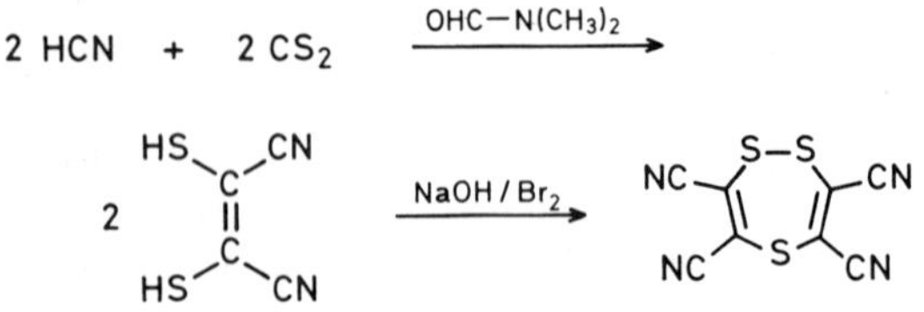

$$2 \text{ HCN} \quad + \quad 2 \text{ CS}_2 \quad \xrightarrow{\text{OHC}-\text{N(CH}_3)_2}$$

Deftan-Fog (E. Merck AG)

Use: fumigant and greenhouse fungicide against mildew, *Botrytis*, and *Cladosporium*.[1077] LD_{50}: 10,000 mg/kg rabbit, oral, acute.

Group b.

1. The rhodanines **1** and **2**, which may be regarded as cyclic carbamodithioates,[1078] have not achieved any great significance as fungicides for control of *Alternaria solani*.

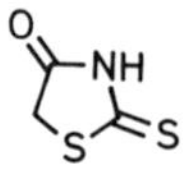

1; 2-Thioxo-4-thiazolidinone (Stauffer Chem. Co.)

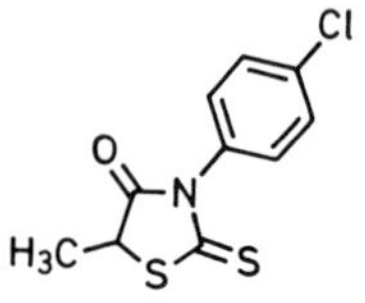

2; 3-(4-Chlorophenyl)-5-methyl-2-thioxo-4-thiazolidinone (Stauffer Chem. Co.)

2. 5-Ethoxy-3-(trichloromethyl)-1,2,4-thiadiazole.[1079]

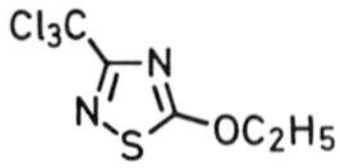

Terrazole® (Olin Mathieson)

Use: seed dressing and soil treatment against *Pythium*, *Rhizoctonia*, and *Fusarium* spp.

3. *N*-2-Benzothiazolyl-*N*'-(1-methylethyl)urea.[1080]

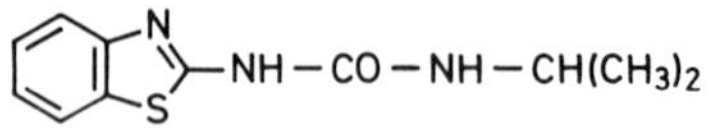

Bentaluron (Ciba-Geigy AG)

Use: against powdery mildews, rust, and *Botrytis*.

4. 5-Methyl-1,2,4-triazolo[3,4-*b*]benzothiazole.[1081]

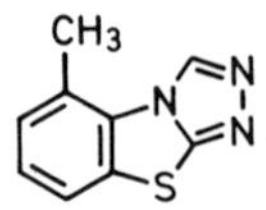

Tricyclazole (Eli Lilly Co.)

High systemic activity against *Pyricularia oryzae*.

5. 3-(2-Propenyloxy)-1,2-benzisothiazole-1,1-dioxide.[1082]

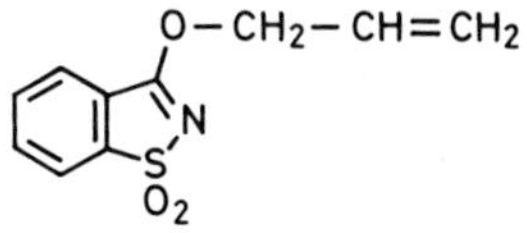

$$O-CH_2-CH=CH_2$$

Oryzaemate

Experimental fungicide in rice.

3.2.5.2. Hydrocarbons

In this group the most significant products are aliphatic hydrocarbons bearing halogen, nitro, or nitroso groups that, owing to their volatility, serve as soil fumigants against soil-borne fungi, and isocyclic hydrocarbons that are used as dressings.

3-Bromo-1-chloro-1-propene is used for control of soil-borne fungi and nematodes. It is manufactured by chlorination of a propene–allyl chloride mixture and

$$Cl-CH=CH-CH_2-Br$$

CBP-55 (Shell Development Co., 1952)

halogen exchange of the product, 1,3-dichloropropene,[1083] with alkali metal bromides.[1084] The intermediate 1,3-dichloropropene (Telone®) is also used as a fumigant.

1-Chloro-2-nitro-propane is a widely active soil fungicide for control of damping-off in cotton.

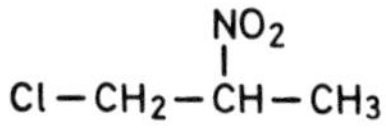

$$Cl-CH_2-\overset{\overset{\displaystyle NO_2}{|}}{CH}-CH_3$$

NJA 5961, Lanstan® (FMC Corp., Niagara Chem. Div.)[1085]

Trichloronitromethane, highly active against soil-borne fungi and with nematicidal and herbicidal properties, is unpleasant to handle because it is a severe irritant to skin, eyes, and respiratory tract.

$$Cl_3C-NO_2$$

Chloropicrin, Picfume® (Dow Chem. Co.)[1086]

Biphenyl **(1)** serves as a post-harvest fungicide for prevention of molds on fruit, particularly citrus,[1087] and is also used for impregnation of packaging materials.

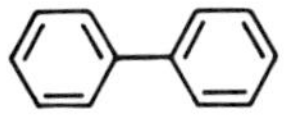

1

LD$_{50}$: 3280 mg/kg rat, oral, acute

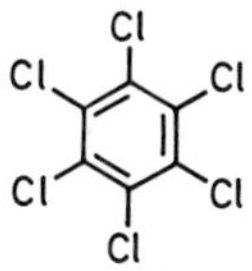

2; HCB

Hexachlorobenzene **(2)** was introduced in 1940 as a seed dressing against, for example, stinking smut of wheat (*Tilletia caries*).[1088] The former worldwide use is now limited because of high soil persistence.

2,3,5,6-Tetrachloro-nitrobenzene developed as a selective fungicide against dry rot of potato (*Fusarium solani* var. *coeruleum*), is relatively untoxic (mice fed 250 mg/kg/day showed no symptoms).

Tecnazene (IG Farbenindustrie, Hoechst), TCNB, Myfusan®, Fusarex®, ®Folosan DB-905

Pentachloronitrobenzene, introduced as a seed dressing, is also used in combination with other fungicides as a dressing owing to its effect on bunt, or stinking

PCNB, Quintozene, Brassicol®, Tritisan®, Folosan®, Botrilex®, Terrachlor®, Tilcarex® (IG Farbenindustrie, 1930);
LD$_{50}$: 12,000 mg/kg rat, oral, acute

smut, of wheat. It is widely used as a substitute for mercury fungicides. *Quintozene* also controls *Rhizoctonia* spp., *Sclerotinia* spp., club root of cabbage (*Plasmodiophora brassicae*), and *Botrytis* diseases.

"Chemagro 2635," a mixture of the two trichlorodinitrobenzenes **3** and **4**, is active against *Rhizoctonia* spp.

3; 80%   4; 20%

1,3,5-Trichloro-2,4,6-trinitrobenzene is highly active against *Cladosporium fulvum*. The halogenated nitrobenzenes all have as a common feature the reactivity

Bulbosan® (Hoechst AG);
LD$_{50}$: 10 000 mg/kg, rat, oral, acute

of a chlorine atom to mercapto groups.[1089] They all attack at the same site of action and hence show cross resistance.[1090]

3.2.5.3. Alcohols and Phenols

Among the aliphatic alcohols, apart from methanol (wet seed dressing Ustilgon®), chlorinated alcohols have fungicidal properties, for example:

1. 1,1,1,3,3,3-Hexacloro-2-propanol.

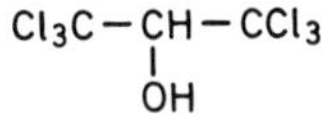

$$Cl_3C-CH-CCl_3$$
$$|$$
$$OH$$

Use: against fruit scab and brown rot of fruit.[1091] Technical synthesis: sodium trichloroacetate and chloral (CO_2 cleavage).[1092]

2. 1,1,1-Trichloro-3-nitro-2-propanol.

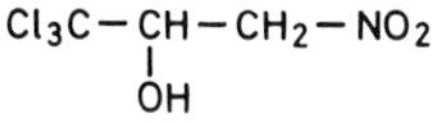

$$Cl_3C-CH-CH_2-NO_2$$
$$|$$
$$OH$$

Systemic fungistatic,[1093] bactericide, and herbicide.[1094]

In spite of high fungicidal activity, the phytotoxicity of the biphenols and halogenated phenols has led to their being used mainly for the preservation of technical materials. However, the esters of dinitro-alkyl-phenols[1095] play a significant role as powdery mildew fungicides with acaricidal side effects.

2-Phenyl-phenol[1095] finds application as a fungicide for the post-harvest dip treatment of citrus fruit but is also widely used as a disinfectant for dyes, leather, textiles, and other technical materials.

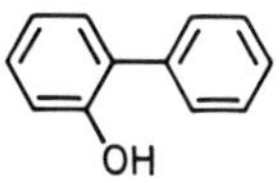

Dowicide 1®, Torsite®, Orthoxenol® (Dow Chem. Co.);
LD$_{50}$: 2700 mg/kg rat, oral, acute; As sodium salt Dowicide A (Dow)

1,4-Dichloro-2,5-dimethoxy-benzene[1096] is used as a seed dressing against damping-off, *Rhizoctonia*, *Sclerotium*, and *Pythium* spp. (mainly for seeds of cotton, beans, sugar beet, and soya).

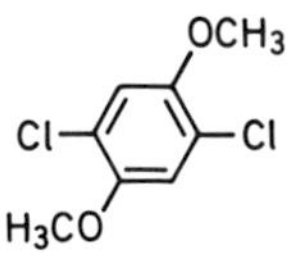

Chloroneb, Demosan®, Tersan® SP (Du Pont, 1966);
LD$_{50}$: 11,000 mg/kg rat, oral, acute

Pentachlorophenol, marketed widely by many firms as wood preservative, insecticide, and herbicide (also in the form of its salts), is fungicidal toward wood-destroying fungi and also kills termites.

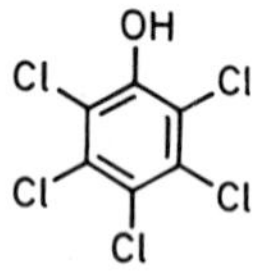
OH
Cl
Cl
Cl
Cl
Cl

PCP, Santobrite®, Dowicide®;
LD_{50}: 210 mg/kg rat, oral, acute

2,2′-Methylenebis[3,4,6-trichlorophenol] serves as a seed dressing for soil-borne fungi, particularly in cotton.

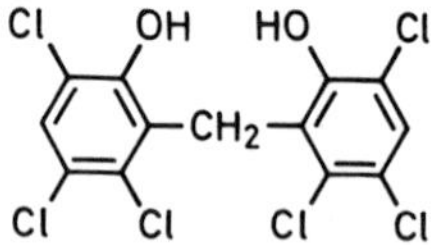
Cl
OH
HO
Cl
CH2
Cl
Cl
Cl
Cl

Hexachlorophene, G 11®, Nabac®, Isobac®;
LD_{50}: 320 mg/kg rat, oral, acute

2-Isooctyl-4,6-dinitrophenyl 2-butenoate, known as an acaricide since 1934, has a particularly high protective and eradicative activity against powdery mil-

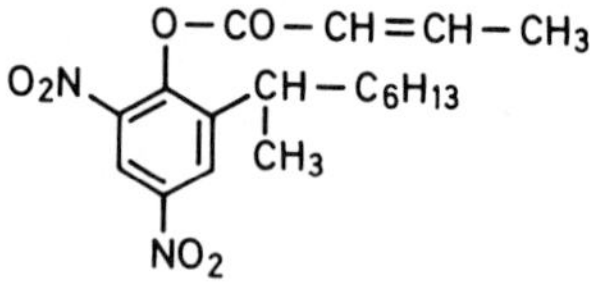
O–CO–CH=CH–CH3
O2N
CH–C6H13
CH3
NO2

Dinocap, Karathane®, Arathane® (Rohm and Haas);
LD_{50}: 980 mg/kg rat, oral, acute

dews[1097,1098] and is used widely against these fungi in fruit, winegrapes, and ornamentals. *Dinocap* is synthesized from isooctylphenol by nitration and subsequent esterification.[1099]

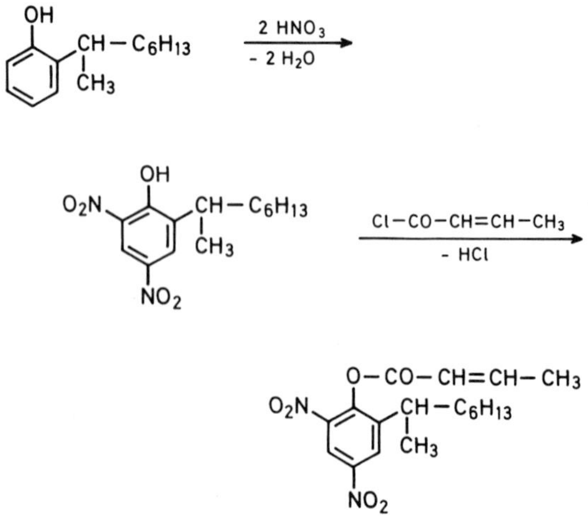
OH
CH–C6H13
CH3
2 HNO3
- 2 H2O
OH
O2N
CH–C6H13
CH3
NO2
Cl–CO–CH=CH–CH3
- HCl
O–CO–CH=CH–CH3
O2N
CH–C6H13
CH3
NO2

Within the class of dinitrooctylphenols[1100] the 2,6-dinitro-4-isooctyl isomer of *dinocap* also has high activity against powdery mildews.

The mechanism of action of the dinitrophenols involves the uncoupling of oxidative phosphorylation. The dinitrophenols thus interfere in the energy system of the cell. This mechanism has been correlated with the fungicidal properties.[1101,1102]

Further products on the market are:

1. 2-(1-Methylpropyl)-4,6-dinitrophenyl 3-methyl-2-butenoate.

Binapacryl, Acricid®, Morocide® (Hoechst AG)
Use: fungicide and acaricide in orchard crops

2. 1-Methylethyl 2-(1-methylpropyl)-4,6-dinitrophenyl carbonate.

Dinobuton, Acrex® (KenoGard AB, Sweden)

3. 2,6-Dinitro-4-nonylphenol (**1**) and methyl 2,6-dinitro-4-(4- or 3-octyl)-phenyl carbonate (**2**).

1; Brandol® mildew fungicide

2; Dinocton-4 (Murphy Chem. Ltd) in orchard crops

3.2.5.4. Oxo Compounds

Of the aliphatic oxo compounds, halogenated aldehydes and ketones and their Schiff bases and aminals have achieved significance as fungicides. Formaldehyde is used to a slight extent but, owing to its phytotoxicity, only as a seed and soil disinfectant without protective action.[1103,1104] Apart from 1,3-dichloro-tetrafluoro-acetone, which has a systemic fungicidal action, and its addition products with dodecylguanidine,[1105] it is mainly the aminals of chloral that are of importance.[1106] These have a systemic action against powdery mildews, scab and rust.

1. *N,N'*-[1,4-Piperazinediylbis(2,2,2-trichloroethylidene)]bis(formamide).[1107]

$$\text{OHC-NH-CH-N} \underset{}{\overset{\text{CCl}_3}{|}} \quad \text{N-CH-NH-CHO} \overset{\text{CCl}_3}{|}$$

Triforine, Funginex®, Saprol® (Boehringer and Sohn, Celamerck GmbH);
Use: systemic soil and foliar fungicide in cereals, cucurbits, apple (against powdery mildews and rusts);
LD_{50}: 6000 mg/kg rat, oral, acute

2. *N*-{2,2,2-trichloro-1-[(3,4-dichlorophenyl)amino]ethyl}formamide.[1108,1109]

$$\text{Cl}_3\text{C-CHO} \quad + \quad \text{H}_2\text{N-CHO} \longrightarrow$$

$$\text{HO-CH-NH-CHO} \overset{\text{CCl}_3}{\underset{}{|}} \xrightarrow{- CO_2}$$

Cloraniformethan (Bayer AG), systemic action against mildews;[1108]
LD_{50}: 2500 mg/kg rat, oral, acute

3. *N*-{2,2,2-trichloro-1-[(3-pyridyl)amino]ethyl}-formamide:

displays systemic activity against mildews when applied as a drench at 5 ppm.[1110]

The cyclic ketone (**1**), made from two molecules of hexachlorocyclopentadiene, has fungicidal action against apple scab and powdery mildew,[1111] in addition to its insecticidal properties.

$$\xrightarrow{\text{H}_2\text{O}}_{- 2 \text{ HCl}}$$

1; Kepone®, GC-1189

Among the aromatic oxo compounds and quinones are to be found derivatives of *p*-quinone used as seed dressings and foliar fungicides as well as pentachloro-benzaldoxime (Mikonol®), a rice fungicide, and 1-phenyl-1-oxo-2-methyleneal-kanes.[1112] Quinones play a significant role as electron acceptors in the respiratory chain owing to their reversible redox reactions.[1113] They react with mercapto and amino groups in biological systems and thus can interfere in biological processes at various points. Their high sublimation tendency, sensitivity to hydrolysis, and photochemical instability[1116] make the majority of fungicidal quinones unsuitable for foliar application.

2,3,5,6-Tetrachloro-2,5-cyclohexadiene-1,4-dione,[1117] available by oxidation of either tetrachlorophenol with chromic acid or of phenol with hydrochloric acid–potassium chlorate, was introduced to the market as a seed dressing with application rates higher than those of the mercury compounds.

Chloranil, Spergon® (US Rubber Co.);
LD$_{50}$: 4000 mg/kg rat, oral, acute

4-(Hydroxyimino)-2,5-dimethyl-2,5-cyclohexadien-1-one[1118] is also a seed dressing.

D-198; LD$_{50}$: 4000 mg/kg rat, oral, acute

Benzoyl [4-(hydroxyimino)-2,5-cyclohexadien-1-ylidene]hydrazide[1119] is highly effective against *Pythium* spp. and is used to treat the seeds of cotton, beet, peas,

Benquinox; COBH, Bayer 15080, Ceredon® (Bayer AG)
LD$_{50}$: 100 mg/kg rat, oral, acute

and beans. *Benquinox* is also used in combination with *thiram* (Ceredon T, Bayer AG) and phenylmercury chloride (Ceredon Spezial, Bayer AG) for control of damping-off diseases.

Quinone-semicarbazone (Boots Co. Ltd.) likewise serves as a seed dressing.[1120]

The better photostability of 2,3-dichloro-1,4-naphthalenedione compared to that of *chloranil* permits its use not only as a seed dressing but also as a foliar fungicide against apple scab and in coffee and citrus crops. Apart from a relatively high fish toxicity, *dichlone* is also somewhat phytotoxic; thus, its application in apples is limited to certain varieties.

Dichlone, Phygon® (US Rubber Co.);
LD$_{50}$: 1300 mg/kg rat, oral, acute

5,10-Dihydro-5,10-dioxo-naphtho[2,3-*b*]-1,4-dithiin-2,3-dicarbonitrile, available by reaction of *dichlone* with 2,3-dimercapto-2-butenedinitrile[1121]

Dithianon, Delan® (Merck AG), Stauffer MV-119;
LD$_{50}$: 1015 mg/kg rat, oral, acute

is used as a foliar fungicide against fruit scab; in coffee, cacao, and citrus growing; and as a seed dressing.

3.2.5.5. Carboxylic Acids and Derivatives

Although a great number of aliphatic carboxylic acids have fungistatic or fungicidal properties,[1122] only a few are used on a technical scale. Propanoic acid and sorbic acid are used in the food industry as food preservatives. Sorbic acid is degraded by the mammalian organism just like other fatty acids[1123] and serves to protect fruit and vegetables from mold and yeast fungi. The technical synthesis of sorbic acid involves condensation of ketene with crotonaldehyde in the presence of boron trifluoride and subsequent treatment with dilute sulfuric acid.

$$H_2C=C=O \quad + \quad H_3C-CH=CH-CHO \xrightarrow[\text{2. dil. } H_2SO_4]{\text{1. } BF_3}$$

$$H_3C-CH=CH-CH=CH-COOH$$

LD$_{50}$: 7360 mg/kg rat, oral, acute

Other carboxylic acid derivatives that have been reported in crop protection are:

1. 2-Chloro-*N*-(2-cyanoethyl)-acetamide.[1124]

$$Cl-CH_2-CO-NH-CH_2-CH_2-CN$$

Udonkor® (Nippon Soda K.K., Japan);
Use: systemic fungicide against powdery mildews on cucurbits and gooseberries (since 1967 in Japan);
LD$_{50}$: 411 mg/kg rat, oral, acute

2. 2-Butynediamide.

$$H_2N-CO-C\equiv C-CO-NH_2$$

Cellocidin®, Cellomate®, Seromoto®;
Use: against *Pellicularia sasakii* (in Japan)

3. 5,5'-Oxybis[3,4-dichloro-2(5H)-furanone].[1125]

Mucochloric anhydride, GC-2466 (Allied Chem. Corp., 1957);
Use: against apple scab; seed protectant

4. 3-Acetyl-6-methyl-2H-pyran-2,4(3H)-dione.[1126]

Dehydroacetic acid, preservative

3.2.5.5.1. Aromatic Carboxylic Acid Derivatives.
Halogen and nitro-substituted aromatic carboxylates and nitriles in particular have high, wide-spectrum activity as protective fungicides.

2,4,5,6-Tetrachloro-1,3-benzenedicarbonitrile[1127] was introduced as a broad-spectrum, protective, foliar- and soil-applied fungicide against *Alternaria*, *Botrytis*,

Chlorothalonil, Daconil® 2787, Forturf®, Bravo® (Diamond Shamrock Co., 1966);
LD_{50}: 10,000 mg/kg rat, oral, acute

Curvularia, *Erysiphe*, *Rhizoctonia*, *Podosphaera*, and *Pythium* spp. in vegetables, corn (maize), peanuts, turf, and ornamentals. It is phytotoxic to pome fruits and grapes.

Other highly active derivatives of isophthalic and nitrobenzoic acids are summarized in Table 20.

4,5,6,7-Tetrachloro-1(3H)-isobenzofuranone is used as a fungicide against *Pyricularia oryzae* in rice.

Rabcide; LD_{50}: 20,000 mg/kg rat, oral, acute

TABLE 20. Derivatives of Isophthalic and Nitrobenzoic Acids

Name	Structure	Activity	Manufacturer	Reference
2,4,6-Trichloro-5-nitro-1,3-benzenedicarbonitrile	DU 1 3710	*Pithomyces chartrum, Venturia inaequalis*	N.V. Philips Gloeilampenfabriken	Neth. P. 6613164 (1966), N.V. Philips Gl.; *C.A.* **69**, 86660z (1968)
Diisopropyl 5-nitro-1,3-benzenedicarboxylate	BAS 3000 F	*Podosphaera leucotricha*	BASF	DAS 1218792 (1964), BASF, Inv.: Polster, Pommer; *C.A.* **65**, 9660c (1966)
4-Bromo-2,4-dinitrobenzoate		*Puccinia recondita, Septoria nodorum*	—	K. Lehtonen et al. *Pestic. Sci.* **3**, 443 (1972)

3.2.5.5.2. Carboxamides. The class of carboxamides embraces a large number of commercial and experimental fungicides with systemic activity, particularly against diseases caused by Basidiomycetes. All of these products have the following basic structure in common. They are α,β-unsaturated carboxylic acid derivatives in which

the C=C double bond either forms part of a benzene ring or is conjugated with electron-donating atoms such as O, N, or S.

The β-carbon in the position *cis* to the carboxy group must be substituted by an alkyl (particularly methyl) group or, in the case of the benzanilides, by a group with a similar spatial requirement (e.g., I, C_2H_5, Br, CH_3, Cl, OH). The substituent in the anilide moiety (X_n) should be an electron-donating group if possible. The carboxanilides act by inhibition of succinate dehydrogenase. This action has been correlated with the activity against *Rhizoctonia solani*[1128] –see Table 21. Work at Ciba-Geigy Ltd. and the Chevron Chem. Co. has led to the discovery of the class of acylated anilines[1128a] containing new, systemic fungicides highly active against Oomycetes.

Methyl *N*-(2,6-dimethylphenyl)-*N*-(methoxyacetyl)-DL-alanine.[1128b] Ridomil®,

Metalaxyl (proposed), CGA 48988, Ridomil® (Ciba-Geigy Ltd.);
LD_{50}: 669 mg/kg rat, oral, acute

a fungicide with a unique combination of residual and systemic properties, is highly active both *in vitro* and *in vivo* against Oomycetes, especially Peronosporales such as *Phytophthora*, *Pseudoperonospora*, *Peronospora*, *Plasmopara*, *Sclerospora*, *Bremia*, and *Pythium* spp. and other species causing downy mildews, late blight, damping-off, and root, stem, and fruit rots. A high activity, at low rates of foliar or soil application, against diseases caused by air- or soil-borne Oomycetes offers promise and performance for use in various crops, for example, potatoes, grape, tobacco, cereals, hops, and vegetables.

Methyl *N*-(2,6-dimethylphenyl)-*N*-(2-furanylcarbonyl)-DL-alanine.[1128c] Fongarid® is also a protective fungicide with systemic properties, showing high activity

Furalaxyl (proposed); CGA 38140, Fongarid® (Ciba-Geigy Ltd.);
LD_{50}: 940 mg/kg rat, oral, acute

TABLE 21. Structure–Activity Relationships among the Carboxamides

Name	Structure	ED_{50} (M) R. Solani	I_{50} (M) Succinate Dehydrogenase	Manufacturer, Trade Name	Use, Activity	LD_{50} mg/kg Rat, Oral, Acute	Reference
Salicylanilide	OH, CO—NH—C$_6$H$_5$	100	85	ICl Shirlan®	Textile preserv.	—	—
Mebenil	CH$_3$, CO—NH—C$_6$H$_5$	10	18	BASF Basfungin®	Soil and foliar fungicide	6000	—
2-Methyl-N-[3-(1-methylethoxy)phenyl]benzamide	CH$_3$, O—CH(CH$_3$)$_2$, CO—NH—	—	—	Kumiai Chem. Ind.	Fung. against *Pellicularia sasakii*	—	DOS 2434430 (1974), Kumiai, Inv.: Chiyomaru et al.; C.A. **82**, 155895n (1975)
Benodanil	I, CO—NH—C$_6$H$_5$	—	1.0	BASF BAS 3170 F	Rusts on cereals, veg., coffee	6400	Belg. P. 714356 (1967), BASF, Inv.: Osieka et al.; C.A. **71**, 49619p (1969)
Cyclafuramid	CH$_3$, CO—NH—, H$_3$C	—	6.0	BASF BAS 3270 F	Basidiomycetes, cereal seed dressing	6400	DOS 2019535 (1970), BASF, Inv.: Distler et al.; C.A. **76**, 46068a (1973)
Furcarbanil	CH$_3$, CO—NH—C$_6$H$_5$, H$_3$C	—	10.0	BASF BAS 3191 F	Basidiomycetes, loose smut of barley	6400	DOS 2006471 (1970); Uniroyal, Inv.: Davis, v. Schmeling; C.A. **73**, 108740n (1970)

Name	Structure			Trademark (Producer)	Uses		References
Methfuroxam, Furavax (proposed)	furan ring; H_3C, CH_3, CH_3, $CO-NH-C_6H_5$	—	—	Uniroyal UDI H 719 Arbosan® (in combination with *imazalil* and *thiabendazole*, Ciba-Geigy)	Smuts, cereal seed dressing	—	DOS 2006472 (1970), Uniroyal, Inv.: Felauer, Kulka; *C.A.* 73, 98933m (1970)
Pyracarbolid	dihydropyran ring; O, CH_3, $CO-NH-C_6H_5$	6.0	—	Hoechst Sicarol®	Basidiomycetes, seed dressing and wettable powder, cereals, coffee	5000	Belg. P. 727245 (1968) Hoechst, Inv.: Scherer, Heubach; *C.A.* 72, 55257f (1970)
Carboxin	O, CH_3, S, $CO-NH-C_6H_5$	0.5	1.0	Uniroyal Vitavax®	Seed dressing, cereals, cotton; smuts, and rusts	3820	USP 3249499 (1965) Uniroyal, Inv.: v. Schmeling et al.; *C.A.* 65, 7190g (1966)
Oxycarboxin	O, CH_3, S, O_2, $CO-NH-C_6H_5$	9.0	13.6	Uniroyal Plantvax®	Soil and foliar fungicide, rusts	2000	See *Carboxin*, above
2,4-Dimethyl-*N*-phenyl-5-thiazole-carboxamide	thiazole ring; CH_3, N, H_3C, S, $CO-NH-C_6H_5$	3.0	2.4	Uniroyal	Rusts, cereals	5600	Neth. P. 6716466 (1966) Uniroyal, Inv.: v. Schmeling et al.; *C.A.* 74, 139879q (1971)
2-Amino-4-methyl-*N*-phenyl-5-thiazole-carboxamide	thiazole ring; CH_3, N, H_2N, S, $CO-NH-C_6H_5$	0.6	3.2	Uniroyal Seedvax®	*Rhizoctonia solani*, seed dressing, cotton	1410	Neth. P. 6716466 (1966) Uniroyal, Inv.: v. Schmeling et al.; *C.A.* 74, 139879q (1971)

against the Peronosporales that cause downy mildews of hops, lettuce, and various other horticultural crops, and against stem and root rots in ornamentals. It can be used, with high protective and curative efficacy, as a soil-drench or soil-incorporated treatment or as a foliar spray against the above pathogens.[1128d]

2-Chloro-*N*-(2,6-dimethylphenyl)-*N*-(tetrahydro-2-oxo-3-furanyl)-acetamide.[1128e] RE 20615 is another new protective, curative, and systemic fungicide

RE 20615 (Chevron Chem. Co.)

from this class, active against Peronosporales pathogens. It has high activity, either alone or in combination with other suitable fungicides, against downy mildews of grapes, hops, potatoes, and other agricultural and horticultural crops.[1128f]

3.2.5.5.3. Carboximides. Apart from 1H-isoindole-1,3(2H)-dithione,[1129] developed as a fungicide for emulsion paints, it is the *N*-(3,5-dichlorophenyl)hetero-

Dithiophthalimide, Fungitrol® (Tenneco Chem. Co.)

cycles in the carboximide class, systemically active against *Botrytis cinerea*, *Cochliobolus miyabeanus*, *Pellicularia sasakii*, and *Sclerotinia sclerotiorum*, which have achieved significance. For example, 1-(3,5-dichlorophenyl-2,5-pyrrolidinedione, a

Dimetachlone (proposed); Ohric®;
LD$_{50}$: 500 mg/kg rat, oral, acute

protective and curative fungicide, is registered in Japan for use in rice, fruit, and vegetables. It is active against *Helminthosporium* spp., *Pellicularia sasakii*, *Pyricularia oryzae*, and powdery mildews.

Derivatives of pyrrolidinedione, oxazolidinedione, and imidazolidinedione[1130] have been reported as structurally related fungicides, some of which are still in the experimental stage (see Table 22).

3.2.5.6. Nitrogen Compounds

3.2.5.6.1. Amines, Amidines, Guanidines. Apart from the nitrogen compounds mentioned above, some amines and ammonium compounds also have fungicidal or bactericidal properties. Owing to problems of phytotoxicity, it is mainly the long-chain aliphatic (particularly with a chain length of C_{10}—C_{13}) and cycloaliphatic amines and their derivatives that have found application in crop protection as foliar fungicides, principally against powdery mildews and scab diseases. The action of the amines, amidines, and guanidines is supposed to depend on their surface-active properties that lead to a destruction of the fungal cell membranes.[1131]

The economically most important compound from this class is dodecylguanidine acetate, which was introduced as a protective and curative locosystemic agent

$$H_3C-(CH_2)_{11}-NH-\overset{NH_2}{\underset{\overset{\|}{NH}}{C}} \cdot H_3C-COOH$$

Dodine, Melprex®, Cyprex® (American Cyanamid, 1956);
LD_{50}: 1000 mg/kg rat, oral, acute

against apple scab.[1132] *Dodine* can be used as a pre-blossom spray in sensitive varieties to avoid problems of phytotoxicity.

2-Heptadecyl-4,5-dihydro-1H-imidazole monoacetate was also found to be a protective and curative fungicide against apple scab. This compound, which is not free from phytotoxicity problems, acts upon nuclein and protein biosynthesis.[1133] *Glyodin* is also used as a corrosion inhibitor.

$$\underset{H}{\overset{N}{\diagup}}\overset{N}{\diagdown}\underset{\overset{|}{CH_3}}{CH}-C_{15}H_{31} \cdot H_3C-COOH$$

Glyodin (Law, Wellman, and McCallan, 1946)

Other amines and amine derivatives that are known to be commercial or experimental mildew fungicides and seed dressings are listed in Table 23.

The aromatic amine, 2,6-dichloro-4-nitro-benzenamine was introduced in 1959 for soil treatment and foliar application against *Botrytis* spp. (lettuce rot) and *Sclerotinia* spp. (fruit rot).

$$\underset{NO_2}{\overset{NH_2}{\underset{}{Cl \diagdown \diagup Cl}}}$$

Dicloran, Allisan® (Boots Co.), Botran® (Upjohn Co.);
LD_{50}: 10,000 mg/kg rat, oral, acute

3.2.5.6.2. Hydrazo and Azo Compounds.

Numerous phenylhydrazines have marked fungitoxic properties, possibly owing to their ability to form hydrazones with essential aldehydes and ketones in the fungal organism.[1134] Phenylhydrazones, phenylazo compounds, and diazonium salts are used in crop protection as seed dressings and in fruit growing. On isolated fungal mitochondria they act as inhibitors of NADH oxidation (*fenaminosulf*).[1135]

The systemically active sodium [4-(dimethylamino)phenyl]diazenesulfonate was

TABLE 22. Carboximides: Experimental Fungicides

Name	Structure	Manufacturer, Trade Name	Use, Activity	Reference
1-(2,5-Dichlorophenyl)-3-methylene-2,4-pyrrolidinedione		Sumitomo	—	S. Afr. P. 2820/69 (1969), Sumitomo, Inv.: Fujinami et al.; *C.A.* **73**, 3667r (1970)
MK 23		Mitsubishi, Taiwan Ihara Co.	Citrus scab	GBP 1324910 (1971), Mitsubishi and Taiwan Ihara, Inv.: Kawada et al.; *C.A.* **80**, 3438r (1974)
Vinchlozolin		BASF Ronilan®	*Botrytis cinerea*	DOS 2207576 (1972), BASF, Inv.: Mangold et al.; *C.A.* **73**, 122203w (1970)
Dichlozoline		Hokko Chem. Co., Sumitomo Sclex®	*Sclerotinia sclerotiorum, Cochliobolus miyabeanus, Botrytis cinerea*	DOS 1811843 (1967), Hokko and Sumitomo, Inv.: Sato et al.; *C.A.* **71**, 101838a (1969)
3-(3,5-Dichlorophenyl)-4-imino-5,5-dimethyl-2-oxazolidinone		Sumitomo	See Sclex®, above	DOS 2002410 (1969), Sumitomo, Inv.: Fujinami et al.; *C.A.* **73**, 87912c (1970)

Compound	Structure	Company / Trade name	Target organisms	References
(1-Cyano-1-methylethyl) 3,5-dichlorophenyl-carbamate		Sumitomo	See Sclex®, above	DOS 2021327 (1970), Sumitomo, Inv.: Fujinami et al.; *C.A.* **67**, 94015g (1967)
3-(3,5-Dichlorophenyl)-1-propyl-2,4-imidazol-idinedione		Sumitomo	*Sclerotinia sclerotiorum*, *Pellicularia sasakii*	DOS 1958183 (1968), Sumitomo, Inv.: Fujinami et al.; *C.A.* **73**, 35376m (1970)
Isovaledione (proposed)		Sumitomo	*Botrytis cinerea*, *Pellicularia sasakii*	DOS 2833767 (1978), Sumitomo, Inv.: Takayama et al.; *C.A.* **91**, 57002k (1979)
RP-26019 *Iprodione*		Rhône Poulenc SA Rovral®	*Botrytis cinerea*, *Monilia, Sclerotinia*	DOS 2149923 (1970) Rhône Poulenc, Inv.: Sauli; *C.A.* **77**, 19647c (1972)
Procymidone		Sumitomo Sumisclex® Sumilex®	See Sclex®, above	DOS 2012652 (1970), Sumitomo, Inv.: Fujinami et al.; *C.A.* **75**, 5516f (1971)

TABLE 23. Fungicidal Amine Derivatives

Name	Structure	Manufacturer, Trade Name	Use	LD_{50} mg/kg Rat, Oral, Acute	Reference
2-Butanamine		Eli Lilly Tutane® Frucote	Control of *Penicillium*, post-harvest treatment of citrus fruit	380	Brit. P. 994125 (1962), Eli Lilly, Inv.: Lindberg; C.A. 63, 3566b (1965)
Piperalin		Eli Lilly Pipron®	Powdery mildew on ornamentals	2500	
Dodemorph		BASF F 238®	Mildew of rose		DAS 1198125 (1961) BASF, Inv.: König et al.; C.A. 58, 4581f (1963)
Tridemorph		BASF Calixin®	Powdery mildew, cereals; application rate 0.5–0.75 liter/ha, eradicative	650	K. H. König, E. H. Pommer, W. Sanne; Angew. Chemie 77, 327 (1965)
Fenpropemorph		BASF Hoffman-La Roche Corbel®	Powdery mildew and rust, cereals; application rate 0.56–0.75 kg/ha	3500	DOS 2656747 (1976), BASF, Inv.: Himmele, Pommer, Goetz; C.A. 89, 109522k (1978)

Name	Structure	Company / Trade name	Use		Reference
Guazatine	$HN=C(NH_2)-NH-(CH_2)_8-NH-(CH_2)_8-NH-C(=NH)-NH_2$	Murphy Chem. Co. Panoctine®	Seed dressing against *Tilletia tritici, Fusarium nivale*	550	USP 3378438 (1962), Geigy, Inv.: Grätzi; *C.A.* **69**, 43816n (1968)
GS 16306	ring structure with $NH-C_{10}H_{21}$, $(H_3C)_3C$ substituent · HCl	Geigy	Experimental product, mildew and scab, apple	–	
Urolocid®	benzyl $-CH_2-\overset{\oplus}{N}(CH_3)_2-CH_2-CO-NH-(CH_2)_{11}-CH_3$	Patchem AG	Fungicide and disinfectant	100	E. H. Frear, Pesticide Index 1969
Propamocarb (proposed)	$(H_3C)_2N-(CH_2)_3-NH-CO-O-(CH_2)_2-CH_3$	Schering AG Previcur®	Fungicide against Oomycetes	–	DOS 1567169 (1966), Schering AG, Inv.: Hoyer, Pieroh; *C.A.* **76**, 21956a (1973)

introduced as a seed dressing and soil treatment with exceptional activity against Phycomycetes such as *Pythium*, *Aphanomyces*, and *Phytophthora* spp.[1136]

$$(H_3C)_2N-\langle\rangle-N=N-SO_2-ONa$$

Fenaminosulf (Bayer AG, 1959);
LD_{50}: 60–75 mg/kg rat, oral, acute

Of the phenylazo and phenylhydrazone derivatives **1-4** used in fruit growing, *drazoxolon* has been on the market the longest:

$$Cl-CH_2-CH=N-NH-\langle\rangle-NO_2 \qquad H_5C_6-NH-NH$$

1; Chloroacetaldehyde-(2,4-dinitrophenyl)-hydrazone; OM 1763 (Olin Mathieson)

2; 1-Phenyl-2-(tetrahydro-3-thienyl)hydrazine S,S-dioxide; Fisons NC 918; systemic against rusts

$$O_2N-\langle\rangle-N=N-C\overset{CH_3}{\underset{CH-COOCH_3}{\mid}}$$

3; Methyl 3-[(4-nitrophenyl)azo]-2-butenoate; NIA 19165 (FMC Chem. Corp., Niagara Div.)

4; 3-Methyl-4,5-isoxazoledione, 4-[(2-chlorophenyl)hydrazone]; *Drazoxolon*;
LD_{50}: 126 mg/kg rat, oral, acute

Drazoxolon is recommended by ICI Plant Protection Division as a seed dressing and foliar fungicide with protective and curative effect for control of mildew and scab infections.[1137] The acetyl compound and the 3-chloro derivative are known to have been experimental products.

3.2.5.6.3. Nitrogen Heterocycles [including the o-Phenylene-bis-thiophanates].

In this group are found important, highly active, systemic heterocyclic fungicides from the classes of pyridines, oxazoles, imidazoles, triazoles, benzimidazoles [and the related *o*-phenylene-bis-thiophanates], pyrimidines, quinoxalines, and triazines. Some are very widely active, but the major activity is frequently centered on powdery mildews and scabs. They are used as partly systemic, protective, curative, and eradicative foliar fungicides and seed dressings.

Pyridines. Some experimental products are found in this class that, owing to labile halogen atoms, are able to react with enzymes containing SH groups and so are active as fungicides. However, some tend to cause skin irritation, and such derivatives cannot be used as foliar sprays (see Table 24).

Parinol is highly active systemically against powdery mildews on fruits and vegetables.[1138]

Parinol, EL-241, Parnon® (Eli Lilly Co.);
LD$_{50}$: 5000 mg/kg rat, oral, acute

Derivatives of quinoline, particularly of 8-hydroxyquinoline (see *oxine-copper*, Section 3.2.2.2) are also known as fungicides, for example:

1. 7-Bromo-5-chloro-8-quinolinyl 2-propenoate.[1139]

Halacrinate (proposed)

2. 2-(8-Hydroxy-5-quinolinyl)ethanone sulfate (2:1).

Quinacetol sulfate;
Use: cereal seed dressing, post-harvest fungicide in citrus

3. 1-(Cyaniminomethyl)-1,2,3,4-tetrahydroquinoline.[1141]

Use: protective fungicide against *Fusicladium*, *Erysiphe*, and *Pythium* spp.

4. 1,2,5,6-Tetrahydro-4H-pyrrolo[3,2,1-*ij*]quinoline-4-one.[1141a]

Pyroquilon (proposed), 4-Lilolidone, CG-114, CGA 49104;
LD$_{50}$: 321 mg/kg rat, oral, acute;
Use: systemic rice fungicide against *Pyricularia oryzae*

TABLE 24. Fungicidal Pyridine Heterocycles

Name	Structure	Activity	Use	LD$_{50}$ (mg/kg Rat, Oral, Acute)	Reference
Dowco® 269 *Pyroxychlor*		*Phytophthora* spp.	Soil and foliar fung. systemic	1500	USP 3 244 722 Dow, Inv.: Johnston, Tomita; *C.A.* **65**, 8884b (1966)
Dowco® 263		Soil-borne fungi	Seed dressing	562 (skin irritant)	USP 3 325 503 (1965), 3 578 434 (1968), Dow, Inv.: Noerske, Tobol; *C.A.* **75**, 19194m (1971)
Ethyl [(2,3,5,6-tetrachloro-4-pyridinyl) sulfonyl]-acetate		Like *thiram*		—	Z. H. Kuprina, E. Shomova et al, Fiziol. Aktiv. Vest-chestva, 1973, **5**, 98–9; *C.A.* **81**, 100516p (1974)
Pyridinitril®		Apple scab, *Alternaria* spp. *Botrytis* spp.	Foliar fungicide	5000	DAS 1 182 896 (1963), E. Merck AG, Inv.: Mohr et al.; *C.A.* **62**, 4012k (1965)

310

Imidazoles and triazoles. This new class of highly active, protective, curative, and partly systemic fungicides first became known through the work of Büchel and Grewe (Bayer AG).[1142] The main focus of activity of these derivatives lies in the control of powdery mildews, rusts, *Cercospora* and *Venturia* spp., cereal diseases caused by loose and covered smuts, and *Helminthosporium* spp.—sometimes at extremely low application rates. Bayleton® is the first product from the azole class to win a significant market share. They find use both as foliar fungicides and as cereal seed dressings.

Buchenauer[1142a] has shown that the mechanism of action of these compounds involves inhibition of ergosterol biosynthesis in the fungi. Commercial and experimental products are listed in Table 25.

Benzimidazoles and phenylene-bis-thiophanates. A class of compounds with a very wide activity spectrum and high systemic action became known with the discovery of the benzimidazole fungicides. Merck & Co. developed *thiabendazole* as a systemic foliar and soil fungicide against *Cercospora*, *Fusarium*, and *Sclerotinia* spp.; Bayer AG developed *fuberidazole* as a cereal seed dressing against *Fusarium* spp. having a higher intensity of action than the organomercurials. Thereafter, Du Pont achieved a breakthrough with the discovery of the benzimidazole carbamates—highly active systemic fungicides particularly against *Botrytis, Ceratocystis, Cercospora, Erysiphe, Fusarium, Pyricularia, Podosphaera,* and *Venturia* spp. Like the other benzimidazole carbamates and the thiophanates, *benomyl* (Du Pont) breaks down to *carbenazim*, fungicidally an equally active metabolite, in the field so that resistance phenomena that have appeared with *benomyl* lead to cross-resistance to other benzimidazole carbamates and thiophanates. These resistance phenomena have limited the scope for application of *benomyl*. Investigations on the mechanism of action of this class of products have shown that the benzimidazoles interfere both in DNA, RNA, and protein biosynthesis and in the course of oxidative phosphorylation (Sisler,[1143] Decallone[1144]). See Table 26 for commercial and experimental products.

Pyrimidines. Pyrimidine derivatives, for example, *methirimol, ethirimol,* and *triarimol*, are also systemic fungicides highly active against powdery mildews and are used in cereals, fruit, and vegetables. Recent trials on *methirimol* and *ethirimol*, introduced by ICI[1145] in 1968, have indicated the appearance of resistance phenomena. (See Table 27.)

Quinoxalines. Quinoxalines were introduced by Bayer AG in 1965 as foliar fungicides. They have a high protective and curative action against powdery mildews in fruit, grapes, tobacco, tea, and ornamentals and have an acaricidal side effect. Starting from *o*-phenylenediamines, they may be prepared by reaction with oxalic acid, chlorination, and reaction with sodium sulfide[1146] (see Table 28). 2,3,5,6- and 2,3,5,8-Tetrachloroquinoxalines have been reported as experimental fungicides.[1147]

Quinazolines. 4-Methyl-4,5-dihydrotetrazolo[1,5-*a*]quinazoline-5-one,[1147a] PP-389 (ICI), is a systemic rice fungicide against *Pyricularia oryzae*.

TABLE 25. Imidazole and Triazole Fungicides

Name	Structure	Manufacturer	Use	LD$_{50}$ (mg/kg Rat, Oral, Acute)	Reference
Fluotrimazol Persulon®		Bayer AG	Mildew control in fruit, cereals, ornamentals	5000	DOS 1795249 (1968), Bayer, Inv.: Büchel et al.; *C.A.* **74,** 100062t (1971)
Triadimefon Bayleton®		Bayer AG	Control of mildew and rust in cereals, vegetables, pome and stone fruit, coffee, ornamentals; systemic	500	DOS 2201063 (1972), Bayer, Inv.: Meiser et al.; *C.A.* **79,** 105257y (1973)
Triadimenol Baytan®		Bayer AG	Widely active seed dressing, powdery mildews, covered smuts	1161	DOS 2324010 (1973), Bayer, Inv.: Krämer et al.; *C.A.* **82,** 156325p (1975)
Bitertanol (proposed) Baycor®		Bayer AG	Foliar fungicide, protective and curative, scab, powdery mildews, *Cercospora, Mycosphaerella, Puccinia* spp.	5000	See *triadimenol* (above)
Diclobutrazol (proposed) Vigil®		ICI Ltd.	Powdery mildews, cereal rust	4000	DOS 2737489 (1977), ICI, Inv.: Balasubramanyan et al., *C.A.* **88,** 184647n (1978)

Name	Structure	Company	Target / Uses	No.	References
Imazalil		Janssen Pharm. NV	*Helminthosporium, Penicillium* on citrus fruits	–	DOS 2063857 (1970) Janssen, Inv.: Godefroi, Schuermang; *C.A.* **75**, 118319n (1971)
CGA 64250 Tilt®		Ciba-Geigy AG	Powdery mildews, cereal rust	1517	DOS 2551560 (1975), Janssen, Inv.: Reet, Heeres, Wals; *C.A.* **85**, 94368f (1976)
CGA 64251 Vanguard®		Ciba-Geigy AG	Powdery mildews, scab, *Cercospora* spp.	1343	See Tilt® (above)
Fenapronil Sisthane®		Rohm and Haas	Powdery mildew, cereals, rusts, *Helminthosporium*	1500	DOS 2604047 (1976), Rohm & Haas, Inv.: Miller, Carley, Chan; *C.A.* **86**, 1093v (1977)
Prochloraz		Boots Co. Ltd.	Powdery mildew, cereals	1600	DOS 2429523 (1974) Boots, Inv.: Brookes et al.; *C.A.* **82**, 15603h (1975)
Butrizol Indar®		Rohm and Haas	*Puccinia rec.*	–	–

313

TABLE 26. Fungicidal Benzimidazoles and Thiophanates

Name	Structure	Manufacturer, Trade Name	Use	LD_{50} (mg/kg Rat, Oral, Acute)	Reference
Thiabendazole		Merck & Co. Mertect®	Seed dressing, ornamentals; foliar fungicide, sugar beet, lawn grass, post-harvest in citrus	3,330	USP 3017415 (1960) Merck & Co., Inv.: Sarett, Brown; *C.A.* **56**, 15517c (1962)
Fuberidazole		Bayer AG Voronit®	Cereal seed dressing	1,100	DAS 1209799 (1964), Bayer, Inv.: Frohberger, Wiegand; *C.A.* **64**, 14900g (1966)
Carbendazim	–NH–COOCH₃	BASF Hoechst AG Bavistine® Derosal®	Foliar fungicide in wine grape, fruit, vegetables, cereal; cereal seed dressing	15,000	USP 3010968 (1959), Du Pont, Inv.: Loux; *C.A.* **58**, 1466g (1963). DOS 1932297 (1969), Bayer, Div.: Widdig, Kühle; *C.A.* **74**, 764265 (1971)
Benomyl	CO–NH–C₄H₉ / NH–COOCH₃	Du Pont Benlate®	Foliar fungicide in wine grape, fruit, vegetables, citrus; cereal seed dressing	>9,590	USP 3541213 (1966), Du Pont, Inv.: Klopping; *C.A.* **72**, 21692d (1970)
Mecarbinzid(e)	CO–NH–CH₂–CH₂–S–CH₃ / NH–COOCH₃	BASF	Experimental product, as *benomyl*	>1,000	DOS 1812100 (1968), BASF, Inv.: Osieka et al.; *C.A.* **73**, 35374j (1970)

Name	Structure	Manufacturer	Use		References
Cypendazole	CO—NH—(CH₂)₅—CN ; benzimidazole with NH—COOCH₃ ($CO-NH-(CH_2)_5-CN$; $NH-COOCH_3$)	Bayer AG	Foliar fungicide; experimental product (activity like *carbendazim* and against **Phycomycetes**)	2,500	DOS 1812005 (1968), Bayer, Inv.: Daum et al.; *C.A.* **73**, 56097m (1970)
Thiophanate	$NH-C(=S)-NH-COOC_2H_5$; $NH-C(=S)-NH-COOC_2H_5$ (o-phenylene)	Nippon Soda Topsin®	Soil and foliar fungicide, fruit, vegetables, coffee, sugar beet	15,000	DOS 1806123 (1967), Nippon Soda, Inv.: Noguchi et al.; *C.A.* **71**, 70347h (1969)
Thiophanateethyl	$NH-C(=S)-NH-COOCH_3$; $NH-C(=S)-NH-COOCH_3$ (o-phenylene)	Nippon Soda Topsin M®	Soil and foliar fungicide, see Topsin®	7,500	See *thiophanate* (above)
NF 48	NH_2 ; $NH-C(=S)-NH-COOCH_3$ (o-phenylene)	Nippon Soda	Cereal seed dressing, experimental product	—	DOS 1930540 (1968), Nippon Soda, Inv.: Noguchi et al.; *C.A.* **73**, 14523s (1970)

TABLE 27. Fungicidal Pyrimidines

Name	Structure	Manufacturer, Trade Name	Use	LD_{50} (mg/kg Rat, Oral, Acute)	Reference
Methirimol	C_4H_9, OH, $N(CH_3)_2$, H_3C, N, N	ICI Milcurb®	Foliar fungicide against powdery mildews on vegetables, systemic	4000	1145
Ethirimol	C_4H_9, OH, $NH-C_2H_5$, H_3C, N, N	ICI Milstem® Milgo®	Seed dressing, spray for cereals against powdery mildews, systemic	1145	
Bupirimate	C_4H_9, $O-SO_2-N(CH_3)_2$, $NH-C_2H_5$, H_3C, N, N	ICI Nimrod®	Experimental product, systemic	—	DOS 2246645 (1971), ICI, Inv.: Cole et al.; *C.A.* 78, 159654y (1973)

Triarimol		Eli Lilly Elanocide®	Foliar fungicide against apple mildew and scab	600	Neth. P. 6806106 (1967) Eli Lilly, Inv.: Davenport et al.; *C.A.* **72**, 100745b (1970)
Fenarimol		Eli Lilly Rubigan®	Foliar fungicide against powdery mildews, scab, *Cercospora* spp., *Fusarium* spp.	4500	See *triarimol* (above)
Nuarimol		Eli Lilly Trimidal®	Seed dressing, *Helminthosporium*, *Ustilago*, *Fusarium*, powdery mildews	—	See *triarimol* (above)

317

TABLE 28. Fungicidal Quinoxalines

Name	Structure	Manufacturer	Use	LD$_{50}$ (mg/kg Rat, Oral, Acute)	Reference
Chinomethionat Oxythioquinox		Bayer AG Morestan®	Foliar fungicide, powdery mildews, fruit, wine grapes, vegetables, ornamentals; acaricidal	9500	DAS 1100372 (1958), Bayer, Inv.: Grewe et al.; C.A. **55**, 26353h (1961)
Chinothionat Thioquinox		Bayer AG Eradex®	Foliar fungicide, fruit, acaricidal	1800	
Chlorquinox		Fisons Lucel®	Foliar fungicide, powdery mildews in cereals	6000	DOS 2107031 (1970), Fisons, Inv.: Barker; C.A. **76**, 21953x (1972). Belg. P. 641214 (1962), Fisons, C.A. **63**, 1175c (1965)

1,2-Oxazoles, 1,3,5-triazines. 3-Hydroxy-5-methyl-1,2-oxazole **(1)** has been developed as a soil fungicide particularly for paddy rice and as a seed dressing for sugar beet against *Fusarium*, *Aphanomyces*, and *Pythium* spp.[1148]

4,6-Dichloro-*N*-(2-chlorophenyl)-1,3,5-triazin-2-amine **(2)** is a widely active foliar fungicide. Synthesized from cyanuric chloride and 2-chloroaniline,[1149] it is used in vegetables, potatoes, and ornamentals for control of *Alternaria*, *Fusarium*, *Plasmopara*, *Puccinia*, and *Rhizoctonia* spp.

1; Hymexazol, Tachigaren®;
LD_{50}: 3112 mg/kg rat, oral, acute

2; *Anilazine* (Ethyl Corp.) Dyrene® (Mobay Ag. Chem. Div./Chemagro);
LD_{50}: 2710 mg/kg rat, oral, acute

3.3. Bactericides and Viricides

Since bacteria and viruses grow intracellularly in plants, their control with chemicals is extremely difficult. At the present time crop protection measures against viruses are directed mainly at control of the vectors. In the case of bacterial infections antibiotics are used—as, for example, *streptomycin*, in human medicine. This

antibiotic, isolated by Waksman,[1150,1151] is marketed as its sulfate or hydrochloride for the control of *Xanthomonas stewartii* in corn (maize), of *Erwinia amylovora* in apples and pears, and of bacterial infections of tobacco (*Pseudomonas tabaci*). It is one of the most active and most widely used bactericides, although its usefulness is limited by cost and phytotoxicity.

The antibiotic *chloramphenicol* is also used together with disodium methyl-arsonate for control of *Xanthomonas* spp. (*Xanthomonas oryzae* in rice).

Shiragen® (Sankyo K.K./Japan)

Of the inorganic compounds used as fungicides, the following copper salts also have bactericidal properties and may be used for control of *Xanthomonas* spp.: copper cyclopentanecarboxylate (see Section 3.2.2.2), copper linoleate (Section 3.2.2.2), copper oxide (Section 3.2.1.2.1), copper oxychloride (Section 3.2.1.2.1), and copper zinc sulfate.

Derivatives of thiocarbonic acids, already discussed as fungicides, are also used for control of *Xanthomonas* spp., particularly in rice, for example, dimethyl-carbamodithioatonickel (Sankel®, see Section 3.2.5.1.3.1) and cadmium propyl xanthogenate.

Cadenax® (Toa Ag. Chem. Co., Japan)

A mixture of 2,4-xylenol and *m*-cresol is marketed under the name Bacticin® (Upjohn) for treatment of bacterial tumors on the trunks of fruit trees.

The cyclic amidine **1** is used as a bactericide against fire blight (*Erwinia amylovora*) of apple, pear, and cherry.

1; 1,4,5,6-Tetrahydro-1-[n]dodecyl-2-methyl-pyrimidine; TD-225 FS, Pennwalt TD-225 (Pennwalt)

Celdion® (**2**) and the *N*-oxide (**3**)[1152,1153] have been recommended as rice bactericides.

2; 4-Phenyl-3-[(phenalmethylene)amino] thiazole-2(3H)-thione; Celdion® (Takeda Seiyaku K.K., Japan); LD$_{50}$: 10,000 mg/kg rat, oral, acute

3; Phenazine-5-oxide; Phenazin W.P.® (Meiji Seika K.K., Japan); nematicidal, acaricidal, fungicidal properties; LD$_{50}$: 12,150 mg/kg rat, oral, acute

The experimental bactericide **4** has a high activity against *Xanthomonas oryzae*.

4;   2-Amino-1,3,4-thiadiazol; TF-128 (Takeda Chem. Ind.)[1154]

2-[(2,3-Dichlorophenyl)aminocarbonyl]-3,4,5,6-tetrachlorobenzoic   acid[1154a] (*techlofthalam*—proposed common name) has a high activity against *Xanthomonas oryzae* as a spray or soil drench.

2-[(4-Chloro-3,5-dimethyl)phenoxy]ethanol was reported as a protective viricide in peaches and a systemic fungicide for control of carnation wilt.[1155]

Experimental Chemotherapeutant 1182 (Union Carbide Co.)

_____ *Chapter Four*

Herbicides

_____

G. JÄGER
Bayer AG, Leverkusen

4.1. Introduction

A *herbicide*, in the broadest sense of the word, is any compound that is capable of
either killing or severely injuring plants and may thus be used for elimination of
plant growth or the killing off of plant parts. The designation *weed* covers a variety
of meanings according to the particular situation. According to the general definition a weed is any plant—either a wild or cultivated variety—that is undesired in
that particular place. In agriculture and horticulture weeds are thus any plants other
than the specific crop being grown. On railway tracks, industrial sites, airports,
paths, open spaces, and the like, the entire vegetation can be regarded as weeds.
Weeds are conveniently divided into dicotyledonous plants, termed *broadleaf
weeds*, and monocotyledonous plants, termed *grass weeds*. (See Table 29.)

Weed control by chemical means has undergone rapid expansion since the introduction of the selective organic herbicides in the period after the last world war.

A steep upward development is particularly noticeable in the last 15 years.
Today a major proportion of the agricultural land devoted to the important crops
in the intensive agricultural industries of Europe, North America, and parts of Asia
is treated with herbicides. The reasons for the increased use of chemical agents are
manifold. Apart from the purely economic aspects, global political questions concerning food also play a significant role in view of the current world food-supply
situation. (See Table 30.)

Strong competitive pressures in the growing of agricultural produce have made
rationalization necessary in the planting, care, and harvesting of cash crops. The
trend to large monocultures has led to an increased danger of weed infestation as
the earlier farming techniques of crop rotation and intensive cultural practices are
being reduced. Simultaneously the shortage and cost of agricultural labor prevent
economic control of weeds either by hand or machine; thus the new chemical pro-
cedures have become necessary to ensure high crop yields.

TABLE 29. Important Broadleaf and Grass Weeds[a]

M *Agropyron repens*	Quackgrass/Couch[b]
M *Alopecurus myosuroides*	Slender foxtail/Blackgrass
D *Amaranthus* spp.	Amaranth, pigweed
D *Anthemis* spp.	Chamomile
M *Apera spica-venti*	Windgrass/Silky bentgrass
D *Atriplex* spp.	Orach(e), saltbush
M *Avena fatua*	Wild oat
M *Carex* spp.	Sedge
D *Chenopodium album*	Common lambsquarters/Fat-hen
D *Cirsium arvense*	Canada thistle/Creeping thistle
D *Convolvulus* spp.	Bindweed, morning glory
M *Cynodon dactylon*	Bermudagrass
M *Cyperus* spp.	Flatsedge, nutsedge
M *Digitaria sanguinalis*	Large crabgrass/Hairy fingergrass
M *Echinochloa crus-galli*	Barnyardgrass
P *Equisetum* spp.	Horsetail
M *Festuca* spp.	Fescue
D *Galeopsis* spp.	Hempnettle
D *Galinsoga* spp.	Galinsoga, gallant soldier
D *Galium* spp.	Bedstraw
D *Linaria vulgaris*	Yellow toadflax/Common toadflax
D *Lolium* spp.	Ryegrass
D *Matricaria* spp.	Chamomile, mayweed
M *Panicum* spp.	Panicum, panic grass
M *Poa* spp.	Bluegrass/Meadowgrass
D *Polygonum* spp.	Smartweed, knotweed
M *Setaria* spp.	Foxtail/Bristlegrass
D *Sinapsis arvensis*	Wild mustard/Charlock
D *Solanum nigrum*	Black nightshade
D *Sonchus* spp.	Sowthistle
M *Sorghum halepense*	Johnsongrass
D *Stellaria media*	Common chickweed
D *Veronica* spp.	Speedwell

[a]D: Dicotyledoneae, M: Monocotyledoneae, P: Pteridophyta.
[b]U.S. term/British term.

The reduction in yield per acre (hectare) due to weed infestation is of no mean significance. Weeds compete with the crop plants for water, light, food, and lebensraum above and below the soil surface. The wild mustard plant, for example, requires twice as much nitrogen and phosphorus and four times as much potassium and water as an oat plant. A serious infestation of the crop with wild oat can reduce the roots of cereal plants to two-thirds of their normal extent, the disturbance to food intake thus caused being reflected in the yield and quality of the cereal harvested.

The value of harvest losses by insect pests, plant diseases, and weeds in spite of

TABLE 30. Some Important Cultivated Plants

Allium spp.	Onion, leek, garlic, etc.
Ananas sativus	Pineapple
Arachis hypogaea	Groundnut, peanut
Avena sativa	Oat
Beta vulgaris spp.	Beta beets (sugar-, fodder-, redbeet, mangel)
Brassica spp.	Brassicas (cabbage, etc.)
Brassica napus	Rape
Citrus spp.	Citrus
Coffea arabica	Coffee tree
Fragaria spp.	Strawberry
Glycine max (*Soja hispida*)	Soybean
Gossypium hirsutum	Cotton
Hordeum vulgare	Barley
Ipomoea batatas	Sweet potato
Lunum usitatissimum	Common flax
Malus spp.	Apple tree
Medicago sativa	Alfalfa, lucerne
Nicotiana tabacum	Tobacco
Oryza sativa	Rice
Panicum milliaceum	Common millet
Phaseolus spp.	Bean
Pisum sativum	Pea
Pyrus spp.	Pear tree
Saccharum officinarum	Sugarcane
Secale cereale	Rye
Solanum lycopersicum	Tomato
Solanum tuberosum	Potato
Sorghum spp.	Sorghum
Triticum spp.	Wheat
Vitis vinifera	Grape
Zea mays	Corn, maize

modern crop protection measures is estimated worldwide as 35% of the potential total harvest. About 9–10% of the reduced yield is caused by competition from weeds.[1156]

Resistance phenomena, as have been observed to a great extent with insecticides and fungicides, have not yet been reported from the use of herbicides. However, a shift in the weed spectrum, due to selection of a few weed species that are difficult to control with the present range of herbicides available, does present a problem in many places.

Apart from reducing the yield losses due to weeds, the use of herbicides also contributes substantially to rationalization of crop care and harvesting. In potato and beet growing whole operations such as hoeing and earthing-up have been made superfluous by the application of suitable herbicides. An economic harvesting of large-scale cash crops is only possible today with machines, which, in some crops, necessitates a special chemical treatment before such machines can be used. For example, cotton is treated with a defoliant and potato is treated with a dessicant to make mechanical harvesting convenient and problem free.

Even to summarize all the applications of herbicides here would exceed the limits imposed on this chapter. The most important areas of use are mentioned in the discussion of the various classes of herbicidal compounds. First, important concepts in the use and mode of action of herbicides are explained in general terms in the following section.

4.2. Use and Mode of Action of Herbicides

4.2.1. CLASSIFICATION OF HERBICIDES ACCORDING TO PRACTICAL CONSIDERATIONS

From the viewpoint of the user or the biologist important points of differentiation for the classification of a herbicide are as follows: time of application, manner in which the agent is absorbed by the plant, the mode of action, and the range of application.

A rough classification can be made on the basis of the manner of absorption and the mode of action. A herbicidal agent taken up by the plant via the roots is termed a *soil herbicide*, in contrast to a *foliage herbicide*, which enters via the green, aerial plant parts. Foliage herbicides that exert their effect directly on the plant parts that have been contacted with the agent are designated *contact herbicides*. All agents that are translocated within the plant after absorption, that is, for which site of absorption and site of action are nonidentical, are called *systemic* or *translocated* herbicides.

Apart from the quantity of herbicide applied, the timing of the application is also important for an optimal control of weeds. Three timings are differentiated according to the state of development of the crop plants when the herbicide is applied.

1. Pre-sowing.
2. Pre-emergence.
3. Post-emergence.

The type of herbicide used is selected according to the nature and state of development of the target weed. Soil-applied herbicides are used for weeds not yet emerged; foliar- or soil-applied herbicides are used post-emergence.

Herbicides are split into two groups according to their range of application. The so-called total herbicides can be used for unspecific extermination of vegetation, for suppression of the entire plant growth on industrial sites, railway tracks, pathways, and so on. Naturally, a long residual action of the herbicide is desirable for this indication to avoid the necessity of multiple treatments at short intervals.

In the use of total herbicides on agricultural land before or after the growing of the crop, however, a correspondingly short duration of action is required to avoid damage to the subsequent crop.

The so-called semitotal herbicides form a subgroup of the total herbicides. These have a high, unspecific weed action but are well tolerated by woody plants. The most important areas of application of such agents are fruit and wine grapes, tropical plantation cultures (citrus, banana, coffee, cacao, palm, tea), forests, and tree and shrub nurseries.

By far the largest and economically most important group of agents used today is that of the selective herbicides, which find application in all crops. A selective herbicide has high activity against the target weed but is well tolerated by the crop plant.

Because it is seldom possible to control all the target weed species simultaneously with one agent, various combinations are used in practice to achieve consistent (as far as possible) control success. According to the type of problem typical soil-applied herbicides (to prevent new growth), foliar-applied agents (to kill off emerged weeds), or mixtures of both are used in this operation, particularly in total weed control.

The characteristics described here for the classification of herbicides, although useful in practice, are not suitable for a clear division since various agents do not adhere strictly to them. Many herbicides are absorbed by the plant both via the roots and via the foliage. They can act as contact agents and can also be translocated, or they can be applied at different states of development. Even the division between total and selective herbicides is not clear-cut. Almost total weed control can also be achieved by raising the application rates of selective agents.

4.2.2. SELECTIVITY OF HERBICIDAL ACTION

The causes of the differences in sensitivity of plants to herbicides are very complex. Although a precise knowledge of the relationships between structure, physicochemical properties, and selective herbicidal action of a chemical would be of supreme importance for the chemist synthesizing new compounds, the state of knowledge in this area is still very fragmentary. A complex interaction of numerous properties of the chemical, the plant contacted with it (morphology, physiology, biochemistry), and its environment (soil factors, humidity, light, temperature) is responsible for the selectivity of herbicides.

According to investigations to date, the causes of herbicidal selectivity may be divided into two groups, the physical and the biochemical.

4.2.2.1. Physical Causes

Physical causes of herbicidal selectivity include application techniques that lead to a spatial separation of the herbicide and the crop plant. For undercanopy spray applications in emerged crops, special techniques are employed to ensure that the herbicide has contact only with the weeds.

The different germination or seed depths of weed and crop plant can lead to a superficially applied soil herbicide only being absorbed by weeds germinating in the upper soil level for a corresponding crop seed depth. A very low migration tendency of the herbicide in the soil is, of course, a precondition.

Fixing of the agent can be caused by its low water solubility, which prevents washing down into deeper soil levels, but interactions between the agent and the soil play a significant role. Depending on their structure, herbicides can be adsorbed in the soil to a high or low degree. The clay and humus content is decisive for the sorptivity of a soil, which is closely related to its cation-exchange capacity.[1157-1159]

Soil-applied herbicides that act as photosynthesis inhibitors must be translocated to their site of action in the green plant parts after uptake in the root system. Differences in the rate of transport in weeds and crop lead to different concentrations in shoots and leaves and thus contribute to selectivity.

Morphologico-anatomical differences among the various plants also influence the selectivity of a herbicide to a great extent. The disposition of the leaves and the structure of the leaf surface affect adhesion, duration of contact, and thus the uptake of the agent.

4.2.2.2. Biochemical Causes

In many cases species-specific detoxification mechanisms are responsible for the high tolerance of various plants toward certain herbicides. Inactivation can occur by rapid metabolism of the agent (nonenzymatic and enzymatic hydroxylation, oxidation, demethylation) within the plant to nonphytotoxic derivatives or by formation of nontoxic conjugates of the agent with plant products (glycosidation, conjugation with amino acids and peptides). The cases in which the chemical applied is first metabolized to the actual active agent within the plant also belong to this group of biochemical factors determining selectivity. The presence or absence of such an activation mechanism thus decides the sensitivity or resistance of the particular plant species. This biochemical inactivation or lack of activation is responsible in many cases for the high tolerance of crop stands toward organic herbicides but also explains the occasional high degree of resistance of some weeds.

Since biochemical selectivity factors are intimately related to the chemical properties of the agents, special biochemical causes of selectivity are discussed in more detail in the corresponding sections on the individual organic herbicides.

4.2.3. MECHANISMS OF ACTION

The precondition for herbicidal activity of a chemical is its ability to interfere in such a way in basic biochemical phytoprocesses that the disturbances thus caused to vital functions lead to severe injury and finally to death of the plant.

The range of possibilities open to a herbicide to disrupt biochemical functions is very wide; the knowledge of the primary processes taking place is, in contrast, very fragmentary. Today the influences on respiration and photosynthesis, two primary metabolic processes closely related to the energy balance of the plant,

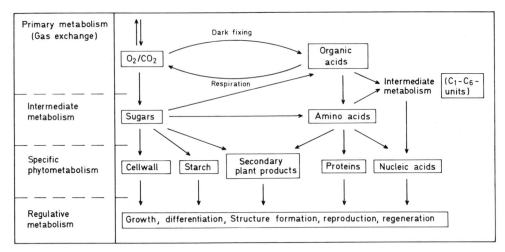

FIGURE 4. Simplified scheme of phytometabolic processes.

belong to the effects most thoroughly investigated. Reliable experimental results on the disruption of elemental biochemical processes by herbicides in the intermediate metabolism, the specific phytometabolism, and the regulative metabolism (Figure 4) are limited to isolated cases. Often only morphological changes or deviations from normal physiological behavior by the plant can be observed as consequences of disturbances in the primary metabolism. Differentiation, reproduction, and regeneration are affected negatively by disruptions in nucleic acid or protein synthesis.

Furthermore, the ability of the whole organism to survive can be severely reduced by disorganization of cellular structures (e.g., destruction of cell membranes) or by inhibition of the synthesis of essential secondary metabolic products (e.g., lipids, carotinoids).

In many cases the total phytotoxic effect of a herbicide cannot be ascribed to interference with one bioprocess alone. Often the herbicide interferes simultaneously in several completely different metabolic processes—admittedly with variable effectiveness and in a manner dependent on concentration. However, it is normally the case that one of the functional disruptions caused by the chemical is so dominant that it can be regarded as the actual cause of the herbicidal effect.

4.2.3.1. Photosynthesis

In photosynthesis, green plants, algae, and some bacteria synthesize carbohydrates from carbon dioxide and water under the influence of light and evolve oxygen as a by-product. The whole process, which is made up of a series of complex biophysical and biochemical operations, is, in the final analysis, a conversion of light, that is, electromagnetic energy, into chemical energy.

Although the principle of "assimilation of carbon dioxide" by plants was recognized in the mid-19th century, it is only in the last few decades that the individual reaction steps and their functions have been studied in depth. Photosynthesis, which takes place in the chloroplasts—highly organized cell organella of higher plants—may be divided into several phases.

1. Activation of the pigment systems in the chloroplasts (normally composed of chlorophyll a and b) by absorption of light quanta.

2. Conversion of the energy stored in the "activated pigments" into chemical energy via electron transport processes in the two Photosystems I and II. The reduction equivalents created by the photolysis of water serve to convert $NADP^\oplus$ into NADPH.

$$NADP^\oplus + H_2O \xrightarrow{h\nu} NADPH + 1/2\,O_2 + H^\oplus$$

Coupled with this is the formation of ATP from ADP and inorganic phosphate Pa (phosphorylation).

$$ADP + P_a \xrightarrow{h\nu} ATP$$

3. Input of the energy contained in NADPH and ATP into subsequent energy-consuming reactions that are not directly light-dependent: fixation and reduction of carbon dioxide.

In the electron-transport chain (Figure 5), made up of Photosystems I and II and various electron carriers (e.g., plastoquinone, cytochrome, plastocyanin, ferredoxin), the electrons released by the photolysis of water are transferred to $NADP^\oplus$ Two electrons flowing through the chain are required to reduce one $NADP^\oplus$ to NADPH and contribute simultaneously to the formation of ATP. Both photosystems are involved in this ATP synthesis, designated noncyclic photophosphorylation, in which, according to latest investigations, not 1,[1160] but $\sim \frac{4}{3}$ATP/2 e,[1161] is formed. In contrast, only Photosystem I is involved in cyclic photophosphorylation, which also occurs under the influence of light but independent of the "open electron-transport system."

Practically all chemicals whose herbicidal activity is due to inhibition of photosynthesis disrupt the light reactions and block the photosynthetic electron transport. Among the inhibitors that interfere with light reaction II are found ureas,[1162–1165] 1,3,5-triazines,[1166–1168] 1,2,4-triazinones,[1169–1171] uracils,[1172,1173] pyridazinones,[1174,1175] 4-hydroxybenzonitriles,[1176,1177] N-arylcarbamates,[1178,1181] and some acylanilides[1163,1182]—representing a series of economically important herbicide classes. According to results available to date—detailed studies have been carried out particularly on *diuron* [*DCMU*, *N'*-(3,4 dichlorophenyl)-*N,N*-dimethylurea]—the sites of inhibition for these agents are probably localized in the electron-transport chain between the electron acceptor of Photosystem II (A_{II}) and the plastoquinone. The bipyridylium herbicides ("quats") *diquat* and *paraquat* interfere with light reaction I in that they function as electron acceptors in place of natural ferredoxin and thus prevent the reduction of $NADP^\oplus$. The hydrogen peroxide synthesized by the reoxidation of the radical cations thus formed (cf. Section 4.3.2.7)

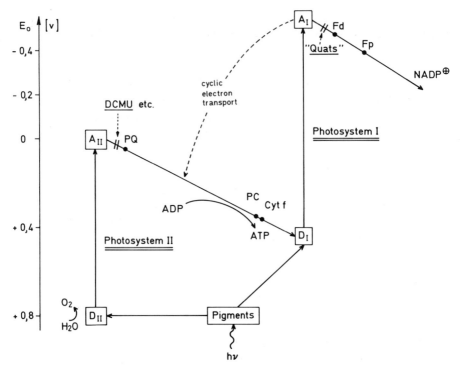

FIGURE 5. Redox potential diagram of the photosynthetic electron transport in chloroplasts. NADP$^+$: Nicotinamide–adenine–dinucleotide–phosphate cation. ADP: Adenosine diphosphate. ATP: Adenosine triphosphate. D$_{I,II}$: Electron donors in Photosystems I and II. A$_{I,II}$: Electron acceptors in Photosystems I and II. Cyt f: Cytochrome f (hemoprotein; E_0 + 0.36 V). Fd: Ferredoxin (nonhemo-ferroprotein of mol. wt. ~11,000; E_0 – 0.42 V). Fp: Fd–NADP$^+$–oxidoreductase (flavoprotein; E_0 ~–0.4 V). PC: Plastocyanin (copper protein of mol. wt. ~22,000; E_0 + 0.36 V). PQ: Plastoquinone.

PQ

is highly cytotoxic (oxidative degradation of unsaturated fatty acids in chloroplasts and cell membranes by radical chain reactions) and to be regarded as the actual cause of the rapid death of plants contacted with "quats.[1183,1184]"

Hill showed in 1937 that photolysis of water also takes place in isolated chloroplasts if an electron acceptor A (Hill reagent: potassium ferricyanide, NADP$^\oplus$, reducible dye) is added to the chloroplast preparations.

$$H_2O \ + \ A \quad \xrightarrow{h\nu\,/\,Chloroplast} \quad AH_2 \ + \ 1/2\,O_2$$

Since then the Hill reaction has contributed greatly to a more precise understanding of primary photosynthetic processes. As a simple *in vitro* method it permits the quantitative determination of the inhibitory properties of photosynthesis blockers on chloroplast systems by measurement of O_2 evolution (oxygen electrode, Warburg manometer) or by following photometrically the reduction of acceptor A in the presence of the inhibitor. The degree of inhibition caused by a given agent X is often expressed as a pI_{50} value, the negative logarithm of the molar concentration of X that effects a 50% inhibition of the Hill reaction.

The ease of availability of quantitative biological data on inhibitors of the Hill reaction, in the form of pI_{50} values, can be used to relate changes in these biological values to changes in measurable molecular or substituent parameters (e.g., distribution coefficient, electronic and steric factors) within various classes of herbicides. These techniques, first introduced by Hansch,[1185-1188] permit predictions of the biological effect of an unknown substance within the class under study. They indicate the molecular properties essential for biological activity and thus put the chemist in the happy position to develop a theoretically based synthesis plan. Such techniques have been applied recently by various research teams to classes of herbicidal compounds such as the *N*-acylcarbamates, ureas, benzimidazoles, 1,2,4-triazin-5-ones, and 1,3,5-triazines.[1169,1170,1189-1192]

4.2.3.2. Respiration

In the process of respiration, which occurs both in the animal and the vegetable organism, the chemical energy present in the form of protein, carbohydrate, and fat is transformed into another form of chemical energy (phosphate bonds in ATP), which is then generally useful for powering biochemical reactions in that organism.

The simplified total picture of respiration (Figure 6) can be arranged into four sections each of which is composed of a sequence of complex single processes.

I. Conversion of protein, carbohydrate, and fat via their simple components (amino acids, hexoses, pentoses, fatty acids, glycerine) to acetyl-coenzyme A (Acetyl-CoA).

II. Oxidation of Acetyl-CoA in the citric acid cycle (Krebs cycle) with formation of reduced nicotinamide–adenine–dinucleotide ($NADH_2$).

III. Reoxidation, that is, transfer of electrons from $NADH_2$ via a series of electron carriers in an electron-transport chain to oxygen.

IV. Coupling of the electron transfer in the energy gradient of the transport chain with storage of the energy released by formation of ATP from ADP and inorganic phosphate (P_a). Three molecules of ATP are formed per molecule of $NADH_2$. This process is called oxidative phosphorylation.

Only the last two sections are of importance with regard to the influence of herbicides on respiration. Like the citric acid cycle, these are confined to certain subcellular organella, the mitochondria. The herbicide can act by blocking the mitochondrial electron transport (A), thus inhibiting respiration, or by preventing coupling of oxidative phosphorylation to the electron-transport chain (B). In the latter case respiration is not inhibited but takes place without production of energy.

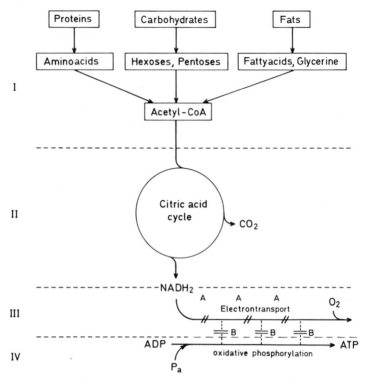

FIGURE 6. Simplified representation of respiration.

Most herbicides interfering in the process of respiration are uncouplers. The dinitrophenols and halophenols act in this manner.[1193] Blocking of the electron-transport chain has only been observed for a few commercial herbicides such as *nitrofen* [2,4-dichloro-1-(4-nitrophenoxy)benzene], which blocks the mitochondrial electron-transport chain and acts simultaneously as a photosynthesis inhibitor by light-activated degradation of xanthophyll pigments.[1194,1195]

4.2.3.3. Miscellaneous Mechanisms of Action

Apart from the inhibitors of photosynthesis and respiration, there are various classes of herbicides that interfere in specific and regulative phytometabolic processes. Generalized growth disturbances are caused by the aryloxyalkylcarboxylic acids and numerous benzoic acids.

These compounds, sometimes referred to as growth regulator herbicides, correspond to the natural phytohormone auxin (3-indolylacetic acid) in their biological effect and may be regarded as "artificial auxins." The morphological changes (abnormal growth, rapid ageing, and abnormal physiological behavior) observed after treatment of plants with these compounds are the same as those to be observed after application of higher quantities of natural auxin.[1196] The primary processes that are responsible for these effects remain practically unidentified in spite of many investigations. It is assumed today that the growth regulator herbicides simulate the naturally occurring 3-indolylacetic acid in their molecular geometry and charge distribution and interact with the same receptor. Experiments indicate that auxin and its synthetic analogues stimulate the construction of the plant cell wall.[1197]

Reasons for the highly selective properties of the growth regulator herbicides— they act almost exclusively against dicotyledons (monocotyledons are only affected by very high doses)—are apparently not to be sought among differences in uptake, translocation, metabolism, or deactivation. They are probably caused by differences in the receptor structures of monocotyledonous and of dicotyledonous plants.[1196]

Derivatives of fluorenylcarboxylic acid ("morphactins"), *naptalam* {2-[(1-naphthalenylamino)carbonyl]benzoic acid} and *TIBA* (2,3,5-triiodobenzoic acid), are representatives of a small class of herbicides and growth regulators whose phytotoxicity is caused by competition with auxin transport in the plant.[1198,1199]

A larger number of commercial herbicides are classed as inhibitors of cell and nucleus division. Of these the alkyl *N*-arylcarbamates also exert a definite inhibitory action on photosynthesis, but the main cause of their herbicidal efficacy is their powerful inhibitory effect on cell division.[1200–1203] The dinitroanilines also inhibit cell and nucleus division.[1204] The appearance of cellular anomalies as a result of disruption of cell division processes has also been demonstrated for other compounds such as maleic hydrazide,[1205] *chlorthal-dimethyl*[1206] (dimethyl 2,3,5,6-tetrachloro-1,4-benzenedicarboxylate), and *propyzamide* [3,5-dichloro-*N*-(1,1-dimethylpropynyl)benzamide]. The molecular biological processes responsible remain largely unelucidated but probably involve a disturbance to certain fundamental biochemical reactions (nucleic acid and protein synthesis).[1202,1203]

Inhibition of protein synthesis has been demonstrated unequivocally for MDMP (**1**) and herbicides of the chloracetamide type.[1207]

$$H_3C-\underset{NO_2}{\overset{NO_2}{\underset{|}{\bigcirc}}}-NH-\underset{CH_3}{\overset{CH_3}{\underset{|}{CH}}}-CO-NH-CH_3$$

1; *N*-Methyl-2-[(4-methyl-2,6-dinitrophenyl)amino]propanamide (SD 15 179, Shell)

MDMP inhibits incorporation of leucine into protein in soybean and wheat.[1208,1209]

A collapse of the plant organism can also be caused by interference with other biosynthetic reactions in which secondary products are formed that are necessary for the maintenance of normal metabolic functions in the plant.

Carotenoids are present in very high concentrations as protective pigments in the chloroplasts where they protect chlorophyll from a photosensitized, oxidative degradation. Some herbicides such as maleic hydrazide, *dichlormate, pyrichlor*, and *ortho*-substituted diphenylethers, which prevent normal carotenoid synthesis in the plant,[1210] thus interfere indirectly with photosynthesis via the consequent rapid degradation of the chlorophyll.

S-Alkyl dialkylcarbamodithioates[1211] and probably also some aliphatic chloro-carboxylic acids inhibit the synthesis of lipids that, among other things, are required for formation of the protective wax layer on the plant surface. It is an open question whether this is the main reason for the herbicidal activity of these compounds (e.g., via increased loss of water) or whether other, as yet unknown processes are responsible.

The mechanisms of action of numerous herbicides remain practically unknown, for example, of almost all inorganics and the organics *dichlobenil, chlorthiamid, bentazone, benzoylpropethyl,* and *diphenamid.*

4.3. Herbicidal Agents

For the discussion of the herbicides (which, with a few exceptions, is restricted to commercial products), subdivision has been made according to chemical considerations since only this method permits a systematic classification. However, a strict adherence to the normal chemical classification has not been maintained in order that biological relationships be preserved as far as possible. Thus herbicides that are first converted in the plant to the actual active compounds are treated together with the latter even when they belong to different classes. In the division into compound classes, the highest priority has been given to the functional groups and partial structures responsible for the herbicidal activity, independent of the nature of the substitution.

Literature references have been limited to the minimum necessary; comprehensive literature citations may be found in recent standard works. Trade names, manufacturers, and the common names laid down by various bodies for the characterization of herbicides (International Organization for Standardization—ISO, British Standards Institution—BSI, Weed Science Society of America—WSSA) are given. The values given for acute toxicity (LD_{50}) have been taken from various handbooks and manufacturers' technical bulletins.

4.3.1. INORGANIC HERBICIDES

The development of weed control by chemical means is closely connected with the use of inorganic chemicals. For a long time—from the beginning of chemical weed control in the second half of the last century with the use of iron(III) sulfate, sulfuric acid, and copper(II) sulfate[1212]—inorganic chemicals dominated in the control of undesired plant growth.

With a few exceptions these acids and salts, produced very cheaply, served as nonselective herbicides. Although even today inorganics are still used to a considerable extent as total herbicides and for certain special indications, they are steadily losing ground to the organics. Apart from the insufficient selectivity for use in crop stands and the high application rates usually necessary, chemical and toxicological properties in many cases also hinder wider application in modern weed control. Relatively few studies have been reported on the mechanisms of action of inorganic herbicides (see Table 31).

4.3.2. ORGANIC HERBICIDES

Starting with the introduction of the phenoxycarboxylic acids about 35 years ago, organic herbicides made the decisive breakthrough into *selective* chemical weed control. Today they are dominant in almost all areas where the control of undesired plant growth is practical. Compounds having herbicidal properties are found in numerous classes whereby the most important commercial, organic herbicides are limited to a few product groups—1,3,5-triazines, carbamates, arylureas, aryloxyfatty acids, and benzoic acid derivatives. Table 32 summarizes the shares of the various herbicide types in the total herbicide turnover for 1974, shows the forecast for 1980 and also reflects the significance of North America in this market.

At the present time corn (maize), soybean, cotton, wheat, and rice (in that order) are the intensively grown, large-scale crops that account for almost 80% of all organic herbicides applied (see Table 33).

4.3.2.1. Hydrocarbons, Alcohols, Aldehydes, Ketones, and Quinones

Only a few herbicidal compounds that have found practical application are to be found in this group. Hydrocarbons are only phytotoxic at high application rates.

High-boiling mineral oils (bp $> 230°$) that contain acidic (phenols) and basic (anilines) components in addition to a high proportion of aromatics have been used in crop protection since about 1920. They are used for control of moss and lichen on fruit trees and find limited application for selective weed control in Umbelliferae (carrot, celery, fennel).

Of the chlorinated hydrocarbons only an isomer mixture of trichlorobenzyl chlorides (*TCBC*; LD_{50}: 3075 mg/kg rat, oral, acute) has achieved any significance as a component of Randox T® (Monsanto; *TCBC* + *allidochlor*), which is used for pre-emergence control of annual grass weeds and certain broadleaf weeds in corn (maize).

Of the alcohols, derivatives of benzyl ether and 2-phenoxyethanol are on the market. As their activity probably depends upon oxidation to the corresponding benzoic acids and phenoxyacetic acids, these are discussed together with the acids in Sections 4.3.2.4.2 and 4.3.2.4.3.

Of the aldehydes and ketones only a few, simple representatives are herbicidal:

Chloral hydrate (LD_{50}: 500 mg/kg rat, oral, acute) has a grass weed-killing effect (panicum varieties) and can be used for pre-sowing treatment in winter rape. The herbicidal properties are apparently due to oxidation to trichloroacetic acid.

Acrolein [Aqualin® (Shell); LD_{50}: 46 mg/kg rat, oral, acute] is used as an aquatic herbicide and algicide.[1223] Its high toxicity toward mammals and fish prohibits wider application.

Hexachloroacetone (HCA-Weed Killer; Allied Chem Corp., 1954; LD_{50}: 1550 mg/kg male rat, oral, acute), a nonselective foliar-applied herbicide, is used in the United States mainly as a defoliant in crops not intended for food or fodder purposes.[1225]

The use of the trihydrate of hexafluoroacetone (HFA; Allied, 1967; LD_{50}: 190 ± 14 mg/kg rat, oral, acute) as a total herbicide is known.[1226]

TABLE 31. Inorganic Herbicides

Formula, Common Name	Trade Name, Manufacturer	Use	LD_{50} (mg/kg Rat, Oral, Acute)	Remarks, References
Sulfuric acid		Total herbicide; potato haulm destruction; selective at low concentrations (onion, cabbage)		Used since 1874;[1212] seldom used today
Iron (II) sulfate		Moss control	5000	[1212]
Ammonium sulfamate ($NH_2-SO_3^{\ominus}NH_4^{\oplus}$) *AMS*	Ammate® X (Du Pont, 1945) Amcide® (Albright & Wilson Ltd.)	Contact herbicide with systemic action for control of woody weeds	3900	[1213]
Potassium cyanate	Aero® Cyanate (American Cyanamid)	Mainly used in USA; selective (onion, rose)	841 (mouse)	Introduced by American Cyanamid
Calcium cyanamide (CaNCN)	Aero Cyanamid® (American Cyanamid)	Main use: nitrogen fertilizer (since 1905). Pre- and post-emergence herbicide; in spring and fall as soil-applied total herbicide; defoliant (cotton)	1400	[1214, 1215]
Sodium chlorate		Total herbicide since 1900 (not on agricultural land); dessicant for all green plant parts (potato haulm destruction); cotton defoliant	1200–7000	Fire hazard on contact with organic material; addition of borate, phosphate, carbonate, or sulfate thus recommended[1216]

Sodium borate ($Na_2B_4O_7 \cdot 10H_2O$; Borax)	Borascu® (US Borax Co.)	Total herbicide for railway and industrial sites; also used in combination with chlorates	2660–5140	1217
Disodium octaborate ($Na_2B_8O_{13} \cdot 4H_2O$)	Polybor® (US Borax Co.)	See borax	5300 (guinea pig)	1218 Manufacture: Spray drying of borax and boric acid in the correct ratio–not a genuine compound
Sodium metaborate ($Na_2B_2O_4 \cdot 4H_2O$)	(US Borax Co.)	See borax	2330	1219
Sodium arsenite ($NaAsO_2$)		Introduced in 1901 as a non-selective agent; cotton defoliant; mostly used in mixture with As_2O_3	10–50	Not registered in Germany
Potassium hexafluoroarsenate ($KAsF_6$) *Hexaflurate*	Nopalmate® (Pennwalt Corp., 1970)	Soil-applied herbicide, systemic selective action against *Opuntia* spp.	1200	1220, 1221

2; Kayaphenone® (Nihon, Noyaku, 1971)

Recently an aromatic ketone, (4-methoxy-3-methylphenyl)-(3-methylphenyl) methanone (2), having useful selective herbicidal properties (against broadleaf and grass weeds in paddy rice), was discovered in Japan.[1227]
It is known that halogenated *p*-benzoquinones (e.g., 3[1228] and 2,5-dibromo-6-

TABLE 32. Division of the Herbicide Market According to Product Groups (Consumer Prices in $US Million, Price Base 1974)[1222]

	United States		World	
Group	1974	1980	1974	1980
Aryloxyalkanecarboxylic acids	50	65	152	260
Arylureas	70	91	173	313
Carbamates	108	120	252	363
Triazines	314	419	620	907
Benzoic acid derivatives	100	115	140	213
Miscellaneous	416	713	853	1366
Total	1018	1523	2190	3422

TABLE 33. Division of the Herbicide Market According to Crop (Consumer Prices in $US Million, Price Base 1974)[1222]

	United States		World	
Crop	1974	1980	1974	1980
Corn (maize)	423	574	680	1061
Cotton	97	140	240	420
Wheat	47	71	200	281
Sorghum	33	49	55	81
Rice	23	33	181	309
Miscellaneous cereals	15	24	54	80
Soybean	286	432	410	584
Tobacco	6	9	11	18
Peanut (groundnut)	17	24	32	55
Sugar beet	11	15	81	132
Miscellaneous crops	16	23	52	81
Alfalfa (lucerne)	6	9	16	27
Hay, fodder	5	9	13	21
Pasture	24	36	40	64
Fruit, nuts	19	32	51	86
Vegetables	30	43	74	122
Total	1058	1523	2190	3422

isopropyl-3-methyl-2,5-cyclohexadiene-1,4-dione[1229]) inhibit photosynthesis. However, only the naphthoquinones **3** and **4** have attained any commercial significance to date as aquatic herbicides and algicides.

3; 2,3-Dichloro-1,4-naphthalenedione,
Dichlone, Phygon XL-50® (US Rubber Co.);
LD$_{50}$: 1500 mg/kg rat, oral, acute

4; 2,2-Dichloro-*N*-(3-chloro-1,4-dihydro-
1,4-dioxo-2-napthalenyl)acetamide,
Quinonamid,[1230] Alginex®, Nosprasit®
(Hoechst, 1972);
LD$_{50}$: 15,000 mg/kg rat, oral, acute

4.3.2.2. Phenols

Many agents are to be found in this class that, owing to their diverse biological activities, have found application in other areas of pest control as well as being used as herbicides.

Thus *DNOC* (**5**, 2-methyl-4,6-dinitro-phenol), which was introduced in 1934 as the first organic herbicide with selective properties and still has a certain significance today, is the oldest synthetic insecticide (Antinonnin®, Bayer), being used as early as 1892 for control of the black-arched tussock caterpillar. Subsequently the phytotoxic properties of a series of other substituted phenols have been described in the patent and other literature, of which some 2,4-dinitro-6-alkyl-phenols and halogenated phenols have attained practical importance.

4.3.2.2.1. 2,4-Dinitro-6-alkyl-penols.[1232] These strongly acidic phenols are typical contact herbicides (dessicants), which are hardly translocated within the plant if at all. Their selective properties are determined by purely physical factors. Plants that have a thick layer of wax preventing the penetration of such phenols are not affected. Root-propagated weeds are difficult to control because of the low degree of translocation of these compounds.

According to *in vitro* studies on isolated mitochondria, the main cause of the herbicidal action may be regarded as uncoupling of oxidative phosphorylation.[1193] An effect on photophosphorylation is first observed at relatively high concentrations. Other effects such as the denaturing of protein caused by phenols and the recently discovered inhibition of enzymatic glycerolipid synthesis in wheat,[1233] which can lead to changes in cell membrane structure, probably also contribute to the biological activity.

The influence of the alkyl residue in the 6-position on the herbicidal properties was studied very early on.[1234] The introduction of higher alkyl groups often leads to a reduction in phytotoxicity with improved selectivity. A further increase in the lipophilicity of the molecule by C_4-C_8 alkyl substitutents coupled with *O*-acylation with a longer-chain carboxylic acid brings the fungicidal and acaricidal properties of this class of compounds to the fore.

The most important herbicidal dinitrophenols and their main areas of use are:

1. 6-Methyl-2,4-dinitro-phenol.[1231]

5; *DNOC*, Sinox® (1932)
Use: control of annual broadleaf weeds in cereal (post-emergence); dessicant for potatoes and leguminous seed crops;
LD_{50}: 25–40 mg/kg rat, oral, acute

2. 2-(1-methylpropyl)-4,6-dinitro-phenol and esters.

R = H; *Dinoseb*, Premerge® (Dow, 1945)[1235]
Use: in cereals, pea, alfalfa (post-emergence); pea, bean, potato (pre-emergence); dessicant;
LD_{50}: 58 mg/kg rat, oral, acute

R = CO–CH₃; *Dinoseb acetate*, Aretit® (Hoechst, 1958)[1236]
Use: in cereals, legumes (post-emergence); potato (pre-emergence);
LD_{50}: 60–65 mg/kg rat, oral, acute

3. 2-(1,1-Dimethylethyl)-4,6-dinitro-phenol and ester.

R = H; *Dinoterb* (Pepro Co., 1957)[1235]
Use: as *dinoseb* but less phytotoxic in cereals; also in combination with *mecoprop* in cereals[1237] (post-emergence);
LD_{50}: 25 mg/kg rat, oral, acute

R = CO–CH₃; *Dinoterb acetate* (Murphy Chem. Ltd., 1965)[1238]
Use: cereal, cotton, legumes (pre-emergence); also nematicidal, and ovicidal for aphids;[1238]
LD_{50}: 62 mg/kg rat, oral, acute

4. 3-Methyl-6-(1,1-dimethylethyl)-2-4-dinitro-phenol acetate.

Minoterb acetate (Murphy Chem. Ltd., 1964)[1239]
Use: sugar beet, cotton, legumes (pre-emergence);
LD_{50}: 42 mg/kg rat, oral, acute

4.3.2.2.2. Halogenated Phenols. Some halogenated phenols also act as contact herbicides. Pentachlorophenol[1240] (*PCP*; LD_{50}: 210 mg/kg rat, oral, acute), manufactured by catalytic chlorination of phenol, has been used as a herbicide since 1940. In the form of its sodium salt it has been used particularly for control of broadleaf and grass weeds in paddy rice and soybeans. Further applications are cotton (defoliant) and potato (pre-harvest haulm destruction). Owing to its fungicidal action, it has been used for a long time as a wood preservative.

The herbicidal properties of halo-substituted 4-hydroxy-benzonitriles, **6**, were discovered in 1959.[1241-1243] The most active members of this group are the 3,5-dihalo derivates listed in Table 34, whereby the biological activity decreases from diiodo **6a** via dibromo **6b** to the dichloro compound **6c**.

They act both as uncouplers of oxidative phosphorylation and as inhibitors of photosynthesis (Hill reaction). Replacement of the cyano residue by a formyl **6d** or carboxy group **6e** reduces the phytotoxicity. This is reflected in the inhibitory values for the Hill reaction as determined on isolated chloroplasts.[1193]

In recent times 4-hydroxy-3,5-diiodo-benzonitrile in particular has attained practical significance.[1244] It is manufactured from 4-hydroxy-benzaldehyde.[1245]

6a; *Ioxynil*, Certrol® (Amchem Prod., 1963), Actril® (May and Baker, 1964); LD_{50}: 110 mg/kg rat, oral, acute

Ioxynil is used for post-emergence control of annual weeds in cereals whereby the herbicidal effect develops to the full under the influence of light. The special advantage of *ioxynil* is that it also kills weeds that are unaffected by growth regulator herbicides such as the aryloxy–fatty acids. *Ioxynil* is also available in var-

TABLE 34. Inhibition of the Hill Reaction by Halophenols

6

	Common Name	X	R	pI_{50}
6a	Ioxynil	I	CN	6.00
6b	Bromoxynil	Br	CN	4.75
6c	Chloroxynil	Cl	CN	3.47
6d		I	CHO	5.52
6e		I	COOH	3.00

ious commercial products in combination with growth regulator herbicides to widen the spectrum of activity. The ester with octanoic acid, (Ioxynil octanoate; Tortril®,[1246] May and Baker, 1963; LD$_{50}$: 390 mg/kg rat, oral, acute), is used as a selective post-emergence herbicide in cereals, sugarcane, and onions.

The corresponding bromo compound[1244] [**6b**; *Bromoxynil*; Brominil® (Amchem Prod., 1963), Buctril® (May and Baker); LD$_{50}$: 190 mg/kg rat, oral, acute] is, like *bromoxynil octanoate*,[1246] less active than *ioxynil* in the same area of application. 3,5-Dichloro-4-hydroxy-benzonitril (**6c**; *Chloroxynil*) is irrelevant as a herbicide.

Recently a selective cereal herbicide (contact, post-emergence) having a lower toxicity has been developed: 3,5-dibromo-4-hydroxybenzaldehyde *O*-(2,4-dinitrophenyl)oxime.

Bromofenoxim,[1247] Faneron® (Ciba, 1969);
LD$_{50}$: 1217 mg/kg rat, oral, acute

4.3.2.2.3. Phenolethers.
The phenolethers belong to the newer generation of herbicides. Although the phytotoxicity of simple phenolethers had been known for a long time,[1248] it was only in the 1960s with the nitrated diphenylethers that a class was found with practically useful properties. The essential characteristic of the commercially interesting phenolethers is a certain pattern of substituents—see formula **7**. The diphenylethers **7**, which are prepared by reaction of

7

R^1 = H(OCH$_3$, O—CO—CH$_3$)
R^2 = Cl, NO$_2$
R^3 = Cl, CF$_3$
R^4 = H(F, Cl)

chloronitrobenzenes with the corresponding phenoxides, interfere in photosynthesis. Their maximum effect is reached under the influence of light.[1195]

Important commercial products are:

1.   2,4-Dichloro-1-(4-nitrophenoxy)benzene.[1249]

Nitrofen, Tok E-25®, Tokkorn® (Rohm and Haas, 1964)
Use: contact and soil-applied herbicide (pre- and post-emergence), against broadleaf and grass weeds in cereal, rice, cotton, potato and vegetables;
LD$_{50}$: 3050 ± 500 mg/kg rat, oral, acute

2. 1,3,5-Trichloro-2-(4-nitrophenoxy)benzene.[1250]

CNP, MO 338 (Mitsui Chem. Co., 1966)
Use: pre- and post-emergence in rice (Japan)

3. 1,5-Dichloro-3-fluoro-2-(4-nitrophenoxy)benzene.[1251]

MO 500 (Mitsui Chem. Co., 1968)
Use: in rice (Japan); reported to have better activity against weeds (*Echinochloa crusgalli, Eleocharis palustris*) in rice than *nitrofen* or *CNP*

4. 2-Nitro-1-(4-nitrophenoxy)-4-(trifluoromethyl)benzene.[1252, 1253]

Fluorodifen, Preforan® (Ciba, 1968)
Use: soybean (pre-emergence), rice (pre- and post-emergence or after planting), cotton, corn (maize);
LD_{50}: 9000 mg/kg rat, oral, acute

5. 2,4-Dichloro-1-(3-methoxy-4-nitrophenoxy)benzene.

(Nihon Nohyaku Co.; Code no. X-52)
Use: in rice;
LD_{50}: 33,000 mg/kg mouse, oral, acute

6. Methyl 5-(2,4-dichlorophenoxy)-2-nitrobenzoate.[1254, 1255]

Bifenox, Modown® (Mobil Chem. Co., 1970)
Use: pre- and post-emergence against important broadleaf and some grass weeds in soybean, corn, sorghum, rice;
LD_{50}: 6400 mg/kg rat, oral, acute

7. 2-Chloro-1-(3-ethoxy-4-nitrophenoxy)-4-(trifluoromethyl)benzene.[1255a]

$$F_3C-\underset{Cl}{\bigcirc}-O-\underset{OC_2H_5}{\bigcirc}-NO_2$$

Oxyfluorfen, Goal® (Rohm and Haas, 1976)
Use: pre- and post-emergence against annual broadleaf and grass weeds in soybean, rice, corn, cotton, peanut, tree fruit, nuts, and many tropical plantation crops and ornamentals;
LD_{50}: >5000 mg/kg rat, oral, acute (technical product)

8. Sodium 5-[(2-chloro-4-(trifluoromethyl)-phenoxy]-2-nitrobenzoate.[1255b]

$$F_3C-\underset{Cl}{\bigcirc}-O-\underset{COONa}{\bigcirc}-NO_2$$

Acifluorfen, Blazer® (Rohm and Haas, 1977)
Use: pre- and post-emergence control of annual broadleaf and grass weeds in soybean, peanut, and other large-seeded legumes;
LD_{50}: 1300 mg/kg male rat, oral, acute (technical product)

4.3.2.3. Derivatives of Carbonic, Carbonothioic, and Carbonodithioic Acids

Among the derivatives of carbonic and carbonothioic acids are to be found the N-substituted carbamates, the S-alkyl-N,N-dialkylcarbamidothioates, and the ureas—three classes of herbicides of immense importance.

4.3.2.3.1. Carbamates. The development of carbamate herbicides began with the discovery of Sexton and Templemann during World War II of the inhibitory action of N-phenylcarbamates on germination. This was also the time when the fundamental work on the growth regulator herbicides was being carried out. The results, first published in 1945,[1200,1256] led to intensive chemical and biological work on this group of compounds, which, in turn, has led to the development of numerous selective agents up to the present day.[1257] At first useful products were only found in the N-arylcarbamate series. Only from 1963 on were N-methylcarbamates with useful activity and selectivity brought onto the market. Commensurate with the significance of the carbamates, numerous studies have been made to elucidate their mechanism of action. Admittedly the N-arylcarbamates interfere, like the ureas and 1,3,5-triazines, with photosynthesis,[1178] but the inhibition values for the Hill reaction,[1258] measured *in vitro* (pI_{50}–*IPC*: 3.40; *CIPC*: 3.92; *Swep*: 5.42) are much lower than those of the ureas. Inhibitory values corresponding to those of typical photosynthesis inhibitors were only found for biscarbamates such as *phenmedipham* (pI_{50}: 6.82).[1179] The fact that carbamates are powerful inhibitors of cell division (mitosis) is probably one of the main reasons for their phytotoxicity. At high concentrations in the biophase the inhibitory effect of the carbamates can even lead to complete suppression of cell division.[1201] Other disturbances to plant cell metabolism have also been observed.[1259,1260]

Of the herbicidal N-methylcarbamates only 2,6-bis(1,1-dimethylethyl)-4-methyl-phenyl methylcarbamate (8)[1261] and—to a slight extent—3,4-dichlorobenzene-methanol methylcarbamate (9) are of interest as soil-applied herbicides with grass weed-killing effect.[1260, 1262]

8; *Terbucarb*, Azak® (Hercules, 1964);
pre- and post-emergence herbicide in cotton, corn, soybean, tomato;
LD_{50}: 34,600 mg/kg rat, oral, acute

9; *Dichlormate*, Rowmate® (Union Carbide, 1966);
pre-emergence in cotton, potato, soybean; post-emergence in rice;
LD_{50}: 2000 mg/kg rat, oral, acute

The *N*-arylcarbamates 10 are much more important and are often characterized by good activity against grass weeds (see Table 35).

The first member of this group to be used in practice, 1-methylethyl phenyl-carbamate (10a, 1946), is available under a variety of trade names as an inhibitor of sprouting in stored potatoes and is used in combination with other carbamates for pre-emergence control of grass weeds in sugar beet, cabbage, and lettuce.

The esters of (3-chlorophenyl)carbamic acid have a higher herbicidal activity. Thus the 1-methylethyl ester (10b, 1951) is used as a potato sprouting inhibitor and as a soil-applied herbicide (pre- and post-emergence) in soybean, cotton, onion, and carrot crops. Its selectivity is ascribed to a different metabolism (ring hydroxylation) in resistant plants (e.g., cotton: →6-hydroxy-*CIPC*) from sensitive plants (→4-hydroxy-*CIPC*). The rate of hydroxylation and the formation of glycosides of the phenolic metabolites probably also determine the detoxification rate.[1273]

The main application of the 4-chloro-2-butynyl ester 10c (Carbyne®, 1958) is in the control of wild oat (*Avena fatua*) in wheat, barley, rape, alfalfa, beans, and sugar beet (post-emergence). In combination with *cycluron*, the 1-methyl-2-propynyl ester 10d is marketed under the trade name Alipur® (BASF, 1963) as a selective herbicide for broadleaf and grass weeds in sugar beet and various vegetable crops. Methyl (3,4-dichlorophenyl)carbamate 10e (FMC, 1960) is mainly used in rice (paddy rice) and legumes.

Further products to be mentioned here are:

1. *N*-Ethyl-2-{[(phenylamino) carbonyl]oxy}proponamide, R-(−) isomer (10f). (See Table 35)

Legurame® (Rhône-Poulenc, 1966)
Use: pre- and post-emergence for control of grass and some broadleaf weeds in winter rape and lettuce

TABLE 35. *N*-Arylcarbamates[1263]

$$Y-\underset{X}{\underset{|}{\bigcirc}}-NH-COOR$$

10

No.	R	X	Y	Common Name	LD$_{50}$ (rat p.o.) mg/kg	Reference
10a	$(CH_3)_2CH-$	H	H	*Propham, IPC*	5,000	1256
10b	$(CH_3)_2CH-$	Cl	H	*Chlorpropham, CIPC*	5,000–7,500	1264, 1265
10c	$ClCH_2-C\equiv C-CH_2-$	Cl	H	*Barban*	1,300	1266, 1267
10d	$HC\equiv C-CH(CH_3)-$	Cl	H	*Chlorbufam*	2,380 (mouse)	1268, 1269
10e	H_3C-	Cl	Cl	*Swep*	552	1270, 1271
10f	$H_5C_2-NH-CO-CH(CH_3)-$	H	H	*Carbetamid*	11,000	1272

2. 2-Propanone O-[(phenylamino) carbonyl] oxime.[1274]

$$H_5C_6-NH-CO-O-N=C(CH_3)_2$$

Proximpham
Use: pre-emergence control of annual broadleaf weeds in beet;
LD$_{50}$: 1540 mg/kg rat, oral, acute

3. Methyl [(4-aminophenyl) sulfonyl]carbamate.

$$H_2N-\bigcirc-SO_2-NH-COOCH_3$$

Asulam, Asulox® (May and Baker, 1968)[1275, 1276]
Use: as alkali metal salt for pre- and post-emergence control of grass weeds in sugar cane;
Mechanism: absorbed by leaves and roots; not a photosynthesis inhibitor; interferes with cell division;
LD$_{50}$: >2000 mg/kg rat, oral, acute

The dicarbamates **11**, which act by inhibiting photosynthesis,[1181] are foliar herbicides with contact action of particular value in post-emergence weeding of sugar and fodder beet because they are well tolerated by these crops.

$$\underset{NH-COOR^2}{\bigcirc}-O-CO-NH-\underset{}{\overset{R^1}{\bigcirc}}$$

11a; $R^1 = R^2 = CH_3$; *Phenmedipham*,[1277] Betanal® (Schering, 1968);
LD$_{50}$: >8000 mg/kg rat, oral, acute

11b; $R^1 = H$, $R^2 = C_2H_5$; *Desmedipham*,[1278] Betanal AM® (Schering, 1971);
LD$_{50}$: >9600 mg/kg rat, oral, acute

Their selectivity depends on the balance between rates of uptake and metabolism (i.e., detoxification) in the various plants.[1279] Thus *phenmedipham* taken up by beet is very rapidly inactivated by metabolism. Highly sensitive weeds such as *Galinsoga parviflora* (frenchweed, gallant soldier) absorb the herbicide rapidly but metabolize it slowly. The resistance of other weeds (chamomile varieties) that also have a slow metabolic detoxification rate is explained by their poor ability to absorb the agent.

4.3.2.3.2. S-Alkyl Carbamothioates.

Compared with the *N*-arylcarbamates, derivatives of carbamothioic acid are relatively late arrivals in weed control. *S*-Ethyl dipropylcarbamothioate (12b, *EPTC*, Table 36) was the first member of this group to reach practical use. In subsequent years a whole series of other carbamothioates (12) was developed (particularly by the Stauffer Chem. Co.) for various applications.

Only the compounds derived from secondary aliphatic amines have sufficient herbicidal activity. They are typical soil-applied herbicides used for pre-emergent control. Because of their relatively high volatility, mechanical incorporation into the soil is necessary. This technique, first applied to these compounds, has since proved its value in the application of other herbicide types. The carbamothioates are active against monocotyledonous and various dicotyledonous plants, but their major use is in control of grass weeds.

Recently antidotes have been found that prevent damage to the crop plants from treatment with carbamothioates and thus improve the selectivity of the latter in sensitive varieties of corn (maize). This can be achieved by coating the crop seed with naphthalic anhydride (13a)[1286] or by application of carbamothioates mixed with 2,2-dichloro-*N*,*N*-di-2-propenyl acetamide (13b).[1287]

$(H_2C=CH-CH_2)_2N-CO-CHCl_2$

13a; Protect® (Gulf Oil, 1972)

13b; e.g., with *EPTC*, Eradicane® (Stauffer, 1973);

LD_{50}: 2000 mg/kg rat, oral, acute

The antidote effect of 13b—and other dichloroacetamides—is due to their stimulation of the formation of glutathione (GSH) and GSH-*S*-transferase in the corn roots.[1288] Carbamoylation of GSH by the phytotoxic carbamothioate-*S*-oxides[1289,1290]—the primary metabolic sulfoxidation products of the *S*-alkyl esters 12—leads to rapid detoxification.

Little is known about the mechanism of action of the carbamothioates. They do not inhibit photosynthesis, and their interference in phytometabolic processes appears to be of a highly complex nature. Some studies indicate that disruption of nucleic acid metabolism or protein biosynthesis[1291,1292] and an influence on lipid biosynthesis (reduced formation of wax)[1211,1293] are partly responsible for the herbicidal effects.

TABLE 36. S-Alkyl Dialkylcarbamothioates

$$R^2\text{-}N\text{-}\underset{\underset{R^3}{|}}{\overset{\overset{O}{||}}{C}}\text{-}S\text{-}R^1$$

12

No.	Common Name	R^1	R^2	R^3	Trade Name, Manufacturer	Use	LD_{50} (mg/kg rat, oral, acute)	Reference
12a	*Ethiolate*	C_2H_5	C_2H_5	C_2H_5	(Prefox®) Gulf Oil, 1972	Corn (in combination with *cyprazine* = Prefox®)	400	1280
12b	*EPTC*	C_2H_5	C_3H_7	C_3H_7	Eptam® Stauffer, 1954	Potato, alfalfa, bean, carrot flax; particularly against *Agropyron repens, Cyperus* spp.	1,630	1281
12c	*Butylate*	C_2H_5	*a*	*a*	Sutan® Stauffer, 1966	Corn	4,660	1281
12d	*Cycloate*	C_2H_5	C_2H_5	C_6H_{11}	Ro-Neet® Stauffer, 1966	Sugar beet, spinach	3,600	1282
12e	*Molinate*	C_2H_5	$-(CH_2)_6-$		Ordram® Stauffer, 1965	Rice (especially against *Echinochloa* spp.)	500–720	1283
12f	*Vernolate*	C_3H_7	C_3H_7	C_3H_7	Vernam® Stauffer, 1965–1966	Soybean, tobacco, tomato, peanut, ornamentals	1,780	1281

12g	*Pebulate*	C_3H_7	C_2H_5	C_4H_9	Tillam® Stauffer, 1960	Sugar beet, tomato, tobacco	1,120	1282
12h	*Di-allate*	$CHCl=CHCl-CH_2-$	$(CH_3)_2CH$	*b*	Avadex® Monsanto, 1961	Beet, potato, corn, beans, pea, winter rape; especially against wild oat, windgrass, slender foxtail	395	1284
12i	*Tri-allate*	$CCl_2=CHCl-CH_2-$	$(CH_3)_2CH$	*b*	Avadex BW®, Far-Go® Monsanto, 1961	Spring barley, spring wheat, beet, pea	1,675–2,165	1284
12j	*Thiobencarb*	$Cl-$⟨benzene ring⟩$-CH_2$	C_2H_5	C_2H_5	Saturn® Kumiai Bolero® Chevron	Rice, soybean, peanut, potato, beet, tomato	1,300	1285
12k	*Tiocarbenil Tiocarbazil* (proposed)	$H_5C_6-CH_2-$	$H_5C_2-CH(CH_3)-$	$H_5C_2-CH(CH_3)-$	Drepamon® Montecatini Edision S.p.A.	Selective rice herbicide (especially against *Echinochloa* spp.); pre- or post-emergence or as rice seed dressing	>10,000	1285a

a $(H_3C)_2CH-CH_2-$.
b $(H_3C)_2CH$.

4.3.2.3.3. Carbamodithioates. The only representative of this group having practical importance is 2-chloro-2-propenyl diethylcarbamodithioate **(14)**, which is used as a pre-emergence soil-applied herbicide principally against grass weeds in various vegetable crops (brassicas, celery, spinach, beans), in corn (maize), and in ornamentals.[1294]

$$(H_5C_2)_2N-C\underset{\displaystyle S-CH_2-\underset{\underset{Cl}{|}}{C}=CH_2}{\overset{\overset{S}{\|}}{\Big\backslash}}$$

14; *Sulfallate*, Vegadex® (Monsanto, 1954);
LD_{50}: 850 mg/kg rat, oral, acute

4.3.2.3.4. Ureas. The effect exerted on plant growth by ureas was first described in 1946,[1295] albeit without these results finding immediate application in the practice of weed control. A systematic study of this group first started in 1951 when the herbicidal properties of N'-(4-chlorophenyl)-N,N-dimethylurea (*monuron*) were announced[1265] and has led to the development of numerous products for various applications up until the present day. Some members of this group find application as total herbicides; others as selective weed control agents in many different crops.

   With the aryloxy–fatty acids and the 1,3,5-triazines, the urea derivatives belong to the commerically most important classes of herbicides. Their herbicidal activity depends on inhibition of photosynthesis by blocking of electron transport in Photosystem II (see Figure 5, Section 4.2.3.1). Being typical inhibitors of phyto-specific bioprocesses, they are characterized by low mammalian toxicities. Herbicidal ureas are mainly absorbed via the plant roots and are translocated with the transpiration current to their site of action in the green plant parts. Often purely physical factors such as uptake and transport in the plant determine the selectivity of the ureas. In addition, some plants such as cotton are able to degrade herbicidal ureas relatively rapidly to nonphytotoxic compounds.

   The maximum herbicidal activity in ureas is found among the tri-substituted derivatives. The phytotoxicity of tetra-substituted ureas is too low to suffice for practical purposes. Of the di-substituted ureas only three are of practical significance:

   1. N-(2-Methylcyclohexyl)-N'-phenylurea.[1296]

$$H_5C_6-NH-CO-NH-\overset{}{\underset{H_3C}{\bigcirc}}$$

Siduron, Tupersan® (Du Pont, 1964)
Use: pre-emergence soil-applied herbicide against annual grass weeds in lawn grass and some dicotyledonous crops;
LD_{50}: >7500 mg/kg rat, oral, acute

   2. N-2-Benzothiazolyl-N'-methylurea.[1297]

Benzthiazuron, Gatnon® (Bayer AG, 1966)
Use: pre-emergence on sugar and fodder beet against broadleaf and grass weeds, mainly
in combination with *lenacil* (see Section 4.3.2.8) as Merpelan® (Bayer AG);
LD_{50}: 1280 mg/kg rat, oral, acute

3. *N*-(2-Methylpropyl)-2-oxo-1-imidazolidinecarboxamide.[1298]

Isocarbamid (Bayer AG)
Use: with *lencacil* (see Section 4.3.2.8) as Merpelan AZ® (Bayer AG, 1974);
LD_{50}: >2500 mg/kg rat, oral, acute

Ureas bearing purely aliphatic substituents are relatively weak herbicides used exclusively as components of combination products, for example:

1. *N'*-Cyclooctyl-*N*,*N*-dimethylurea.[1299]

Cycluron (BASF, 1958)
Use: as a mixture with *chlorbufam* (see Section 4.3.2.3.1, Table 35) for pre-emergence
weed control in sugar beet and spinach;
LD_{50}: 2600 mg/kg rat, oral, acute

2. *N*,*N*-Dimethyl-*N'*-[1-+2-(2,3,3*a*,4,5,6,7,7*a*-octahydro-4,7-methano-1H-
indenyl)]urea.[1300]

Isonoruron (BASF, 1967)
Use: in combination with *brompyrazon* in potatoes and winter cereals;
LD_{50}: >500 mg/kg rat, oral, acute

The most important groups of tri-substituted ureas are those of the *N'*-aryl-*N*,*N*-
dimethyl derivatives **15** and the *N'*-aryl-1-methoxy-1-methyl derivatives **16**, which
are listed in Tables 37 and 38.

   A 1 : 1 mixture of *monuron* **(15b)** and trichloroacetic acid (*monuron*-TCA; LD_{50}:
2300 mg/kg rat, oral, acute) is marketed as Urox® (Allied Chem Corp., 1956) as a
total herbicide for nonagricultural land.[1301]

   N-(3,4-Dichlorophenyl)-*N*-[(dimethylamino)carbonyl]benzamide **(17a)** is of significance as a soil-applied herbicide in wine grape and fruit growing.[1313]

TABLE 37. *N*-Aryl-*N'*,*N'*-dimethylureas

$$Ar-NH-CO-N(CH_3)_2$$

15

No.	Common Name	Ar	Trade Name, Manufacturer	Use	LD$_{50}$ (mg/kg Rat, Oral, Acute)	Reference
15a	*Fenuron*		Dybar® Du Pont, 1957	Against deep-rooted weeds; control of woody plants; pre- and post-emergence	6,400	1265, 1301
15b	*Monuron* (*CMU*)		Telvar® Du Pont, 1952	Total herbicide with long residual action; selective in cotton, sugarcane, asparagus at low application rates	3,600	1265, 1301
15c	*Diuron* (*DCMU*)		Karmex® Du Pont, 1954	Total herbicide; selective in cotton, sugarcane, vegetables, fruit, asparagus at low application rates	3,400	1265, 1302
15d	*Fluometuron*		Cotoran® Ciba, 1960	Selective in cotton, sugarcane, citrus (pre-emergence); post-emergence in cereals	8,000	1303
15e	*Isoproturon*		Arelon® Hoechst	In wheat and barley against annual grass weeds	1,800 males, 2,400 females	1304
15f	*Metoxuron*		Dosanex® Sandoz, 1968	Selective in cereals and carrots (pre- and post-emergence) particularly against slender foxtail, windgrass, wild oat, ryegrass	3,200	1305, 1306

	Structure	Trade name	Use	LD$_{50}$	No.
15g *Chlorotoluron*		Dicuran® Ciba, 1969	Root and foliage herbicide against grass weeds and shallow-germinating seed-propagated broadleaf weeds in winter wheat and barley (pre- and post-emergence)	10,000	1307
15h *Karbutylate*		Tandex® FMC, 1967	Total herbicide (railway tracks, industrial sites, paths, open spaces)	3,000	1308
15i *Chloroxuron*		Tenoran® Ciba, 1960	Root- and foliage-absorbed herbicide selective in leek, celery, onion, carrot and strawberry	3,000	1309
15j *Thiochlormethyl*		Clearcide® Bayer	Selective rice herbicide (direct-drilled and transplanted rice) mainly effective against broadleaf weeds and algae (experimental product)	432 males, 336 females	1310
15k *Difenoxuron*		Lironion® Ciba-Geigy	Root- and foliage-absorbed herbicide for control of weeds in onions (post-emergence)	>7,750	

TABLE 38. *N*-Aryl-*N'*-methoxy-*N'*-methylureas

$$Ar-NH-CO-N \begin{smallmatrix} OCH_3 \\ CH_3 \end{smallmatrix}$$

16

No.	Common Name	Ar	Trade Name, Manufacturer	Use	LD$_{50}$ (mg/kg Rat, Oral, Acute)	Reference
16a	*Monolinuron*		Aresin® Hoechst, 1958	Pre-emergence against broadleaf and grass weeds in potato, winter cereals, asparagus, corn, beans, wine grape, and ornamental shrubs	2250	1311
16b	*Metobromuron*		Patoran® Ciba, 1963	Pre-emergence against annual grass and broadleaf weeds in potato, kidney bean, tobacco, peanut	2500	1312
16c	*Linuron*		Afalon® Hoechst, 1960; Lorox® Du Pont, 1960	Pre- and post-emergence in soybean, cotton, potato, corn, beans, pea, winter wheat, asparagus, carrot, and in fruit and wine grape growing	4000	1311
16d	*Chlorbromuron*		Maloran® Ciba, 1961	Pre-emergence in soybean, potato, carrot	>5000	1312

17a; *Phenobenzuron*, Benzomarc® (Pepro, 1963);
LD_{50}: 5000 mg/kg rat, oral, acute

Replacement of one of the methyl groups in the ureas of type **15** (see Table 37) by larger residues leads to a fall in herbicidal activity with a simultaneous improvement in selectivity, for example:

1. *N*-Butyl-*N'*-(3,4-dichlorophenyl)-*N*-methylurea.[1264,1265,1314]

Neburon, Kloben® (Du Pont, 1957)
Use: selective root-absorbed herbicide for control of annual broadleaf and grass weeds in wheat, alfalfa, strawberry, and woody ornamentals;
LD_{50}: >11,000 mg/kg rat, oral, acute

2. *N'*-(4-Chlorophenyl)-*N*-methyl-*N*-(1-methyl-2-propynyl)urea.[1315]

Buturon, Eptapur® (BASF, 1966)
Use: Mainly against grass weeds (windgrass, slender foxtail) in winter cereals and corn (maize);
LD_{50}: 3000 mg/kg rat, oral, acute

The *N'*-aryl-*N*-methoxy-*N*-methylureas **16**, of which *monolinuron* **(16a)** and *linuron* **(16c)** are the most important representatives, have better selective properties and higher herbicidal activity than the *N'*-aryl-*N*,*N*-dimethylureas (**15**, Table 37). They are mainly absorbed via the roots but are also taken up by the foliage to a greater or lesser degree. In the soil they are much more rapidly degraded by microbial processes than the ureas of type **15** (Table 37). The greatly reduced residual action or persistence decreases the danger of carryover damage to the next crop.

Urea derivatives having heterocyclic substituents have been studied increasingly in recent years with respect to their herbicidal activity. Of the numerous compounds described in journals and patent literature as herbicidal, derivatives of

2-amino-benzo-1,3-thiazole and substituted 2-amino-1,3,4-thiadiazoles deserve a special mention.

Of great significance today is *N*-2-benzothiazolyl-*N*,*N'*-dimethylurea **(17b)**, a root- and foliage-absorbed herbicide that has proved its value for the control of broadleaf and grass weeds in cereals, peas, and beans.[1316,1317]

17b; *Methabenzthiazuron*, Tribunil® (Bayer, 1968);
LD_{50}: 2500 mg/kg male rat, oral, acute

Recently, various companies have developed 1,3,4-thiadiazolylureas[1318] **(18)** (see Table 39) that are characterized by a very wide spectrum of activity and are used as nonselective agents for total weed control on railway tracks, industrial sites, paths, and open spaces. Urea derivatives such as *buthidazole*[1318a] are also used for the same purpose.

Buthidazole, Ravage® (Velsicol Chem. Co.);
LD_{50}: 1581 mg/kg male rat, oral, acute

N-Phenyl-*N'*-(1,2,3-thiadiazol-5-yl)urea (*thidiazuron*)[1318b] is a plant growth regulator used as a defoliant in cotton.

Thidiazuron, Dropp® (Nor-Am Ag. Prod. Inc.; Schering AG, 1976);
LD_{50}: >4000 mg/kg rat, oral, acute

TABLE 39. *N*,*N'*-Dimethyl-*N*-(1,3,4-thiadiazol-2-yl)ureas

18

No.	Common Name	R	Trade Name, Manufacturer	LD_{50} (mg/kg Rat, Oral, Acute)	Reference
18a	*Ethidimuron*	C_2H_5	Ustilan® Bayer, 1973	>5000	1319, 1320
18b	*Thiazafluron*	CF_3	Erbotan® Ciba-Geigy, 1973	278	1321, 1322
18c	*Tebuthiuron*	$(CH_3)_3C$	Spike® Elanco, 1973	644	1323, 1324

4.3.2.4. Carboxylic Acids and Derivatives

Numerous carboxylic acids and their salts, esters, and amides are to be found in the range of herbicides used today. Owing to their mechanism of action, the majority of organic acids and their esters that are used for weed control may be classed with the synthetic plant growth hormones of the auxin type. Among these growth regulator herbicides, the aryloxyalkanecarboxylic acids occupy a special position owing to the significant role they played in the development of modern weed control and because today they still account for a large part of the total herbicide production both in terms of tonnage and turnover (see Table 33).

The carboxamides, whose biological properties are often fundamentally different from those of the acids, salts, and esters, are dealt with in Section 4.3.2.4.5. Depending on the type of amide, their phytotoxic properties are mainly caused by disruption of protein synthesis in the plant or an inhibition of photosynthetic electron transport in light reaction II.

4.3.2.4.1. Aliphatic and Araliphatic Carboxylic Acids. Among the aliphatic carboxylic acids only monochloro- and trichloroacetic acid and 2,2-dichloropropanoic acid are of interest as herbicides. They are normally used as their sodium salts for control of grass weeds, broadleaf weeds being affected only slightly.

The biochemical causes of their phytotoxic properties remain unknown. It is presumed that they deactivate enzyme systems by interacting with proteins.[1325,1326] Whether the inhibition of plant lipid metabolism, which has been demonstrated,[1327,1328] makes a decisive contribution to the herbicidal effect of these chlorinated fatty acids remains an open question.

Monochloroacetic acid *(MCA)*,[1329] which is also reported to have a defoliant action, is used in the form of its sodium salt [*SMCA*, Monooxone® (ICI Plant Protection, 1956) LD_{50}: 650 mg/kg rat, oral, acute] as a post-emergence contact herbicide in some vegetable crops (leeks, onions).

Trichloroacetic acid is of greater importance. The sodium salt [*TCA*, Nata® (Du Pont, Dow, 1947); LD_{50}: 3200–5000 mg/kg rat, oral, acute] is used to control grass weeds—particularly quackgrass (couch), slender foxtail, and wild oat—in rape, beet, potato, and spring cereals prior to planting and for removal of grasses on paths, tracks, and open spaces.

Whereas *TCA* is mainly absorbed via the roots, sodium 2,2-dichloropropanoate [*dalapon-sodium*,[1330] Dowpon® (Dow, 1953); LD_{50}: 9330 mg/kg male rat, oral, acute; 7570 mg/kg female] is taken up by the roots and the foliage. The range of applications of *dalapon-sodium* corresponds to that of *TCA*. Owing to its low toxicity toward fish, *dalapon* is additionally used for control of reeds, sedges, and rushes in drainage canals and ditches.

The glycol ester of trichloroacetic acid {1,2-ethanediyl[bis(trichloroacetate)]},[1331] Glytac® (Hooker Chem. Corp., 1965); LD_{50}: 7000 mg/kg rat, oral, acute} is a nonselective agent for suppression or control of grass weeds and tall-growing broadleaf weeds and for spot treatment in cotton and soybean.

Arylacetic acids belong to the group of growth regulator herbicides. Following the isolation by Kögl in 1934[1332] of the natural plant growth regulator "heteroauxin" **(20)**, 1-naphthaleneacetic acid (1-NAA) **(19)** was one of the first synthetic analogues for which growth regulating properties were observed[1333],[1334] and for which a selective herbicidal action could be demonstrated at higher concentrations.[1335]

19; 1-NAA; LD$_{50}$: 1000 mg/kg rat, oral,
acute

20; β-Indolylacetic acid

1-NAA (19), esters of which were used just after World War II to prevent premature sprouting in potatoes, is only of historical interest and has no commercial significance. Many compounds from the class of phenylacetic acids and phenylacetonitriles have been described as herbicidal in the patent literature. The only one from this class that has a certain significance is 2,3,6-trichlorobenzeneacetic acid[1336],[1337] (21). The sodium salt is a pre- and post-emergence selective herbicide used against grass and some broadleaf weeds in corn (maize), sugarcane, and asparagus (also for control of aquatic weeds in the United States).

21; *Fenac* (WSSA), *Chlorfenac* (ISO), Fenac® (Amchem. Prod., 1965);
LD$_{50}$: 1780 mg/kg rat, oral, acute

Among the longer-chain alkanoic acids, methyl (±)-α,4-dichlorobenzenepropanoate (22) is characterized by high selectivity.[1338],[1339]

22; *Chlorofenprop-methyl*, Bidisin® (Bayer AG, 1968);
LD$_{50}$: 1190 mg/kg rat, oral, acute

Bidisin® is a specific contact herbicide for the control of wild oat (*Avena fatua*) in beet, barley, wheat, and potato. Few experimental data are available concerning its primary mechanism of action and the reasons for its remarkable selectivity.

Bidisin® induces autolysis in the leaves of *Avena fatua*. Through activation of hydrolytic cell enzymes, high-molecular-weight cellular components (proteins, starch, nucleic acids) are degraded, thus causing the death of the plant.[1340] According to recent studies the herbicidal activity of the L(−) enantiomer is twice as high as that of the racemate, the D(+) form being biologically almost inactive.[1341]

Derivatives of 9-hydroxy-9H-fluorene-9-carboxylic acid,[1342] grouped together under the name "morphactins," belong to the plant growth regulators and are discussed in Chapter 5. However, the butyl ester should be mentioned here.

HO COOC$_4$H$_9$

Fluorenol, Flurecol-butyl
Use: with *MCPA-isooctyl* as Amiten® (E. Merck AG, 1964) or with *ioxynil* (see Section 4.3.2.2.2) for post-emergence control of weeds in winter and spring cereals;[1343]
LD$_{50}$: 5000 mg/kg rat, oral, acute

4.3.2.4.2. Aryloxyalkanoic Acids. The search for synthetics that could influence the growth behavior of plants in the same manner as auxin (β-indolylacetic acid, **20**), which occurs naturally in the regulation system of plants, led during World War II to the discovery of the selective herbicidal properties of ring-chlorinated phenoxyacetic acids.

Independent of one another, the phytotoxic activity[1335,1344] of (4-chloro-2-methylphenoxy)acetic acid[1345] **(24)** was discovered in Britain and of (2,4-dichloro-phenoxy)acetic acid[1346] **(23)** in the United States between 1941 and 1945.

23; *2,4-D*; LD$_{50}$: 375 mg/kg rat, oral, acute

24; *MCPA*; LD$_{50}$: 700 mg/kg rat, oral, acute

The fundamental discovery that these compounds, at higher application rates, not only act as growth inhibitors but can also be used for destruction of dicotyledonous plants and are well tolerated by many monocotyledonous crop plants revolutionized agriculture and was the starting point for the development of modern weed control. Within a few years *MCPA* **(24)** and *2,4-D* **(23)** were in widespread use as selective herbicides, particularly in cereals. Their great advantages here were cheap manufacture (starting materials are *o*-cresol for *MCPA* and phenol for *2,4-D*, chlorine, and acetic acid) and relatively low mammalian toxicity.

(2,4,5-Trichlorophenoxy)acetic acid **(25)**,[1347,1348] another product from this group, was also introduced in the mid-1940s.

25; *2,4,5-T*; LD$_{50}$: 500 mg/kg rat, oral, acute

2,4,5-T has particularly high activity against woody plants and is used—usually in combination with other herbicides—for control of trees and shrubs, release of forest conifers, and control of otherwise intractable broadleaf weeds (*Anthemis, Matricaria, Galeopsis* spp.).

A teratogenic effect has been observed in animal experiments with the technical product, which was traced back to contamination with 2,3,7,8-tetrachlorodibenzo-1,4-dioxin **(26)**.[1349] This by-product is formed during the alkaline hydrolysis of

1,2,4,5-tetrachlorobenzene to 2,4,5-trichlorophenol. The content of **26** in *2,4,5-T* can be reduced to below 0.5 ppm by modifications to the manufacturing process and by special purification techniques.[1350]

26

Derivatives of 2-phenoxyethanols should be classed with the phenoxyacetic acids since their activity apparently depends on oxidation to the corresponding bioactive acid in the plant or in the soil.

1. Sodium 2-(2,4-dichlorophenoxy)ethyl sulfate.[1351,1352]

Disul-sodium, Sesone® (Union Carbide)
Use: pre-emergence in strawberry, asparagus, and peanut

2. 2-(2,4,5-Trichlorophenoxy)ethyl 2,2-dichloropropanoate.[1353]

Erbon, Baron®, Erbon® (Dow); Total herbicide; LD_{50}: 1120 mg/kg rat, oral, acute

3. Mixture of tris[2-(2,4-dichlorophenoxy)ethyl]phosphite and bis[2-(2,4-dichlorophenoxy)ethyl]phosphonate.[1354]

2,4-DEP, Falone® (US Rubber Co.); Action: similar to *2,4-D* (**23**); LD_{50}: 850 mg/kg rat, oral, acute

In addition to the aryloxyacetic acids, the corresponding ring substituted 2-(aryloxy)propanoic acids **27** and 4-(aryloxy)butanoic acids **28** are also characterized by high herbicidal activity.

In contrast, the phytotoxicity of the 3-(aryloxy)propanoic acids is very low. Of the 2-substituted propanoic acids **27**, (±)-2-(2,4-dichlorophenoxy)propanoic **(27a)**[1355] and (±)-2-(4-chloro-2-methylphenoxy)propanoic acids **(27b)** are mainly used for weed control in cereals. (±)-(2,4,5-Trichlorophenoxy) propanoic acid[1355] **(27c)** has, like *2,4,5-T* (see Section 4.3.2.4.2), good activity against woody plants and is used for weed control in fields and pastures (United States) and as a selective

27

27a;   X = Cl, Y = H; *Dichlorprop*; LD$_{50}$: 800 mg/kg rat, oral, acute
27b;   X = CH$_3$, Y = H; *Mecoprop*; LD$_{50}$: 930 mg/kg rat, oral, acute
27c;   X = Y = Cl; *Fenoprop, Silvex*; LD$_{50}$: 650 mg/kg rat, oral, acute

28

28a;   X = Cl; *2,4-DB*; LD$_{50}$: 700 mg/kg rat, oral, acute
28b;   X = CH$_3$; *MCPB*; LD$_{50}$: 680 mg/kg rat, oral, acute

herbicide in rice and sugarcane. Only the D-(+) form of the propanoic acid derivatives 27 is biologically active, although they are used in practice as racemates.

The 4-(phenoxy)butanoic acids 28a and 28b,[1357,1358] readily available from the corresponding phenoxides and γ-butyrolactone, were introduced about 10 years after the phenoxyacetic acids. They should be regarded as precursors of *2,4-D* and *MCPA* as they are converted into the latter in the plant by β-oxidation.[1359] Herein lies the particular importance of the 4-(phenoxy)butanoic acids. Owing to the biochemical activation necessary to release the herbicidal action, they can also be used for weed control in some dicotyledonous crop plants such as legumes (alfalfa, clover, peas) in which β-oxidation does not take place.

The aryloxy–fatty acids are translocated and are normally used in the form of salts (alkali metal or ammonium salts) or esters [methyl, isopropyl, butyl, octyl, 2-(butyloxy)ethyl, or propylene glycol esters]. The salts are mainly taken up by the roots; the more lipophilic esters by the aerial plant parts. In practice, mixtures of different aryloxy–fatty acids or combinations with suitable herbicides from other classes are used to obtain an optimal spectrum owing to the different degrees of effectiveness of individual agents against the various broadleaf weeds.

Numerous investigations have been made on the reasons for herbicidal activity among the phenoxy–fatty acids. *2,4-D* (24) and analogues show typical auxin properties in various test systems. The morphological changes (thickening and distortion of the stems, formation of aerial roots) observed after treatment of the plants with phenoxyacetic acids are indications of excessive auxin activity. The primary biochemical processes responsible are, however—as with auxin itself—largely unknown.

An activity spectrum completely different from that of the classical phenoxy–fatty acids is found with the 4-(aryloxy)phenoxypropanoates, a class of herbicides discovered recently in the laboratories of Hoechst AG. These are selective herbicides that are suitable for control of grass weeds such as wild oat, slender foxtail, and panicum varieties in almost all broadleaf crops,[1359a] particularly post-emergence.

$$Cl—\text{(ring)}—O—\text{(ring)}—O—\overset{\underset{|}{CH_3}}{CH}—COOR$$

X

a;   X = Cl, R = CH$_3$; *Dichlofop-methyl*, Hoelon®, Hoe-Grass®, Illoxan® (Hoechst AG, 1977);
LD$_{50}$: 580 mg/kg rat, oral, acute (technical product)

b;   X = H, R = (H$_3$C)$_2$CH–CH$_2$–; *Chlofop-isobutyl*, Alopex® (Hoechst, 1977); LD$_{50}$: 723
mg/kg female rat, oral, acute

4.3.2.4.3. Aromatic Carboxylic Acids.

Many halogenated—particularly the multiply chlorinated—benzoic acids are phytotoxic.[1360] In 1954 a benzoic acid derivative, 2,3,6-trichlorobenzoic acid,[1361] was introduced that has similar growth regulator properties to the phenoxy–fatty acids and is used for control of broadleaf weeds in cereals and grass seed crops. The technical product, made by oxidation of the mixture of 2,3,6- and 2,4,5-trichlorotoluene obtained by chlorination of 2-chlorotoluene, contains ~40% of the isomeric 2,4,5-trichlorobenzoic acid (30), which has little herbicidal activity.

2,3,5,6-Tetrachlorobenzoic acid (29b) is also powerfully phytotoxic but has not achieved any commercial significance.[1362]

29a;   X = H; *2,3,6-TBA*, Trysben® (Du   30
Pont); LD$_{50}$: 1500 mg/kg rat, oral, acute
29b;   X = Cl; 2,3,5,6-TBA

Replacement of the chlorine in the 2-position of *2,3,6-TBA* by a methoxy group does not lead to loss of activity. 3,6-Dichloro-2-methoxybenzoic acid[1363] (31a) is used as a growth regulator herbicide in combination with *MCPA* (24) or *2,4-D* (23) for control of broadleaf weeds in cereals and grassland.

$$\xrightarrow[\text{- NaCl}]{\text{NaOH}} \qquad \xrightarrow{CO_2, \text{ Pressure}}$$

$$\xrightarrow{(H_3CO)_2SO_2}$$

31a;   X = H; *Dicamba*, Banvel® (Velsicol, 1965);
LD$_{50}$: 2900 ± 800 mg/kg rat, oral, acute
31b;   X = Cl; *Tricamba*, Banvel T® (Velsicol);
LD$_{50}$: 970 mg/kg rat, oral, acute

3,5,6-Trichloro-2-methoxybenzoic acid[1364] **(31b)**, which can be used as a post-emergence herbicide in cereals, is of little practical relevance.

It is appropriate that 2,3,5-triiodobenzoic acid *(TIBA)* be mentioned here. Although reported to have herbicidal properties, *TIBA* is used today solely as a plant growth regulator (see Chapter 5).

2,6-Dichlorobenzonitrile[1365,1366] **(32)** and 2,6-dichlorobenzenecarbothi-amide[1367,1368] **(33)**, which was introduced later and is converted in the soil to the nitrile **32**, are two herbicidal halobenzoic acid derivatives whose mechanism of action is virtually unknown. Hydrolysis to 2,6-dichlorobenzoic acid or the amide (both are also phytotoxic) has been excluded as the reason for their high herbicidal activity.

32; *Dichlobenil*, Casoron® (N. V. Philips Dupahr, 1960); LD$_{50}$: 3160 mg/kg rat, oral, acute

33; *Chlorthiamid*, Prefix® (Shell, 1963); LD$_{50}$: 757 mg/kg rat, oral, acute

Dichlobenil **(32)** is absorbed via roots and leaves and translocated. It is suitable for weed control in rice. Like *chlorthiamid* **(33)**, it is used for post-emergence weed control in fruit and other crops and for total weed control on non-crop land at high application rates. (See Section 4.3.2.2.2 for the halogenated 4-hydroxybenzonitriles.)

A derivative of an aromatic dicarboxylic acid, dimethyl 2,3,5,6-tetrachloro-1,4-benzenedicarboxylate[1369] **(34)**, is a selective pre-emergence agent in vegetables and ornamentals against annual grass weeds, common chickweed, and *Chenopodium* spp. Recent studies have shown one of the causes of the herbicidal activity to be the formation of multinuclear cells owing to inhibition of cell division.[1206]

34; *Chlorthal-dimethyl* (ISO), *DCPA* (WSSA), Dacthal® (Diamond Shamrock, 1959); LD$_{50}$: >300 mg/kg rat, oral, acute

4.3.2.4.4. Amino acids.
Only a few of the amino acid derivatives with herbicidal activity have achieved any practical significance.

3-Amino-2,5-dichlorobenzoic acid[1370] **(35)**, also used in the form of its methyl ester and amide, inhibits the root development of emerging weeds and is employed for control of dicotyledonous weeds and some grasses in soybean, corn (maize), tomato, carrot, and cucurbits (pre-emergence). A stable, inactive *N*-glycoside is formed in tolerant crop plants such as soybean.

Approximately 1500 tonnes of *chloramben* were used in the United States in 1970–about 5% of the 2,4-D quantity applied in the United States in the same year.

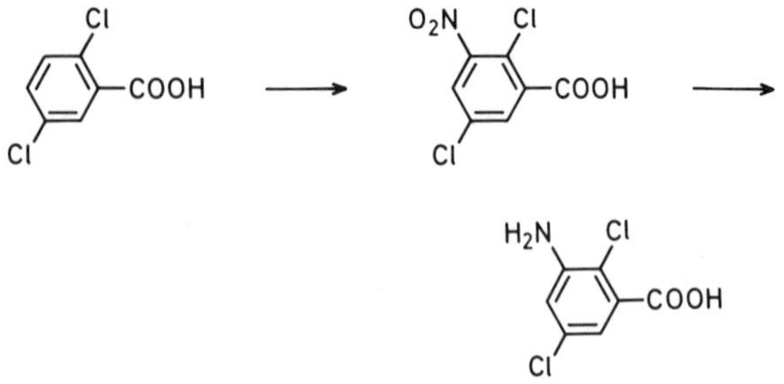

35; *Chloramben*, Amiben® (Amchem, 1968);
LD_{50}: 5620 mg/kg rat, oral, acute

4-Amino-3,5,6-trichloropyridine-2-carboxylic acid[1371] **(36)**, a highly active herbicide—especially against broadleaf weeds—was introduced in 1963. High activity against deep-rooted weeds and woody plants makes it particularly suited for application in forest crops. Owing to the lack of tolerance shown by many crop plants, it is of little significance in agriculture (wheat, corn). *Picloram* **(36)** is manufactured according to the following scheme.[1372]

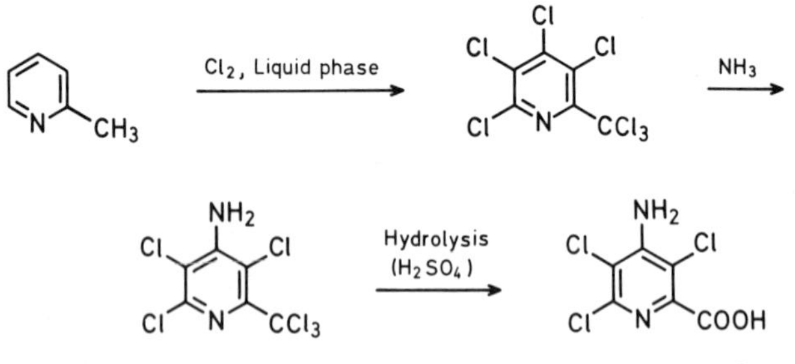

36; *Picloram*, Tordon® (Dow);
LD_{50}: 8200 mg/kg rat, oral, acute

4-Chloro-2-oxo-3(2H)-benzothiazoleacetic acid **(37)**,[1373, 1374] which may be regarded as an amino acid derivative, is a herbicide with growth regulator properties mainly used in combination with other growth regulator herbicides (phenoxyfatty acids, *dicamba*) for post-emergence control of *Galium aparine*, *Stellaria media*, and *Matricaria* spp. in cereals, grass, clover, and rape.

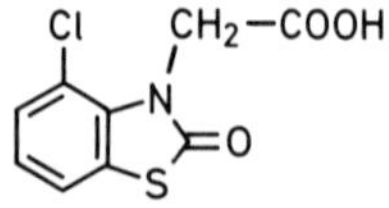

37; *Benazolin*, Ben-Cornox® (Boots, 1965);
LD_{50}: >3000 mg/kg rat, oral, acute

Two derivatives of 2-aminopropanoic acid have been introduced recently specifically for control of wild oat in cereals.

Whereas ethyl *N*-benzoyl-*N*-(3,4-dichlorophenyl)-DL-alaninate **(38a)** is only tolerated sufficiently well by wheat,[1375, 1376] the ester **(38b)** can also be used for control of *Avena fatua* in barley.

38a; X = Cl, R = C_2H_5; *Benzoylprop-ethyl*, Suffix® (Shell, 1969);
LD$_{50}$: 1555 mg/kg rat, oral, acute

38b; X = F, R = $CH(CH_3)_2$; *Flamprop-isopropyl*, Barnon® (Shell, 1972);
LD$_{50}$: >3000 mg/kg rat, oral, acute

It has been demonstrated that the corresponding free carboxylic acids and not the esters are responsible for the herbicidal activity. The herbicidal activity and selectivity are thus mainly determined by the different rates of ester hydrolysis in wild oat and the crop grasses.[1377, 1378]

4.3.2.4.5. Carboxamides. Most of the carboxamide-type herbicides used in practice may be divided into two groups according to their chemical structure and the mechanism of action mainly responsible for their herbicidal activities. The anilides **39** (Table 40) belong to the first group. They are inhibitors of photosynthesis (light reaction II) and disrupt the primary metabolism of the plant. One exception is 2-[(1-naphthalenylamino)carbonyl]benzoic acid **(40)**,[1379, 1380] which interferes with auxin transport in the plant in a manner similar to the "morphactins."[1381]

40; *Naptalam*, Alanap® (Uniroyal, 1950);
Use: pre-emergence in potato, soybean, peanut;
LD$_{50}$: 8200 mg/kg rat, oral, acute

The activity of the chloracetamides **41** (see Table 41) seems to depend on inhibition of protein synthesis.

Various detoxification mechanisms have been detected that account for the selectivities of anilides **39** (Table 40) and the chloracetamides **41** (Table 41). Thus *propanil* **39a**, for example, is cleaved relatively rapidly by an acylanilide hydrolase in the rice plant.[1393] The 3,4-dichloroaniline is then detoxified by glycoside formation. The resistance of corn (maize) to *allidochlor* **41a** and *propachlor* **41b** has been traced back to the formation of herbicidally inactive conjugates of these agents with glutamylcysteine and glutathione.[1394]

Whereas the amides **39, 41a, 41b, 41c** are synthesized in the normal way by acylation of the corresponding primary or secondary amine, *alachlor* **41d** and

TABLE 40. Anilides

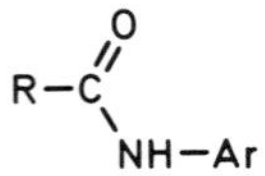

$$R-\overset{\displaystyle O}{\overset{\displaystyle \|}{C}}-NH-Ar$$

39

No.	Common Name	R	Ar	Trade Name, Manufacturer	Use	LD$_{50}$ (mg/kg Rat, Oral, Acute)	Reference
39a	*Propanil*	C_2H_5	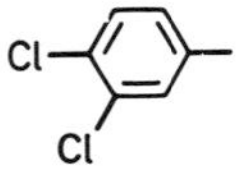	Surcopur® (Bayer); Stam F-34® (Rohm and Haas); Rogue® (Monsanto); 1960	Contact herbicide of brief persistence; control of broadleaf and grass weeds in rice and potato (post-emergence)	1,400	1382
39b	*Pentanochlor*	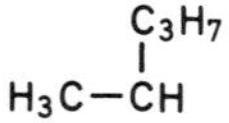 $H_3C-CH-C_3H_7$	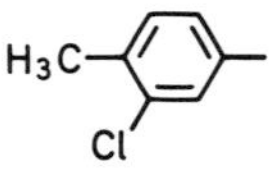	Dutom®, Solan® FMC, 1958	Post-emergence in celery, tomato, carrot, parsley, strawberry; absorbed via the foliage	>10,000	1383
39c	*Monalide*	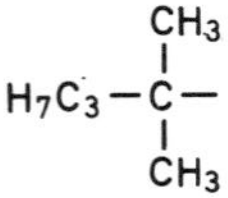 $H_7C_3-\overset{CH_3}{\underset{CH_3}{C}}-$	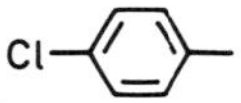	Potablan® Schering AG, 1963	Post-emergence weed control in umbelliferous crops; absorbed via root and leaf	>4,000	1384, 1385
39d	*Chloranocryl*	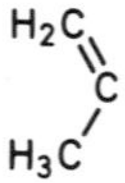 $H_2C=\overset{}{\underset{H_3C}{C}}$	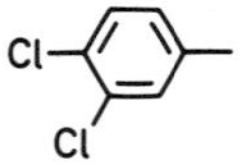	Dicryl® FMC, 1960	In cotton against grass and some broadleaf weeds; post-emergence contact herbicide	1,800–3,000	1386, 1387

366

39e	*NCA*		Karsil® FMC	10,000	1384	Contact herbicide; post-emergence in celery, carrot, strawberry
39f	*Cypromid*		Clobber® Gulf Oil, Spencer Chemical Division, 1965	218	1388	Contact herbicide in corn, cereals, cotton, and umbelliferous crops (post-emergence)
39g	*Mefluidide*		Embark® 3M Co., 1977	>4,000	1388a	Growth regulator for suppression of growth of turf grasses; enhancement of sucrose content of sugarcane; as post-emergence grass weed herbicide in soybean

TABLE 41. Chloracetamides

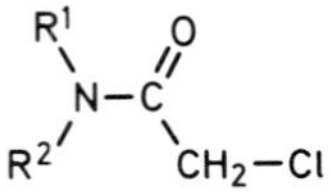

$$R^1 \underset{R^2}{\overset{}{>}} N-\overset{\overset{O}{\|}}{C}\diagdown_{\displaystyle CH_2-Cl}$$

41

No.	Common Name	R^1	R^2	Trade Name, Manufacturer	Use	LD_{50} (mg/kg Rat, Oral, Acute)	Reference
41a	*Allidochlor*	$H_2C=CH-CH_2$	$H_2C=CH-CH_2$	Randox® Monsanto	Pre-emergence against germinating grass weeds in corn, soybean, potato, sugarcane, and various vegetable crops	700	1389, 1390
41b	*Propachlor*	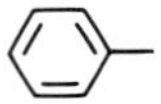 phenyl	$(H_3C)_2CH$	Ramrod® Monsanto	Pre-emergence mainly against annual grass and some broadleaf weeds in corn, soybean, cotton, sugarcane, peanut, and some vegetables	1200	1389
41c	*Prynachlor*	phenyl	$HC\equiv C-CH\underset{\underset{CH_3}{\|}}{}$	Butisan®, Basamaize® BASF	Selective herbicide in soybean, rape, onion, and brassicas	1177	—

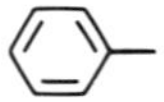

41d	*Alachlor*	C₂H₅ / C₂H₅ ring structure; H₃CO—CH₂	Lasso® Monsanto, 1965	Pre-emergence against annual grass and some broadleaf weeds in corn, soybean, cotton, sugarcane, peanut, rape, and brassicas	1200	1391, 1392
41e	*Butachlor*	C₂H₅ / C₂H₅ ring structure; H₉C₄O—CH₂	Machete® Monsanto, 1969	Pre-emergence against annual grass and some broadleaf weeds in rice, wheat, barley, sugar beet, cotton, peanut, and brassicas	3300	1392
41f	*Metolachlor*	CH₃ / C₂H₅ ring structure; CH₃ / H₃COCH₂—CH	Dual®, Bicep® (in combination with *atrazine*) Ciba-Geigy	Selective herbicide (pre-emergence and pre-plant incorporated) with activity against grasses; used in corn, soybean, peanut, and white potato	2750	1392a
41g	*Dimethachlor*	CH₃ / CH₃ ring structure; H₃COCH₂CH₂—	Teridox® Ciba-Geigy	Selective herbicide against annual broadleaf and grass weeds in rape	1600	1392b

butachlor **41e** are manufactured according to the following scheme:

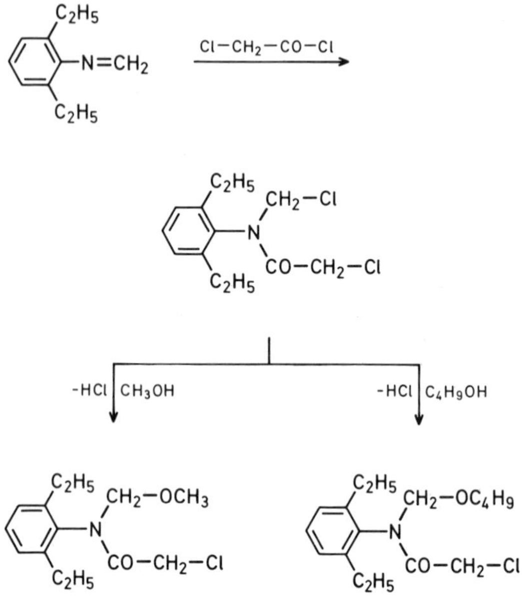

Various other carboxamide herbicides occupy a special position according to their structures and mechanisms of action. 3,5-Dichloro-*N*-(1,1-dimethylpropynyl)benzamide[1395,1396] **(42)** is used as a pre-emergence herbicide against grass and some broadleaf weeds in soybean, cotton, clover, alfalfa, and ornamental trees and shrubs. It is absorbed via the roots and acts as a mitosis poison.

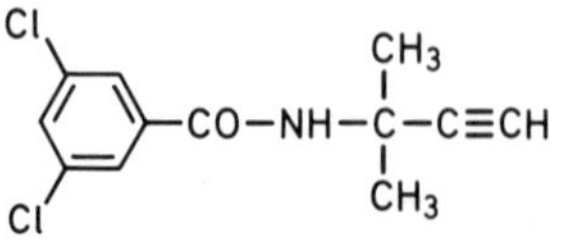

42; *Propyzamide*, Kerb® (Rohm and Haas, 1965);
LD_{50}: 5620 mg/kg female rat, oral, acute; 8350 mg/kg male

Disruption of cell division is also assumed[1398] to be the mechanism of action of *diphenamid* **(43)**.[1397]

Napropamide **(44)**[1399] is absorbed from the soil and inhibits root growth.

[(Benzoylamino)oxy]acetic acid **(45)**[1400,1401] is used for post-emergence control of kochia in sugar beet.

$$(H_5C_6)_2CH-CO-N(CH_3)_2$$

43; Dymid® (Eli Lilly, 1960), Enide® (Upjohn);
Use: pre-emergence in tomato, potato, peanut and ornamentals against grasses and certain broadleaf weeds. Absorbed via the roots;
LD_{50}: 970 mg/kg rat, oral, acute

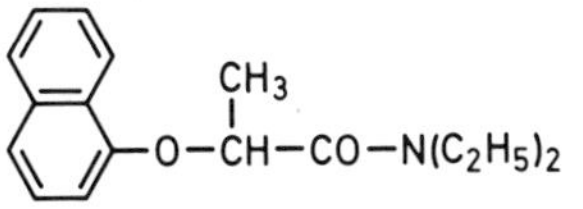

44; Devrinol® (Stauffer, 1971);
LD$_{50}$: 5000 mg/kg rat, oral, acute

$$H_5C_6-CO-NH-O-CH_2-COOH$$

45; *Benzadox*, Topcide® (Gulf Oil, 1965);
LD$_{50}$: 5600 mg/kg rat, oral, acute

4.3.2.5. 2,6-Dinitroanilines

Tri-substituted N,N-dialkylanilines of the following structure belong to those groups of herbicides that have been discovered in recent years. Their herbicidal properties

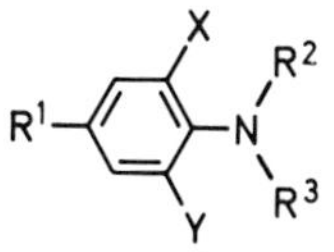

46

are closely connected with the nature and position of the substituent on the aryl ring (see Table 42). High herbicidal activity is always found when the amine function is flanked by two electron-withdrawing groups (X, Y). The substituent in the 4-position (R^1), also essential for activity, may be an alkyl group or an electronegative residue. Introduction of a further substituent in the 3-position generally has no effect on the phytotoxicity. Of all the herbicidal agents from this class only the derivatives of 2,6-dinitroaniline (**46**; X, Y = NO$_2$) have achieved commercial importance to date (Table 42). They are used as soil-applied pre-emergence herbicides (direct incorporation in the soil before sowing) for selective weed control—mainly against grasses—in various crops, the most important of which are cotton and soybean. The dinitroanilines are mitosis poisons. Their herbicidal activity is apparently chiefly caused by interference with nuclear and cell division, which leads to the production of multinuclear cells and formation of abnormal cell nuclei. A marked reduction in root and shoot growth in affected plants is the typical macroscopic symptom.

2,6-Dinitro-N,N-dipropyl-4-(trifluoromethyl)benzenamine (**46a**), the first and today commercially the most significant herbicide from this class, was introduced in 1960.

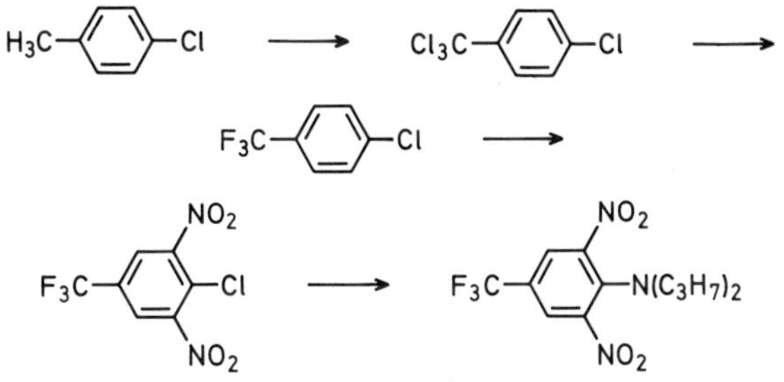

46a

TABLE 42. 2,6-Dinitroanilines

46

No.	Common Name	Structure	Trade Name, Manufacturer	Use	LD$_{50}$ (mg/kg Rat, Oral, Acute)	Reference
46a	*Trifluralin*	F_3C— ring with NO$_2$, NO$_2$, N(C$_3$H$_7$)$_2$	Treflan® Eli Lilly, 1960	Against annual grasses and certain broadleaf weeds in cotton, soybean, sugar beet, beans, pea, peanut, and tomato; pre-sowing	>10,000	1402, 1403
46b	*Benfluralin* *Benefin*	F_3C— ring with NO$_2$, NO$_2$, N(C$_4$H$_9$)(C$_2$H$_5$)	Balan®, Bonalan®, Quilan® Eli Lilly, 1965	Against grasses and broadleaf weeds in tobacco, peanut, alfalfa, and some vegetables; pre-sowing and pre-emergence	>10,000	1402, 1403
46c	*Profluralin*	F_3C— ring with NO$_2$, NO$_2$, N(CH$_2$—cyclopropyl)(C$_3$H$_7$)	Pregard®, Tolban® Ciba-Geigy, 1970	In cotton and soybean; pre-sowing and pre-emergence	~10,000	1404
46d	*Dinitramine*	F_3C— ring with H$_2$N, NO$_2$, NO$_2$, N(C$_2$H$_5$)$_2$	Cobex® US Borax, 1971	In cotton, soybean, sunflower, beans, peanut, carrot; pre-sowing and pre-emergence	3,000	1405

	Name	Structure	Trade name / Company		Reference
46e	Nitralin	H_3C-SO_2 benzene ring with NO_2, NO_2, $N(C_3H_7)_2$	Planavin® Shell, 1966	In winter rape, cotton, soy-bean, peanut, tobacco, tomato; pre-sowing and pre-emergence	>6,000 — 1406, 1407
46f	Oryzalin	H_2N-SO_2 benzene ring with NO_2, NO_2, $N(C_3H_7)_2$	Surflan®, Dirimal® Eli Lilly	In soybean, rice, and potato; pre-emergence	>10,000 — 1408
46g	Isopropalin	$(H_3C)_2CH$ benzene ring with NO_2, NO_2, $N(C_3H_7)_2$	Paarlan® Eli Lilly, 1969	In tobacco, pepper, and paprica; pre-sowing	>5,000 — 1403
46h	Fluchloralin	F_3C benzene ring with NO_2, NO_2, N with $(CH_2)_2-CH_3$ and CH_2-CH_2-Cl	Basalin® BASF	Against annual grasses and broadleaf weeds in cotton and soybean; pre-plant incorporated	1,550 — 1408a
46i	Ethalfluralin	F_3C benzene ring with NO_2, NO_2, N with C_2H_5 and $CH_2-C=CH_2$ (CH_3)	Sonalan® Eli Lilly	In drybean and cotton; pre-emergence	>10,000 — 1408b

N-(1-Ethylpropyl)-3,4-dimethyl-2,6-dinitrobenzenamine[1409] **(47)**, a divergent from the *N,N*-dialkylamines **46**, was the first *N*-monoalkyl-2,6-dinitroaniline to be marketed. It is mainly used for control of grasses in corn (maize), cereals, and rice (pre-emergence) and in cotton, soybean, peanut, and beans (pre-sowing, direct incorporation).

47; *Pendimethalin*, Stomp®, Prowl® (American Cyanamid);
LD_{50}: 1050 mg/kg female rat, oral, acute; 1250 mg/kg male

Another monoaklyl derivative, **47a**, finds application for control of annual broadleaf weeds and grasses in soybean and cotton, and as a selective herbicide in direct-seeded rice.

47a; *Butralin*, Amex 820® (Amchem. Prod.);
LD_{50}: 12,600 mg/kg rat, oral, acute

4.3.2.6. Five-Membered Heterocyclics

Although the high herbicidal activity of numerous five-membered heterocyclics (e.g., substituted pyrrolidones; 1,3-oxazolidinediones, pyrazoles, benzo-1,2,3-thiadiazoles, imidazoles, and benzimidazoles[1410-1414]) is known from journals and the patent literature, few such compounds have been introduced to date as commercial products.

One of the oldest heterocyclics used for total weed control is 1H-1,2,4-triazol-3-amine,[1415] a nonselective translocated herbicide absorbed by roots and leaves with particular activity against deep-rooted weeds. It is often used as a synergistic component in many total herbicidal mixtures (with ureas, 1,3,5-triazines).

Amitrole, Aminotriazole, ATA, Weedazol® (Amchem, 1954);
LD_{50}: 1100–2500 mg/kg rat, oral, acute

The herbicidal activity of *amitrole* is considerably enhanced by the addition of ammonium thiocyanate, itself only weakly phytotoxic[1416] (*amitrole T*, Weedazol TL®, Amchem). Among other things *amitrole* inhibits the biosynthesis of carotenoids,[1210] which are apparently necessary for the maintenance of chloroplast structure and as protectants against the photooxidative degradation of chlorophyll.

Bleaching-out (chlorosis) and the almost white color of newly formed plant parts are the external characteristics of plants treated with *amitrole*.

Nothing is known to date about the mechanism of action of 3-[2,4-dichloro-5-(1-methylethoxy)phenyl]-5-(1,1-dimethylethyl)-1,3,4-oxadiazol-2(3H)-one **(48)**.

48; *Oxadiazon*,[1417] Ronstar® (Rhône-Poulenc, 1969)
Use: pre- and post-emergence in rice, wine grape, and orchard fruit
LD_{50}: >8000 mg/kg rat, oral, acute

2-(3,4-Dichlorophenyl)-4-methyl-1,2,4-oxadiazolidine-2,5-dione **(49)** is a selective pre-emergence herbicide for control of grasses and seed-propagated broadleaf weeds in potato and cotton. It is also used in citrus and grape.[1418, 1419]

49; *Methazole*, Probe® (Velsicol, 1970)
LD_{50}: 1350 mg/kg rat, oral, acute

Two recently introduced products should also be mentioned here.

1. 2-Ethoxy-2,3-dihydro-3,3-dimethyl-5-benzofuranol methanesulfonate.[1420]

Ethofumesate, Nortron® (Fisons)
Use: pre-emergence for control of grasses and certain broadleaf weeds in sugar beet;
LD_{50}: >6400 mg/kg rat, oral, acute

2. 1,2-Dimethyl-3,5-diphenyl-1H-pyrazolium methyl sulfate.[1421, 1422]

Difenzoquat, Avenge® (American Cyanamid, 1974)
Use: post-emergence for control of wild oat in wheat and barley;
LD_{50}: 470 mg/kg rat, oral, acute

4.3.2.7. Pyridines and Pyridinium Salts

Among the herbicides available commercially, pyridines are not represented to any great extent. Although herbicidal activity has been ascribed to many in the patent

literature, only 2,3,5-trichloro-4-pyridinol (50)[1423] and the pyridyloxyacetic acid
50a, apart from *picloram* (36), have actually been marketed.

50; *Pyrichlor*, Daxtron® (Dow, 1965);
LD$_{50}$: 80–130 mg/kg rat, oral, acute

50a; *Trichlopyr*, Garlon 3A® (Dow, 1970);
Use: control of broadleaf weeds in cereals, of weeds and woody plants on non-crop land;
LD$_{50}$: 713 mg/kg rat, oral, acute

The practical applications of *pyrichlor* are severely restricted by its unfavorable
toxicological properties. It has high activity against grasses and is used as a total
herbicide (industrial sites, tracks, roadsides, etc.) and as a selective agent in some
crops (sugarcane, cotton, peanut).

Certain pyridinium salts are of much greater economic importance. Within this
class of herbicides the bipyridylium salts **51** and **52**, prepared by alkylation of 2,2'-
and 4,4'-bipyridyl with 1,2-dibromoethane and chloromethane, respectively, occupy
a special position.

51; *Diquat*,[1424–1427] Reglone® (ICI, 1955);
LD$_{50}$: 231 mg/kg rat, oral, acute

52; *Paraquat*,[1425,1427] Gramoxone® (ICI, 1960);
LD$_{50}$: 150 mg/kg rat, oral, acute

Studies on structure–herbicidal activity relationships among bipyridylium compounds have shown that only quaternary salts of 2,2'- and 4,4'-bipyridyl are sufficiently active whereby—particularly with the 2,2'-bipyridyl derivatives—the nature
of the alkyl residue is crucial for the phytotoxic activity.

Diquat (**51**) and *paraquat* (**52**) are rapidly acting contact herbicides and dessi-
cants that only affect aerial plant parts. In the soil the "quats" are strongly absorbed
and inactivated.

These properties have opened up special application areas for *diquat* and particularly for *paraquat*. Whereas *diquat* mainly acts against dicotyledonous plants and is
suitable for such applications as potato haulm destruction before harvesting, *para-*

quat may be used against both broadleaf and grass weeds. The main use of *paraquat* is for total nonresidual weed control. The advantage here is that the treated areas can be shown even when the old vegetation is still dying, without any risk of damage to the next crop. This technique has brought about a reform in some agricultural areas and obviated the necessity for erosion-promoting ploughing and the like (no-tillage technique[1427a]).

Paraquat (52) is also used for weed control in coffee, orchard fruit, and grape crops and for control of aquatic weeds (owing to its low toxicity to fish).

Another herbicidal 4,4'-bipyridylium salt is *morfamquat*[1428] (53, ICI), which, however, has achieved little commercial significance.

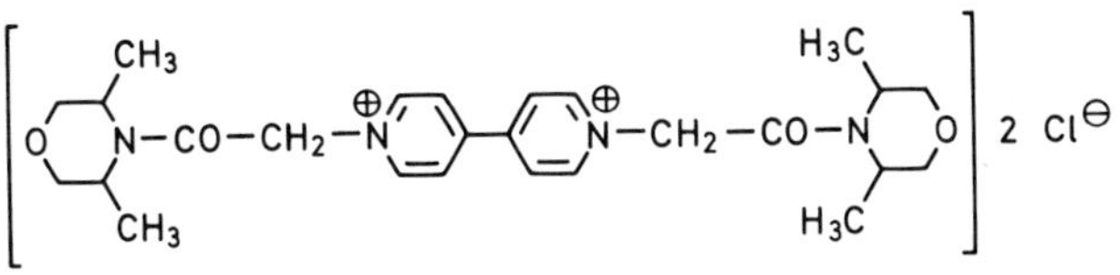

53

The mechanism of action of the "quats" (disruption of light reaction I in the process of photosynthesis, see Figure 5)[1183,1184] concords with the marked dependence of the herbicidal activity on light. In the dark or in the absence of oxygen the phytotoxicity of the "quats" is much reduced.

4.3.2.8. Pyridazines and Pyrimidines

At the beginning of the 1960s a new class of compounds—the pyridazine-3-ones—was discovered in the laboratories of BASF. Some compounds from this class have since achieved practical importance as selective agents. The first, 5-amino-4-chloro-2-phenyl-3(2H)-pyridazinone[1429,1430] (54), was introduced in 1962 and subsequently became of great value as a special herbicide in beet crops. It is manufactured according to the following scheme.

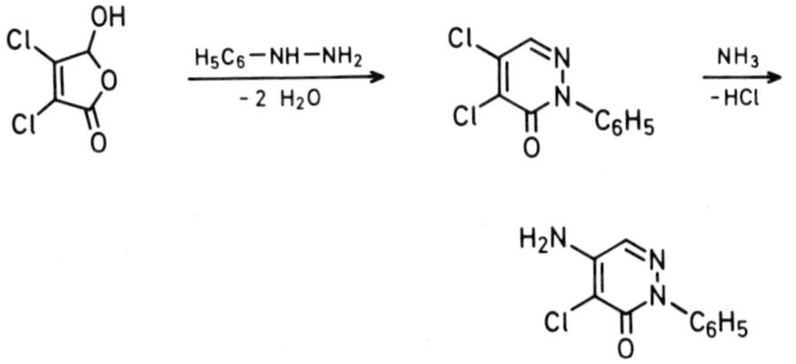

54; *Pyrazon*, Pyramin® (BASF);
LD$_{50}$: 3600 mg/kg rat, oral, acute

Pyrazon is a soil-applied herbicide mainly absorbed via the roots and translocated. Generally it is applied pre-emergence directly after sowing. *Pyrazon* is particularly effective against shallow-germinating weeds but ineffective in the control of some grasses. Its activity is probably mainly due to inhibition of photosynthesis (it inhibits the Hill reaction). The selectivity of *pyrazon* is biochemical in origin; in beet plants it is rapidly deactivated by formation of a herbicidally inactive *N*-glucoside.[1431]

The analogous bromo compound[1430] **55** is a component of the combination product Basanor® (see Section 4.3.2.3.4).

55; *Brompyrazon* (BASF, 1968); LD$_{50}$: >6400 mg/kg rat, oral, acute

56; *Norflurazon*, Zorial® (Sandoz); LD$_{50}$: >8000 mg/kg rat, oral, acute

57; *MH* (US Rubber, 1949); LD$_{50}$: 6950 mg/kg rat, oral, acute

4-Chloro-5-(methylamino-2-[3-(trifluoromethyl)phenyl]-3(2H)-pyridazinone (*norflurazon*, **56**)[1432] was introduced in 1971 as a selective herbicide for control of grasses and most annual broadleaf weeds in cotton.

Maleic hydrazide (**57**)[1433] is mainly used as a growth regulator (see Chapter 5). Apart from this growth inhibition in grasses[1434] and prevention of sucker development in tobacco, *MH* is also used in combination with "growth regulator herbicides" (components to control broadleaf weeds) for specialized applications (retardation of plant growth on roadsides, highway central reservations, etc.).

The 3-aryloxypyridazines **58** are available from maleic hydrazide (**57**). Their herbicidal properties were first recognized in Japan,[1435] for example:

58a; 3-(2-methylphenoxy)pyridazine, *Credazine*, Kusakira® (Sankyo, 1970); LD$_{50}$: 3090 mg/kg rat, oral, acute

Of this group only *credazine* is in use at the present, in Japan, as a soil herbicide for control of annual grasses and certain broadleaf weeds in tomato and strawberry.

2-Aryloxypyrimidines **59** are also herbicidal,[1436] but no compound from this group has got beyond the experimental stage to date.

59; e.g.: Ar = C_6H_5, R = CH_3, Ar = 2,6-$Cl_2C_6H_3$, R = H

Of the pyrimidines, only derivatives of uracil[1437] have achieved any practical significance. The herbicidal activity of these compounds is due to inhibition of photosynthesis[1172, 1173] (they are potent inhibitors of the Hill reaction).

3-Cyclohexyl-6,7-dihydro-1H-cyclopentapyrimidine-2,4(3H, 5H)-dione[1438] **(60)**, available by the following synthetic sequence is a selective soil-applied herbicide for

60; *Lenacil*, Venzar® (Du Pont, 1974);
LD_{50}: >11,000 mg/kg rat, oral, acute

control of annual grasses and broadleaf weeds in sugar and fodder beet, spinach and strawberry by pre-plant incorporation or pre-emergence treatment. Combinations of *lenacil* with *benzthiazuron* (see Section 4.3.2.3.4; as Merpelan®, Bayer AG) and *isocarbamid* (see Section 4.3.2.3.4; as Merpelan AZ®, Bayer AG) are also used in the same crops.

The major area of use of the 3-alkyl-5-halo-6-methyl-uracils **61**, synthesized from acetoacetate and alkylurea, is as total and semitotal herbicides.

61a; R = -CH(CH_3)C_2H_5, X = Br; *Bromacil*, Hyvar X® (Du Pont, 1952)[1439, 1440];
LD_{50}: 5200 mg/kg rat, oral, acute

61b; R = -CH(CH_3)_2, X = Br; *Isocil*,[1439] Hyvar® (Du Pont, 1962); (of little commercial interest);
LD_{50}: 3250 mg/kg rat, oral, acute

61c; R = -C(CH_3)_3, X = Cl; *Terbacil*,[1439] Sinbar® (Du Pont, 1966);
LD_{50}: 5000 mg/kg rat, oral, acute

Bromacil (**61a**), the most important member of this group, is used—mainly in combination with other agents (*diuron, amitrole*)—for total weed control on non-crop land and as a semitotal herbicide in established citrus plantations.

Terbacil (**61b**) is also used as a component in total herbicides and for control of many annual and some perennial weeds in citrus, sugarcane, and orchard fruit.

The 3-alkyltetrahydro- (e.g., **62**) and -decahydroquinazoline-2,4-diones[1441,1442] (e.g., **63**), chemically related to the uracils, are also herbicidal, but no product from this group has yet found acceptance in practice.

3-(1-Methylethyl)-(1H)-2,1,3-benzothiadiazin-4(3H)one 2,2-dioxide (**64**),[1442,1443] closely related to **62**, is synthesized from methyl anthranilate by the following procedure.

64; *Bentazone*, Basagran® (BASF, 1968);
LD$_{50}$: 1100 mg/kg rat, oral, acute

It is a post-emergence contact herbicide (absorbed via the leaves) for control of broadleaf weeds (e.g., *Matricaria* and *Anthemis* spp., *Gallium aparine*, *Stellaria media*) in winter and spring cereals. The mechanism of action of *bentazone* is as yet unknown.

4.3.2.9. 1,3,5-Triazines

The marked herbicidal activity of trisubstituted 1,3,5-triazines was discovered in the Geigy laboratories in 1955.[1444,1445] Thus began the development of a group of herbicides that has subsequently achieved great practical significance. After the phenoxy–fatty acids, the 1,3,5-triazine derivatives are probably the most important class of herbicides in economic terms today.

The starting material for the manufacture of all herbicidal 1,3,5-triazines is cyanuric chloride (2,4,6-trichloro-1,3,5-triazine), the trimerization product from cyanogen chloride. The stepwise replacement of two of the three chlorine atoms by identical or different C_2-C_4-alkylamino residues leads to the first generation of triazine herbicides **65** (see Table 43), of which *simazine* **(64a)** and *atrazine* **(65b)** are the most important.

Although the 1,3,5-triazine derivatives **66** are highly active herbicides, their instability precludes any practical application.[1446] In contrast, the replacement of one or two chlorine atoms in cyanuric chloride by arylamino groups leads to herbicidally inactive compounds.

Triazine herbicides of the second generation ["triatones", **67** (Table 44); "triatrynes", **68** (Table 45)] are available from *atrazine* **(65b)** and so on according to the following scheme.

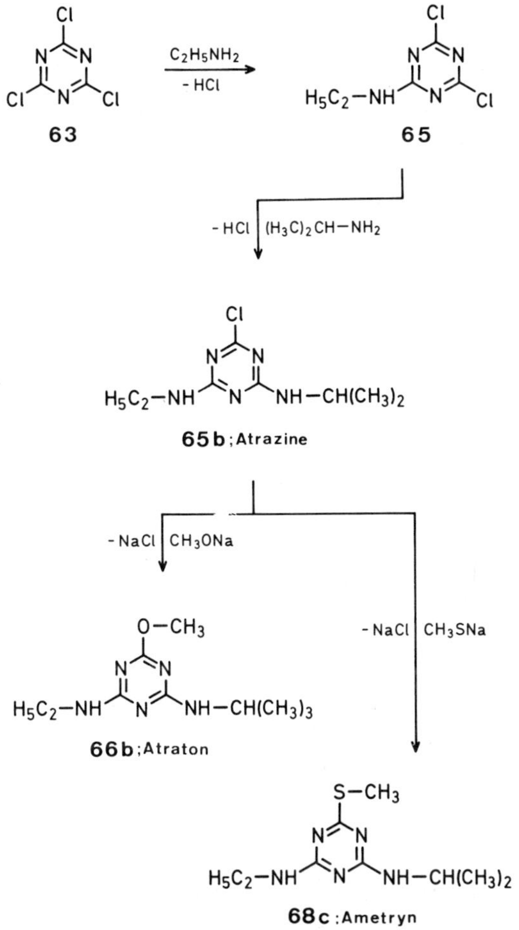

65b;Atrazine

66b;Atraton

68c:Ametryn

Introduction of an ethylthio residue as the third substituent—for example, in *dipropetryn* **(69)**—also leads to compounds of high herbicidal activity.

TABLE 43. 6-Chloro-1,3,5-triazine Derivatives ("Triazines")

$$\underset{65}{\underset{R^1\diagdown N \diagup N \diagdown R^2}{\overset{Cl}{\overset{|}{\underset{N}{\bigtriangleup}}}}}$$

No.	Common Name, Trade Name (Manufacturer)	R^1	R^2	Use	LD$_{50}$ (mg/kg Rat, Oral, Acute)	Reference
65a	*Simazine* Gesatop® (Geigy, 1955)	H$_5$C$_2$—NH	H$_5$C$_2$—NH	Selective in corn, citrus, olive, asparagus, grape, coffee, tea, cocoa (pre-emergence); against aquatic weeds and algae in ponds, fish hatcheries; at higher application rates as total herbicide on non-crop land	>5000	1447
65b	*Atrazine* Gesaprim® (Geigy, 1958)	H$_5$C$_2$—NH	(H$_3$C)$_2$CH—NH	Selective in corn, sorghum, sugarcane, and pineapple; algicide	3080	1447
65c	*Propazine* Gesamil® (Geigy, 1960)	(H$_3$C)$_2$CH—NH	(H$_3$C)$_2$CH—NH	Mainly in combination with other herbicides (*atrazine, amitrole, 2,4-D*) for pre-emergence total weed control on non-crop land	>5000	1447
65d	*Trietazine* Gesafloc® (Geigy, 1960)	H$_5$C$_2$—NH	(H$_5$C$_2$)$_2$N	Marketed since 1972 as a component of combination herbicides (*linuron, simazine*)	2830 male, 4000 female	1447
65e	*Terbuthylamine* Gardoprim® (Geigy, 1966)	H$_5$C$_2$—NH	(H$_3$C)$_3$C—NH	Similar to *simazine* and *atrazine* but with a longer residual action	2160	1447
65f	*Cyprazine* Outfox® (Gulf Oil, 1970)	(H$_3$C)$_2$CH—NH	▷—NH	Selective in corn against broadleaf and certain grass weeds (post-emergence)	1200 ± 200	1448
65g	*Cyanazine* Bladex® (Shell, 1971)	H$_5$C$_2$—NH	NC—C(CH$_3$)$_2$NH	Selective (short residual action) in corn, pea, beans, winter wheat, barley, sugarcane, soybean, and peanut (pre- and post-emergence)	182	1449

TABLE 44. 6-Methoxy-1,3,5-triazines ("Triatones")

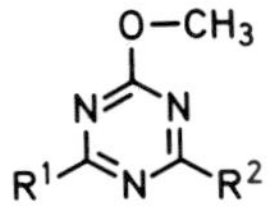

67

No.	Common Name, Trade Name (Manufacturer)	R^1	R^2	Use	LD_{50} (mg/kg) Rat, Oral, Acute	Reference
67a	*Simeton* Gesadural® (Geigy)	H_5C_2-NH	H_5C_2-NH	Soil- and foliar-applied herbicide (pre- and post-emergence); of little commercial significance	535	1447
67b	*Atraton* Gesatamin® (Geigy, 1959)	H_5C_2-NH	$(H_3C)_2CH-NH$	Selective in flax and cotton	1465	1447
67c	*Prometon* Primatol® (Geigy, 1958)	$(H_3C)_2CH-NH$	$(H_3C)_2CH-NH$	Total herbicide on non-crop land	2980	1447
67d	*Secbumeton* Etazin® (Geigy, 1966)	H_5C_2-NH	$H_5C_2-CH-NH$ $\vert$ CH_3	Selective against grass and broadleaf weeds, both annual and perennial, in alfalfa and sugarcane; also component of total herbicides	2680	1447
67e	*Terbumeton* Caragard® (Geigy, 1966)	H_5C_2-NH	$(H_3C)_3C-NH$	In citrus, apple orchards, vineyards, and forestry —often in combination with *terbuthylazine* (pre- and post-emergence); absorbed via leaves and roots	485	1447

383

TABLE 45. 6-Methylthio-1,3,5-triazines ("Triatrynes")

$$R^1 \underset{N}{\overset{N}{\diagdown}} \underset{\substack{N \\ \\ 68}}{\overset{S-CH_3}{\diagup}} R^2$$

No.	Common Name, Trade Name (Manufacturer)	R^1	R^2	Use	LD_{50} (mg/kg Rat, Oral, Acute)	Reference
68a	*Simetryn* — (Geigy)	H_5C_2-NH	H_5C_2-NH	No commercial significance	535–1830	1450
68b	*Desmetryn* Semeron® (Geigy, 1964)	H_3C-NH	$(H_3C)_2CH-NH$	Selective post-emergence herbicide against broadleaf and grass weeds in brassicas	1390	1450
68c	*Ametryn* Gesapax®, Evik® (Geigy, 1964)	H_5C_2-NH	$(H_3C)_2CH-NH$	Pre- and post-emergence in pineapple, sugarcane, banana, citrus, pome and stone fruit, and coffee and for potato haulm destruction; aquatic herbicide	1405	1450

68d	*Terbutryn* Igran® (Geigy, 1966)	H_5C_2—NH	$(H_3C)_3C$—NH	Selective pre- and post-emergence herbicide in winter cereals (slender foxtail, windgrass, broadleaf weeds), sunflower, and potato	2400–2980	1451
68e	*Prometryn* Gesagard®, Caparal® (Geigy, 1962)	$(H_3C)_2CH$—NH	$(H_3C)_2CH$—NH	Selective pre- and post-emergence herbicide against broadleaf weeds in leek, celery, onion, cotton, potato, and sunflower	3150–3750	1450
68f	*Methoprotryne* Gesaran® (Geigy, 1965)	$(H_3C)_2CH$—NH	H_3CO—$(CH_2)_3$—NH	Against grass weeds in winter cereals (post-emergence, spring application), also in combination with *sinazine* and *mecoprop*	>5000	1452
68g	*Dimethametryn* C 18 898 (Ciba 1968)	H_5C_2—NH	$(H_3C)_2CH$—CH—NH CH_3	Selective against broadleaf weeds in rice (direct-drilled and transplanted); in combination with *piperophos* (Avirosan®) also against grass weeds	3000	1453
68h	*Aziprotryn* Mesoranil® (Ciba, 1967)	N_3	$(H_3C)_2CH$—NH	Selective in brassicas, rape (pre-emergence) against annual broadleaf and certain grass weeds; against weeds in pea (post-emergence)	3600–5830	1454
68i	*MPMT* Lambast® (Monsanto, 1966)	H_3CO—$(CH_2)_3$—NH	H_3CO—$(CH_2)_3$—NH	Selective in peanut, soybean (post-emergence), and pea (pre-emergence)	1400	1455

$$S-C_2H_5$$

$$(H_3C)_2CH-NH \qquad NH-CH(CH_3)_2$$

69; *Dipropetryn*, Cotofor® (Ciba-Geigy, 1971)
Use: pre-emergence in cotton;

LD_{50}: 4050 mg/kg rat, oral, acute

The differences, often very marked, in range of activity, duration of activity, selectivity, and stability on soil—the "triatrynes" **(68)**, for example, are generally degraded much more rapidly than the "triazines" **(65)** and the "triatones" **(67)**—enable the whole range of triazine herbicides to be used for all the various applications in agriculture. They are used as total, as semitotal, and as selective herbicides in numerous crops. Of special significance is the use of 1,3,5-triazines in tropical crops and corn (maize) growing. The main area of application in Europe is in cereals. *Atrazine* **(65b)** and *simazine* **(65a)** are the major corn herbicides. *Atrazine*, which in contrast to *simazine* is also very effective under dry conditions, is not very effective in the control of panic grass. This problem has been practically solved by combination with certain grass weed herbicides such as *alachlor* **(41d)**.

As with the urea herbicides, the main reason for the phytotoxic properties of the 1,3,5-triazines is the inhibition of photosynthesis (light reaction II).[1166-1168] However, it has been shown that the triazines also influence other biochemical processes in plants. Thus inhibition of growth of nonphotosynthetic tissue in the dark has been observed after treatment with *atrazine* **(65b)** and *simazine* **(65a)**; an effect that is not found with the herbicidally inactive 2-hydroxy analogues, **70b** and **70a**, of these agents—even at higher concentrations.[1456]

$$OH$$

$$R^1 \qquad N \qquad R^2$$

a : $R^1 = R^2 = H_5C_2-NH$

b : $R^1 = H_5C_2-NH$; $R^2 = (H_3C)_2CH-NH$

c : $R^1 = R^2 = (H_3C)_2CH-NH$

70

An inhibition of oxidative phosphorylation in plant mitochondria has also been detected experimentally with certain 1,3,5-triazines such as *prometryn* **(68e)**.[1457] *Simazine* stimulates the activity of nitrate reductase in various crop plants (wheat, corn, rice). An enhanced uptake of nitrate by the roots coupled with an increased protein content is a positive and desirable side effect found with use of this herbicide.[1458] If ammonium salts serve as a source of nitrogen for the plants then no increase in protein content is observed.

Both physical and biochemical factors must be taken into consideration as reasons for the selectivity of the 1,3,5-triazines. The lack of solubility of some 1,3,5-triazines in water (**65b**: 5 ppm/20–22°C; **65c**: 8.6 ppm/20–22°C), coupled with strong adsorption on various soil types, leads to these agents remaining in the upper soil layers, where they are taken up by germinating weeds. Deep-rooted crop plants do not then come into contact with the herbicide. Cotton, which is not sensitive to many 1,3,5-triazines such as *prometryn* (68a), absorbs the agents but has the ability to store them in certain parts of the plant distant from the site of action (i.e., the chloroplasts) and inactivate them by complex formation with plant products.

The high tolerance of corn (maize) toward 6-chloro-1,3,5-triazines ("triazines"; Table 43) was previously ascribed solely to a nonenzymatic, plant-specific degradation whereby the conversion into the inactive 6-hydroxy-1,3,5-triazines (70) with the aid of the glucoside 71[1459,1460] represented the first and decisive step. (This glucoside has been isolated from corn and other plants insensitive to the triazines 65.)

71

Although this mechanism of detoxification is important for the selectivity in corn, it has been shown recently that the formation of herbicidally inactive conjugates of the "triazines" 65 with amino acids and peptides containing SH groups—for example, of *atrazine* (65b) with glutathione and cysteine—also makes a significant contribution to the selectivity of this group of triazine herbicides in corn.[1461] The "triatones" (67) and "triatrynes" (68), which cannot be inactivated in this way in corn, have insufficient selectivity in this crop.

Other mechanisms of degradation and detoxification of the herbicidal 1,3,5-triazines are the conversion of a methylthio group to hydroxy—for example, the degradation of *prometryn* (68e) to the 6-hydroxy-1,3,5-triazine (70c) in carrots—and the elimination of alkyl groups. Such reactions are important in microbial metabolic degradation.

Recently the high herbicidal activity of other derivatives of 1,3,5-triazine, such as the tetrahydro-1,3,5-triazine-2,4-diones, has been discovered. One commercial development from this group is 3-cyclohexyl-6-(dimethylamino)-1-methyl-1,3,5-triazine-2,4 (1H,3H)-dione[1462] (72), used as a total herbicide on non-crop land on sites not adjacent to deciduous trees and other desirable plants. The very short residual action of this product—a disadvantage in this application—can be compensated by mixing with other soil herbicides.

72; *Hexazinone*, Velpar® (Du Pont, 1975);
LD_{50}: 1690 mg/kg rat, oral, acute

4.3.2.10. 1,2,4-Triazines

Recently a new class of herbicides was discovered in the laboratories of Bayer AG—the 4-amino-1,2,4-triazin-5-ones.[1463] Some derivatives from this class are characterized by high herbicidal activity coupled with pronounced selectivity. The synthesis of some compounds of this class such as the 4-amino-3-(methylthio)-1,2,4-triazin-5(4H)-ones, 73 (Bayer AG, 1964),[1464] starts from thiocarbohydrazide.

a; R = C_6H_5; Aglypt® (Bayer AG, 1969);
Use: flax;

b; R = $(CH_3)_3C$; *Metribuzin*, Sencor® (Bayer AG, 1971; Mobay), Lexone® (Du Pont);
LD_{50}: 2200 mg/kg male rat, oral, acute; 2345 mg/kg female rat

4-Amino-3-(methylthio)-6-phenyl-1,2,4-triazin-5(4H)-one **(73a)**[1463] was developed first but is of little practical significance owing to its limited range of applications.   4-Amino-6-(1,1-dimethylethyl)-3-(methylthio)-1,2,4-triazin-5(4H)-one **(73b)**[1463,1465] is, however, of great interest, its major applications being the control of broadleaf and grass weeds in soybean[1466] and potato (pre-emergence). It is also used in tomato, asparagus, alfalfa, and sugarcane.

The hydrazone **74**, available by condensation of *metribuzin* **(73b)**[1467] with 2-methyl-propanal, is a selective herbicide used for control of broadleaf and grass weeds in winter barley and spring barley and wheat.

74; *Isomethiozin*, Tantizon® (Bayer AG, 1975);
LD_{50}: >10,000 mg/kg rat, oral, acute

Closely related to Aglypt® **(73a)** is 4-amino-3-methyl-6-phenyl-1,2,4-triazin-5(4H)-one **(75)**,[1468-1470] which, owing to its high activity against broadleaf and grass weeds, its long application period (pre- and post-emergence), and its high selectivity to sugar and fodder beets, is used with great success in these crops. Goltix® is synthesized by condensation of phenylglyoxylate with acethydrazide.

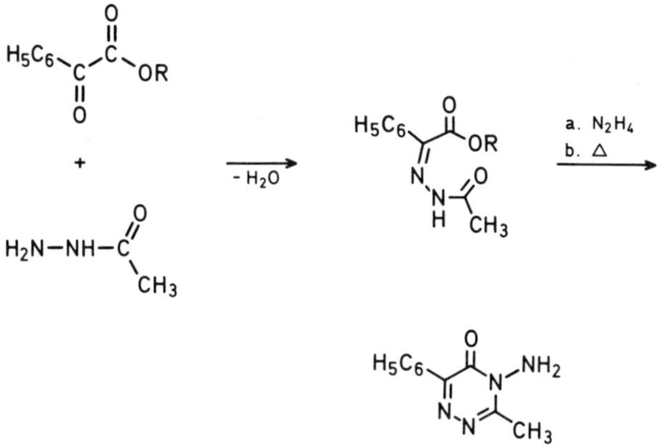

75; *Metamitron*, Goltix® (Bayer AG, 1975); LD$_{50}$: 3340 mg/kg rat, oral, acute

The herbicidal action of these 4-amino-1,2,4-triazin-5(4H)-ones has been traced to an inhibition of photosynthetic electron transport within the area of light reaction II.[1169-1171] The following pI_{50} values for inhibition of the Hill reaction have been determined on isolated chloroplasts.

73a 6.63
73b 6.7
74   6.13

Little is known of the reasons for the pronounced selectivity of these compounds. With *metribuzin* (73b), it has been shown that differences in the translocation of the agent in resistant and sensitive plants could play a role. Specific detoxification reactions in beet at the 4-amino or 3-methyl group are presumed to be the reason for the high selectivity of *metamitron* (75) in these plants.[1470] Studies on structure–activity relationships for this heterocyclic system revealed that the 4-amino group can be replaced by a methylamino residue without significant loss of activity. However, deamination leads to weakly active compounds; the introduction of a dimethylamino group in position 4 causes an almost complete loss of activity. Even the immediate precursor of *metribuzin* (73b) is herbicidally inactive.

4.3.2.11. Organophosphorus and Organoarsenic Compounds

Organophosphorus compounds have been used for about 25 years now as selective herbicides and harvest aids. *S,S,S*-Tributyl phosphorotrithioate[1471] (76) and tributyl phosphorotrithioite[1472] (77), both translocated foliar herbicides, belong to the oldest products in this group.

$$(H_9C_4-S)_3PO$$

76; DEF® (Mobay Ag. Div./Chemagro, 1956); LD$_{50}$: 325 mg/kg rat, oral, acute

$$(H_9C_4-S)_3P$$

77; *Merphos*, Folex® (Mobil Oil); LD$_{50}$: 1270 mg/kg rat, oral, acute

Both compounds are used as defoliants in cotton.

Together with the phosphite mixture *2,4-DEP* mentioned earlier (Section 4.3.2.4.2), *O*-(2,4-dichlorophenyl) *O*-methyl isopropylphosphoramidothioate was one of the first selective herbicides based on phosphorus to be introduced.[1473] This product, of little commercial significance, is mainly used for control of broadleaf weeds in lawn grass.

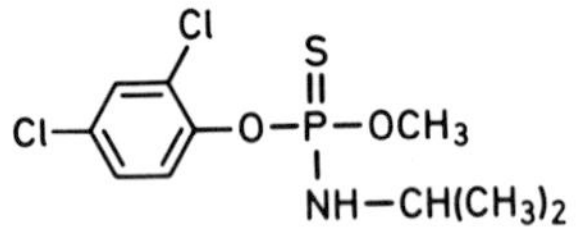

DMPA, Zytron® (Dow)

Nitrophenyl esters of phosphoramidothioic acid have recently achieved significance in Japan. *O*-(4-Methyl-2-nitrophenyl) *O*-methyl isopropylphosphoramidothioate 78 is used as a selective herbicide in many vegetable crops and in dry paddy rice (pre-emergence). As with the 5-methyl derivative 79, the main activity is control of grass weeds.

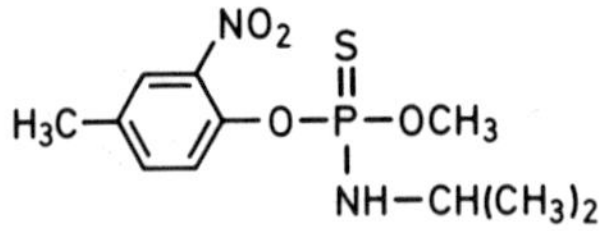

78; *Amiprofos-methyl*, Tokunol M® (Nitokuno, 1976); LD$_{50}$: 600 mg/kg rat, oral, acute

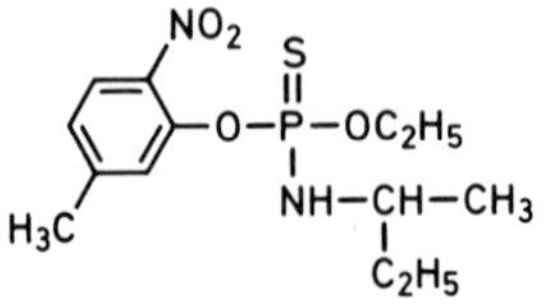

79; Cremart® (Sumitomo, 1976)

Little has been published concerning the mechanism of action of these two compounds. It is assumed that Cremart® (79) interferes with cell division processes in higher plants and algae.[1474]

The phosphorodithioate 80, which is absorbed via roots, coleoptila, and leaves of the plants, is active against annual grasses and is selective in rice.

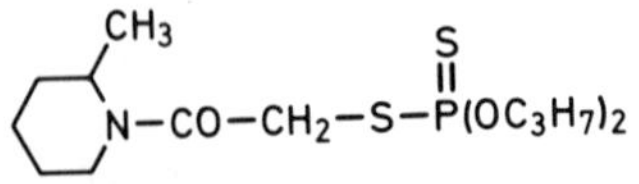

80; *Piperophos*,[1475] Rilof®—also marketed in combination with *dimethametryn* as Avirosa® (Ciba-Geigy, 1969);
LD$_{50}$: >2150 mg/kg rat, oral, acute

$$\langle\underset{=}{=}\rangle-SO_2-NH-(CH_2)_2-S-\overset{\overset{\text{S}}{\|}}{P}[O-CH(CH_3)_2]_2$$

81; *Bensulide*, Betasan®, Prefar® (Stauffer, 1969);
LD_{50}: 770 mg/kg rat, oral, acute

In order to control broadleaf weeds as well, combinations with *dimethametryn* (Table 45) and—in tropical areas—with "growth regulator herbicides," are on the market.

Bensulide (**81**)[1476] is used for pre-emergence control of broadleaf and grass weeds in rice, melon, and lettuce. The mechanism of action of this compound has not yet been elucidated.

In 1971 the first phosphonic acid derivative—*N*-(phosphonomethyl)glycine[1477] (**82**)—was introduced as an experimental herbicide.

$$Cl-CH_2-\overset{\overset{\text{O}}{\|}}{P}(OH)_2 \quad + \quad HOOC-CH_2-NH_2 \quad \xrightarrow[-\,HCl]{NaOH}$$

$$HOOC-CH_2-NH-CH_2-\overset{\overset{\text{O}}{\|}}{P}(OH)_2$$

82; *Glyphosate*, Roundup® (Monsanto);
LD_{50}: 4320 mg/kg rat, oral, acute

Today it is used in the form of the isopropylamine salt as a nonselective, non-residual post-emergence herbicide. It is very effective against deep-rooted perennial species and annual and biennial species of grasses, sedges, and broadleaf weeds.

The carbamoylphosphonate **82a** finds application as a foliar herbicide for control of woody plants on non-crop land and in forestry.[1477a]

$$H_5C_2O-\overset{\overset{\text{O}}{\|}}{\underset{\underset{O^{\ominus}\;NH_4^{\oplus}}{|}}{P}}-CO-NH_2$$

82a; *Fosamine-ammonium*, Krenite® (Du Pont, 1974);
LD_{50}: 24,000 mg/kg rat, oral, acute

Although the use of arsenicals in crop protection is no longer permitted in many countries, the worldwide turnover of organic and organic arsenic herbicides is still considerable. For example, in 1974 the turnover of arsenic herbicides amounted to about 10% of that of the carbamate and thiocarbamate herbicides in financial terms. According to current projections these relationships will only change slowly worldwide in the next few years.

Among the organoarsenicals, the salts of methylarsonic acid and dimethylarsinic acid (cacodylic acid) play a prominent role. The methylarsonates **83** and **84** have the advantage over the sodium arsenite **85** of lower toxicity and better selectivity. The salts are contact herbicides with some systemic properties used for control of grass weeds in cotton and on non-crop land.

$$\underset{\underset{OH}{|}}{\overset{\overset{O}{\|}}{H_3C-As-ONa}}$$

$$\underset{\underset{OH}{|}}{\overset{\overset{O}{\|}}{H_3C-As-ONH_4}}$$

$$\underset{\underset{ONa}{|}}{\overset{\overset{O}{\|}}{H_3C-As-ONa}}$$

83; MSMA, Ansar® (An-
sul, 1956);
LD$_{50}$: 900 mg/kg rat, oral,
acute

84; *MAMA*, Ansar® (An-
sul, 1956);
LD$_{50}$: 1800 mg/kg rat, oral,
acute

85

Cacodylic acid **(86)** is used as its sodium salt as a total herbicide on non-crop land (post-emergence), as a dessicant and defoliant in cotton, and for killing un-wanted trees (by direct injection). It is virtually inactivated on contact with soil.

$$\overset{\overset{O}{\|}}{(H_3C)_2As-ONa}$$

86; Phytar® (Ansul, 1958);
LD$_{50}$: 1350 mg/kg rat, oral, acute

For bibliography, see ref. 1478.

_____ *Chapter Five*

Plant Growth Regulators

_____

W. DRABER
Bayer AG
Wuppertal-Elberfeld

5.1. Introduction

Defining plant growth regulators in a way that distinguishes them from herbicides is a difficult matter. This may be seen in the frequent, although undesirable, use of the term "growth regulator herbicide"[1479] (hormone-type herbicide), which may be justified on historical grounds in that the types of compounds thus described—phenoxyacetic acids, halogenated benzoic acids, and araliphatic carboxylic acids—originate from work on analogues of the first known natural "growth regulator," the phytohormone 3-indolylacetic acid.[1480]

In this chapter plant growth regulators are defined as compounds that influence the growth of plants but without any lethal effect on the target plant being intended in their use. The "growth regulator herbicides" are thus clearly excluded by this definition as are the cotton defoliants (such as DEF® and Folex®), regrowth after defoliation not being desired. Some products that play a double role as herbicides and growth regulators (*MH, chlorflurecol, endothal*) are dealt with here. The fact that typical growth regulators and even the endogenic phytohormone ethylene can kill plants in high doses, whereas sublethal doses of herbicides, on the other hand, can occasionally produce desirable effects on growth and physiology (e.g., enhanced protein content with photosynthesis inhibitors), does not mean that these are exceptions to the definition. These are simply well-known phenomena to be observed with all biologically active compounds.

The scientific study of growth regulators began with Went's discovery of auxin activity[1481] (1928) and Kögl's structural elucidation[1482] (1934) of the first natural phytohormone, 3-indolylacetic acid (IAA). The phytophysiological effect of gibberellic acid was also recognized in the 1920s by Japanese researchers.[1483] Structural elucidation and proof of hormone character followed, however, about

35 years later.[1484,1485] More recent is the discovery of the other natural phyto-hormones—the cytokinines[1486] (1964), abscisic acid[1487] (1965), and ethylene[1488] (1962) as endogenic maturity hormone. Parallel to this, numerous synthetic compounds are being found in increasing numbers that exert phytophysiological effects, some analogous and some antagonistic to those of the natural phytohormones. A considerable amount of research work has been carried out in industrial laboratories, for example, at ICI (the gibberellins). On the other hand, in a few cases (*chlormequat*) industry has drawn benefit from the results of pure phytophysiological studies. Great optimism was widespread in the 1960s concerning the early commercial applicability of growth regulators, at least in highly developed agricultural industries. In 1968 a turnover of $50–75M (manufacturers' prices) was predicted in the United States for 1975.[1489] This prediction can hardly have been verified. The actual U.S. turnover in growth regulators (excluding defoliants) was $15M (manufacturers' prices) in 1974. Another estimate[1490] predicted the world turnover in 1974 as $77M (consumers' prices). The same source gives a prediction of $40M (United States) and $118M (world) in 1980, based on consumer prices in 1974.

At the moment the future chances for growth regulators must be assessed conservatively in spite of great scientific interest. The great potential for improvements in yield and quality and for rationalization in production is confronted by equally great difficulties in the development of new products. The reason is that, in chemical interference in the plant system, one can hardly expect to be able to change one variable (in a desirable manner) and hold all others constant. Different varieties of the same plant species can react very differently. Environmental conditions (e.g., climate, fertilizer) exert a considerable influence on the effect. Biological variability necessitates a much better statistical back-up of results than when, as for herbicides, only lethal effects are being assessed. The test programs must therefore be much more detailed, and the development of a growth regulator is thus much more resource intensive than that of a herbicide.[1491] A large number of laboratory tests for growth regulating effects exist, and they are relatively simple to perform.[1492] Their relevance to practice is, however, frequently doubtful. In spite of all the problems most of the companies in the United States, Europe, and Japan engaged in the crop protection business are also active in the growth regulator field.[1491] Considerable research capacity also exists in the iron curtain countries.[1493]

5.2. Economic Significance and Use

At the present moment, the United States is probably the largest market for growth regulators, a highly developed and sophisticated agricultural industry being a prerequisite for their use. The turnover in America is only composed of a few products. Between 1969 and 1974 only *ethephon* (Ethrel®), shown in Table 46, entered the list as a new and successful product.[1494,1495,1495a]

TABLE 46. U.S. Turnover in Growth Regulators (Consumer Prices)

Product	Use	Turnover in $M 1978	1974	1969
Maleic hydrazide	Tobacco, potato	10.2	14.3	15.5
Ethephon (Ethrel®)	Grape, vegetables, fruit, hevea	3.8	6.0	–
Daminozide (Alar®)	Apple, grape, cherry	7.8	2.2	0.9
Gibberellic acid	Grape, citrus, vegetables	1.6	2.2	1.3
Off-Shoot T®	Tobacco	5.9	–	–
Miscellaneous		4.1	3.5	1.5
Total		33.4	28.4	18.4

In addition to the growth regulators, about $20M was turned over in the United States in 1974 with defoliants (mainly for cotton) and dessicants (1978: $30M). Compared with a herbicide turnover of $1014M in the same year, these figures are modest indeed (1978: $2100M).

One important product that has not achieved any significance thus far in the U.S. market is *chlormequat* (CCC®). Approximately 50% of the total wheat acreage in Europe is treated with *chlormequat*[1496] to shorten the stem and thus prevent lodging caused either physiologically or by fungal infection (*Cercosporella herpotrichoides*).

5.3. Commercial and Important Experimental Products

The compounds used as growth regulators are listed in Table 47. Most of them were introduced a long time ago; some are in an advanced stage of development. The products in the table have been listed alphabetically within the following groups.

Carboxylic acids and esters.
Onium compounds.
Heterocycles.
Miscellaneous.

Phenylacetic acids—although they act as growth regulators in low concentrations—have not been included, because they are used practically only as herbicides (see Section 4.3.2.4.1). Typical cotton defoliants (e.g., DEF®, Folex®; see Section 4.3.2.11) are also excluded. The naturally occurring phytohormones, in so far that they are not included in the table, are discussed in Section 5.5, since their activity in most cases depends on a disturbance to the balance of the natural hormones.

TABLE 47. Growth Regulators (Commercial Products)

No.	Common Name, Trade Name, Code	Structure	Manufacturer, Year of Introduction	LD$_{50}$ (mg/kg), Animal	Properties and Use	Patent	Reference
			A. Carboxylic Acids and Esters				
1	*Chlorflurecol-methyl* Maintain® CF 125 IT 3456	(fluorene structure with Cl, HO, COOCH$_3$)	E. Merck AG 1965	3100 rat	Inhibits growth and causes numerous morphological changes; in combination with *MH* as growth retarder in turf, with *MCPA* as selective herbicide (Aniten®)	DAS 1301173 (1962/69), E. Merck; Inv.: Jacobi et al.; *C.A.* **62,** 11086d (1965)	1497–1501
2	*Daminozide* Alar®, B-Nine® B-995	CH$_2$—CO—NH—N(CH$_3$)$_2$ | CH$_2$—COOH dimethylalkylamine salt	Uniroyal 1962	8400 rat	Growth regulator, retards vegetative growth in fruit trees and ornamentals, hastens ripening in cherry, peach, nectarine, improves fruit set in grape, causes compact growth and induces flowering in house plants	DAS 1267464 (1961/68), Uniroyal; Inv.: Hagemann et al.; *C.A.* **57,** 10281d (1962)	1489, 1502–1504
3	*Endothal, Endothall* Ripenthol® Accelerate®	(bicyclic structure with COOH, COOH, O)	Pennwalt 1954	51 rat	Water-soluble pre- and post-emergence herbicide, delays degradation of saccharose in sugarcane by preventing ripening—hence increase in yield, also used as cotton defoliant (Accelerate)	USP 3482959 (1967/69), Pennwalt; Inv.: Nickell et al.; *C.A.* **72,** 11572a (1970)	1505, 1506
4	*Ethephon* Ethrel®, Cepha® Amchem 66-329	O ‖ Cl—CH$_2$—CH$_2$—P(OH)$_2$ Na salt	Amchem 1968 GAF	4200 rat	Growth regulator, releases the endogenous ripening hormone ethylene, stimulates flowering, accelerates ripening of pineapple, tomato, apple, cherry, stimulates latex flow in *Hevea*; harvest aid in walnut	DOS 2053967 (1969/71), Amchem; Inv.: Randall, Vogel; *C.A.* **75,** 36343b (1971)	1507

No.	Name	Structure	LD₅₀	Manufacturer, Year	Uses	Ref.
5	GA Gibberellic acid, Pro-Gibb GA$_3$	(steroid-type structure with OH, HO, H$_3$C, COOH, CH$_2$, C=O)	6300 rat	ICI, Elanco, Nihon, Nohyaku ~1955	Endogenous growth hormone, used for increasing yield and improving quality in grape, cherry, citrus; accelerating flowering in house plants; stimulating germination of malt barley	1508, 1509
6	Glyphosine Polaris®	HOOC—CH$_2$—N(CH$_2$—P(OH)$_2$, O)(CH$_2$—P(OH)$_2$, O)	3900 rat	Monsanto 1979	Hastens ripening and increases level of sugar in sugarcane	USP 3455675 (1965/69), Monsanto; Inv.: Irani; C.A. 71, 91654w (1969) — 1510
7	IBA 4-(Indol-3-yl)-butyric acid Hormolin®	indole—(CH$_2$)$_3$—COOH	100 mouse	Merck & Co.	Auxin (is metabolized to IAA), for rooting of cuttings	
8	NAA 1-Naphthyl-acetic acid Fruitone-N®	naphthalene—CH$_2$—COOH	1000 rat	Amchem 1946	Growth regulator (IAA analogue) for thinning of apple, pear, induction of flowering in pineapple, prevention of pre-harvest fruit drop, stimulation of rooting of cuttings	1511
9	Naptalam Analap®, NPA, ACP 322	HOOC—(benzene)—NH—CO—NH—(naphthalene); Na salt	2000 rat	Uniroyal 1950	Inhibitor of seed germination, anti-auxin, for thinning of peach	USP 2556664 (1949/51), Uniroyal; Inv.: Hoffmann, Smith; C.A. 45, 8194c (1951) — 1512

TABLE 47. (*Continued*)

No.	Common Name Trade Name, Code	Structure	Manufacturer, Year of Introduction	LD$_{50}$ (mg/kg), Animal	Properties and Use	Patent	Reference
10	*Pydanon* H 1244		C.F. Spiess 1970	2100 mouse	Growth and development inhibitor, does not lead to habitual changes, use in nursery crops	DOS 1620359 (1965/69), C.F. Spiess; Inv.: Knoevenagel, Himmelreich; *C.A.* **71**, 3394d (1969)	1513
11	*Dicamba-methyl* Racuza®		Velsicol 1974	2700 rat	Yield increase in sugarcane, enhancement of sugar content in sugar beet, melon, grapefruit, the free acid is used as a herbicide (*dicamba*)	USP3013054 (1958/61) Velsicol; Inv.: Richter; *C.A.* **56**, 10049d (1962)	1491
12	*TIBA* Regim-8		Amchem 1968 Mobay Ag.	813 rat	Growth regulator (anti-auxin?), yield increase in soybean: thicker growth, better pod set, only effective in certain varieties, however; yield increase in apple		1514, 1515
13	4-CPA Tomatotone® Tomato Fix®		Dow 1950	850 rat	Growth regulator, to improve bloom set in tomato and to thin peach		

B. Onium Compounds

No.	Name	Structure	Company/Year	Tox.	Description	Ref.
14	Chlormequat chloride CCC®, Cyclocel®, Cyogan®	$[Cl-CH_2-CH_2-\overset{\oplus}{N}(CH_3)_3]\,Cl^{\ominus}$	Cyanamid 1959 BASF Österr. Stickstoffw. Makteshim-Agan	670 rat	Growth regulator, inhibits GA biosynthesis, shortens and thickens the lower stem internodes, main use: wheat, also ornamentals (azaleas, poinsettias), improves fruit set in pear	DAS 1294734 (1959/69), Research Corp. USA; Inv.: Tolbert; C.A. **61**, 6298c (1964)	1516–1519
15	Chlorphonium Phosphon®	$\left[\underset{Cl}{\overset{Cl}{\bigcirc}}-CH_2-\overset{\oplus}{P}(C_4H_9)_3 \right] Cl^{\ominus}$	Virginia Carolina Chem. Co. 1959 Mobil	178 rat	Growth regulator, GA antagonist? height retardant only used to date in ornamentals (lilies, chrysanthemums)	USP 3103431 (1961/63), Virginia Carol. Chem. Co.; Inv.: Wilson; C.A. **59**, 13282b (1963)	1520, 1521

C. Heterocyclic Derivatives

| 16 | Ancymidol A. Rest®, Reducymol® EL 531 | | Elanco 1971 | 4500 rat | Growth inhibitor; reduces internode elongation, excellent results with glasshouse plants such as chrysanthemum, poinsettia, dahlia, tulip | USP 3868244 (1974/75), Eli Lilly; Inv.: Taylor et al.; C.A. **72**, 100745b (1979) | 1522, 1523 |
| 17 | Cycloheximide Acti-aid®, Acti-dione® | | Upjohn 1948 | 2.5 rat, 65 monkey, 133 mouse | Growth regulator and fungicide; inhibits protein biosynthesis; promotes abscission of orange and olive (harvest aid); by-product of streptomycin manufacture | | 1524, 1525 |

TABLE 47. (Continued)

No.	Common Name Trade Name, Code	Structure	Manufacturer, Year of Introduction	LD$_{50}$ (mg/kg), Animal	Properties and Use	Patent	Reference
18	*MH* Maleic hydrazide MH-30®		US Rubber 1948	6900 rat, noncarcinogenic	Growth regulator; inhibits cell division but not extension; translocated, for prevention of sucker development in tobacco; inhibition of growth of grass, bushes, and trees; component of total herbicides		1526, 1527
19	Release® ABC 3030		Abbott 1974		Promotes ripening and abscission; is not translocated; harvest aid in citrus (orange)	USP 3869274 (1972/75), Abbott; Inv.: Crovetti et al.; *C.A.* **82**, 98005w (1975)	1528
20	Tomacon®		Takeda	1130 rat	Growth regulator, to improve bloom set on tomato; no longer commercially available	BP 1022655 (1963/65), Takeda; Inv.: Toyozato et al.: *C.A.* **61**, 16073f (1974)	

No.	Name	Structure	Producer/Year	LD50	Use	References	Ref. No.
21	Dikegulac-sodium Atrinal® Ro 07-6145		Hoffmann-La Roche 1973 Dr. R. Maag Ltd.	18,000 rat	Systemic growth regulator; reduces apical dominance and increases side branching and flower-bud formation on ornamentals (chemical pinching); temporarily retards longitudinal growth on bushes and woody plants		1528a

D. Miscellaneous Compounds

No.	Name	Structure	Producer/Year	LD50	Use	References	Ref. No.
22	Fluoridamid Sustar® MBR 6033		3M Co. 1971	2600 mouse	Yield enhancement in sugarcane; retards growth of grass	USP 3639474 (1969/72), 3M Co.; Inv.: Harrington et al.; C.A. 72. 121183g (1970)	1529, 1530
23	Off-Shoot T® n-Octanol (+n-Decanol)	$H_3C-(CH_2)_7-OH$ $H_3C-(CH_2)_9-OH$	Procter and Gamble 1968	10,000 rat	Nonspecific growth inhibitor; for prevention of sucker development in tobacco		1531
24	Orthonil Kanionil®, PRB-200® PRB 8		PRB 1974	920–1010 rat	Growth regulator with activity similar to auxin; for chemical thinning of apple, peach, plum, and for yield enhancement in sugar beet and potato	DOS 1768815 (1967/71), PRB; Inv.: Brepoels, Busschots; C.A. 75, 139671m (1971)	1532–1535
25	Pik-Off® CGA-2291 Enthanedial dioxime	HON=CH–CH=NOH	Ciba-Geigy 1974	180 rat	Harvest aid (abscission agent) in orange, olive	DOS 2338010 (1972/74), Ciba-Geigy; Inv.: Wilcox, Taylor; C.A. 80, 132803a (1974)	1491

5.4. Chemistry

The growth regulators listed in Table 47 belong to all different classes of compounds. Most of them are available by simple synthetic sequences. At the other extreme two are, in fact, natural products, gibberellic acid and *cycloheximide.*

Gibberellic acid, also known as GA_3, is one of numerous gibberellins isolated to date from plants and fungi, mainly from *Gibberella fujikuroi (Fusarium monili-forme)*, which are formally derived from the hydrocarbon 1, designated "gibberellane", but in fact biosynthesized from the hydrocarbon kaurene 2, itself formed via

1   2

mevalonic acid and geranyl-geranyl-pyrophosphate. Ring B of the gibberellins is formed by contraction of the middle 6-ring in kaurene with simultaneous generation of the carboxy group, which is present in all gibberellins. Following an agreement between the British and Japanese research groups, which have carried out most of the work, the gibberellins are called GA_n.

Counting had reached GA_{45} in 1975, which was reported by J. MacMillan et al. (University of Bristol).[1536] The Japanese group works mainly under N. Takahashi (University of Tokyo). Both groups are also working on partial and total syntheses in the gibberellin series, which are, however, of no practical relevance.

The gibberellins used in practice, GA_3 and the less important GA_7, are produced by fermentation. *Cycloheximide*, which has a much simpler structure than GA_3 and was synthesized as early as 1947,[1524, 1525] is also obtained by fermentation. *Cycloheximide* is a by-product in the manufacture of streptomycin.

The instability of metallated pyrimidine makes extremely low temperatures necessary in the synthesis of *ancymidol* (Table 47, 16).

Ancymidol

Chlorflurecol-methyl (Table 47, **1**) is also available via an organometallic reaction.

The benzylic acid rearrangement of 2-chloro-9,10-phenanthrenedione is a synthetic alternative. *Ethephon* (Table 47, **4**) acts by releasing ethylene in the plant. This may be observed *in vitro* about pH 4.1 but is possibly catalyzed by enzymes in the plant. Other ethylene-releasing agents such as (2-hydroxyethyl)hydrazine and *N*-aminomorpholine[1537] cause similar phytophysiological effects to *ethephon* (promotion of abscission).

5.5. Mechanisms of Action

In any discussion of the mechanisms of action of synthetic plant growth regulators mention must be made of the endogenous phytohormones since the majority of the synthetic regulators interfere in some way or other with the equilibria regulated by these hormones.

The structures and most important biological functions of the five types of phytohormones known to date are shown in Table 48.

Numerous synthetic growth regulators interfere with the auxins and gibberellins, as may be seen from the table. There is no contradiction in the fact that the structural analogues of IAA such as NAA, 4-CPA, or *2,4-D* promote growth at lower doses but have a herbicidal effect at higher concentrations. These structural analogues, particularly *2,4-D*, are much less easily metabolized than IAA and thus throw the equilibria out of balance, among other things, by stimulation of the endogenic ethylene synthesis, the "chaotropic hormone."[1538]

Reciprocation of synthetic regulators with cytokinines has not yet been reported. However, there exist numerous synthetic analogues of zeatine with high biological activity, not one of which has yet achieved the breakthrough into commercial use.

This also holds true for abscisic acid, which has interesting and completely reversible properties as an inhibitor of germination and growth. Its practical use remains out of the question on the grounds of cost and lack of metabolic stability.

The product with the greatest turnover, *MH*, acts as an inhibitor of mitosis[1539,1540] but does not influence cell extension. It thus reduces apical growth and flower-bud formation (chemical pinching).

TABLE 48. Naturally Occurring Growth Regulators (Phytohormones)

Group, Major Representative	Typical Biological Effects	Reciprocation with Synthetic PGRs
Auxins CH_2-COOH on indole ring 1H-Indole-3-acetic acid IAA	Growth promoter; necessary for longitudinal growth in all plant parts including the roots; stimulates ethylene synthesis at higher concentrations	*Chlorflurecol-methyl*—transport IBA—generator NAA—structural analogue *Naptalam*—transport Orthonil—analogue *TIBA*—transport Tomacon®—structural analogue Tomatotone®—structural analogue
Gibberellins GA₃	Growth promoter; initiates enzyme syntheses in germinating seeds; normalizes growth of dwarf mutants; promotes longitudinal and foliar growth; induces flowering	*Chlormequat chloride*—synthesis *Chlorphonium*—synthesis *Daminozide*—synthesis (?)
Cytokinines Zeatin	Inhibits senescence; promotes cell division and cell elongation; delays the degradation of chlorophyll and other ageing processes	
Abscisic acid ABA	Growth and development inhibitor (phytotranquilizer); partial antagonistic activity towards IAA and GA₃; accelerates ripening and senescence	*Pydanon*—analogue (?)
Ethylene $H_2C=CH_2$	Promoter of ripening and senescence ("chaotropic hormone"); accelerates ripening of fruit, degradation of chlorophyll and leaf drop; inhibits growth	*Ethephon*—generator Ethanedial dioxime—releases $CH_2=CH_2$ (?) Release®—releases $CH_2=CH_2$ (?)

Higher fatty acids and alcohols such as Off-Shoot® (Table 47, **23**) have a similar physiological effect. Activity among the alcohols varies with chain length, C_8-C_{10} representing an optimum.[1541]

Nothing is known to date about the mechanism of action of Racuza® (*dicamba-methyl*, Table 47, **11**) and Sustar® (*fluoridamid*, Table 47, **22**), both of which enhance the yield from sugarcane.

Endothal (Table 47, **3**), which is also used for this purpose, is reported to delay the enzymatic degradation of saccharose,[1542] possibly via inhibition of protein synthesis since *de novo* enzyme syntheses are also necessary for ageing processes. The high mammalian toxicity of *endothal* (LD_{50}: 51 mg/kg rat, oral, acute) and of *cycloheximide* indicates that a mechanism nonspecific to plants is operating.

Ancymidol has not been investigated to date. However, it has been found for the structurally related fungicide *triarimol* [α-(2,4-dichlorphenyl)-α-phenyl-5-pyrimidinemethanol, see Table 27,] that this pyrimidine derivative interferes in the lipid and sterol biosynthesis of fungi.[1543] *Triarimol* is also a growth inhibitor at higher concentrations. Thus, a common mechanism may be assumed.

Glyphosine (Table 47, **6**) inhibits the synthesis of aromatic amino acids such as tyrosine and phenylalanine *in vitro*.[1544] Whether this result is relevant to the *in vivo* action and how the phytophysiological effect (delay of saccharose degradation) is then to be explained (negative effect on enzyme synthesis?) is an open question.

For bibliography, see ref. 1545.

Chapter Six

Formulation Aids

H. NIESSEN
Bayer AG, Leverkusen

6.1. Introduction

Formulation of a crop protection agent is the art of bringing it into a form in which it can be applied with suitable equipment. Frequently the effective use of modern crop protection chemicals is, in the first instance, a problem of distribution. This may be illustrated with the following example.

In the instructions for the use of Morestan® (Bayer AG) for control of spider mites and powdery mildew on fruit, a concentration of 0.03% product in the spray solution is recommended. As the formulated product contains only 25% active ingredient this signifies a concentration of only 0.0075% active ingredient, or, for a spray solution of 2000 l/ha, 150 g active ingredient per hectare. In order for the product to be effective, it must be distributed evenly over the whole area of foliage present in 1 ha of land, and this is a multiple of the land area. Assuming a factor of 10, one can calculate a concentration of 1.5×10^{-7} g active ingredient/cm^2 foliage!

In order to achieve such a distribution, many crop protection agents are diluted with water before use. In field crops it is normal for 200–600 l spray mixture to be sprayed per hectare with the usual equipment. In tree and small fruit crops up to 2500 l/ha is sprayed according to the specific situation. In aerial crop spraying the quantity can be reduced to below 25 l/ha. Specially equipped aircraft frequently work with 0.5–5 l/ha using ultra-low-volume techniques. Since the quantity of active ingredient per hectare must be held practically constant, one naturally has vastly different concentrations of active ingredient in the spraying liquid.

A review of the most important water-miscible formulations has been published elsewhere.[1546] Furthermore, many products are formulated as ready to use or are diluted with substances other than water. The most important of such products are summarized in Table 49. For a short review of market formulations of crop protection chemicals, their composition, and the technology of their preparation, see ref. 1547.

TABLE 49. Formulation of Crop Protection Agents That Are Not Diluted with Water

Formulation	Quantity to Be Distributed	Manner of Distribution
Dustable powder	20–40 kg/ha	Wide
Granule	5–200 kg/ha	Wide, row or spot
Seed treatment	1–6 kg/ton seed	Together with the seed
Fogging concentrate		Specialized applications, e.g., in forests, or closed buildings
ULV products	0.5–5 l/ha	Wide, mainly from aircraft
Oil dispersible powder Oil miscible liquid	0.5–5 kg/ha	Mainly wide, dissolved or dispersed in 5–50 l/ha mineral oil or fatty oil
Smoke tin Aerosol dispenser	10–50 g/100 m^3	Fumigation in greenhouses according to instructions for use

An entire series of quality requirements is specified for every formulation, apart from even distribution: it must be easy to use, it must permit the active ingredient to deploy its biological activity to the full, and it must be well tolerated by the plants. It is frequently possible to compensate for inherently unfavorable properties of an active ingredient to a certain degree by choice of suitable formulation aids and by the proper preparation of the formulation.[1548] The storage shelf life of formulations is of special importance. In many cases a slow decomposition of the active ingredient is observed during long storage, particularly at higher temperatures. Unsuitable formulation aids can catalyze such decomposition reactions or even function as a reaction partner for the active ingredient. With the exception of true solutions, all formulations are thermodynamically unstable systems, and thus diverse physical changes can also occur during storage, generally leading to a further worsening of the properties of the product. Thus, in the search for suitable formulations, intensive checks on storage stability are necessary in addition to studies on the biological properties.[1549]

6.2. *Legal Requirements Concerning Formulation Aids*

In most countries of the world, the permit to market a crop protection agent is dependent upon official registration. However, as the examination of the biological properties of such an agent is always carried out on the formulated product, the toxicology of which is also thoroughly examined, a special registration of formulation aids is generally not regarded as necessary. Only in the United States does the law concerning crop protection agents require that, in principle, a tolerance has to be established for *every* substance that finds application in crop protection. In special cases certain substances, which must be named, can be excluded from this requirement. Many formulation aids fall under this clause. The corresponding decrees have been published continuously in the last few years in the *U.S. Federal Register* (see also ref. 1550). Generally the active ingredients used in crop protection are far more toxic than the formulation aids, most of which are used in other sectors, not just crop protection. It is actually self-evident that highly toxic compounds are not used as formulation aids. Substances for which threshold limit values have been laid down are only used to the extent that these values cannot be exceeded during correct handling and application of the product.

6.3. Formulation Aids

6.3.1. SURFACTANTS (TENSIDES)

Practically all formulations that are diluted with water before use contain surfactants as emulsifiers, dispersants, and wetters. They have been discussed in detail elsewhere.[1546]

6.3.2 POWDERED MINERALS

Dustable powders generally contain 1–10% a.i. mixed with powdered minerals as carriers and diluents, the particle size of which may vary according to the method of application and so on but normally lies below 75 μm. The dusts are most often made up with calcium carbonate, talc, kaolin, montmorillonite, and attapulgite.[1551] The selection is made according to the compatibility with the active ingredient, price, and availability.

For *seed treatments* and *dispersible powders* the same diluents and carriers can be used as for dusts. However, for dispersible powders a finer particle-size spectrum is necessary, the proportion above 40 μm not exceeding a few percent.

The *surface acidity* of a mineral powder is an important parameter for the compatibility of the powder with the active ingredient to be formulated. A scale of so-called pK_a values may be set up by observation of color changes of a series of indicator dyestuffs during adsorption onto the powder surface. This determination of the pK_a value of a mineral powder is an important criterion in the selection of a particular powder.[1552]

6.3.3. SYNTHETIC HYDRATED SILICAS

When preparing *dispersible powders* it is usually necessary to add highly dispersed synthetic silicas to improve the physical properties of the powder. The nature and quantity are mainly determined by the properties of the active ingredient and its concentration in the formulated product.

If a liquid active ingredient is desired in a powder form, this is achieved by mixing with approximately the same quantity of such a highly sorptive material. By addition of wetters and dispersants and thorough homogenization one can obtain, for example, a powder dispersible in water.

Even with solid active ingredient one can achieve improvements in grindability and pourability and prevent clumping during the storage of powder formulations by addition of such silicas. Often small quantities suffice.

We differentiate between precipitated silicas and those prepared by pyrolysis. The first group is made by precipitation of water glass with acids. The dried powder is composed of agglomerates of primary particles with diameters of ~25 nm and specific surface areas of 50–250 m^2/g (BET). Co-precipitates of free silica and silicates may be prepared by addition of calcium or magnesium or both. The pH values, measured in aqueous suspension, may thus be regulated from acidic into the highly alkaline region.

The second group is prepared by flame hydrolysis of silicon(IV) chloride. They differ from the precipitated silicas in their high chemical purity, pore-free particle surface, and the low number of silanol groups per unit surface area.[1553] Hydrophobic silicas are made by reaction of a proportion of these silanol groups with dimethyldichlorosilane.

6.3.4. GRANULAR CARRIERS

Two types must be differentiated in products of granular form.[1547]

A. Products made by impregnation (soaking or coating) of granular carriers.

B. Products made by granulation of powdery mixtures of active ingredient and formulation aids.

These two types differ in the active ingredient concentrations attainable. For type A, an active ingredient content of the order of 20% is the maximum technically feasible. For type B, active ingredient contents of the same order as those in dispersible powders are attainable in principle.

6.3.4.1. Carriers for Granular Formulations of Type A

The particle-size spectrum plays a decisive role in the use of all granular formulations. The spreading equipment requires even, free-flowing granules. The precondition for this is a limited range of carrier particle sizes. Granulations in which the upper and lower limits of granule size are not further apart than the ratio of 1:2 are technically feasible and fulfill all requirements. The particle diameter of most granular formulations in crop protection lies between 0.3 and 3 mm.

Porous carriers with not too large a specific surface area are ideal for the formulation of liquid active ingredient. In the formulation process the active ingredient is sprayed onto the stirred carrier. Examples of carriers of this nature are: pumice, brick, certain calcium carbonates, sepiolite, and bentonite, as well as specially manufactured carriers made by spray granulation and subsequent calcination of various clays (fuller's earth, florex, kaolin). Furthermore, it is possible to coat a nonsorptive nucleus with a sorptive layer that can then be impregnated.[1554]

Apart from liquid active ingredient, solutions of active ingredient can also be formulated into granules in the manner just described. However, with solid active ingredients it is often better to coat them onto a nonporous carrier. In its simplest form a powdery active ingredient or an active ingredient premix is stuck onto the carrier with a suitable liquid. Vice versa, one can spread a liquid active ingredient onto the carrier and "dry" it on with a solid such as synthetic silica. According to the natures of active ingredient and formulation aids it is possible to create coatings of various thicknesses and mechanical strength.

Cheap carrier materials such as calcite and sand are the first to be considered for granular formulations of this nature, whereby not only the particle size but also the shape plays a part. Round granules are less subject to abrasion than angular ones, have better flow properties, and cause less wear and tear on the spreading equipment. In addition to sand and calcite, a variety of pregranulated materials of organic and inorganic origins are used. The selection ranges from dolomite to ground and sieved nutshells and corncobs. Nonresinifying oils such as mineral oils, emulsifiers, or fat oils, or resinifying oils such as linseed oil and solutions or dispersions of plastics or rubbers (latexes) are suitable for use as stickers.

A special aspect is the impregnation of granular fertilizers with crop protection agents. What seems to be an ideal combination at first glance, however, proves to be less attractive when scrutinized. The application rates for granulated crop protection agents of the type described differ from those of fertilizers by a factor of 10. This means that impregnated fertilizers contain only 1% of the a.i. concentration of normal granular formulations, and therefore 10 times as much material must be subjected to a formulation process. Depending on the type of fertilizer, stability problems may be encountered, particularly with sensitive active ingredients. In some countries such as West Germany, such combinations are not permitted. By these means the legislative authorities wish to prevent the indiscriminate use of crop protection agents when the pest levels make it unnecessary and only a fertilizer is required. For these reasons the combination of crop protection chemicals with fertilizers is limited to a few specialized applications in a few countries.

6.3.4.2. Carriers for Granular Formulations of Type B

Various techniques are available for the granulation of powdery mixtures of a.i. and formulation aids.

Disc granulation.

Extrusion processes.

Spray-drying of solutions or suspensions.

In each case a careful sieving of the end product is necessary. As a rule, the granules that are too large (the oversize) are ground, and those too small (the undersize) are recycled.

In principle, the same carrier materials can be used as for dispersible powders, that is, mineral powders of various types and surfactants. By some means or other an adequate mechanical stability of the granules must be achieved. Additional stickers of very different natures are necessary when this cannot be achieved by sintering of the granular particles.

For granulation techniques involving addition of water, those water-soluble substances are suitable that are partially dissolved and stick together during the subsequent drying process. Examples are sodium sulfate, sugar, and lignin sulfonates. Polymeric compounds are, of course, also suitable as stickers, examples being polyvinyl alcohols and acetates, cellulose derivatives, and even compounds that polymerize during the granulation process, such as the formaldehyde–urea system.

Under certain circumstances it is possible to create special effects regarding rate of release of a.i. in the field by including various quantities of plastics or resins in the formulation. A very strict cost–benefit analysis is necessary in view of the relatively high costs involved.

6.3.5. SOLVENTS

The selection of solvents for the preparation of liquid formulations (emulsifiable concentrates, soluble concentrates, ULV products) is made according to the following criteria.

1. The solubility of the active ingredient
2. Chemical inertness toward the active ingredient
3. Phytotoxicity.
4. Toxicology.
5. Availability and price.
6. Flammability.
7. Volatility.

Aliphatic hydrocarbons are poor solvents for most active ingredients and are therefore normally only used for very dilute formulations. Aromatic hydrocarbons are used frequently, the fractions C_8 to C_{12} being favored, that is, from technical xylene mixtures to substituted naphthalenes. Apart from the toxicity of benzene, their high flammability prohibits the use of lower aromatics. On the other hand, phytotoxicity frequently increases with increasing molecular weight. Ketones are excellent solvents for many organic compounds. Most of them are, however, at least partially water soluble. They are therefore mainly suitable for the formulation of water-soluble and for liquid-active ingredients. When they are used for the formulation of emulsifiable concentrates of solid water-insoluble active ingredient, one frequently observes a crystallizing-out of the active ingredient during the preparation of the spray mix. This is caused by the solvent passing into the aqueous phase whereupon the active ingredient is precipitated in the "oil droplets" of the emulsion. Here ketones can normally only be used together with other, water-insoluble solvents as cosolvents.

This is also valid for alcohols and glycols and their ethers and esters as well as for highly polar aprotic solvents such as dimethylsulfoxide and dimethylformamide.

Chlorinated solvents are only used to a very slight extent. Chlorobenzene is used occasionally; dichlormethane is used more frequently as a relatively nontoxic, highly volatile, and nonflammable solvent for hygiene and stored-product agents and for aerosols.

The most important groups of solvents have thus been named. Others can be used when certain considerations make it necessary.

6.3.6. AIDS FOR THE CREATION OF SPECIAL EFFECTS

As already mentioned, it is frequently possible to balance out unfavorable properties of an active ingredient to a certain extent by careful formulation. The problems encountered are specific to the particular active ingredient, and therefore the number of compounds that may be considered as potential formulation aids is very large. Only a few groups of aids can be discussed here together with examples.

6.3.6.1. Stabilizers

One aim of formulation is often to improve the stability of the active ingredient in order to achieve an adequate storage life for the finished product. Decomposition can be initiated by adsorption of the a.i. onto active sites on the surface of powdered or granular carriers. Alcohols or glycol derivatives, for example, can be added as *deactivators*.

pH-Regulators such as organic acids and bases can prevent decomposition reactions that are favored in certain pH ranges. Certain epoxidized compounds are suitable as proton traps.

In many cases it is necessary to find *inhibitors* for certain highly specific reactions. This frequently makes detailed kinetic studies necessary. Compounds suitable for use as additives can thus belong to all classes of chemicals.

6.3.6.2. Defoamers

If the surfactants added to a crop protection formulation cause a persistent foaming during preparation and spraying of the spray mixture then the addition of *defoamers* is necessary. Silicones are the first choice, owing to their chemical inertness, but other compounds are also used.

6.3.6.3. Viscosity Regulators

Viscosity plays a leading role in ULV products (see Section 6.1) and suspension concentrates. For aqueous suspensions the viscosity may be raised by addition of swellable clays such as bentonite or of water-soluble polymers: it may be lowered by certain dispersing agents. In organic systems even small quantities of synthetic silica can cause a drastic increase in viscosity.[1553]

6.3.6.4. Dust Binders

For certain application forms, for example, seed treatments, powders are required that produce as little dust as possible. This can be achieved by addition of oils and waxes or of hygroscopic substances like glycerine.

6.3.6.5. Hydrophobing Agents

Tracking powders for rodent control or floating dustable powders should be hydrophobic. The oils and waxes used as dust binders are also suitable for this purpose, as are synthetic hydrophobic silicas or adsorbable polar compounds that coat the surface in such a way that the hydrophobic part of the molecule faces outward. An example of the latter is magnesium stearate.

6.4. *Additives and Adjuvants*

Under this heading fall those aids that are added to the spray mixtures of crop protection agents in the tank mix technique and improve the properties of the agents during or after application. Six such aids are currently registered in West Germany for this purpose.[1555] Additives and adjuvants can, according to composition and purpose, fulfill the following functions.

1. Improvement in wetting of drops.
2. Alteration of the volatility of the spray mixture.
3. Improvement in rain-fastness (i.e., resistance of a spray coating toward washing-off).
4. Improvement in the penetration of the active ingredient into the plant.
5. Regulation of the pH of the spray mix.
6. Improvement in distribution of the active ingredient over the plant.
7. Improvement in the compatibility of various crop protection agents in the tank mix.
8. Reduction of drift during spraying.

In principle, the same compounds may be considered as such additives that are also used as formulation aids—mainly surfactants, polymers, oils, and buffers. Although various improvements can be achieved with additives and adjuvants, their use is not without problems and is not appropriate under all conditions. Scrupulous prior checks are advisable.[1548]

References

Chapter 1. Introduction

1. R. Wegler (Ed.), *Chemie der Pflanzenschutz- und Schädlingsbekämpfungsmittel*, Vol. 1 and 2, Springer Verlag, Berlin–Heidelberg–New York (1970), Vol. 3 (1976), Vol 4 (1977).

2. H. H. Cramer, *Ullmanns Encyklopädie der technischen Chemie*, Keyword Insektizide, Wirtschaftliches (1977).

3. J. J. Lipa, *Ochr. Rosl.* **18**, 18-21 (1974).

4. G. Haug, Pflanzenschutzforschung der Industrie, in: R. Wegler (Ed.), *Chemie der Pflanzenschutz- und Schädlingsbekämpfungsmittel*, Vol. 3, Springer Verlag, Berlin–Heidelberg–New York (1976).

5. K. Lürssen, *Der Biologieunterricht* **9**, Issue 4 (1973).
   G. Mohr, Pflanzliche Wuchsstoffe und Wachstumsregulatoren, *Chem. Ztg.* **97**, 409 (1973).

6. "World Pesticide Markets," *Farm Chemicals* **138**, 45 (1975).

7. A. E. Wechsler, J. E. Harrison, and J. Neumeyer, EPA Study for the Office of Pesticide programs, A. D. Little, Cambridge, Mass. (1975).

8. H. H. Cramer, Pflanzenschutz und Welternte, *Pflanzenschutz-Nachr.* **20**, 1 (1967).

9. O. E. Fischnich, *Arch. DLG* **37**, 16 (1966).

10. W. Bartels and S. v.Eicken, *Chem. Labor Betr.* **23**, 64-266 (1972).

11. H. H. Cramer, Pflanzenschutz und Welternte, Brochure of *GIFAP*, Brussels (1974) and *Chemie + Fortschritt*, VCI-Schriftenreihe, Issue 3, (1972).
    H. H. Cramer, Ökonomisch-ökologische Wechselwirkungen in: *Pflanzenschutz- und Schädlingsbekämpfungsmittel*, Vol. 3, Springer Verlag, Berlin–Heidelberg–New York (1976).

12. Vector Control, *WHO Chronicle* **25**, 5 (1971).

12a. K. H. Büchel, *Nachr. Chem. Techn. Lab.* **28**, 719 (1980).

12b. K. H. Büchel, *Chem. Internat.* **1980**, 17.

13. H. H. Cramer, "Ökonomisch-ökologische Wechselwirkungen, in: R. Wegler (Ed.), *Chemie der Pflanzenschutz- und Schädlingsbekämpfungsmittel*, Vol. 3, Springer Verlag, Berlin–Heidelberg–New York (1976).

14. H. H. Cramer, *Bayer Ber.* **30**, 27 (1973).

15. F. Coulston, in: *Environ. Qual. Saf.* **1**, 40 (1972) and **2**, 125 (1973), Georg Thieme Verlag, Stuttgart.

16. G. Haug, *Bayer Ber.* **28**, 7 (1972).

Chapter 2.   Agents for Control of Animal Pests

SECTION 2.1.   NATURALLY OCCURRING INSECTICIDES AND SYNTHETIC ANALOGUES

17. C. B. Gnadinger, *Pyrethrum Flowers*, 2. Ed., McLaughlin Gromley King, Minneapolis 1936.

18. L. Crombie and M. Elliot, *Fortschr. Chem. Org. Naturst.* **19**, 120 (1961).

19. U. Claussen, in: R. Wegler, *Chemie der Pflanzenschutz- und Schädlingsbekämpfungsmittel*, Vol. I, 88–97, Springer Verlag, Berlin–Heidelberg–New York (1970).

20. J. E. Casida, *Pyrethrum, the Natural Insecticide*, Academic Press, New York–London (1973).

21. W. Perkow, *Die Insektizide, Chemie, Wirkungsweise und Toxizität*, p. 77–84, Dr. Alfred Hüthig Verlag, Heidelberg (1956).

22. R. S. Cahn, C. K. Ingold, and V. Prelog, *Angew. Chem. Int. Ed. Engl.* **5**, 388 (1966).

22a. M. Elliott and N. F. Janes, in: J. E. Casida, *Pyrethrum, the Natural Insecticide*, pp. 61–63, Academic Press, New York–London (1973).

23. H. Staudinger and L. Ruzicka, *Helv. Chim. Acta* **7**, 177, 201, 212, 236, 245 and 448 (1924).

 H. Staudinger, O. Muntwyler, L. Ruzicka, and S. Seibt, *Helv. Chim. Acta* **7**, 390 (1924).

24. F. B. La Forge and W. F. Barthel, *J. Org. Chem.* **12**, 199 (1947).

24a. M. Elliott and N. F. Janes, *Chem. Soc. Reviews* **7**, 473 (1978).

25. I. G. Campbell and S. H. Harper, *J. Chem. Soc. (London)* **1945**, 283.

26. M. Elliot and N. F. Janes, in: J. E. Casida, *Pyrethrum, the Natural Insecticide*, pp. 65–70, Academic Press, New York–London (1973).

27. M. Elliot and N. F. Janes, in: J. E. Casida, *Pyrethrum, the Natural Insecticide*, p. 90, Academic Press, New York–London (1973).

28. N. Ohno, K. Fujimoto, Y. Okuno, T. Mizutani, M. Hirano, N. Itaya, T. Honda, and H. Yoshioka, *Agric. Biol. Chem.* **38**, 881 (1974).

29. E. E. Kenaga and C. S. End, *Commercial and Experimental Organic Insecticides*, Entomological Society of America (1974).

30. M. Elliot, A. W. Farnham, N. F. Janes, P. H. Needham, D. A. Pulman, and J. H. Stevenson, *Nature* **246**, 169 (1973).

31. M. S. Schechter, N. Green, and F. B. La Forge, *J. Amer. Chem. Soc.* **71**, 1717, 3165 (1949).

32. US P. 2886485 (1957) U.S. Government, Inv.: W. F. Barthel, B. H. Alexander, J. B. Gahan, and P. G. Piquett; *C.A.* **53**, 19,922d (1959).

33. T. Kato, K. Ueda and K. Fujimoto, *Agric. Biol. Chem.* **28**, 914 (1964).

34. Y. Katsuda, in: A. S. Tahori, Proceedings of the 2nd Int. IUPAC-Congress of Pesticide Chemistry, Vol. I, p. 443, Gordon and Breach Science Publishers, New York (1972).

35. M. Elliot, A. W. Farnham, P. H. Needham, and B. C. Pearson, *Nature* **213**, 493 (1967).

36. L. Velluz, J. Martel, and G. Nomine, *C.R. Acad. Sci., Paris* **268**, 2199 (1969).

36a. J. Lhoste and F. Rauch, *Pestic. Sci.* **7**, 247 (1976).

36b. K. Fujimoto, N. Itaya, Y. Okuno, T. Kadota, and T. Yamaguchi, *Agric. Biol. Chem.* **37**, 2681 (1973).

36c. M. Elliott, N. F. Janes, and C. Potter, *Ann. Rev. Entomol.* **23**, 443 (1978).

36d. DOS 2231312 (1972), Sumitomo Chem. Co., Inv.: T. Matsuo, N. Itaya, Y. Okuno, T. Mitzutani, and N. Ohno; *C.A.* **78**, 84,072w (1973).

36e. M. H. Breese, *Pestic. Sci.* **8**, 264 (1977).

36f. DOS 2326077 (1973), National Research Development Corp., Inv.: M. Elliott, N. F. Janes, and D. A. Pulman; *C.A.* **80**, 132,901f (1974).

37. M. Elliot, A. W. Farnham, N. F. Janes, P. H. Needham, and D. A. Pulman, *Nature* **248**, 710 (1974).

38. J. M. Barnes and R. D. Verschoyle, *Nature* **248**, 711 (1974).

38a. N. Ohno, K. Fujimoto, Y. Okuno, T. Mizutani, M. Hirano, N. Itaya, T. Honda, and H. Yoshioka, *Pestic. Sci.* **7**, 241 (1976).

38b. DOS 2605828 (1976), American Cyanamid Co., Inv.: R. W. Addor and M. S. Schrider; *C.A.* **85**, 105,432a, 123,660c (1976).

38c. DOS 2802962 (1978), ICI Ltd., Inv.: R. K. Huff; *C.A.* **89**, 197,045k (1978).

38d. G. Holan, D. F. O'Keefe, C. Virgona, and R. Walser, *Nature* **272**, 734 (1978).

38e. DOS 2757066 (1977), American Cyanamid Co., Inv.: G. Berkelhammer and V. Kameswaran; *C.A.* **90**, 186,606p (1979).

38f. W. K. Whitney, *Proceedings of the 10th British Crop Protection Conference–Insecticides and Fungicides*, Brighton 1979, p. 387.

38g. J. Lhoste and C. Piedallu, *Pestic. Sci.* **8**, 254 (1977).

38h. C. N. E. Ruscoe, *Pestic. Sci.* **8**, 236 (1977).

39. R. D. O'Brien, *Insecticides, Action and Metabolism*, Academic Press, New York (1967).

40. I. Yamamoto, in: R. D. O'Brien and I. Yamamoto, *Biochemical Toxicology of Insecticides*, pp. 193–200, Academic Press, New York (1970).

41. W. F. Barthel, in: J. E. Casida, *Pyrethrum, the Natural Insecticide*, Academic Press, New York (1973).

42. C. Eagleson, *Soap Chem. Spec.* **18**, 125 (1942).

43. H. L. Haller, F. B. La Forge, and W. N. Sullivan, *J. Org. Chem.* **7**, 185 (1942).

44. U. Claussen, in: R. Wegler, *Chemie der Pflanzenschutz- und Schädlingsbekämpfungsmittel*, Vol. I, pp. 98–102, Springer Verlag, Berlin–Heidelberg–New York (1970).

45. Summary on the mechanism of action: I. Yamamoto, in: J. E. Casida, *Pyrethrum, the Natural Insecticide*, pp. 195–210, Academic Press, New York (1973).

46. H. Martin, in: *The Scientific Principles of Crop Protection*, 5th Ed., Arnold, London (1964).

47. L. Crombie, *Fortschr. Chem. Org. Naturst.* **21**, 275 (1963).

48. U. Claussen, in: R. Wegler, *Chemie der Pflanzenschutz- und Schädlingsbekämpfungsmittel*, Vol. I, pp. 103–105, Springer Verlag, Berlin–Heidelberg–New York (1970).

49. E. E. Kenaga, *Bull. Entomol. Soc. Am.* **12**, 161 (1966).

50. U. Claussen, in: R. Wegler, *Chemie der Pflanzenschutz- und Schädlingsbekämpfungsmittel*, Vol. I, pp. 105–108, Springer Verlag, Berlin–Heidelberg–New York (1970).

51. E. F. Rogers, F. Koninszy, I. Shavel, and K. Folkers, *J. Amer. Chem. Soc.* **70**, 3086 (1948).
    K. Wiesner, Z. Valenta, and J. A. Findlay, *Tetrahedron Lett.* **1967**, 221.

52. L. Feinstein and M. Jacobson, *Fortschr. Chem. Org. Naturst.* **10**, 452–455 (1953).

53. A. M. Heimpel, *Ann. Rev. Entomol.* **12**, 287 (1967).

54. M. Jakobson, *J. Amer. Chem. Soc.* **71**, 366 (1949).

55. F. Korte, K. H. Büchel, and A. Zschocke, *Chem. Ber.* **94**, 1952 (1961).

56. K. Konishi, IUPAC-Congress, *Pesticide Chem.*, Tel Aviv, Febr. 1971.

SECTION 2.2. INSECTICIDAL CHLOROHYDROCARBONS

57. A. v. Baeyer, O. Zeidler, J. Weiler, O. Fischer, E. Jäger, E. Ter Mer, W. Hemilian, and E. Fischer, *Ber.* **7**, 1181 (1874).

58. P. Müller, *DDT–Das Insektizid Dichlordiphenyltrichloräthan und seine Bedeutung*, Vol. 1, Birkhäuser Verlag, Basel–Stuttgart (1955).

59. Swiss, P. 226180 (1940), J. R. Geigy AG, *C.A.* **43**, 6358g (1949).

60. *The Place of DDT in Operations Against Malaria and Other Vectorborne Diseases*, Off. Rec. W.H.O., No. 190, 176 (1971).

61. S. H. Mosher, M. R. Cannon, E. A. Conroy, R. E. van Strien, and D. P. Spalding, *Ind. Eng. Chem.* **38**, 916 (1946).

62. M. S. Schechter and H. L. Haller, *J. Amer. Chem. Soc.* **66**, 2129 (1944).

63. M. Beroza and M. C. Bowman, *Anal. Chem.* **37**, 291 (1965).

64. D. W. Johnston, *Science* **186**, 841 (1974).

65. L. J. Mullins, *Chem. Rev.* **54**, 289 (1954).

66. G. Holan, *Nature* **221**, 1025 (1969).

67. Neth. Appl. 6412298 (1965), Monsanto Chemicals; *C.A.* **63**, 9862g (1965).

68. J. R. Corbett, *The Biochemical Mode of Action of Pesticides*, p. 169 ff., Academic Press, London (1974).

69. S. S. Perry and W. M. Hoskins, *J. Econ. Entomol.* **44**, 850 (1951).

70. Y. H. Atallah and W. C. Nettles Jr., *J. Econ. Entomol.* **59**, 560 (1966).

71. O. Johnson, CW Report "Pesticides 72," *Chem. Week* **111**, (4), 17 (1972).

72. P. Läuger, H. Martin, and P. Müller, *Helv. Chim. Acta* **27**, 892 (1944).

73. DBP 871979 (1943), Farbwerke Hoechst, Inv.: K. Pfaff, M. Erlenbach, and W. Finkenbrink, *Chem. Zentralblatt* **1953**, 7169.

74. Swiss P. 237581 (1943), J. R. Geigy AG, *C.A.* **43**, 58,899b (1949).

75. USP 2516186 (1948), Purdue Research Foundation, H. B. Hass and R. T. Blickenstaff; *C.A.* **44**, 11,011a (1950).

76. W. L. Barrett, *J. Econ. Entomol.* **45**, 90 (1952).

77. M. Faraday, *Philosophical Transactions* (1825).

78. L. van der Linden, *Ber.* **45**, 231 (1912).

79. USP 2010841 (1933), Great Western Electro-Chemical Co., Inv.: H. Bender, *Chem. Zentralblatt* **1936** I, 1112.

80. A. Dupire and M. Raucourt, *C.R. Seances Acad. Agric. Fr.* **20**, 470 (1942).

81. R. Slade, *Chem. Ind.* (*London*) **40**, 314 (1945).

82. K. Schwabe and P. P. Rammelt, *Z. Physik. Chem.* (*Leipzig*) **204**, 310 (1955).

83. USP 2942035 (1960), Columbia-Southern Chemical Corp., Inv. F. Strain, E. Bissinger; *C.A.* **54**, 22,495f (1960).

84. US 2765272 (1956), Columbia-Southern Chemical Corp., Inv.: J. A. Neubauer, F. Strain, F. E. Kung, and E. Bissinger; *C.A.* **51**, 11,645f (1957).

85. USP 2758077 (1956), Inv.: A. LaLande, G. M. Knorr, and M. E. Aeugle; *C.A.* **51**, 664i (1957).

86. Brit. P. 637412 (1946), Solvay & Cie., *C.A.* **44**, 7872b (1950).

87. USP 2797195 (1952), Columbia-Southern Chemical Corp., Inv.: J. A. Neubauer, F Strain, F. E. Kung, and F. C. Dehn; *C.A.* **51**, 16,532h (1957).

88. DBP 1024959 (1958), BASF, Inv.: H. Schlecht; *C.A.* **54**, 7589b (1960).

89. K. C. Kauer, R. B. Du Vall, and R. L. Alquist, *Ind. Eng. Chem.* **39**, 1335 (1947).

90. O. Hassel, *Quart. Rev. Chem. Soc.* **7**, 221 (1953).

91. H. D. Orloff, *Chem. Rev.* **54**, 347 (1954).

92. M. S. Schechter and I. Hornstein, *Anal. Chem.* **24**, 544 (1952).

93. G. R. Raw, *CIPAC-Handbook*, Vol. I, Collaborative International Pesticides Analytical Council Ltd., Harpenden, U.K. (1970).

94. E. W. Balson, *Trans. Faraday Soc.* **43**, 54 (1947).

95. E. Ulmann, *Lindan, Monographie eines Insektizids*, K. Schillinger Verlag, Freiburg (1972).

96. USP 2546174 (1945), Hercules Powder Co., Inv.: D. Stonecipher; *C.A.* **45**, 5358b (1951).

97. W. Le Roy Parker and J. G. Beacher, *Bulletin of the Delaware Univ. Agricultural Experiment Station* 264 (1947).

98. J. E. Casida, Lecture, The Third International Congress of Pesticide Chemistry, Helsinki, (1974).

99. M. L. Anagnostopoulos, H. Parlar, and F. Korte, *Chemosphere* **3**, 65 (1974).

100. A. J. Graupner, C. L. Dunn, *J. Agric. Food Chem.* **8**, 286 (1960).

101. W. Gruch and P. Steiner, *Mitt. Biol. Bundesanst. Land.- Forstwirtsch.*, Berlin–Dahlem **102**, 64 (1960).

102. P. M. Mehrle and F. L. Mayer, *Science* **188**, 343 (1975).

103. F. Straus, L. Kollek, and W. Heyn, *Chem. Ber.* **63**, 1868 (1930).

104. Brit. P. 703202 (1951), N. V. de Bataafsche Petroleum Maatschappij, Inv.: R. E. Lidov; *C.A.* **49**, 6995h (1955).

105. H. E. Ungnade and E. T. Mc Bee, *Chem. Rev.* **58**, 249 (1958).

106. USP 2616825 (1951), Allied Chem. & Dye Corp., Inv.: E. E. Gilbert and S. L. Giolito; *C.A.* **47**, 2424d (1953).

107. E. T. Mc Bee, C. W. Roberts, J. D. Idol, and R. H. Earle, *J. Amer. Chem. Soc.* **78**, 1511 (1956).

108. WHO Monogr. Ser. No. **443** (1970).

109. *Chem. Week* **117**, (8) (1975).

110. USP 3393223 (1963) ≡ Neth. Appl. 6402964 (1964), Allied Chem. & Dye Corp., *C.A.* **62**, 7658c (1965).

111. USP 3402209 (1965), Allied Chem. & Dye Corp., Inv.: E. E. Gilbert, P. Lombardo, E. J. Rumanowski, and B. Sukornick; *C.A.* **70**, 11,226z (1969).

112. E. E. Gilbert, P. Lombardo, E. J. Rumanowski, and G. L. Walker, *J. Agric. Food Chem.* **14**, 111 (1966).

113. H. J. Prins, *Rec. Trav. Chim. Pays-Bas* **65**, 455 (1946).

114. USP 2671043 (1954), Allied Chem. & Dye Corp., Inv.: E. E. Gilbert; *C.A.* **48**, 10,290c (1954).

115. K. L. E. Kaiser, *Science* **185**, 523 (1974).

116. W. Furness, *Proc. 6th Brit. Insectic. Fungic. Conf.* **2**, 541 (1971).

117. USP 2519190 (1950), Velsicol Corp., Inv.: J. Hyman; *C.A.* **45**, 647f (1951).

118. C. W. Kearns, L. Ingle, and R. L. Metcalf, *J. Econ. Entomol.* **38**, 661 (1945).

119. R. Riemschneider and A. Kühnl, *Pharmazie* **3**, 115 (1948).

120. USP 2598561 (1949), Velsicol Corp., Inv.: M. Kleiman; *C.A.* **47**, 1190i (1953).

121. P. B. Polen, *Chlordane, Composition, Analytical Considerations and Terminal Residues*, presented to a meeting of the IUPAC Commission, Commission on Terminal Residues, Genf, 1966.

122. J. G. Saha and Y. W. Lee, *Bull. Environ. Contam. Toxicol.* **4**, 285 (1969).

123. R. Riemschneider and A. Kühnl, *Monatsh. Chem.* **86**, 879 (1953).

124. C. Vogelbach, *Angew. Chem.* **63**, 378 (1951).

125. R. B. March, *J. Econ. Entomol.* **45**, 452 (1952).

126. K. H. Büchel, A. E. Ginsberg, and R. Fischer, *Chem. Ber.* **99**, 421 (1966).

127. S. J. Christol, *Adv. Chem. Ser.* **1**, 184 (1950).

128. W. P. Cochrane, M. Forbes, and A. S. Y. Chau, *J. Assoc. Off. Agric. Chem.* **53**, 769 (1970).

129. E. P. Ordas, *J. Agric. Food Chem.* **4**, 444 (1956).

130. A. S. Chau and W. P. Cochrane, *J. Assoc. Off. Agric. Chem.* **52**, 1092 (1969).

131. P. Steiner and W. Gruch, *Mitt. Biol. Bundesanst. Land. Forstwirtsch.*, Berlin–Dahlem **95**, 66 (1959).

132. Brit. P 714869 (1952), Arvey Corp., *C.A.* **50**, 402a (1956).

133. USP 2661378 (1953), Shell Development Co., Inv.: T. G. McKenna, S. B. Soloway, R. E. Lidov, and J. Hyman; *C.A.* **49**, 378e (1955).

134. USP 2606910 (1946), Velsicol Corp., Inv.: S. H. Herzfeld, R. E. Lidov, and H. Bluestone; *C.A.* **47**, 8775b (1953).

135. W. M. Rogoff and R. L. Metcalf, *J. Econ. Entomol.* **44**, 910 (1951).

136. P. C. Polen and P. Silverman, *Anal. Chem.* **24**, 733 (1952).

137. D. M. Coulson and L. A. Cavanagh, *Anal. Chem.* **32**, 1245 (1960).

138. B. Davidow and J. Radomski, *J. Pharmacol. Exp. Ther.* **107**, 259 (1953).

139. A. S. Perry, A. M. Mattson, and A. J. Buchner, *J. Econ. Entomol.* **51**, 346 (1958).

140. R. Kaul, W. Klein, and F. Korte, *Tetrahedron* **26**, 331 (1970).

141. J. R. W. Miles, C. M. Tu, and C. R. Harris, *J. Econ. Entomol.* **64**, 839 (1971).

142. K. H. Büchel, A. E. Ginsberg, and R. Fischer, *Chem. Ber.* **99**, 405 (1966).

143. DBP 1020346 (1957), Ruhrchemie AG, Inv.: H. Feichtinger, H. Tummes, and S. Puschhof; *C.A.* **53**, 19,922f (1959).

144. DBP 960284 (1955), Ruhrchemie AG, Inv.: H. Feichtinger and H. Tummes; *C.A.* **53**, 17,149b (1959).

145. DBP 959229 (1957), Ruhrchemie AG, Inv.: H. Feichtinger; *C.A.* **53**, 13,674a (1959).

146. DBP 1026325 (1958), Ruhrchemie AG, Inv.: H. Feichtinger and H. W. Linden; *C.A.* **54**, 11,049h (1960).

147. DBP 1015797 (1957), Farbwerke Hoechst AG, Inv.: H. Frensch and H. Goebel; *C.A.* **53**, 16,023h (1959).

148. DBP 963282 (1957), Farbwerke Hoechst AG, Inv.: H. Frensch, W. Staudermann, and W. Finkenbrink; *C.A.* **53**, 19,923f (1959).

149. USP 2799685 (1957), Farbwerke Hoechst AG, Inv.: H. Frensch, W. Staudermann, H. Goebel, and W. Finkenbrink; *C.A.* **52**, 2062i (1958).

150. W. Finkenbrink, *Nachrichtenbl. Dtsch. Pflanzenschutzdienst (Braunschweig)* **8**, 183 (1956).

151. Brit. P 909588 (1960), Hooker Chemical Corp., Inv.: R. H. Kimball, E. Leon, E. J. Geering, and S. J. Nelson; *C.A.* **58**, 6848a (1963).

152. S. E. Forman, A. J. Durbetaki, V. Cohen, and R. A. Olofson, *J. Org. Chem.* **30**, 169 (1965).

153. H. Maier-Bode, *Residue Rev.* **1968**, 22.

154. H. Maier-Bode, *Arch. Pflanzenschutz* **3**, 201 (1967).

155. K. Ballschmiter and G. Tölg, *Angew. Chem.* **78**, 775 (1966).

156. DBP 1002341 (1957), Farbwerke Hoechst AG, Inv.: H. Frensch and W. Finkenbrink; *C.A.* **54**, 7589c (1960).

157. DBP 1114057 (1953), Farbwerke Hoechst AG, Inv.: H. Frensch, W. Staudermann, W. Finkenbrink; *C.A.* **56**, 10,001c (1962).

158. C. W. Kearns, C. Weinman, and G. C. Decker, *J. Econ. Entomol.* **42**, 127 (1949).

159. USP 2875256 (1959), Shell Development Co., Inv.: J. Hyman, E. Freireich, and R. E. Lidov; *C.A.* **53**, 13,082d (1959).

160. USP 2635977 (1953), Shell Development Co., Inv.: R. E. Lidov; *C.A.* **48**, 2769h (1954).

161. E. S. Goodwin, R. Goulden, A. Richardson, and J. G. Reynolds, *Chem. Ind. (London)* **1960**, 1220.

162. A. A. Danish and R. E. Lidov, *Anal. Chem.* **22**, 702 (1950).

163. J. A. Moss and D. E. Hathway, *Biochem. J.* **91**, 384 (1964).

164. W. L. Hayes, *Ann. Rev. Pharmacol.* **1**, 27 (1965).

165. G. T. Brooks, *World Rev. Pest Control* **5**, 62 (1966).

166. USP 2676131 (1954), Shell Development Co., Inv.: B. Soloway; *C.A.* **48**, 8473f (1954).

167. Shell Chemical Corp., *Handbook of Aldrin, Dieldrin and Endrin Formulations*, 2nd ed., Agricultural Chemicals Division, New York (1959).

168. USP 2676132 (1954), Shell Development Co., Inv.: H. Bluestone; *C.A.* **48**, 8474b (1954).

169. USP 2813915 (1957), Shell Development Co., Inv.: J. M. Howald and C. D. Marshall; *C.A.* **52**, 5460d (1958).

170. D. D. Phillips, G. E. Pollard, and S. B. Soloway, *J. Agric. Food Chem.* **10**, 217 (1962).

171. J. D. Rosen, D. J. Sutherland, and G. R. Lipton, *Bull. Environ. Contam. Toxicol.* **1**, 133 (1966).

172. J. T. Snelson, 4th Internat. Symp. on Toxicological Aspects of Environmental Quality, Munich, September 1975.

173. G. T. Brooks, *Nature* **186**, 96 (1960).

174. *Nachr. Chem. Tech.* **22**, 370 (1974).

175. R. Wegler, *Chemie der Pflanzenschutz- und Schädlingsbekämpfungsmittel*, Vol. 1, Springer Verlag, Berlin–Heidelberg–New York (1970).

H. Martin, *Pesticide Manual, British Crop Protection Council*, Worcester (1972).

W. Perkow, *Wirksubstanzen der Pflanzenschutz- und Schädlingsbekämpfungsmittel*, Verlag Paul Parey, Berlin (1971).

G. T. Brooks, *Chlorinated Insecticides I, Technology and Application*, CRC Press Inc., Cleveland (1974).

H. Maier-Bode, *Pflanzenschutzmittel-Rückstände*, E. Ulmer–Verlag, Stuttgart (1965).

R. D. O'Brien, *Insecticides, Action and Metabolism*, Academic Press, New York (1967).

H. Martin, *The Scientific Principles of Crop Protection*, 5th Ed., Arnold, London (1964).

R. L. Metcalf, *Organic Insecticides*, Wiley Interscience, New York–London–Sydney–Toronto (1955).

G. Zweig, *Analytical Methods for Pesticides, Plant Growth Regulators and Food-Additives*, Vol. II, *Insecticides*, Academic Press, New York (1964).

O. R. Klimmer, *Pflanzenschutz- und Schädlingsbekämpfungsmittel. Abriß einer Toxikologie und Therapie von Vergiftungen*, Hundt–Verlag, Hattingen (1971).

J. R. Corbett, *The Biochemical Mode of Action of Pesticides*, Academic Press, London–New York (1974).

SECTION 2.3. ORGANOPHOSPHORUS INSECTICIDES

176. G. Schrader, *Die Entwicklung neuer insektizider Phosphorsäureester*, p. 3, Verlag Chemie, Weinheim/Bergstraße (1963).

177. V. M. Clark, D. W. Hutchinson, A. J. Kirby, and S. G. Warren, *Angew. Chem.* **76**, 704 (1964).

178. C. Fest and K.-J. Schmidt, *The Chemistry of Organophosphorus Pesticides*, p. 43, Springer Verlag, Berlin–Heidelberg–New York (1973).

179. H. McCombie, B. C. Saunders, and G. J. Stacey, *J. Chem. Soc.* **1945**, 380.

180. P. Nylen, *Ber.* **57**, 1023 (1924).

181. T. Milobendzki and A. Sachnowski, *Chem. Polski* **15**, 34 (1917).

182. DAS 1079022 (1958), Bayer AG, Inv.: H. Schliebs; *C.A.* **55**, 14,307i (1961).

183. G. O. Doak and L. D. Freedman, *Chem. Rev.* **61**, 31 (1961).

184. F. Cramer, *Agnew. Chem.* **72**, 236 (1960).

185. A. J. Kirby and S. G. Warren, *The Organic Chemistry of Phosphorus*, p. 37, Elsevier Publishing Co., Amsterdam–London–New York (1967).

186. DBP 835145 (1949), Bayer AG, Inv.: H. Jonas; *C.A.* **49**, 12,529f (1955).

187. T. Mukaiyama and T. Fujisawa, *Bull. Chem. Soc. Japan* **34**, 812 (1961).

188. N. N. Melnikow, J. A. Mandelbaum, and P. G. Zaks, *Ž. obšč. Chim.* **29**, 522 (1959).

189. J. H. Fletcher, J. C. Hamilton, I. Hechenbleikner, E. I. Hoegberg, B. J. Sertl, and J. T. Cassaday, *J. Amer. Chem. Soc.* **72**, 2461 (1950).

190. DAS 1191369 (1963), Knapsack-Griesheim, Inv.: H. Niermann, J. Cremer, and H. Harnisch; *C.A.* **63**, 8199e (1965).

191. G. M. Steinberg, *J. Org. Chem.* **15**, 637 (1950).

192. A. Michaelis, *Leibigs Ann. Chem.* **326**, 175, 194 (1903).

193. DAS 1119860 (1958), Bayer AG, Inv.: R. Schliebs and H. Kaiser; *C.A.* **58**, 6863c (1963).

194. J. P. Komkow, K. W. Karawanow and S. Iwin, *Ž. obšč. Chim.* **28**, 2963 (1953).

195. A. M. Kinnear and E. A. Perren, *J. Chem. Soc.* **1952**, 3437.

196. D. G. Coe, B. J. Perry, and R. K. Brown, *J. Chem. Soc.* **1957**, 3604.

197. Collomp, "Les Trilons" (1949) Bull. Inf. Scient. Min. Guerre, Paris, No. 23, gl.

198. USP 2662917 (1951), Continental Oil Co., Inv.: W. L. Jensen; *C.A.* **48**, 13,711g (1954).

199. Ph. de Clermont, *C.R.* **39**, 338 (1854).

200. A. D. F. Toy, *J. Amer. Chem. Soc.* **70**, 3882 (1948).

201. USP 2495220 (1946), Eastman Kodak Co., Inv.: A. Bell; *C.A.* **44**, 3203e (1950).

202. USP 2486658 (1947), Monsanto Chemical Co., Inv.: G. M. Kosolapoff; *C.A.* **44**, 1644c (1950).

203. DBP 848812 (1950), Bayer AG, Inv.: G. Schrader and R. Mühlmann, *C.A.* **47**, 5425i (1953).

204. DBP 918603 (1941), Bayer AG, Inv.: G. Schrader and H. Kükenthal; *C.A.* **40**, 4448a (1956).

205. DAS 1210835 (1964) ≡ Neth. P. 6508556 (1966), Bayer AG, Inv.: G. Schrader, W. Lorenz, G. Unterstenhöfer, and I. Hammann; *C.A.* **65**, 7058b (1966).

206. USP 3309266 (1965), Chevron Research Co., Inv.: Ph.S. Magee; Corresp. Patent: Neth. 6602588.

207. G. Hilgetag and H. Teichmann, *Agnew. Chem.* **77**, 1001 (1965).

208. USP 2586655 (1948), American Cyanamid Co., Inv.: E. O. Hook and P. H. Moss; *C.A.* **46**, 8144e (1952).

209. DBP 917668 (1952), Bayer AG, Inv.: W. Lorenz and G. Schrader; *C.A.* **49**, 12,529a (1955).

210. USP 2793224 (1954), Stauffer Chemical Co., Inv.: L. W. Fancher; *C.A.* **51**, 14,196i (1957).

211. DBP 836349 (1950), Bayer AG, Inv.: G. Schrader, *C.A.* **49**, 3786a (1955).

212. DBP 818352 (1949), Bayer AG, Inv.: G. Schrader, *C.A.* **47**, 5959i (1953).

213. DBP 926488 (1953), Bayer AG, Inv.: G. Schrader and W. Lorenz, *C.A.* **50**, 2653f (1956).

214. DBP 830509 (1950), Bayer AG, Inv.: G. Schrader; *C.A.* **47**, 1727f (1953).

215. DBP 961083 (1954), Bayer AG, Inv.: G. Schrader; *C.A.* **51**, 15,549h (1957).

216. T. R. Fukuto and R. L. Metcalf, *J. Amer. Chem. Soc.* **76**, 5103 (1954).

217. DBP 947368 (1954), Bayer AG, Inv.: R. Mühlmann, W. Lorenz, and G. Schrader; *C.A.* **51**, 4413f (1957).

218. DBP 949229 (1955), Bayer AG, Inv.: W. Lorenz, R. Mühlmann, G. Schrader, and K. Tettweiler; *C.A.* **51**, 12,957c (1957).

219. DBP 961670 (1953), Pest Control Ltd., Inv.: D. W. J. Lane and D. F. Heath; *C.A.* **54**, 813a (1960).

220. DAS 1116217 (1958), Sandoz AG, Inv.: J. P. Leber and K. Lutz; *C.A.* **56**, 11,447f (1962).

221. DBP 917668 (1952), Bayer AG, Inv.: W. Lorenz and G. Schrader; *C.A.* **49**, 12,529a (1955).

221a. Neth. P. 6714776 (1967), Nihon Nohyaku Co., Inv.: Anon.; *C.A.*, not abstracted.

222. DAS 1123863 (1959) ≡ USP 2943974 (1960), Rhône–Poulenc, Inv.: J. Métivier; *C.A.* **55**, 391b (1961).

223. H. Finger, *J. Prakt. Chem.* **37**, 431 (1888).

224. DBP 927270 (1953), Bayer AG, Inv.: W. Lorenz; *C.A.* **52**, 2908e (1958).

225. Belg. P. 673390 (1964) ≡ Neth. P. 6516125 (1966), Hercules Powder Co., Inv.: Anonym; *C.A.* **65**, 16,910d (1966).

226. DAS 1125929 (1958) ≡ USP 2984669 (1961), Chemische Werke Albert, Inv.: B. Brähler, R. Zimmermann, and J. Reese; *C.A.* **55**, 22,341h (1961).

227. Fr.P 1335755 (1962), Geigy, Inv.: Anonym; *C.A.* **60**, 1764g (1964).

228. Fr.P 1133785 (1955), C. H. Broehringer, Sohn, Inv.: Anonym; *Chem. Zentralblatt* **1959**, 605.

229. DAS 1011416 (1955), Bayer AG, Inv.: G. Schrader; *C.A.* **54**, 24,551e (1960).

230. Fr.P 1153596 (1956) ≡ USP 2947662 (1960), Montecatini, Inv.: R. Fusco, G. Losco, and M. Perini; *C.A.* **54**, 25,545b (1960).

231. USP 2494283 (1948), American Cyanamid Co., Inv.: J. T. Cassaday, E. I. Hoegberg, and B. D. Gleißner; *C.A.* **44**, 3516a (1950).

232. USP 3047459 (1955), Montecatini, Inv.: M. Perini and G. Speroni; *C.A.* **58**, 1349b (1963).

233. Fr.P 1234879 (1959) ≡ DBP 1146486 (1958), C. H. Boehringer, Sohn, Inv.: R. Sehring, and K. Zeile; *C.A.* **59**, 8790g (1963).

234. USP 2959610 (1959), American Cyanamid Co., Inv.: R. W. Young and G. Berkel-hammer; *C.A.* **55**, 7288b (1961).

235. DAS 1163310 (1962), Bayer AG, Inv.: G. Oertel and H. Malz; *C.A.* **60**, 10,549b (1964).

236. USP 3007845 (1958), American Cyanamid Co., Inv.: R. I. Hewitt and E. Waletzky; *C.A.* **56,** 10,638g (1962).

237. Brit. P. 814587 (1957), Sandoz Ltd., Inv.: Anonym; *C.A.* **53,** 22,716h (1959).

238. Fr.P 1285498 (1961) ≡ DBP 1138977 (1962), Ciba, Inv.: E. Beriger; *C.A.* **58,** 11,224e (1963).

239. DAS 1148806 (1960) ≡ Brit. P. 900557 (1962), Sandoz Ltd., Inv.: K. Lutz and M. Schuler; *C.A.* **57,** 16,404g (1962).

240. Ital.P 625074 (1959), Montecatini, Inv.: G. Losco, G. Rossi and G. Michieli; *C.A.* **57,** 8445b (1962).

241. Fr.P 1206931 (1958), Murphy Chemical Co., Inv.: M. Pianka and D. J. Polton; *Chem. Zentralblatt* **1962,** 9870.

242. USP 2578652 (1950), American Cyanamid Co., Inv.: J. T. Cassaday; *C.A.* **46,** 6139c (1952).

243. USP 2725328 (1954), Hercules Powder Co., Inv.: W. R. Diveley and A. D. Lohr; *C.A.* **50,** 5972h (1956).

244. USP 2864826, Hercules Powder Co., Inv.: W. R. Dively; *C.A.* **53,** 9058a (1959).

245. DBP 814152 (1948), Bayer AG, Inv.: G. Schrader; *Chem. Zentralblatt* **1952,** 600.

246. G. Schrader, *Z. Naturforsch.* **18b,** H. 11,965 (1963).

247. Bayer–Pflanzenschutz: E 605® (Company brochure) Leverkusen 1953.

248. J. B. McPherson and G. A. Johnson, *J. Agric. Food Chem.* **4,** 42 (1956).

249. USP 2503390 (1948), DuPont de Nemours & Co., Inv.: A.G. Jelinek; *C.A.* **44,** 6435d (1950).

250. J. Drábek and J. Pelikán, *Chem. Prům.* **6/31,** 293 (1956).

251. Belg. P. 596091 (1960) ≡ Brit. P. 919874 (1963), Bayer AG, Inv.: W. Lorenz; *C.A.* **59,** 7431f (1963).

252. USP 3135780 (1960), Sumitomo Chemical Co., Inv.: Sh. Suzuji, K. Fujii, Y. Nishizawa, and T. Kadota; Corresp. Pat: Jap. P. 6215147 (1959); *C.A.* **59,** 6934a (1963).

253. DBP 921870 (1952), Bayer AG, Inv.: G. Schrader; *C.A.* **50,** 1083a (1956).

254. USP 2664437 (1950), American Cyanamid Co., Inv. J. H. Fletcher; *C.A.* **48,** 13,717h (1954).

255. USP 2761806 (1954), Virginia Carolina Chem. Co., Inv.: W. P. Boyer; *C.A.* **51,** 663h (1957).

256. DAS 1050768 (1957), Bayer AG, Inv.: G. Schrader; *C.A.* **56,** 2473c (1962).

257. USP 2811480 (1956), Dow Chemical Co., Inv.: M. G. Norris and L. L. Wade; *C.A.* **52,** 6710c (1958).

258. DAS 1099530 (1959), Bayer AG, Inv.: H. Schlör, F. Schegk, and G. Schrader; *C.A.* **56,** 8748e (1962).

259. DAS 1174104 (1961) ≡ Belg. P. 625198 (1963), C. H. Boehringer, Sohn, Inv.: R. Sehring and K. Zeile; *C.A.* **60,** 13,187a (1964).

260. USP 3459836 (1965). Velsicol Chem. Co., Inv.: S. B. Richter; *C.A.* **71,** 81,519s (1969).

261. Belg. P. 579237 (1959) ≡ USP 2929762 (1958), Dow Chemical Co., Inv.: J. L. Wasco, L. L. Wade, and J. F. Landram; *C.A.* **54,** 18,439d (1960).

262. R. L. Metcalf and T. R. Fukuto, *J. Agric. Food Chem.* **4,** 930 (1956).

263. C. Fest and K.-J. Schmidt, *The Chemistry of Organophosphorus Pesticides*, p. 187, Springer Verlag, Berlin–Heidelberg–New York (1973).

264. USP 3042703 (1960), Bayer AG, Inv.: E. Schegk and G. Schrader; Corresp. Pat. Brit. 819689.

265. DAS 1063177 (1957), Bayer AG, Inv.: D. Delfs and K. Wedemeyer; *C.A.* **55,** 15,420g (1961).

266. DAS 1121882 (1958) ≡ Brit. P. 880297 (1959), Bayer AG, Inv.: H. Kayser, and G. Schrader; *C.A.* **56,** 8635b (1962).

267. Neth. P. 6508899 (1965), Bayer AG, Inv.: Sh. Kishino, A. Kudamatsu, I. Takase, K. Shiokawa, and Sh. Yamaguchi.

268. Belg. P. 648531 (1963), American Cyanamid Co., Inv.: J. B. Lovell and R. W. Baer; *C.A.* **63,** 11,433e (1965).

269. USP 3150040 (1961), Sumitomo Chemical Co., Inv.: S. Kuramoto, Y. Nishizawa, H. Sakamoto, and T. Mizutani; *C.A.* **64,** 2008h (1966).

270. Fr.P 1302593 (1961) ≡ DBP 1214040 (1966), Sumitomo Chemical Co., Inv.: S. Kuramoto, Y. Nishizawa, and H. Sakamoto, and T. Mizutani; *C.A.* **65,** 4579c (1966).

271. Fr.P 1600932 (1968), Bayer AG, Inv.: G. Schrader, I. Hammann, and W. Stendel; *C.A.* **74,** 125,186z (1971).

272. USP 3005004 (1959), American Cyanamid Co., Inv.: G. Berkelhammer; *C.A.* **56,** 4680b (1962).

273. M. Eto and Y. Oshima, *J. Agric. Biol. Chem.* **26,** 452 (1962).

274. M. Eto, Y. Kinoshito, T. Kato, and Y. Oshima, *J. Agric. Biol. Chem.* **27,** 789 (1963).

275. USP 2988474 (1960), Stauffer Chemical Co., Inv.: K. Szabo, J. G. Brady, and T. B. Williamson; *C.A.* **55,** 23,918i (1961).

276. DBP 814297 (1948) ≡ Brit. P. 670030 (1952), Bayer AG, Inv.: G. Schrader; *C.A.* **47,** 5438i (1953).

277. DBP 881194 (1951) ≡ Brit. P. 713142 (1954), Bayer AG, Inv.: G. Schrader; *C.A.* **50,** 411f (1956).

278. USP 3257416 (1963), Cooper, McDougall & Robertson Ltd., Inv.: N. C. Brown and D. T. Hollinshead; Corresp. Pat. Fr. AD 84110.

279. USP 3284455 (1963), Bayer AG, Inv.: C. Fest, G. Schrader; Corresp. Pat. Belg. 638710.

280. Fr.P 1360901 (1962), Dow Chemical Co., Inv.: R. H. Rigterink; *C.A.* **61,** 16,052b (1964).

280a. USP 2759937 (1955), American Cyanamid Co., Inv.: S. DuBreuil; *C.A.* **51,** 2885a (1957).

281. DBP 910652 (1952), Geigy AG, Inv.: H. Gysin and A. Margot; *C.A.* **50,** 10,770e (1956).

281a. DOS 2209554 (1972), Sandoz AG, Inv.: K. H. Milzner and F. Reisser; *C.A.* **77,** 152,218e (1972).

281b. H. J. Knutti and F. Reisser, *Proceedings of the 8th British Insecticide and Fungicide Conference,* 1975.

281c. Brit. P. 1019227 (1963), ICI Ltd., Inv.: G. V. McHattie; *C.A.* **64,** 14,197b (1966).

Brit. P. 1204552 (1968), ICI Ltd., Inv.: S. P. Sharpe and B. K. Snell; *C.A.* **74,** 22,875u (1971).

282. DAS 1156274 (1958) ≡ USP 2918468 (1959), American Cyanamid Co., Inv.: J. K. Dixon, Sh. DuBreuil, N. L. Boardway, and F. M. Gordon; *C.A.* **54,** 9971b (1960).

283. DAS 1545817 (1965) ≡ Neth. P. 6607054 (1966), Bayer AG, Inv.: K.-J. Schmidt and I. Hammann; *C.A.* **66,** 95,085f (1967).

283a. DAS 1767970 (1968), Hoechst AG, Inv.: W. Finkenbrink and L. F. Emmel; *C.A.* **73,** 44,282e (1970).

283b. DOS 2150098 (1970), Agripat S. A., Inv.: D. Dawes and B. Böhner; *C.A.* **77,** 34,528h (1972).

DAS 1299924 (1967), Hoechst AG, Inv.: O. Scherer and H. Mildenberger; *C.A.* **71**, 101,861c (1969).

284. DBP 962608 (1954), Bayer AG, Inv.: W. Lorenz and R. Wegler; *C.A.* **51**, 15,588a (1957).

285. DAS 1238902 (1965) ≡ Neth. P. 6605907 (1966), Bayer AG, Inv.: W. Lorenz, C. Fest, I. Hammann, M. Federmann, W. Flucke, and W. Stendel; *C.A.* **67**, 22,013s (1967).

286. R. W. Addor, *Org. Chem.* **29**, 738 (1964).

287. Fr.P 1327386 (1962) ≡ Belg. P. 618155 (1962), American Cyanamid Co., Inv.: R. W. Addor, and J. B. Lovell; *C.A.* **59**, 10,066f (1963).

287a. W. K. Whitney and J. L. Ashton, *Proceedings of the 8th British Insecticide and Fungicide Conference 1979*, p. 625.

288. DBP 975092 (1952), Ciba AG, Inv.: R. Sallmann, *C.A.* **59**, 5715f (1963).

289. DBP 1003720 (1954), Bayer AG, Inv.: W. Lorenz; *C.A.* **53**, 12,175d (1959).

290. DBP 974569 (1952), Ciba AG, Inv.: R. Sallmann; *C.A.* **56**, 10,640i (1962).

291. USP 3116201 (1952), Shell Oil Co., Inv.: R. R. Whetstone and D. Harman; *C.A.* **60**, 6748f (1964).

292. USP 2685552 (1952), Shell Development Co., Inv.: A. R. Stiles; *C.A.* **48**, 12,365c (1954).

293. USP 3068268 (1961), Shell Oil Co., Inv.: Ch. H. Tiemann and A. R. Stiles; Corresp. Pat. Belg. 613828.

294. DAS 1098939 (1956) ≡ USP 2908605 (1959), Ciba AG, Inv.: E. Beriger and R. Sallmann; *C.A.* **54**, 15,819a (1960).

294a. DOS 1768315 (1968), Hoechst AG, Inv.: O. Scherer, H. Röchling, and H. Mildenberger; *C.A.*, not abstracted.

DOS 1643608 (1967), Agripat SA, Inv.: B. Böhner and K. Rüfenacht; *C.A.* **70**, 106,068p (1969).

295. K.-J. Schmidt, *Chemical Aspects of Organophosphate Pesticides in View of the Environment*, in: *Environmental Quality and Safety*. Ed.: F. Coulston and F. Korte, Vol. 4, p. 96–108, Georg Thieme Verlag, Stuttgart (1975).

296. C. Fest and K.-J. Schmidt, *Insektizide Phosphorsäureester*, in: *Chemie der Pflanzenschutz- und Schädlingsbekämpfungsmittel*, Ed.: R. Wegler, Vol. 1, pp. 248–438, Springer Verlag, Berlin–Heidelberg–New York (1970).

297. M. Eto, *Organophosphorus Pesticides: Organic and Biological Chemistry*, CRC Press, Cleveland, Ohio (1974).

298. J. H. Ruzicka, J. Thomson, and B. B. Wheals, *J. Chromatogr.* **31**, 37 (1967).

299. Y. P. Sun, *J. Econ. Entomol.* **61**, 949 (1968).

300. A. R. Main, *Science* **144**, 992 (1964).

301. A. L. Lehninger, *Biochemie*, p. 156/157, Verlag Chemie, Weinheim/Bergstraße, 1975.

302. M. Uchiyama, T. Yoshida, K. Homma, and T. Hongo, *Environmental Quality and Safety*, (Ed.) F. Coulston and F. Korte, Vol. 4, p. 109f., George Thieme Verlag, Stuttgart (1975).

303. M. Eto, *Organophosphorus Pesticides: Organic and Biological Chemistry*, p. 157, CRC Press, Cleveland, Ohio (1974).

304. B. B. Brodie, J. R. Gillette, and B. N. LaDu, *Ann. Rev. Biochem.* **27**, 427 (1958).

305. T. Nakatsugawa and P. A. Dahm, *J. Econ. Entomol.* **58**, 500 (1965).

306. J. B. Knaak, M. A. Stahmann, and J. E. Casida, *J. Agric. Food Chem.* **10**, 154 (1962).

307. R. E. Menzer, *Res. Rev.* **48**, 79 (1973).

308. R. Anliker, E. Beriger, M. Geiger, and K. Schmid, *Helv. Chim. Acta* Vol. **XLIV**, 1635 (1961).

309. D. A. Lindquist and D. L. Bull, *J. Agric. Food Chem.* **15**, 267 (1967).

310. D. L. Bull and D. A. Lindquist, *J. Agric. Food Chem.* **12**, 311 (1964).

311. M. Eto, *Organophosphorus Pesticides: Organic and Biological Chemistry*, p. 166, CRC Press, Cleveland, Ohio (1974).

312. J. S. Bowman and J. E. Casida, *J. Agric. Food Chem.* **5**, 192 (1952).

313. C. Fest and K.-J. Schmidt, *The Chemistry of Organophosphorus Pesticides*, p. 98, Springer Verlag, Berlin–Heidelberg–New York (1973).

314. H. Niessen, H. Tietz, and H. Frehse, *Pflanzenschutz-Nachr. "Bayer"* **15**, (3), 148 (1962).

315. J. G. Leesch and T. R. Fukuto, *Pestic. Biochem. Physiol.* **2**, 223 (1972).

316. J. B. McBain and L. J. Menn, *Pestic. Biochem. Physiol.* **1**, 356 (1971).

317. J. Métivier, *Pesticide Chemistry Proceedings Second International IUPAC-Congress* Vol. 1, p. 325, Gordon & Breach, London (1972).

318. N. F. Janes et al., *J. Agric. Food Chem.* **21**, 121 (1973).

319. C. C. Conaway and Ch. O. Knowles, *J. Econ. Entomol.* **62**, 286 (1969).

320. W. Mücke, K. O. Alt, and H. O. Esser, *J. Agric. Food Chem.* **18**, 208 (1970).

321. F. W. Seume and R. D. O'Brien, *J. Agric. Food Chem.* **8**, 36 (1960).

322. H. Frehse, *Pesticide Terminal Residues*, p. 9, Butterworths, London (1971).

323. W. H. Gutenmann, L. E. St. John, and D. J. Lisk, *J. Agric. Food. Chem.* **19**, 1259 (1971).

324. E. Benjamini, R. L. Metcalf, and T. R. Fukuto, *J. Econ. Entomol.* **52**, 99 (1959).

325. H. Jarczyk, *Pflanzenschutz-Nachr. "Bayer"* **19**, (1), 1 (1966).

326. J. Miyamoto, A. Wakimura, and T. Kadota, *Environ. Qual. Saf.* **1**, 235–39 (1972).

327. R. A. Neal, *Biochem. J.* **103**, 183 (1967).

328. C. J. Whitten and D. L. Bull, *Pestic. Biochem. Physiol.* **4**, 266 (1974).

329. G. N. Smith, B. S. Watson, and F. S. Fischer, *J. Agric. Food Chem.* **15**, 127 (1967).

330. J. J. Menn and J. B. McBain, *J. Agric. Food Chem.* **12**, 163 (1964).

331. R. A. Werner, *J. Econ. Entomol.* **66**, 867 (1973).

332. W. Dedek and H. Schwarz, *Atompraxis* **12**, 603 (1966).

333. C. Fest and K.-J. Schmidt, *The Chemistry of Organophosphorus Pesticides*, p. 247, Springer Verlag, Berlin–Heidelberg–New York (1973).

334. E. Hodgson and J. E. Casida, *J. Agric. Food Chem.* **10**, 208 (1962).

335. J. E. Casida, L. McBride, and R. P. Niedermeier, *J. Agric. Food Chem.* **10**, 370 (1962).

336. F. W. Plapp and J. E. Casida, *J. Agric. Food Chem.* **6**, 662 (1958).

337. W. Hofer, Bayer AG, manuscript in preparation.

338. E. Boyland and L. F. Chasseud, *Adv. Enzymol.* **32**, 132–219 (1969).

339. R. M. Hollingworth, *J. Agric. Food Chem.* **17**, 987 (1969).

340. D. L. Bull, *Res. Rev.* **43**, 1 (1972).

341. A. Morello, A. Vardanis, and E. Y. Spencer, *Can. J. Biochem.* **46**, 885 (1968).

342. N. Motoyama and W. C. Dautermann, *Pestic. Biochem. Physiol.* **2**, 170 (1972).

343. J. Fukami and T. Shishido, *J. Econ. Entomol.* **59**, 1338 (1966).

344. T. van Bao, I. Szabó, P. Ruzcska, and A. Czeizel, *Humangenetik* **24**, p. 39–57, Springer Verlag, Berlin–Heidelberg–New York (1974).

345. H. H. Sauer, *J. Agric. Food Chem.* **20**, 578 (1972).

346. R. D. O'Brien, *Insecticides-Action and Metabolism*, Academic Press, New York–London (1967).

SECTION 2.4. INSECTICIDAL CARBAMATES

347. E. Stedman, *Biochem. J.* **20**, 719 (1926).
348. O. Loewi and E. Nawratil, *Pflüg. Arch. Ges. Physiol.* **214**, 689 (1926).
349. H. Gysin, *Experientia* **8**, 205 (1954).
350. R. Wiesmann, R. Gasser, and H. Grob, *Experientia* **7**, 117 (1951).
351. J. A. Durden, Pers. Comm., Union Carbide Corp.
352. R. Pulver and R. Domenjoz, *Experientia* **7**, 306 (1951).
353. DBP 956638 (1952); DBP 962124 (1953); DBP 964818 (1953) ≡ Brit. P. 743958 (1953), E. Merck AG, Inv.: E. Jacobi, S. Lust, A. van Schoor, O. Zima; *C.A.* **53**, 22718d,e (1959); *C.* **1956**, 6792.
354. M. J. Kolbezen, R. L. Metcalf, and T. R. Fukuto, *J. Agric. Food Chem.* **2**, 864 (1954).
355. J. Fraser, I. R. Harrison, and S. B. Wakerley, *J. Sci. Food Agric. Suppl.* **1968**, 8.
356. L. P. Batjer and B. J. Thompson, *Proc. Am. Soc. Hortic. Sci.* **77**, 1 (1964).
357. R. Wiesmann and C. Kocher, *Z. Angew. Entomol.* **33**, 297 (1951).
358. F. L. C. Baranoyovits and R. Gosh, *Chem. Ind.* (*London*), 1018 (1969).
359. H. L. Haynes, J. A. Lambrech, and H. H. Moorefield, *Contrib. Boyce Thompson Inst.* **18**, 507 (1958).
360. R. C. Back, *J. Agric. Food Chem.* **13**, 198 (1965).
361. G. P. Georghiu and R. L. Metcalf, *J. Econ. Entomol.* **55**, 125 (1962).
362. G. Unterstenhöfer, *Pflanzenschutznachr. Bayer* **15**, 181 (1962).
363. R. L. Metcalf, T. R. Fukuto, and M. Y. Winton, *J. Econ. Entomol.* **55**, 889 (1962).
364. G. Unterstenhöfer, *Meded. Landbouwhogesch. Gent* **28**, 758 (1963).
365. A. J. Lemin, G. A. Boyack, and R. M. McDonald, *J. Agric. Food Chem.* **13**, 124 (1965).
366. A. Jäger, *Z. Angew. Entomol.* **58**, 188 (1966).
367. *Nat. Pestic. Contr. Techn. Release No.* **6**, 2 (1966).
368. J. Fraser, D. Greenwood, I. R. Harrison, and W. H. Wells, *J. Sci. Food Agric.* **18**, 372 (1967).
369. E. E. Kenaga, *Bull. Entomol. Soc. Am.* **12**, 161 (1966).
370. W. R. Steinhausen, *Z. Angew. Zool.* **55**, 108 (1968).
371. R. L. Metcalf, C. Fuertes-Polo, and T. R. Fukuto, *J. Econ. Entomol.* **56**, 862 (1963).
372. F. Bachmann and J. B. Legge, *J. Sci. Food Agric. Suppl.* **1968**, 39.
373. G. Prins, *Bull. Entomol. Res.* **56**, 231 (1965).
374. F. Barlow and A. B. Hadaway, *Pestic. Sci.* **1**, 117 (1970).
375. M. H. J. Weiden, H. H. Moorefield, and L. K. Payne, *J. Econ. Entomol.* **58**, 154 (1965).
376. G. A. Roodhaus and N. B. Joy, *Meded. Landbouwhogesch. Gent* **33**, 833 (1968).
377. J. C. Felton, *J. Sci. Food Agric. Suppl.* **1968**, 32.
378. L. K. Payne, H. A. Stansbury, and M. H. J. Weiden, *J. Med. Chem.* **8**, 525 (1965).
379. D. A. Allison, C. Sinclair, and M. R. Smith, *Proc. 7th Brit. Fungic. Insectic. Conf.* **1973**, 395.
380. M. H. Bresse, *7th. Intern. Crop Protection Congress Paris* **1970**, 117.
381. J. W. Davis and C. B. Gowan, *J. Econ. Entomol.* **67**, 130 (1974).

382. K. Konishi, *Proc. 2. Intern. IUPAC Congr. Pestic. Chem.* **1**, 179 (1972); M. Sakai and Y. Sato, ibid., 445.

383. R. L. Metcalf and T. R. Fukuto, *J. Econ. Entomol.* **58**, 1151 (1965).

384. *National Pestic. Control Ass. Techn. Release No. 4* (1966).

385. *National Pestic. Control Ass. Techn. Release No. 13* (1966).

386. H. F. Schoof, *Bull. Entomol. Soc. America* **12**, 338 (1966).

387. K. S. McKinlay and W. K. Martin, *Can. Entomol.* **99**, 748 (1967).

388. M. H. J. Weiden, A. J. Borasch, E. L. Boyd, A. A. Sousa, and L. K. Payne, *J. Econ. Entomol.* **60**, 873 (1967).

389. C. R. Worthing, *J. Hortic. Sci.* **44**, 235 (1969).

390. P. A. Norman, D. K. Reed, and C. R. Crittenden, *J. Econ. Entomol.* **63**, 1409 (1970).

391. R. M. Sacher and J. F. Olin, *J. Agric. Food Chem.* **20**, 354 (1972).

392. B. Hetnarski and R. D. O'Brien, *J. Agric. Food Chem.* **20**, 543 (1972).

393. J. C. Boling, *J. Econ. Entomol.* **65**, 1737 (1972).

394. R. W. Addor, *J. Agric. Food Chem.* **13**, 207 (1965).

395. R. O. Drummond and W. J. Gladney, *J. Econ. Entomol.* **62**, 934 (1969).

396. R. O. Drummond and T. M. Whetstone, *J. Econ. Entomol.* **67**, 237 (1974).

397. A. R. Friedman and E. G. Goodrich, *J. Agric. Food Chem.* **19**, 856 (1971).

398. J. C. Felton, *J. Sci. Food Agric. Suppl.* **1968**, 32.

399. R. L. Jones, T. R. Fukuto, and R. L. Metcalf, *J. Econ. Entomol.* **65**, 28 (1972).

400. D. D. Rosenfeld and J. R. Kilsheimer, *J. Agric. Food Chem.* **22**, 926 (1974).

401. G. P. Georghiu, R. L. Metcalf, and F. E. Gidden, *Bull. World Health Organ.* **35**, 691 (1966).

402. J. B. Briggs, *East Malling Res. Stn., Maidstone, Engl., Rep.*, 149 (1969).

403. B. M. Glancey, K. F. Baldwin, and C. S. Lofgren, *Mosq. News* **29**, 36 (1969).

404. D. E. Clegg and P. R. Martin, *Pestic. Sci.* **4**, 4 (1973).

405. M. A. H. Fahmy, T. R. Fukuto, R. O. Myers, and R. B. March, *J. Agric. Food Chem.* **18**, 793 (1970).

406. C. H. Schaeffer and W. H. Wilder, *J. Econ. Entomol.* **63**, 480 (1970).

407. A. L. Black, Y. C. Chiu, M. A. H. Fahmy, and T. R. Fukuto, *J. Agric. Food Chem.* **21**, 747 (1973).

408. M. A. H. Fahmy, Y. C. Chiu, and T. R. Fukuto, *J. Agric. Food Chem.* **22**, 59 (1974).

409. DOS 2254359 (1972), Bayer AG, Inv.: P. Siegle, E. Kühle, I. Hammann, and W. Behrenz; *C.A.* **81**, 37,408h (1974).

410. DOS 2425211 (1973), Union Carbide Corp., Inv. J. A. Durden and T. D. J. D'Silva; *C.A.* **83**, 164,240t (1975).

411. L. K. Payne, H. A. Stansbury, and M. H. J. Weiden, *J. Agric. Food Chem.* **14**, 356 (1966).

412. DOS 1963061 (1968), E. I. du Pont de Nemours and Co., Inv.: J. B. Buchanan, J. J. Fuchs, E. W. Raleigh, and H. M. Loux; *C.A.* **74**, 93,100s (1971).

413. E. Casida, K.-B. Augustinsson, and G. Jonson, *J. Econ. Entomol.* **53**, 502 (1960).

414. M. L. Bender and R. B. Homer, *J. Org. Chem.* **30**, 3975 (1965).

415. T. R. Fukuto, M. A. H. Fahmy, and R. L. Metcalf, *J. Agric. Food Chem.* **15**, 273 (1967).

416. T. R. Fukuto, R. L. Metcalf, R. L. Jones, and R. O. Myers, *J. Agric. Food Chem.* **17**, 923 (1969).

J. A. Durden and M. H. J. Weiden, *J. Agric. Food Chem.* **22**, 396 (1974).

417. G. B. Koelle, *Handbuch der experimentellen Pharmakologie*, Vol. 15, Springer Verlag, Berlin–Göttingen–Heidelberg (1963).

418. N. Engelhard, K. Prchal, and M. Nenner, *Angew. Chem.* **79**, 604 (1967).

419. R. L. Metcalf and T. R. Fukuto, *J. Agric. Food Chem.* **13**, 220 (1964).

420. I. B. Wilson, M. Hatch, and S. Ginsburg, *J. Biol. Chem.* **235**, 2312 (1960).

421. A. R. Main, *Can. Med. Assoc. J.* **100**, 161 (1969).

422. K. Hellenbrand, *J. Agric. Food Chem.* **15**, 825 (1967).

423. R. M. Hollingworth, T. R. Fukuto, and R. L. Metcalf, *J. Agric. Food Chem.* **15**, 235 (1967).

424. A. H. Aharoni and R. D. O'Brien, *Biochemistry* **7**, 1538 (1968).

425. N. R. McFarlane, P. J. Jewess, and P. E. Porter, in: N. Aldridge and F. Reiner, *Enzyme Inhibitors as Substrates*, Kap. 8.2, Elsevier Publ. Co., Amsterdam–London–New York (1972).

426. M. E. Eldefrawi, O. Toppozada, and R. D. O'Brien, *Proc. 6. Intern. Congr. Plant Protection*, p. 214, Wien 1967.

427. B. Hetnarski and R. D. O'Brien, *Pestic. Biochem. Physiol.* **2**, 132 (1972); *Biochemistry* **12**, 3883 (1973); *J. Agric. Food Chem.* **23**, 709 (1975).

428. D. Nachmansohn, *Chemical and Molecular Basis of Nerve Activity*, Chap. 10, Academic Press, New York (1959).

429. C. Hansch and E. W. Deutsch, *Biochim. Biophys. Acta* **126**, 117 (1966).

430. M. H. J. Wieden, *J. Sci. Food Agric. Suppl.* **1968**, 19; *Bull. W.H.O.* **44**, 203 (1971).

431. A. L. Black. Y. C. Chiu, T. R. Fukuto, and T. A. Miller, *Pestic. Biochem. Physiol.* **3**, 435 (1973).

432. J. J. K. Boulton, C. B. C. Boyce, P. J. Jewess, and R. F. Jones, *Pestic. Sci.* **2**, 10 (1971).

433. R. L. Metcalf, *Bull. W.H.O.* **44**, 43 (1971).

434. F. Reiner, *Bull. W.H.O.* **44**, 109 (1971).

435. E. Heilbronn, *Biochem. Pharmacol.* **12**, 25 (1963).

436. O. R. Klimmer, *Pflanzenschutz- und Schädlingsbekämpfungsmittel, Abriß einer Toxikologie*, p. 71, Hundt-Verlag, Bonn (1971).

437. USP 3800037 (1970) ≡ DOS 2041986 (1971), Ciba-Geigy, Inv.: H. Martin, G. Pissiotas, and O. Rohr; *C.A.* **75**, 35,491m (1971).

438. J. R. Barnes, *J. Econ. Entomol.* **62**, 86 (1969).

439. J. F. Casida, *J. Agric. Food Chem.* **18**, 753 (1970).

440. C. Hansch, *J. Med. Chem.* **11**, 920 (1968).

441. J. Desmarcheliev, R. I. Krieger, P. W. Lee, and T. R. Fukuto, *J. Econ. Entomol.* **66**, 631 (1973).

442. L. B. Brattsten and R. L. Metcalf, *J. Econ. Entomol.* **63**, 101 (1970).

443. R. M. Sacher, R. L. Metcalf, E. R. Metcalf, and T. R. Fukuto, *J. Econ. Entomol.* **64**, 1011 (1971).

SECTION 2.5. INSECTICIDES FROM MISCELLANEOUS CLASSES

444. R. Wegler, *Chemie der Pflanzenschutz- und Schädlingsbekämpfungsmittel*, Vol. 1, Springer Verlag, Berlin–Heidelberg (1970).

445. W. Perkow, *Wirksubstanzen der Pflanzenschutz- und Schädlingsbekämpfungsmittel*, Verlag Paul Parey, Berlin–Hamburg (1971/1974).

446. H. Martin and C. R. Worthing, *Pesticide Manual*, 4th ed., British Crop Protection Council, Worcester (1974).

447. W. T. Thomson, *Agricultural Chemicals, Book 1: Insecticides, Acaricides, and Ovicides*, 1975 Revision, Thomson Publications, Indianapolis, Indiana (1975).

448. W. T. Thomson, *Agricultural Chemicals, Book 3: Fumigants, Growth Regulators, Repellents, and Rodenticides*, 1974 Revision, Thomson Publications, Indianapolis, Indiana (1974).

449. J. J. van Daalen, J. Meltzer, R. Mulder, and K. Wellinga, *Naturwissenschaften* **59**, 312 (1972).

450. DOS 2123236 (1971 Neth. Prior. 1970). N. V. Philips' Gloeilampenfabriken, Inv.: K. Wellinga, R. Mulder; *C.A.* **76**, 85,578m (1972).

451. K. Wellinga, R. Mulder, and J. J. van Daalen, *J. Agric. Food Chem.* **21**, 348,993 (1973).

452. R. Mulder and M. J. Gijswijt, *Pestic. Sci.* **4**, 737 (1973).

453. L. C. Post, B. J. de Jong, and W. R. Vincent, *Pestic. Biochem. Physiol.* **4**, 473 (1974).

454. DRP 673246 (1934), W. v. Leuthold, *Frdl.* **25**, 1331 (1942).

455. R. Riemschneider, *Monatsh. Chem.* **82**, 600 (1951).

456. DOS 2110056 (1971) ≡ S. Afric. P. 7104236 (1972), Bayer AG, Inv.: W. Meiser, K. H. Büchel, and W. Behrenz; *C.A.* **77**, 136,296y (1972).

457. A. Eben and G. Kimmerle, unpublished work.

SECTION 2.6. NEMATICIDES

458. J. F. Spears, *Agric. Chem.* **14**, (1), 39, 41, 111 (1959).

459. J. F. Spears, *Agric. Chem.* **14**, (2), 36 (1959).

460. B. Homeyer in: R. Wegler, *Chemie der Pflanzenschutz- und Schädlingsbekämpfungsmittel*, Vol. 1, p. 573–583, Springer Verlag, Berlin–Heidelberg–New York (1970).

461. R. L. Metcalf, *Advances in Pest Control Research*, Vol. 3, p. 181, Interscience, New York (1960).

462. A. G. Newhall, *Botan. Rev.* **21**, 189 (1955).

463. A. L. Taylor and C. W. McBeth, *Proc. Helminthol. Soc. Wash.* **7**, 94 (1940).

464. R. C. Baines, F. J. Foote, and J. P. Martin, *Citrus Leaves* **36**, (10), 6–8, 24, 27 (1956); *Calif. Citogr.* **41**, (12), 427, 448 (1956).

465. C. W. McBeth and G. B. Bergeson, *Plant Dis. Rep.* **39**, 223 (1955).

466. USP 2734084 (1953), Dow Chemical Co., Inv.: M. P. Doerner; *C.A.* **50**, 15,581e (1956).
    USP 2766554 (1954), Stauffer Chemical Co., Inv.: S. C. Dorman and A. B. Lindquist; *C.A.* **51**, 4638i (1957).

467. D. C. Torgeson, D. M. Joder, and J. B. Johnson, *Phytopathology* **47**, 536 (1957).

468. DOS 2308807 (1973, US Prior. 1972) Buckman Laboratories, Inc., Inv.: J. D. Buckman and J. D. Pera; *C.A.* **80**, 26,786n (1974).

469. Neth. P. 73274 (1953), A. P. Weber; *C.A.* **48**, 3626g (1954).
    Neth. P. 83819 (1957), Directie van de Staatsmijnen in Limburg; *C.A.* **52**, 1541f (1958).

470. Neth. P. 77330 (1955), Directie van de Staatsmijnen in Limburg; *C.A.* **49**, 16,321c (1955).

471. USP 2761806 (1954), Virginia–Carolina Chemical Corp., Inv.: P. Bayer; *C.A.* **51**, 663g (1957).

472. USP 3042703 (1956) ≡ Brit. P. 819689 (1957), Bayer AG, Inv.: E. Schegk and G. Schrader; *C.A.* **55**, 11751h (1961).

473. USP 2978479 (1959), Bayer AG, Inv.: H. Kayser and G. Schrader; *C.A.* **56**, 8633f (1962).

474. USP 3005749 (1959), Dow Chemical Co., Inv.: C. R. Youngson; *C.A.* **56**, 5164d (1962).

475. DAS 1156274 (1958) ≡ USP 2918468 (1957), American Cyanamid Co., Inv.: J. K. Dixon, S. Du Breuil, N. L. Boardway, and F. M. Gordon; *C.A.* **54**, 9971b (1960).

476. *VII. Int. Congr. Plant Protection*, p. 123, Paris (1970).

477. Neth. P. 6605690 (1966) ≡ USP 3268393 (1965), Mobil Oil Corp., Inv.: J. H. Wilson; *C.A.* **65**, 14,362d (1966).

   Neth. P. 6609632 (1966, DB Prior. 1965), Bayer AG, Inv.: K. Mannes, G. Schrader, and B. Homeyer; *C.A.* **67**, 2758k (1967).

478. DOS 2405288 (1974, US Prior. 1973), Shell Internationale Research Maatschappij B. V., Inv.: C. W. McBeth, L. F. Ward, and D. D. Phillips; *C.A.* **81**, 135,442p (1974).

479. DOS 2263429 (1973, US Prior. 1971), Dow Chemical Co., Inv.: R. H. Rigterink; *C.A.* **79**, 115,430p (1973).

480. DOS 2260015 (1972, Schweiz. Prior. 1971), Ciba-Geigy AG, Inv.: D. Dawes and B. Boehner; *C.A.* **79**, 66,364j (1973).

481. Belg. P. 649260 (1964, DB Prior. 1963), Bayer AG, Inv.: R. Heiss, E. Boecker, W. Behrenz, and G. Unterstenhöfer; *C.A.* **64**, 14,170g (1966).

482. USP 3217037 (1962) ≡ Fr. P. 1377474 (1963), Union Carbide Corp., Inv.: L. K. Payne and M. H. J. Weiden; *C.A.* **63**, 2900a (1965).

483. DOS 1768623 (1968) ≡ S. Afric. P. 6803629 (1968, US Prior. 1967), E. I. du Pont de Nemours and Co., Inv.: J. B. Buchanan; *C.A.* **70**, 114,646r (1969).

484. USP 3193561 (1960), American Cyanamid Co., Inv.: R. W. Addor; *C.A.* **63**, 11,577b (1965).

SECTION 2.7. ACARICIDES

485. R. Wegler, *Chemie der Pflanzenschutz- und Schädlingsbekämpfungsmittel*, Vol. 1, Springer Verlag, Berlin–Heidelberg–New York (1970).

486. H. Martin and C. R. Worthing, *Pesticide Manual*, 4th ed., British Crop Protection Council, Worcester (1974).

487. W. Perkow, *Wirksubstanzen der Pflanzenschutz- und Schädlingsbekämpfungsmittel*, Verlag Paul Parey, Berlin–Hamburg (1971/74).

488. W. T. Thomson, *Agricultural Chemicals*, Book I, *Insecticides, Acaricides, and Ovicides*, 1975 revision, Thomson Publications, Indianapolis, Indiana (1975).

489. *Farm Chemicals Handbook 1975*, Meister Publishing, Willoughby, Ohio (1975).

490. K. Gätzi, Akarizide, in: *Ullmanns Encyklopädie der technischen Chemie*, 4th ed., Vol. 7, pp. 1–9, Verlag Chemie, Weinheim/Bergstraße (1974).

491. DOS 1909868 (1969, US Prior. 1968), 1926366 (1970, US Prior. 1968), The Upjohn Co., Inv.: G. Kaugars, E. G. Gemrich; *C.A.* **72**, 54,978e, 100,323n (1970).

492. R. E. Rice and R. A. Jones, *Calif. Agric.* **26**, 12 (1972); *C.A.* **77**, 84,446a (1972).

493. A. G. Shelhime, C. R. Crittenden, and T. B. Hull, *Fla. Entomol.* **54**, 79 (1971); *C.A.* **75**, 34,463s (1971).

494. *Farm Chemicals Handbook 1972*, C. 154/155, Meister Publishing, Willoughby, Ohio (1972).

495. K. H. Büchel, F. Korte, and R. B. Beechey, *Angew. Chem.* **77**, 814 (1965).

496. R. L. Williamson and R. L. Metcalf, *Science* **158**, 1694 (1967).

497. D. E. Burton, A. J. Lambie, J. C. L. Ludgate, G. T. Newbold, A. Percival, and D. T. Saggers, *Nature* **208**, 1166 (1965).

498. D. T. G. Jones and W. A. Watson, *Nature* **208**, 1169 (1965).

499. K. H. Büchel, F. Korte, A. Trebst, and E. Pistorius, *Angew. Chem.* **77**, 911 (1965).

500. K. H. Büchel, W. Draber, A. Trebst, and E. Pistorius, *Z. Naturforsch.* B **21**, 243 (1966).

501. D. T. Saggers and M. L. Clark, *Nature* **215**, 275 (1967).

502. H. Zimmer, O. A. Homberg, and M. Jayawant, *J. Org. Chem.* **31**, 3857 (1966).

503. DOS 2115666 (1971, US Prior. 1970), Shell Internationale Research Maatschappij N.V., Inv.: C. A. Horne; *C.A.* **76**, 69,178a (1972).

504. D. Asquith, *J. Econ. Entomol.* **66**, 237 (1973);

 S. L. Poe, *J. Econ. Entomol.* **66**, 490 (1973).

SECTION 2.8. OTHER METHODS OF PEST CONTROL

505a. *Review Literature on Hormones and Hormone Mimics*

A. A. Akhrem, I. S Levina, and Yu. A. Titov, The Chemistry of Ecdysones, *Russ. Chem. Rev.* **40**, 760–772 (1971).

M. T. El-Ibrashi, Insect Hormones and Analogues: Chemistry Biology and Insecticidial Potencies, *Z. Angew. Entomol.* **66**, 113–114 (1970).

M. G. Gersch and G.-A. Böhm, *Aktuelle Aspekte und Tendenzen der Hormonforschung bei Insekten, Chap. 3: Das Ecdyson und andere Verbindungen mit Häutungshormonaktivität, Biol. Rundsch.* **9**, 1–11 (1971).

L. I. Gilbert, Invertebrate Hormones. I. The Chemistry of Insect Hormones, *Proc. III. Int. Congr. Endocrin., Mexico*, pp. 340–346 (1968).

H. Hikino and Y. Hikino, Arthropod Molting Hormones, in: *Fortschritte der Chemie Organischer Naturstoffe*, Zechmeister (Ed.), Vol. 28, pp. 257–312, Springer Verlag, Berlin–Heidelberg–New York (1970).

D. H. S. Horn, The Ecdysones, in: *Naturally Occurring Insecticides*, M. Jacobson and D. G. Crosby (Eds.), pp. 333–459, Marcel Dekker, New York (1971).

P. Karlson, Ecdyson, das Häutungshormon der Insekten, *Naturwissenschaften* **53**, 445–453 (1966).

P. Karlson, Insect Hormones and Insect Pheromones as Possible Agents for Insect Control—A Critical Review, *IUPAC-Symposium, Johannesburg, South Africa*, 14.–18. July (1969).

P. Karlson and G. E. Sekeris, Ecdysone, An Insect Steroid Hormone, and Its Mode of Action, in: *Recent Progress in Hormone Research*, G. Pincus (Ed.), pp. 473–502, Academic Press, New York–London (1966).

G. Koller, Die Hormone der wirbellosen Tiere, in: *Fermente, Hormone, Vitamine*, R. W. Dirschere (Ed.), Vol. II, pp. 814–868, Georg Thieme Verlag, Stuttgart (1960).

Y. L. Meltzer, *Hormonal and Attractant Pesticide Technology*, Chap. 5, Ecdysones and Related Products, pp. 95–151, Noyes Data Corporation, New Jersey (1971).

K. Nakanishi, The Ecdysones, *Pure Appl. Chem.* **25**, 167–195 (1971).

M. J. Piechowska, Z. Sinkeewicz, and M. Bielinska, Hormones of insect metamorphosis, *Post. Biochem.* **16**, 449–481 (1970).

H. H. Rees, Ecdysones, in: *Aspects of Terpenoid Chemistry and Biochemistry*, T. W. Goodwin (Ed.), pp. 181–222, Academic Press, New York–London (1971).

F. Schmal, Endocrines of Arthropods, *Chem. Zool.* **6**, 307–345 (1971).

H. A. Schneiderman, Insect Hormones and Insect Control, in: *Insect Juvenil Hormones Chemistry and Action*, J. J. Menn and M. Beroza (Eds.), pp. 3–27, Academic Press, New York–London (1972).

H. A. Schneiderman, A. Krishnakumaran, P. J. Bryant, and F. Sehnal, Endocrinological Strategies in Insect Control, *Agric. Sci. Rev.* **8**, 13–25 (1970).

K. Slama, M. Romanuk, and F. Sorm, Insect Hormones and Bioanalogues, Chap. III, *Chemistry and Physiology of Ecdysoids*, pp. 303–387, Springer Verlag, Berlin–Heidelberg–New York (1974).

G. R. Wyatt, Insect Hormones, in: *Biochemical Actions of Hormones*, G. Litwack (Ed.), Vol. II, pp. 385–490, Academic Press, New York–London (1972).

C. E. Berkoff, The Chemistry and Biochemistry of Insect Hormones, *Quart. Rev. Chem. Soc.* **23**, 372–391 (1969).

P. Ellis, Can Insect Hormones and Their Mimics Be Used to Control Pests?, *Pest Articles and News Summaries* **14**, 329–342 (1968).

J. B. Siddall, Chemical Aspects of Hormonal Interactions, in: *Chemical Ecology*, E. Sondheimer and J. B. Simeone (Eds.), pp. 281–306, Academic Press, New York–London (1970).

C. M. Williams, Hormonal Interactions between Plants and Insects, in: *Chemical Ecology*, E. Sondheimer and J. B. Simeone (Eds.), pp. 103–132, Academic Press, New York–London (1970).

505. C. M. Williams, *Nature* **178**, 212 (1956).

506. S. Kopec, *Biol. Bull.* **42**, 323 (1922).

507. G. Fraenkel, *Nature* **133**, 834 (1934).

508. V. B. Wigglesworth, *Q. J. Microsc. Sci.* **77**, 191 (1934).

509. S. Fukuda, *Proc. Imp. Acad. (Tokyo)* **16**, 417 (1940).

510. C. M. Williams, *Sci. Am.* **217**, 13 (1967).

511. M. Locke, in: *The Physiology of Insects*, M. Rockstein (Ed.), Vol. 3, pp. 379–470, Academic Press, New York (1964).

512. M. Gersch, *Vergleichende Endokrinologie der wirbellosen Tiere*, Akademische Verlagsgesellschaft, Geest & Portig, Leipzig (1964).

513. P. M. Jenkin, *Ann. Endocrinol. (Paris)* **27**, 331 (1966).

514. M. Gersch, *Am. Zool.* **1**, 53 (1961).

515. M. Gersch and G. A. Böhm, *Biol. Rundsch.* **9**, 7 (1971).

516. M. T. El-Ibrashy, *Z. Angew. Entomol.* **66**, 118 (1970).

517. H. A. Schneiderman, A. Krishnakumaran, P. J. Bryant, and F. Sehnal, *Agric. Sci. Rev.* **8**, 21 (1970).

518. H. S. Horn, E. J. Middleton, and J. A. Wunderlich, *Chem. Commun.* **1966**, 339.

519. W. E. Robbins, J. N. Kaplanis, J. A. Svoboda, and M. J. Thompson, *Ann. Rev. Entomol.* **16**, 53 (1971).

520. H. Oberlander, *J. Insect Physiol.* **15**, 1803 (1969).

521. H. Oberlander, *J. Insect Physiol.* **18**, 223 (1972).

522. A. Krishnakumaran and H. A. Schneiderman, *Nature* **220**, 601 (1968).

523. J. E. Wright, *Science* **163**, 390 (1969).

524. H. A. Schneiderman and A. Krishnakumaran, *Agric. Sci. Rev.* **8**, 13 (1970).

525. A. Butenandt and P. Karlson, *Z. Naturforsch.* **96**, 389 (1954).

526. R. Huber and W. Hoppe, *Chem. Ber.* **7**, 2403 (1965).

527. U. Kerb, G. Schulz, P. Hocks, R. Wiechert, A. Furlenmeier, A. Fürst, A. Langemann, and G. Waldvogel, *Helv. Chim. Acta* **49**, 1601 (1966).

528. J. B. Siddall, A. B. Cross, and J. H. Fried, *J. Amer. Chem. Soc.* **88**, 862 (1966).

529. C. M. Williams, *Anat. Rec.* **120**, 743 (1954).

530. F. Hampshire and D. H. S. Horn, *Chem. Commun.* **1966**, 37.

531. D. H. S. Horn, E. J. Middleton, J. A. Wunderlich, and F. Hampshire, ref. 518.

532. P. Hocks and R. Wiechert, *Tetrahedron Lett.* **1966**, 2989.

533. H. Hoffmeister and H. F. Grützmacher, *Tetrahedron Lett.* **1966**, 401.

534. G. Hüppi and J. B. Siddall, *J. Amer. Chem. Soc.* **89**, 6790 (1967).

535. U. Kerb, R. Wiechert, A. Furlenmeier, and A. Fürst, *Tetrahedron Lett.* **1968**, 4277.

536. D. S. King and J. B. Siddall, *Nature* **221**, 955 (1969).

537. J. N. Kaplanis, M. G. Thompson, R. T. Yamamoto, W. E. Robbins, and S. J. Louloudes, *Steroids* **8**, 605 (1966).

538. M. N. Galbraith, D. H. S. Horn, J. A. Thomson, G. J. Neufeld, and R. J. Hackney, *J. Insect Physiol.* **15**, 1225 (1966).

539. J. B. Siddall, Chemical Aspects of Hormonal Interactions, in: *Chemical Ecology*, E. Sondheimer and J. B. Simeone (Eds.), p. 281, Academic Press, New York–London (1970).

540. A. A. Akhrem, I. S. Levina, and Yu. A. Titov, *Russ. Chem. Rev.* **40**, 760 (1971).

   H. Hikino and Y. Hikino, Arthropod Molting Hormones, in: *Fortschritte der Chemie organischer Naturstoffe*, Zechmeister (Ed.), Vol. 28, p. 257, Springer Verlag, Berlin–Heidelberg–New York (1970).

   F. Sorm, The Chemistry of Moulting Hormones, in: *Insect Hormones and Bioanalogues*, K. Slama, M. Romanuk, F. Sorm (Eds.), p. 303, Springer Verlag, Berlin–Heidelberg–New York (1974).

541. T. Takemoto, Y. Hikino, H. Jin, and H. Hikino, *J. Pharm. Soc. Jap.* **88**, 359 (1968).

542. K. Nakanishi, M. Koreeda, S. Sasaki, M. L. Chang, and H. Y. Hsu, *Chem. Commun.* **1966**, 915.

543. M. N. Galbraith and D. H. S. Horn, *Chem. Commun.* **1966**, 905.

544. S. Imai, T. Toyosato, M. Sakai, Y. Sato, S. Fujioka, E. Nurata, and M. Goto, *Chem. Pharm. Bull.* **17**, 335 (1969).

545. T. Takemoto, S. Ogawa, N. Nishimoto, S. Arihara, and K. Bue, *Yakugaku Zasshi* **87**, 1414 (1967).

546. H. Hikino and T. Takemoto, *Naturwissenschaften* **59**, 91 (1972).

   H. Hikino and Y. Hikino, Arthropod Molting Hormones, in: *Fortschritte der Chemie organischer Naturstoffe*, Zechmeister (Ed.), Vol. 28, p. 257, Springer Verlag, Berlin–Heidelberg–New York (1970).

   H. H. Rees, Ecdysones, in: *Aspects of Terpenoid Chemistry and Biochemistry*, T. W. Goodwin (Ed.), p. 180, Academic Press, London–New York (1971).

547. A. Butenandt and P. Karlson, *Z. Naturforsch.* **96**, 389 (1954).

548. P. Karlson, *Vitam. Horm. (N.Y.)* **14**, 227 (1956).

549. P. Karlson, *Ann. Sci. Nat. Zool. Biol. Anim.* **18**(11), 125 (1956).

550. P. Karlson, H. Hoffmeister, W. Hoppe, and R. Huber, *Justus Liebigs Ann. Chem.* **662**, 1 (1963).

551. M. Hori, *Steroids* **14**, 33 (1969).

552. K. Slama, Bioassays for Ecdysoids, in: *Insect Hormones and Bioanalogues*, K. Slama, M. Romanuk, F. Sorm (Eds.), p. 332, Springer Verlag, Berlin–Heidelberg–New York (1974).

553. P. Karlson, *Vitam. Horm. (N.Y.)* **14**, 227 (1956).

554. Y. Sato, M. Sakai, S. Imai, and S. Fuioka, *Appl. Entomol. Zool.* **3**, 49 (1968).

555. T. Ohtaki, R. D. Milkman, and C. M. Williams, *Proc. Nat. Acad. Sci. U.S.A.* **58**, 981 (1967).

556. G. B. Staal, *K. Ned. Akad. Wet.* **70**, 409 (1967).

557. C. M. Williams, *BioScience* **18**, 791 (1968).

558. W. E. Robbins, J. N. Kaplanis, M. J. Thompson, T. J. Shortino, C. F. Cohen, and S. C. Joyner, *Science* **161**, 1158 (1968).

559. A. Krishnakumaran and H. A. Schneiderman, *Nature* **220**, 601 (1968).

560. A. Krishnakumaran and H. A. Schneiderman, *Gen. Comp. Endocrinol.* **12**, 515 (1969).

561. A. Krishnakumaran and H. A. Schneiderman, *Biol. Bull.* (*Woods Hole, Mass.*) **139**, 520 (1970).

562. M. E. Lowe, H. D. S. Horn, and M. N. Galbraith, *Experientia* **24**, 518 (1968).

563. J. E. Wright, *Science* **163**, 390 (1969).

564. A. Krishnakumaran and H. A. Schneiderman, ref. 559.

565. M. Muftig, *Parasitology* **59**, 365 (1969).

566. M. Kobayashi, T. Takemoto, S. Ogawa, and N. Nishimoto, *J. Insect Physiol.* **13**, 1395 (1967).

567. C. M. Williams, *Biol. Bull* (*Woods Hole, Mass.*) **134**, 344 (1968).

568. H. A. Schneiderman, A. Krishnakumaran, P. J. Bryant, and F. Sehnal, ref 517.

569. A. Krishnakumaran, S. J. Berry, H. Oberlander, and H. A. Schneiderman, *J. Insect Physiol.* **13**, 1 (1967).

570. H. A. Schneiderman, Insect Hormones and Insect Control, in: *Insect Juvenile Hormones Chemistry and Action*, J. J. Menn, M. Beroza (Eds.), p. 3, Academic Press, New York-London (1972).

571. C. M. Williams and W. E. Robbins, *BioScience* **18**, 791, 797 (1968).

572. K. Slama, Effect of Ecdysoids in Other Animals and Plants, in: *Insect Hormones and Bioanalogues*, K. Slama, M. Romanuk, F. Sorm (Eds.), p. 367, Springer Verlag, Berlin-Heidelberg-New York (1974).

573. H. Hikino, S. Nabetani, K. Nomoto, T. Arai, T. Takemoto, T. Otaka, and M. Uchiyama, *J. Pharm. Soc. Jap.* **89**, 235 (1969)

574. W. J. Burdette and R. C. Richards, *Nature* **189**, 666 (1961).

575. H. Hikino et al., *Yakugaku Zasshi* **89**, 235 (1969).

576. S. Okui et al., *Chem. Pharm. Bull.* (*Tokyo*) **17**, 75 (1969).

577. H. Hikino and T. Takemoto, *Naturwissenschaften* **59**, 91 (1972).

578. H. Hikino and T. Takemoto, ref. 577.

579. H. Matsuda, K. Kawaba, and S. Yamamoto, *Nippon Yakurigaku Zasshi* **66**, 551 (1970).

580. H. Hikino and T. Takemoto, ref. 577.

581. G. R. Wyatt, Insect Hormones, in: *Biochemical Actions of Hormones*, G. Litwack (Ed.), Vol. 2, p. 400, Academic Press, New York-London (1972).

582. H. H. Rees, Ecdysones, in: *Aspects of Terpenoid Chemistry and Biochemistry*, T. W. Goodwin (Ed.), p. 200, Academic Press, London-New York (1971).

583. H. Hikino and Y. Hikino, Arthropod Molting Hormones, in: *Fortschritte der Chemie organischer Naturstoffe*, Zechmeister (Ed.), Vol. 28, p. 291, Springer Verlag, Berlin-Heidelberg-New York (1970).

584. D. H. S. Horn, The Ecdysones, in: *Naturally Occurring Insecticides*, M. Jacobson and D. G. Crosby (Eds.), p. 431, Marcel Dekker, New York (1971).

585. P. Karlson and C. Bode, *J. Insect Physiol.* **15**, 111 (1969).

586. H. H. Rees, ref. 582.

587. D. H. S. Horn, E. J. Middleton, J. A. Wunderlich, and F. Hampshire, *Chem. Commun.* **1966**, 339.

588. T. Takemoto, Y. Hikino, H. Hikino, S. Ogawa, and N. Nishimoto, *Tetrahedron* **25**, 1241 (1969).

589. T. Ohtaki, R. D. Milkman, and C. M. Williams, ref. 555.

590. T. Ohtaki and C. M. Williams, quoted in: C. M. Williams, *Hormonal Interactions between Plants and Insects*, in: *Chemical Ecology*, E. Sondheimer, J. B. Simeone (Eds.), p. 103, Academic Press, New York–London (1970).

591. M. Koreeda, K. Nakanishi, and M. Goto, *J. Amer. Chem. Soc.* **92**, 7512 (1970).

592. W. E. Robbins, J. N. Kaplanis, M. J. Thompson, T. J. Shortino, C. F. Cohen, and S. C. Joyner, *Science* **161**, 1158 (1968).

593. F. Sorm, Steroid Compounds Interfering with Ecdysis, in: *Insect Hormones and Bioanalogues*, K. Slama, M. Romanuk, and F. Sorm (Eds.), p. 325, Springer Verlag, Berlin–Heidelberg–New York (1974).

594. G. R. Wyatt, ref. 581.

  A. A. Akhrem, I. S. Levina, and Yu. A. Titov, ref. 540, p. 769.

  H. H. Rees, ref. 582, sec 214.

  D. H. S. Horn, ref. 584, pp. 441.

  F. Sorm, Relation between the Structure and Activity of Insect Moulting Hormones, ref. 593, p. 318.

595. P. Karlson, *J. S. Afr. Chem. Inst.* **22**, 41 (1969).

596. T. Ohtaki, R. D. Milkman, and C. M. Williams, *Biol. Bull.* **135**, 322 (1968).

597. P. Karlson and C. Bode, ref. 585.

598. E. Shaaya, *Z. Naturforsch.* **24B**, 718 (1969).

599. K. Nakanishi et al., *Chem. Commun.* **1966**, 915.

600. M. N. Galbraith and D. H. S. Horn, *Chem. Commun.* **1966**, 905.

601. J. Jizba, V. Herout, and F. Sorm, *Tetrahedron Lett.* **1967**, 1689.

601a. *Review Literature on Juvenile Hormones and Juvenoids.*

  C. E. Berkoff, The Chemistry and Biochemistry of Insect Hormones, *Quart. Rev. Chem. Soc.* **23**, 372–391 (1969).

  W. S. Bowers, Juvenile Hormones, in: *Naturally Occurring Insecticides*, M. Jacobson, D. G. Crosby (Eds.), pp. 307–332, Marcel Dekker, New York (1971).

  C. E. Dyte, Evolutionary Aspects of Insecticide Selectivity, *Proc. 5th Br. Insectic. Fungic. Conf.*, 393–397 (1969).

  M. T. El-Ibrashy, Insect Hormones and Analogues: Chemistry, Biology, and Insecticidal Potencies, *Z. Angew. Entomol.* **66**, 113–144 (1970).

  P. Ellis, Can Insect Hormones and Their Mimics Be Used to Control Pests?, *Pest Articles and News Summaries* (*A*) **14**, 329–342 (1968).

  L. I. Gilbert, Invertebrate Hormones, I. The Chemistry of Insect Hormones, *Proc. III. Int. Congr. Endocrin., Mexico*, 340–346 (1968).

  G. W. Irving, Jr., Agricultural Pest Control and the Environment, *Science* **168**, 1419–1424 (1970).

  E. F. Knipling, Alternate Methods of Controlling Insect Pests, *FDA Pap.* **3**, 16–24 (1969).

  J. J. Menn and M. Beroza, *Insect Juvenile Hormones Chemistry and Action*, pp. 1–341, Academic Press, New York–London (1972).

  A. Pfiffner, Juvenil Hormones, in: *Aspects of Terpenoid Chemistry and Biochemistry*, T. W. Goodwin (Ed.), pp. 95–135, Academic Press, London–New York (1971).

  H. Schmutterer, Biotechnische Verfahren in der Schädlingsbekämpfung–Möglichkeiten und Grenzen erläutert am Beispiel der Juvenilhormon-Analoge, *Mitteilungen der Biologischen Bundesanstalt*, Issue 151, 119–134 (1973).

  H. A. Schneiderman, The Strategy of Controlling Insect Pests with Growth Regulators, *Bull. Soc. Ent. Suisse* **44**, 141–149 (1971).

H. A. Schneiderman, A. Krishnakumaran, P. J. Bryant, and F. Sehnal, Endocrinological and Genetic Strategies in Insect Control, *Proc. Symp. Potent. Crop. Protect., N.Y. State Agric. Exp. Sta., Geneva*, 14–25 (1969).

J. B. Siddall, Chemical Aspects of Hormonal Interactions, in: *Chemical Ecology*, E. Sondheimer and J. B. Simeone (Eds.), pp. 281–306, Academic Press, New York-London (1970).

K. Slama, Insect Juvenile Hormone Analogs, *Ann. Rev. Biochem.* **40**, 1079 (1971).

K. Slama, M. Romanuk, and F. Sorm, *Insect Hormones and Bioanalogues*, pp. 1–477, Springer Verlag, Berlin-Heidelberg-New York (1974).

P. J. Thomas and P. L. Bhatnagar-Thomas, Use of a Juvenile Hormone Analogue as Insecticide for Pests of Stored Grain, *Nature* **219**, 949 (1968).

J. De Wilde, New Elements for Integration in Pest Control Schedules, *Meded. Rijksfac. Landbouwwet., Gent* **33**, 597–604 (1968).

J. De Wilde, The Present Status of Hormonal Insect Control, *EPPO Bull.* **1**, 17–23 (1971).

C. M. Williams and W. E. Robbins, Conference on Insect-Plant Interactions, *BioScience* **18**, 791–799 (1968).

C. M. Williams, Third Generation Pesticides, *Sci. Am.* **217**, 13–17 (1967).

C. M. Williams, Juvenile Hormone Insecticides, *Accademia Nazionale dei Lincei, Roma, Quaderno* **128**, 79–87 (1969).

C. M. Williams, Hormonal Interactions between Plants and Insects, in: *Chemical Ecology*, E. Sondheimer and J. B. Simeone (Eds.), pp. 103–132, Academic Press, New York-London (1970).

G. R. Wyatt, Insect Hormones, in: *Biochemical Actions of Hormones*, G. Litwack (Ed.), Vol. II, pp. 432–462, Academic Press, New York-London (1972).

Yu. S. Tsizin and A. A. Drabkina, The Juvenile Hormone of Insects and Its Analogues, *Russ. Chem. Rev.* **39**, 498–509 (1970).

602. C. M. Williams, *Nature* **178**, 212 (1956).

603. C. M. Williams, *Biol. Bull.* **124**, 355 (1963).

604. C. M. Williams, *Sci. Am.* **217**, 13 (1967).

605. P. Schmialek, *Z. Naturforsch.* **16B**, 461 (1961).

606. P. Schmialek, *Z. Naturforsch.* **18B**, 516 (1963).

607. W. S. Bowers, M. J. Thompson, and E. C. Uebel, *Life Sci.* **4**, 2323 (1965).

608. H. Röller, K. A. Dahm, C. C. Sweeley, and B. M. Trost, *Angew. Chem.* **79**, 190 (1967).

609. A. S. Meyer, H. A. Schneiderman, E. Hanzmann, and J. H. Ko, *Proc. Nat. Acad. Sci. U.S.A.* **60**, 853 (1968).

610. K. J. Judy, D. A. Schooley, L. L. Dunham, M. S. Hall, B. J. Bergot, and J. B. Siddall, *Proc. Nat. Acad. Sci. U.S.A.* **70**, 1509 (1973).

611. K. Slama and C. M. Williams, *Proc. Nat. Acad. Sci. U.S.A.* **54**, 411 (1965).

612. W. S. Bowers, H. M. Fales, M. J. Thompson, and E. C. Uebel, *Science* **154**, 1020 (1966).

613. V. Cerny, L. Dolejs, L. Labler, F. Sorm, and K. Slama, *Collect. Czech. Chem. Commun.* **32**, 3926 (1967).

614. W. S. Bowers, *Science* **161**, 895 (1968).

615. W. S. Bowers, *Science* **164**, 323 (1969).

616. K. Slama, *Ann. Rev. Biochem.* **40**, 1079 (1971).

617. K. Slama, Instruction on Some Biological Assays with Juvenoids, in: *Insect Hormones and Bioanalogues*, K. Slama, M. Romanuk, and F. Sorm (Eds.), p. 124, Springer Verlag, Berlin-Heidelberg-New York (1974).

618. G. B. Staal, Biological Activity and Bioassay of Juvenile Hormone Analogs, in: *Insect Juvenile Hormones Chemistry and Action*, J. J. Menn and M. Beroza (Eds.), p. 69, Academic Press, New York–London (1972).

619. W. S. Bowers, ref. 614.

620. L. I. Gilbert and H. A. Schneiderman, *Trans. Amer. Microsc. Soc.* 79, 38-(1960).

621. C. M. Williams, ref. 602.

622. L. I. Gilbert and H. A. Schneiderman, ref. 620.

623. H. A. Schneiderman, A. Krishnakumaran, V. G. Kulkarni, and L. Friedman, *J. Insect Physiol.* 11, 1641 (1965).

624. H. Röller and J. S. Bjerke, *Life Sci.* 4, 1617 (1965).

625. H. Röller, J. S. Bjerke, and W. H. McShan, *J. Insect Physiol.* 11, 1185 (1965).

626. V. B. Wigglesworth, *J. Insect Physiol.* 15, 73 (1969).

627. K. Slama, ref. 616.

628. M. Romanuk and K. Slama, Structure-activity Relationships, in: *Insect Hormones and Bioanalogues*, K. Slama, M. Romanuk, and F. Sorm (Eds.), p. 194, Springer Verlag, Berlin-Heidelberg–New York (1974).

629. K. Slama, M. Suchy, and F. Sorm, *Biol. Bull.* 134, 154 (1968).

630. F. Sehnal and V. J. A. Novak, *Acta Entomol. Bohemoslov.* 66, 137 (1969).

631. C. M. Williams, *Biol. Bull.* 116, 323 (1959).

632. A. Krishnakumaran and H. A. Schneiderman, *J. Insect Physiol.* 11, 1517 (1965).

633. H. A. Schneiderman and L. I. Gilbert, *Science* 143, 325 (1964).

634. W. S. Bowers, *Bull. W.H.O.* 44, 381 (1971).

635. L. M. Riddiford and C. M. Williams, *Proc. Nat. Acad. Sci. U.S.A.* 57, 595 (1967).

636. K. Slama and C. M. Williams, *Nature* 210, 329 (1966).

637. K. Slama, *BioScience* 18, 791 (1968).

638. P. Masner, K. Slama, and V. Landa, *J. Econ. Entomol.* 63, 706 (1970).

639. P. Masner, K. Slama, and V. Landa, *J. Embryol. Exp. Morphol.* 20, 25 (1968).

640. M. Slade and C. F. Wilkinson, *Science* 181, 672 (1973).

641. J. F. Crow, *Ann. Rev. Entomol.* 2, 227 (1957).

642. H. A. Schneiderman, *Mitt. Schweiz. Entomol. Ges.* 44, 141 (1971).

643. K. Slama, *Ann. Rev. Biochem.* 40, 1079 (1971).

644. D. C. Cerf and G. P. Georghiou, *Nature* 239, 401 (1972).

645. C. E. Dyte, *Nature* 238, 84 (1972).

646. G. B. Staal, *Bull. W.H.O.* 44, 39 (1971).

647. G. B. Staal, ref. 618.

648. C. A. Henrick, G. B. Staal and J. B. Siddall, *J. Agr. Food Chem.* 21, 355 (1973).

649. J. B. Davis, L. M. Jackman, P. T. Siddons and B. C. L. Weedon, *J. Chem. Soc.* 1966, 2154.

650. G. Pattenden and B. C. L. Weedon, *J. Chem. Soc.* 1968, 1984.

651. USP 3163669 (1964) ≡ DBP 1109671 (1958), BASF, Inv.: W. Stilz and H. Pommer; *C.A.* 56, 8571a (1962).

652. USP 3177226 (1965) ≡ DBP 1109671 (1968), BASF, Inv.: W. Stilz and H. Pommer; *C.A.* 56, 8571a (1962).

653. A. Franke, G. Mattern, and W. Traber, *Helv. Chim. Acta* 58, 268 (1975).

654. D. E. Leonard, *Ann. Rev. Entomol.* 19, 197 (1974).

655. F. M. Pallos and J. J. Menn, Synthesis and Activity of Juvenile Hormone Analogs, in:

Insect Juvenile Hormones Chemistry and Action, J. J. Menn, M. Beroza (Eds.), p. 303, Academic Press, New York–London (1972).

656. B. A. Pawson, F. Scheidl, and F. Vane, Environmental Stability of Juvenile Hormone Mimicking Agents, in: *Insect Juvenile Hormones Chemistry and Action*, J. J. Menn, M. Beroza (Eds.), p. 191, Academic Press, New York–London (1972).

657. R. W. Bagley and J. C. Bauernfeind, Field Experiences with Juvenile Hormone Mimics, in: *Insect Juvenile Hormones Chemistry and Action*, J. J. Menn and M. Beroza (Eds.), p. 113, Academic Press, New York–London (1972).

658. W. F. Chamberlain and D. E. Hopkins, *Ann. Entomol. Soc. Am.* **63**, 1363 (1970).

659. K. Novak and F. Sehnal, *Acta Entomol. Bohemoslov.* **70**, 20 (1973).

660. K. Novak and F. Sehnal, *Z. Angew. Entomol.* **73**, 312 (1973).

661. L. Varjas and F. Sehnal, *Entomol. Exp. Appl.* **16**, 115 (1973).

661a. W. S. Bowers, T. Ohta, J. S. Cleere, and P. A. Marsella, *Science* **193**, 542 (1976).

W. S. Bowers and R. Martinez-Padro, *Science* **197**, 1369 (1977).

661b. H. Rembold, Ch. Czoppelt, and G. K. Sharma, *Z. Naturforsch.* **34c**, 1261 (1979).

662. J. J. van Daalen, J. Meltzer, R. Mulder, and K..Wellinga, *Naturwissenschaften* **59**, 312 (1972).

663. J. D. Bijloo and R. Mulder, *Gewasbescherming* **5**(1), 13 (1974).

664. R. Mulder and M. H. Gijswijt, *Pestic. Sci.* **4**, 737 (1973).

665. L. C. Post and R. Mulder, *ACS Symposium Ser.* **2**, 136–144, Am. Chem. Soc., Washington (1974).

666. I. Ishaaya and J. E. Casida, *Pestic. Biochem. Physiol.* **4**, 484 (1974).

667. DOS 2123236 (1971; Neth. Prior. 1970), N. V. Philips' Gloeilampenfabriken, Inv.: K. Wellinga and R. Mulder; *C.A.* **76**, 85,578m (1972).

668. H.-H. Nölle, E. J. van Busschbach, and A. Verloop, 40. Deutsche Pflanzenschutztagung in Oldenburg i.O., 6.–10. Oktober 1975, *Wirkungsweise und Anwendung des Insektizids Diflubenzuron.*

669. R. L. Metcalf, P. Y. Lu, and S. Bowlus, *J. Agric. Food Chem.* **23**, 359 (1975).

670. U. Skatulla, *Anz. Schädlingskd. Pflanz. Umweltschutz* **48**, 17 (1975).

671. H. Holst, *Z. Pflanzenkr. Pflanzenschutz* **81**, 1 (1975).

671a. *Review Literature on Chemosterilants.* A. B. Borkovec, Advances in Pest Control Research, Vol. VII, pp. 1–143, Interscience, New York (1966).

M. A. Bulyginskaya and M. D. Vronskikh, Insect chemoserilants, *Russ. Chem. Rev.* **39**, 964–976 (1970).

J. M. Franz and A. Krieg, *Biologische Schädlingsbekämpfung*, p. 1–209, Verlag Paul Parey, Berlin (1972).

R. Heitefuß, *Pflanzenschutz, Grundlagen der praktischen Phytomedizin*, p. 1–270, Georg Thieme Verlag, Stuttgart (1975).

G. E. La Brecque and C. N. Smith, *Principles of Insect Chemosterilization*, p. 1–354, North Holland, Amsterdam (1968).

M. Stüben, Chemosterilantien, *Mitt. Biol. Bundesanstalt, Berlin*, Issue **133**, 1–84 (1969).

R. Wegler, *Chemie der Pflanzenschutz- und Schädlingsbekämpfungsmittel*, Vol. 1, p. 475, Springer Verlag, Berlin–Heidelberg–New York (1970).

672. E. F. Knipling, *J. Econ. Entomol.* **48**, 459 (1955).

E. F. Knipling, *Science* **130**, 902 (1959).

673. E. F. Knipling, *J. Econ. Entomol.* **53**, 415 (1960).

674. A. H. Baumhover, A. J. Graham, B. A. Bitter, E. D. Hopkins, W. D. New, F. H. Dudley, and R. C. Bushland, *J. Econ. Entomol.* **48**, 462 (1955).

675. G. C. La Brecque, P. H. Adcock, and C. N. Smith, *J. Econ. Entomol.* **53**, 802 (1960).

676. USP 2858306 (1957), Olin Mathieson Chemical Corp., Inv.: R. F. W. Rätz and C. J. Grundmann: *C.A.* **54**, 1543h (1960).

677. K. R. S. Ascher, J. Meisner, and S. Nissim, *World Rev. Pest Control* 7(2), 84 (1967).

678. W. Klassen and T. H. Chang, *Science* **154**, 920 (1966).

679. J. W. Hayes, Jr., *Bull. W.H.O.* **31**, 721 (1964).

679a. *Review Literature on Attractants (Pheromones).* M. C. Birch, *Pheromones*, in: *Frontiers of Biology*, Vol. 32, p. 495, North Holland, Amsterdam–London (1974).

M. Boneß, Insektenpheromone und ihre Anwendungsmöglichkeiten, *Naturwiss. Rundsch.* **26**, 515-522 (1973).

K. Eiter, Insektensexuallockstoffe, in: *Fortschritte der Chemie organischer Naturstoffe*, L. Zechmeister (Ed.), Vol. 28, pp. 204-255, Springer Verlag, Berlin–Heidelberg–New York (1970).

K. Eiter, Insekten-Sexuallockstoffe, in: *Chemie der Pflanzenschutz- und Schädlingsbekämpfungsmittel*, R. Wegler (Ed.), Vol. 1, pp. 497-522, Springer Verlag, Berlin–Heidelberg–New York (1970).

J. M. Franz and A. Krieg, *Biologische Schädlingsbekämpfung*, pp. 1-209, Verlag Paul Parey, Berlin (1972).

R. Heitefuß, *Pflanzenschutz, Grundlagen der praktischen Phytomedizin*, pp. 1-270, Georg Thieme Verlag, Stuttgart (1975).

M. Jacobson, *Insect Sex Pheromones*, pp. 1-382, Academic Press, New York–London (1972).

J. G. MacConnel and R. M. Silverstein, Neue Ergebnisse von Insektenpheromonen, *Angew. Chem.* **85**, 647 (1973).

K. Mayer, Die Verwendung von Lockstoffen in der Schädlingsbekämpfung, *Gesunde Pflanz.* **20**, 179 (1968).

O. Vostrowsky, H. J. Bestmann, and E. Priesner, Struktur und Aktivität von Pheromonen, *Nachr. Chem. Tech.* **21**, 501 (1973).

680. P. Karlson and A. Butenandt, *Ann. Rev. Entomol.* **4**, 39 (1959).

681. P. Karlson and M. Lüscher, *Naturwissenschaften* **46**, 63 (1959).

682. D. Schneider, *Ind.-Elektron. Forsch. Fertigung* **5**, 3 (1955).

683. D. Schneider, *Z. Vergl. Physiol.* **40**, 8 (1957).

684. D. Schneider, *Experientia* **31**, 89 (1957).

685. A. Butenandt, R. Beckmann, D. Stamm, and E. Hecker, *Z. Naturforsch.* **14B**, 283 (1959).

686. A. Butenandt, E. Hecker, M. Hopp, and W. Koch, *Justus Liebigs Ann. Chem.* **658**, 39 (1962).

687. E. Truscheit and K. Eiter, *Justus Liebigs Ann. Chem.* **658**, 65 (1962).

688. M. Jacobson, Sexpheromones in insect control, in: *Insect Sex Pheromones*, Chap. XIII, pp. 277-291, Academic Press, New York–London (1972).

689. M. Beroza and E. F. Knipling, *Science* **177**, 19 (1972).

690. H. H. Toba, N. Green, A. N. Kishaba, M. Jacobson, and J. W. Debolt, *J. Econ. Entomol.* **63**, 1048 (1970).

691. W. L. Roelofs and A. Comeau, *Nature* **220**, 600 (1968).

W. L. Roelofs, R. J. Bartell, A. S. Hill, R. T. Carde, and L. H. Waters, *J. Econ. Entomol.* **65**, 1276 (1972).

692. J. H. Tumlinson, D. D. Hardee, R. C. Gueldner, A. C. Thompson, P. A. Hedin, and J. P. Minyard, *Science* **166,** 1010 (1969).

693. M. Boneß, *Naturwiss. Rundsch.* **26,** 515 (1973).

694. H. Z. Levinson, *Umschau* **71,** 945 (1971).

695. J. H. Tumlinson, D. D. Hardee, R. C. Gueldner, A. C. Thompson, P. A. Hedin, and J. P. Minyard, *J. Org. Chem.* **36,** 2616 (1971).

696. K. Eiter, Insekten-Sexuallockstoffe, in: *Chemie der Pflanzenschutz- und Schädlingsbekämpfungsmittel,* R. Wegler (Ed.), Vol. 1, pp. 497–522, Springer Verlag, Berlin–Heidelberg, New York (1970).

697. D. A. Carlson, M. S. Mayer, D. L. Silhacek, J. D. James, M. Beroza, and B. A. Bierl, *Science* **174,** 76 (1971).

698. B. A. Bierl, M. Beroza, and C. W. Collier, *Science* **170,** 87 (1970).

699. N. Green, M. Jacobson, and J. C. Keller, *Experientia* **25,** 682 (1969).

700. P. A. Maddison et al., *J. Econ. Entomol.* **66,** 591 (1973), DOS 2123989 (1971; US. Prior. 1970), Stauffer Chemical Co., Inv.: K. S. Shim and D. J. Martin; *C.A.* **76,** 45,812b (1972).

701. M. Beroza and M. Jacobson, *World Rev. Pest Control* **2,** 591 (1963).

702. M. Beroza and R. Sarmiento, *J. Econ. Entomol.* **59,** 1295 (1966).

703. R. Zurflüh, L. L. Dunham, V. L. Spain, and J. B. Siddall, *J. Amer. Chem. Soc.* **92,** 425 (1970).

703a. *Review Literature on Insect Repellents.* K. H. Büchel, Insekten-Repellents, in: *Chemie der Pflanzenschutz- und Schädlingsbekämpfungsmittel,* R. Wegler (Ed.), Vol. 1, pp. 487–496, Springer Verlag, Berlin–Heidelberg–New York (1970).

K. Mayer, *Pharmazie* **7,** 150–217 (1952).

G. E. Utzinger, *Angew. Chem.* **63,** 430 (1951).

704. R. R. Painter, in: *Pest Control,* W. W. Kilgore, R. L. Doutt (Eds.), p. 267, Academic Press, New York (1967).

705. USP 2408389 (1944), USA by Secy. of Agriculture, Inv.: S. I. Gertler; *C.A.* **41,** 559b (1947).

706. I. H. Gilbert, H. K. Gouck, and C. N. Smith, *J. Econ. Entomol.* **48,** 741 (1955).

707. P. Granett and H. L. Haynes, *J. Econ. Entomol.* **38,** 671 (1945).

708. USP 2138540 (1936), Kilgore Development Co., Inv.: J. H. Ford; *C.A.* **33,** 1759[1] (1939).

709. USP 2293256 (1938), National Carbon Co., Inv.: P. Granett; *C.A.* **37,** 988[7] (1943).

710. B. V. Travis and C. N. Smith, *J. Econ. Entomol.* **44,** 428 (1951).

711. L. D. Goodhue and C. Linnaid, *J. Econ. Entomol.* **45,** 133 (1952).

712. L. D. Goodhue and R. E. Stansbery, *J. Econ. Entomol.* **46,** 982 (1953).

712a. *Review Literature on Biological Methods.* H. D. Burges and N. W. Hussey, *Microbial Control of Insects and Mites,* pp. 1–861, Academic Press, London (1971).

J. M. Franz and A. Krieg, *Biologische Schädlingsbekämpfung,* pp. 1–208, Verlag Paul Parey, Berlin–Hamburg (1972).

K. S. Hagen and J. M. Franz, A History of Biological Control, in: *History of Entomology,* pp. 433–476, Ann. Rev. Inc., Palo Alto/USA (1973).

A. Haisch, Das Autozidverfahren, eine Möglichkeit des biologischen Pflanzenschutzes, *Z. Pflanzenkr. Pflanzenschutz* **75,** 257 (1968).

C. M. Ignoffo, K. D. Biever, and D. Hostetter, Living Insecticides, *Chem. Technol.* **1975,** 396.

E. F. Knipling, The Potential Role of Sterility for Pest Control, in: *Principles of Insect*

Chemosterilisation, G. C. La Brecque and C. N. Smith (Eds.), pp. 7–40, North Holland, Amsterdam (1968).

A. Krieg, Neues über Bacillus thuringiensis und seine Anwendung, *Mitt. Biol. Bundesanst. Berlin*, Issue 125 (1967).

H. Laven, Genetische Methoden zur Schädlingsbekämpfung, *Anz. Schädlingskd.* **41**, 1 (1968).

K. Maramorosch, *Insect viruses*, p. 192, Springer Verlag, Berlin–Heidelberg–New York (1968).

E. Müller-Kögler, *Pilzkrankheiten bei Insekten*, p. 444, Paul Parey Verlag, Berlin–Hamburg (1965).

E. A. Steinhaus, *Insect Pathologie; An Advanced Treatise*, Vol. I and II, p. 661, 689, Academic Press, New York–London (1963).

M. Stüben, Chemosterilantien, *Mitt. Biol. Bundesanst. Berlin*, Issue 133, pp. 1–84 (1969).

H. L. Sweetman, *The Principles of Biological Control–Interrelation of Hosts and Pests and Utilization in Regulation of Animal and Plant Populations*, pp. 1–560, Wm. C. Brown, Dubuque, Iowa (1958).

R. Wegler, *Chemie der Pflanzenschutz- und Schädlingsbekämpfungsmittel*, Vol. 1, p. 31, Springer Verlag, Berlin–Heidelberg–New York (1970).

713. H. L. Sweetman, *The Principles of Biological Control–Interrelation of Hosts and Pests and Utilization in Regulation of Animal and Plant Populations*, pp. 1–560, Wm. C. Brown, Dubuque, Iowa (1958).

714. J. M. Franz and A. Krieg, *Biologische Schädlingsbekämpfung*, pp. 1–208, Verlag Paul Parey, Berlin–Hamburg (1972).

715. C. M. Ignoffo, K. D. Biever, and D. Hostetter, Living Insecticides, *Chem. Technol.* **1975**, 396.

716. J. M. Franz and A. Krieg, ref. 714, p. 59.

717. A. H. Baumhover, A. J. Graham, B. A. Bitter, E. D. Hopkins, W. D. New, F. H. Dudley, and R. C. Bushland, *J. Econ. Entomol.* **48**, 462 (1955).

718. H. Laven, Genetische Methoden zur Schädlingsbekämpfung, *Anz. Schädlingskd.* **41**, 1 (1968).

719. H. Laven, ref. 718, p. 2.

720. F. Greer, C. M. Ignoffo, and R. F. Anderson, The First Viral Pesticide, *Chem. Technol.* **1971**, 342.

SECTION 2.9. MOLLUSCICIDES

721. *Reviews*

W. Foerst, *Ullmanns Encyklopädie der technischen Chemie*, 3rd. Ed., Vol. 15, pp. 140–141, Suppl. Vol., p. 559, Verlag Urban & Schwarzenberg, München–Berlin (1964) and 1970.

F. D. Judge, *J. Econ. Entomol.* **62**, 1393 (1969).

W. Perkow, *Wirksubstanzen der Pflanzenschutz- und Schädlingsbekämpfungsmittel*, Verlag Paul Parey, Berlin–Hamburg (1971/1974).

H. Martin and C. R. Worthing, *Pesticide Manual*, 4th. ed., British Crop Protection Council, Worcester 1974.

W. T. Thomson, *Agricultural Chemicals*. Book 1: *Insecticides, Acaricides, and Ovicides*, 1975 revision, Thomson Publications, Indianapolis, Indiana (1975).

Farm Chemicals Handbook 1975, Meister Publishing, Willoughby, Ohio (1975).

721a. *Naturwiss. Rundsch.* **23**, 113 (1970).

SECTION 2.10. RODENTICIDES

721a. *Reviews*

E. Enders, in: R. Wegler, *Chemie der Pflanzenschutz- und Schädlingsbekämpfungsmittel*, Vol. 1, Springer Verlag, Berlin–Heidelberg–New York (1970).

W. Perkow, *Wirksubstanzen der Pflanzenschutz- und Schädlingsbekämpfungsmittel*, Verlag Paul Parey, Berlin–Hamburg (1971/1974).

H. Martin and C. R. Worthing, *Pesticide Manual*, 4th. ed., British Crop Protection Council, Worcester (1974).

Farm Chemicals Handbook 1975, Meister Publishing, Willoughby, Ohio (1975).

G. Hermann, *Bundesgesundheitsblatt* **12**, 129 (1969).

G. Hermann, *7. Arbeitstagung der Schweizerischen Gesellschaft für Lebensmittelhygiene (SGLH) vom 8.11.1974*, Issue 2, pp. 37–45, Schriftenreihe der SGLH, Zürich (1974).

N. G. Gratz, *Bull. W.H.O.* **48**, 469 (1973).

722. DOS 2424806 (1974, Brit. Prior. 1973), Ward, Blenkinsop & Co. Ltd., Inv.: M. R. Hadler and R. S. Shadbolt; *C.A.* s. Ms. p. 47.

723. M. R. Hadler and R. S. Shadbolt, *Nature* **253**, 275 (1975).

724. DOS 2310636 (1973, Brit. Prior. 1972), Ward, Blenkinsop & Co. Ltd., Inv.: M. R. Hadler; *C.A.* **80**, 23,550 (1974).

725. Anonym, *Chemische Industrie* **26**, 176 (1974).

726. I. Tokumitsu et al., *Botyu-Kagaku* **38**, 202 (1973).

T. Kusano et al., *Botyu-Kagaku* **39**, 70 (1974).

727. DOS 2009964 (1970, US Prior. 1969), Billiton-M & T Chemische Industrie N.V., Inv.: M. Schwarcz, C. B. Beiter, S. Kaplin, O. E. Loeffler, M. E. Martino, and S. Damle; *C.A.* **73**, 120,748d (1970).

728. C. B. Beiter, M. Schwarcz, and G. Crabtree, *Soap Chem. Spec.* **46**, 38 (1970).

729. DOS 2305196 (1973, US Prior. 1972) ≡ Fr. Appl. 2176069 (1973), Rohm & Haas Co., Inv.: J. E. Ware, E. E. Kolbourne, and D. L. Peardon; *C.A.* **80**, 82,687t (1974).

730. D. L. Peardon, *Pest Control* **42**, 14 (1974).

Chapter 3. Fungicides and Bactericides

731. J. G. Horsfall, *Principles of Fungicidal Action*, Chronica Botanica Company, Waltham, Mass. (1956).

732. A. de Bary, *Untersuchungen über die Brandpilze und die durch sie verursachten Krankheiten der Pflanze*, Berlin 1853, cf. G. W. Keitt, History of Plant Pathology, in: *Plant Pathology*, I, pp. 61–67, Academic Press, New York (1959).

733. E. Müller and W. Loeffler, *Mykologie*, Georg Thieme Verlag, Stuttgart (1971).

734. G. M. Hoffmann, F. Nienhaus, F. Schönbeck, H. C. Weltzien, and H. Wilbert, *Lehrbuch der Phytomedizin*, Paul Parey Verlag, Berlin–Hamburg (1976).

735. H. H. Cramer, Pflanzenschutz und Welternte, *Pflanzenschutz Nachr.* **20**(1) (1967).

736. H. Martin, *The Scientific Principles of Crop Protection*, 5th. ed., Arnold, London (1964).

737. A. Millardet, *J. Agric. Prat.* **49**, 513, 801 (1885).

738. E. Gäumann, *Die Pilze*, 2nd. ed., Birkhäuser Verlag, Basel-Stuttgart (1964).

739. G. N. Agrios, *Plant Pathology*, Academic Press, New York–London (1969).

740. H. G. Aach, Die Viren, in: *Handbuch der Biologie,* Vol. I/1, pp. 287–352, Akademische Verlagsgesellschaft Athenaion, Konstanz-Stuttgart (1965).

741. H. A. Kirchner, *Grundriß der Phytopathologie und des Pflanzenschutzes*, p. 22, VEB Gustav Fischer Verlag, Jena (1967).

742. H. Jakob, *Untersuchungen zur Thermotherapie von Steinobstvirosen*, Acta Phytomedica, Paul Parey Verlag, Berlin–Hamburg (1974).

743. H. H. Cramer, Pflanzenschutz und Welternte, *Pflanzenschutz Nachr.* **20**(1), 272, 228 (1967).

744. F. Grossmann, *World Rev. Pest Control* **7**, 167 (1968).

745. J. Kuć, *World Rev. Pest Control* **7**, 42 (1968).

746. F. Grossmann, *World Rev. Pest Control* **7**, 176 (1968).

747. J. Barner and K. Röder, *Z. Pflanzenkr. (Pflanzenpathol.) Pflanzenschutz, Sonderh.* **71**, 210 (1964).

748. H. F. Emden and K. E. Cockshall, *J. Exp. Botan.* **18**, 707 (1967).

749. H. H. Mayr and E. Preschy, *Z. Acker. Pflanzenbau* **118**, 109 (1963).

750. R. Heitfuss and W. H. Fuchs, *Phytopathol. Z.* **37**, 348 (1960).

751. A. Kaars Sijpesteijn, *World Rev. Pest Control* **8**, 138 (1969).

752. M. Gordon, *Can. J. Chem.* **51**, 748 (1973).

753. J. G. Horsfall and G. A. Zentmyer, *Phytopathology* **32**, 186 (1942).

754. F. Grossmann, *Phytopathol. Z.* **45**, 139 (1962).

755. A. Matta, *Riv. Patol. Veg. (3)*, **3**, 99 (1963).

756. R. H. Biffen, *J. Agric. Sci.* **1**, 4 (1905); **2**, 109 (1907).

757. T. Römer, W. J. Fuchs, and K. Isenbeck, *Kühn Arch.* **45**, 427 (1938).

758. E. H. Coons, *Phytopathology* **27**, 622 (1937).

759. M. B. Waite, *Yearb. US. Dep. Agric.* **1925**, 453.

760. W. F. Hanna, Pl. Prot. Conf. 1956, p. 31, Butterworths Scientific Publications, London (1957).

761. J. v. Liebig, *Die Chemie in ihrer Anwendung auf Agricultur u. Physiologie*, 9th. ed., F. Vieweg & Sohn, Braunschweig (1876).

762. G. T. Spints, *J. Agric. Sci.* **5**, 231 (1913).

763. G. Gassner and K. Hassebrauk, *Phytopath. Z.* **3**, 535 (1931).

764. J. Kuć, *World Rev. Pest Control* **7**, 42 (1968).

765. W. F. Bewley and S. G. Paine, *Rep. Exp. Res. Stat. Cheshunt*, p. 22 (1919).

766. H. Müller and E. Molz, *Fühlings Landw. Z.* **66**, 42 (1917).

767. W. B. Tisdale, *J. Agric. Res.* **24**, 55 (1923).

768. L. R. Jones, J. C. Walker, and J. Monteith, *J. Agric. Res.* **30**, 1027 (1925).

769. W. F. Bewley, *Bull. Minist. Agric.*, p. 77 (1939).

770. T. Small, *Ann. Appl. Biol.* **17**, 71 (1930).

771. H. H. McKinney, *J. Agric. Res.* **26**, 195 (1923).

772. G. M. Reed, *Phytopathology* **14**, 437 (1924).

773. C. Rumm, *Ber.* **13**, 189 (1895).

774. W. T. Swingle, *Bull. US-Dep. Agric. Div. Veg. Phys. Path.* **9**, 1–37 (1896).

775. D. C. Erwin, S. D. Tsai, and R. A. Khan, *Proceed. Cotton Disease Council* (1975).

776. R. W. Marsh, *Systemic Fungicides*, John Wiley, New York; *Farm. Chem.* **137**, 29 (1974).

777. E. Müller and W. Loeffler, *Mykologie*, p. 98, Georg Thieme Verlag, Stuttgart (1971).

778. W. Forsyth, *A Treatise on the Culture and Management of Fruit Trees*, 2nd ed., p. 358, Nichols and Son, London (1803).

779. B. Prevost, *Memoire sur la cause immediate de la Carie ou Charbon des Blés*, Bernard, Paris (1807).

780. L. R. Streeter, *Tech. Bull. N.Y. St. Agric. Exp. Sta. Nr.* 125 (1927).

781. A. S. McDaniel, *Ind. Eng. Chem.* **26**, 340 (1934).

782. G. Rupprecht, *Angew. Bot.* **3**, 253 (1921).

783. F. M. Turell, *Plant Physiol.* **25**, 13 (1950).

784. *Gdnrs'. Cron.*, p. 419 (1852).

785. DRP 399137 (1920), Farbenfabriken vorm. Friedrich Bayer & Co., Inv.: H. Karstens and C. Hansen; *Chem. Zentralblatt* **1924 II**, 1396.

786. A. F. Parker-Rhodes, *Ann. Appl. Biol.* **29**, 136 (1942).

787. J. G. Horsfall, *Principles of Fungicidal Action*, Chronica Botanica Co., p. 166, Waltham, Mass. (1956).

788. A. Wöber, *Z. Pflanzenkrankh.* **30**, 51 (1920).

789. A. Millardet and W. Gayon, *J. Agric. Prat.* **49**, 513, 707 (1885); and **51**, 698, 728, 765 (1887).

790. E. Masson, *J. Agric. Prat.* **51**, 814 (1887).

791. W. F. Bewley, *J. Minist. Agric.* **28**, 653 (1921).

792. H. Martin, *Ann. Appl. Biol.* **20**, 342 (1933).

793. J. G. Horsfall and J. M. Hamilton, *Phytopathology* **25**, 21 (1935).

794. USP 2051910 (1933) ≡ Fr. P. 778426 (1934), California Spray Chemical Corporation; *Chem. Zentralblatt* **1935 II**, 3002.

795. H. Martin, *Guide to the Chemicals used in Crop Protection*, 4th ed., Canada Department of Agriculture, London, Ontario 1961, with suppl. by E. Y. Spencer (1964).

796. B. Prevost, *Memoire sur la cause immediate de la Carie ou Charbon des Blés*, Bernard, Paris (1807).

797. J. Branas and J. Dulac, *C.R. Acad. Sci. Paris* **197**, 938, 1245 (1933).

798. B. Delage, *Chim. Ind. (Paris)* **27**, 853 (1932).

799. R. L. Wain and E. H. Wilkinson, *Ann. Appl. Biol.* **30**, 379 (1943).

800. H. Martin, R. L. Wain, and E. H. Wilkinson, *Ann. Appl. Biol.* **29**, 412 (1942).

801. B. F. Lutman, *Phytopathology* **12**, 305 (1922).

802. M. H. Moore, H. B. S. Montgomery, and H. Shaw, *Rep. E. Malling, Res. Sta.* **1936**, 259.

803. J. W. Roberts and L. Pierce, *Phytopathology* **22**, 415 (1932).

804. L. Hiltner, *Prakt. Blätter f. Pflanzenbau u. Pflanzenschutz* **18**, 65 (1915).

805. H. Martin, *Guide to the Chemicals used in Crop Protection*, 4th ed., Canada Department of Agriculture, London, Ontario 1961, with suppl. by E. Y. Spencer (1964).

806. H. Schör and H. Klös, unpublished.

807. DAS 1003733 (1953), Farbenfabriken Bayer AG, Inv.: H. Klös, W. Schacht; *C.A.* **54**, 4390e (1960).

808. K. H. Slotta and K. R. Jacobi, *J. Prakt. Chem.* 2 **120**, 272 (1929).

809. D. E. H. Frear, *Pesticide Index*, College Science Publishers, 3rd State College, Pennsylvania (1965).

810. USP 2452595 (1948), 2481438 (1949), Du Pont, Inv.: D. F. Mowery; *C.A.* **43**, 1805e (1949); **44**, 3666f (1950).

811. USP 2598562 (1951), Velsicol Corp., Inv.: M. Kleimann; *C.A.* **46**, 7278a (1952).

812. W. Schöller, W. Schrauth, and W. Essers, *Ber.* **46**, 2866 (1913).

813. W. Bonrath, *Nachr. Schädlingsbek.* **10**, 23 (1935).

814. W. T. Thomson, *Agriculture Chemicals, IV – Fungicides*, Thomson Publications, Indianapolis, Indiana (1973).

815. E. K. Okaisabor, *Phytopathologische Zeitschrift* **69**, 125 (1970).

816. Swiss P. 287787 (1950), Sandoz Ltd., *C.A.* **48**, 6070e (1954).

817. Swiss P. 451934 (1963), Sandoz Ltd., *C.A.* **63**, 18152c (1965).

818. Swiss P. 234851 (1910), 312281 (1914), Farbenfabriken vorm. Friedrich Bayer & Co.; *Frdl.* **10**, 1272; **13**, 983.

819. G. Wesenberg, *Nachr. Schädlingsbekämpf.* **13**, 103 (1938).

820. I. Riehm, *Mitt. aus d. Kaiserl. Anstalt f. Land- u. Forstwirtschaft*, Issue 14, p. 8, April (1913).

821. D. Hunter et al., *Quart. J. Med.* **9**, 193 (1940).

822. O. Dimroth, *Justus Liebigs Ann. Chem.* **446**, 148 (1926).

823. A. N. Nesmejanow, *Ber.* **62**, 1010 (1929).

824. O. A. Reutow, *Angew. Chem.* **72**, 198 (1960).

825. F. L. Howard, S. B. Locke, and H. L. Keil, *Proc. Am. Soc. Hortic. Sci.* **45**, 131 (1947).

826. D. E. H. Frear, *Pesticide Index*, pp. 315–318, College Science Publishers, State College, Pennsylvania (1969).

827. R. H. Daines, *Phytopathology* **26**, 90 (1936).

828. J. P. Booer, *Ann. Appl. Biol.* **31**, 340 (1944).

829. W. T. Thomson, *Agricultural Chemicals, IV – Fungicides*, Thomson Publications, Indianapolis, Indiana (1973).

830. W. T. Thomson, *Agricultural Chemicals, IV – Fungicides*, Thomson Publications, Indianapolis, Indiana (1973).

831. C. L. Mason, *Phytopathology* **38**, 740 (1948).

832. J. G. A. Luyten and G. J. M. van der Kerk, *Investigations in the Field of Organotin Chemistry*, Tin Research Institute, Greenford/England (1955).

833. W. R. Neumann, *Angew. Chem.* **75**, 225 (1963).

834. DAS 1084722 (1959), Farbwerke Hoechst AG, Inv.: C. Dörfelt and H. Gelbert; *C.A.* **55**, 20,962a (1961).

835. W. T. Thomson, *Agricultural Chemicals, IV – Fungicides*, Thomson Publications, Indianapolis, Indiana (1973).

836. K. Härtel, *Angew. Chem.* **70**, 135 (1958); **76**, 304 (1964).

837. W. T. Thomson, *Agricultural Chemicals, IV – Fungicides*, Thomson Publications, Indianapolis, Indiana (1973).

838. DAS 1215709 (1964), C. H. Boehringer Sohn, Inv.: L. Schröder, K. Thomas, and D. Jerchel; *C.A.* **65**, 5489f (1966).

839. DAS 1219028 (1963), C. H. Boehringer Sohn, Inv.: L. Schröder, K. Thomas, and D. Jerchel; *Chem. Zentralblatt* **1967**, 17-2976.

840. P. Schicke et al., *Pflanzenschutzberichte (Wien)* **38**, 189 (1968).

841. DAS 1227905 (1963), Neth. Appl. 6414757 (1965), C. H. Boehringer Sohn, Inv.: L. Schröder, K. Thomas, and D. Jerchel; *C.A.* **64**, 757f (1966).

842. A. H. McIntosh, *Ann. Appl. Biol.* **69**, 43 (1971).

843. R. R. Ison et al., *Pest. Sci.* **2**, 152 (1971).

844. A. J. Quick and R. Adams, *J. Amer. Chem. Soc.* **44**, 809 (1922).

845. A. v. Baeyer, *Justus Liebigs Ann. Chem.* **107**, 279 (1858).

846. DAS 1079886 (1958), Farbenfabriken Bayer AG, Inv.: P. E. Frohberger and E. Urbschat; *C.A.* **55**, 19121i (1961).

847. F. K. Beilstein, *Handbuch der Organ. Chemie*, 4th ed., Vol. 4, II 574, III 1797, Springer Verlag, Göttingen–Berlin–Heidelberg (1942) resp. (1963).

848. F. Grewe, Rückblick auf 25 Jahre Fungizidforschung, *Pflanzenschutz Nachr.* 18, 45 (1965).

849. C. Fest and K. J. Schmidt, *The Chemistry of Organophosphorus Pesticides*, Springer Verlag, Berlin–Heidelberg–New York (1973).

850. H. Scheinpflug and H. F. Jung, *Pflanzenschutz-Nachr. Bayer* 21, 79 (1968).

851. Jap. P. 72/33 340 (1969) Kumiai Chem. Co., Inv.: H. Sugiyama, K. Takita, and H. Ito; *C.A.* 78, 3952h (1973).

852. *Japan Chem. Week* 24.4.69, p. 4, see also *C.A.* 77, 122,880x (1972).

853. DAS 1213664 (1963); Neth. Appl. 6412154 (1965), Farbenfabriken Bayer AG, Inv.: H. Scheinpflug, H. F. Jung, and G. Schrader; *C.A.* 64, 1296f (1966).

854. W. T. Thomson, *Agricultural Chemicals, IV–Fungicides*, Indianapolis, Indiana (1973).

855. Neth. P. 6717383 (1967) ≡ Fr. P. 1560374 (1969), Sumitomo Kagaku (Japan); *C.A.* 72, 90,043e (1970).

856. M. Kado, *Experimental Approaches of Pesticide Metabolism, Degradation and Mode of Action*, US–Japan Seminar 16–19 August 1967, p. 121, Nikko, Japan.

857. Belg. P. 686048 (1967, DB Prior. 1965) ≡ Neth. Appl. 6611860 (1967), Farbenfabriken Bayer AG, Inv.: G. Schrader, K. Mannes and H. Scheinpflug; *C.A.* 68, 49,297v (1968).

858. DOS 1545790 (1965) ≡ Neth. Appl. 6602131 (1966), Farbwerke Hoechst AG, Inv.: O. Scherer and H. Mildenberger; *C.A.* 66, 37,946b (1967).

859. 21st. Intern. Symposium over Fytopharmacie u. Fytiatrie, Gent, 6.5.1969.

860. DAS 1209355 (1958); Brit. P. 888686 (1962), N.V. Philips Gloeillampenfabriken, Inv.: M. J. Koopmans, J. Meltzer, H. O. Huisman, B. G. van den Bos, and K. Wellinga; *C.A.* 60, 2951h (1964).

861. Brit. P. 1034493 (1964); Belg. P. 661891 (1965), Dow Chemical Co., Inv.: H. Tolkmith, and H. O. Senkbeil; *C.A.* 64, 19,489g (1966).

862. H. Tolkmith and D. R. Musell, *World Rev. Pest Control* 6, 74 (1967).

863. DAS 1072245 (1957); Brit. P. 822476 (1959). Farbenfabriken Bayer AG, Inv.: G. Schrader; *C.A.* 54, 7559a (1960).

864. M. Kado, *Experimental Approaches of Pesticide Metabolism, Degradation and Mode of Action*, US–Japan Seminar 16.–19.8.67, p. 121, Nikko, Japan.

864a. D. J. Williams, B. G. W. Beach, D. Horrière, and G. Marechal, *British Crop Protection Conference–Pests and Diseases Proceedings*, Vol. 2, p. 565 (1977).

865. G. B. Sanford, *Phytopathology* 16, 525 (1926).

866. W. A. Millard and C. B. Taylor, *Ann. Appl. Biol.* 14, 202 (1927).

867. J. E. Machacek, *McDonald College (McGill University) Techn. Bull.* 7 (1928).

868. R. Weindling, *Phytopathology* 22, 837 (1932); 24, 1153 (1934); 27, 1175 (1937).

869. R. Weindling and O. H. Emerson, *Phytopathology* 26, 1068 (1936).

870. P. W. Brian and H. G. Hemming, *Ann. Appl. Biol.* 32, 214 (1945).

871. M. R. Bell et al., *J. Amer. Chem. Soc.* 80, 1001 (1958).

872. A. Fleming, *Brit. J. Exp. Path.* 10, 226 (1929).

873. A. E. Oxford, H. Raistrick, and P. Simonart, *Biochem. J.* 33, 240 (1939).

874. A. J. Whiffen, N. Bohonos, and R. L. Emerson, *J. Bacteriol.* 52, 610 (1946).

875. I. A. M. Cruickshank, *World Rev. Pest Control* 5, 161 (1966).

876. D. M. Spencer, J. H. Topps, and R. L. Wain, *Nature* 179, 651 (1957).

877. B. H. H. Bergmann, J. C. M. Beijersbergen, J. C. Overeem, and A. Kaars Sijpesteijn, *Rec. Trav. Chim. Pays-Bas* 86, 709 (1967).

878. E. C. Kornfeld, R. G. Jones, and T. V. Parke, *J. Amer. Chem. Soc.* **71**, 150 (1949).

879. B. E. Leach, J. H. Ford, and A. J. Whiffen, *J. Amer. Chem. Soc.* **69**, 474 (1947).

880. W. T. Thomson, *Agricultural Chemicals, IV – Fungicides*, Indianapolis, Indiana (1973).

881. H. D. Sisler and M. R. Siegel, *Antibiotics*, Vol. I, Springer Verlag, Berlin-Heidelberg-New York (1967).

882. J. F. Grove et al. *J. Chem. Soc. (London)* **1952**, 3977.

883. A. E. Oxford, H. Raistrick, and P. Simonart, *Biochem. J.* **33**, 240 (1939).

884. G. Stork and M. Tomasz, *J. Amer. Chem. Soc.* **86**, 471 (1964).

885. P. W. Brian, *Ann. Bot. (London)* **13**, 59 (1949).

886. L. P. Miller, *Pest Articles and News Summaries, Sect. B* **14**, 239 (1968).

887. J. F. Anderson, *J. Econ. Entomol.* **59**, 1476 (1966).

888. Jap. P. 63/70 718 ≡ Neth. P. 6415131 (1965), Zaidan Hojin Biseibutsu K.K.; *C.A.* **64**, 11824f (1966).

889. H. Yonehara and N. Otake, *Tetrahedron Lett.* **1966**, 3785.

890. H. Seto et al., *Tetrahedron Lett.* **1966**, 3793.

891. Brit. P. 978632 (1964, Jap. Prior. 1960), Japan Antibiotic Research Assoc. and Kaken Chemical Co.: *C.A.* **62**, 11,117a (1965).

892. Jap. P. 63/25298 (1961), Y. Sumiki and H. Umezawa; *C.A.* **60**, 7415a (1964).

893. K. Isono and S. Suzuki, *Agric. Biol. Chem.* **32**, 1193 (1968).

894. W. T. Thomson, *Agricultural Chemicals, IV – Fungicides*, Indianapolis, Indiana (1973).

895. *Japan Pesticide Information* No. 16, p. 21.

896. DOS 2231979 (1972), Kitasuto Institut, Nihon Tokushu Noyaku Seizo K.K., Inv.: T. Hata, S. Omura, M. Kutagiri, J. S. Awaya, S. Kuyama, S. Higashikawa, K. Yasui, and H. Terada; *C.A.* **78**, 82,937h (1973).

897. S. Horii and Y. Kameda, *Chem. Commun.* **1972**, 747.

898. K. Imai et al., *J. Pharm. Soc. Japan* **76**, 400 (1956).

899. F. Bohlmann and K. M. Kleine, *Chem. Ber.* **95**, 39 (1961).

900. J. C. Lewis et al., *U.S. Dept. Agric., Bur. Agric. Ind. Chem., Mimeo, Circ. Ser. AIC* **231**, 1 (1949).

901. H. Erdtman, Natural Tropolones, in: K. Paech and M. V. Tracey, *Moderne Methoden der Pflanzenanalyse*, vol. III, p. 351, Springer Verlag, Berlin-Heidelberg-New York (1955).

902. H. W. Buston and S. K. Roy, *Arch. Biochem.* **22**, 1 (1949).

903. M. Holden, B. C. Seegal, and H. Baer, *Proc. Soc. Exp. Biol. Med.* **66**, 54 (1947).

904. R. Newton and J. A. Anderson, *Can. J. Res.* **1**, 86 (1929).

905. N. N. Kargapolova, *Sci. Res. Work Inst. Plant Protect., Leningrad* **1935**, 491.

906. G. C. Farkas and Z. Király, *Phytopathol. Z.* **44**, 105 (1962).

907. J. C. Walker and M. A. Stahmann, *Ann. Rev. Plant Physiol.* **6**, 351 (1955).

908. R. Sakai, K. Tomiyama, and T. Takemori, *Ann. Phytopathol. Soc. Japan.* **29**, 120 (1964).

909. H. Nienstaedt, *Phytopathology* **43**, 32 (1953).

910. H. Erdtman, *Naturwissenschaften* **27**, 130 (1939).

911. E. Rennerfeldt, *Medd. Statens Skogsforskningsinst.* **33**, 331 (1944).

912. R. Fischer and F. Stauder, *Pharm. Zentralhalle* **72**, 97 (1932).

913. L. Saint Rat and P. de Luteraan, *C.R. Acad. Sci. Paris* **224**, 1587 (1947).

914. H. Erdtman and J. Gripenberg, *Acta Chem. Scand.* **2**, 625, 639, 644 (1948).

915. E. Rennerfeldt, *Physiol. Plant.* **1**, 245 (1948).

916. B. A. Birdsong, R. Alston, and B. L. Turner, *Can. J. Botany* **38**, 499 (1960).

917. C. J. Shepherd, *Aust. J. Biol. Sci.* **15**, 483 (1962).

918. A. I. Virtanen, *Suom. Kemistil. B* **34**, 29 (1961).

919. A. I. Virtanen and P. K. Hietala, *Acta Chem. Scan.* **9**, 1543 (1955).

920. A. Stoll and E. Seebeck, *Experimentia* **3**, 114 (1947).

921. C. J. Cavallito, J. H. Bailey, and J. S. Buck, *J. Amer. Chem. Soc.* **67**, 1032 (1945).

922. J. M. Segal, *Hindustan. Antibiot. Bull.* **4**, 3 (1961).

923. J. C. Walker, S. Morell, and H. H. Foster, *Am. J. Bot.* **24**, 536 (1937).

924. D. Davis, *Phytopathology* **54**, 290 (1964).

925. J. M. M. Cruickshank, *Ann. Rev. Phytopathol.* **1**, 351 (1963).

926. J. Kuć, *World Rev. Pest Control* **7**, 176 (1968).

928. C. H. Fawcett and D. M. Spencer, Natural Antifungal Compounds, in: D. C. Torgeson, *Fungicides*, Vol. II, p. 637 ff., Academic Press, New York–London (1969).

929. R. M. Letcher et al., *Phytochemistry* **9**, 249 (1970).

930. D. M. Spencer, J. H. Topps, and R. L. Wain, *Nature* **179**, 641 (1957).

931. R. L. Wain, D. M. Spencer, C. H. Fawcett, *Soc. Chem. Ind. (London), Monograph* No. **15**, 109–114 (1961).

932. C. H. Fawcett et al., *J. Chem. Soc. Sect. C.* **1968**, 2455.

933. M. Hiura, *Gifu Daigaku Nogakubu Kenkyu Hokoku* **50**, 1 (1943).

934. N. Zuzuki and S. Toyoda, *Nogyo Gijutsu Kenkyusho Hokoky, Byori Konchu* **8**, 131 (1957).

935. I. Uritani, M. Uritani, and H. Yamada, *Phytopathology* **50**, 30 (1960).

936. N. Bernard, *Ann. Sci. Nat.: Botan. Biol. Vegetale* **[IX]9**, 1 (1909).

937. E. Gäumann and H. Kern, *Phytopathol. Z.* **35**, 347 (1959).

938. M. Schellenbaum, Isolierung und Konstitutionsaufklärung des Orchinols, *Dissertation Eidgenössische Technische Hochschule Zürich*, No. 2977, 555 (1959).

939. H. M. Fish et al., *Phytochemistry* **12**, 437 (1973).

940. A. Böller et al., *Helv. Chim. Acta* **40**, 1062 (1957).

941. N. Mukherjee et al., *Sci. Cult. (Calcutta)* **40**, 198 (1974).

942. J. A. M. Cruickshank and D. R. Perrin, *Aust. J. Biol. Sci.* **16**, 111 (1963).

943. H. Fukui et al., *Agric. Biol.-Chem. (Tokyo)* **37**, 417 (1973).

944. H. D. van Etten, *Phytochemistry* **12**, 1791 (1973).

945. W. A. Andreae, *Can. J. Res. Sect. C* **26**, 31 (1948).

946. J. C. Hughes and T. Swain, *Phytopathology* **50**, 398 (1960).

947. I. Uritani and I. Hoshiya, *J. Agric. Chem. Soc. Jap.* **27**, 161 (1953).

948. E. Sondheimer, *Phytopathology* **51**, 71 (1969).

949. C. H. Fawcett and D. M. Spencer, in: D. C. Torgeson, *Fungicides*, Vol. II, S. 645, 655, Academic Press, New York–London (1969).

950. V. E. Sokolova et al., *Dokl. Akad. Nauk SSSR* **136**, 723 (1961); *C.A.* **55**, 16,668f (1961).

951. R. S. Clark et al., *Phytopathology* **49**, 594 (1959).

952. M. Gordon et al., *Can. J. Chem.* **51**, 748 (1973).

953. R. S. Burden et al., *Phytochemistry* **14**, 221 (1975).

954. USP 1972961 (1931), Du Pont, Inv.: W. H. Tisdale and I. Williams; *Chem. Zentralblatt* **1935 II**, 557.

955. G. E. Harrington, *Science* **93**, 311 (1941).

956. P. J. Anderson, *Plant Dis. Rep.* **26**, 201 (1942).

957. R. R. Kincaid, *Plant Dis. Rep.* **26**, 223 (1942).

958. J. C. Dunegan, *Plant Dis. Rep.* **27**, 101 (1943).

959. M. C. Goldsworthy, E. L. Green, and M. A. Smith, *J. Agric. Res.* **66**, 277 (1943).

960. J. M. Hamilton and D. H. Palmiter, *Farm Res.* (*N.Y.*) **9**, 14 (1943).

961. H. Martin, *J.S.E. Agric. Coll. Wye* **33**, 38 (1934).

962. DRP 642532 (1934), IG Farbenindustrie, Inv.: H. Günzler, F. Heckmanns, and E. Urbschat; Frdl. **23**, 1737.

963. H. S. Cunningham and E. G. Sharvelle, *Phytopathology* **30**, 4 (1940).

964. R. H. Wellman and S. E. A. McCallan, *Contrib. Boyce Thompson Inst.* **14**, 151 (1946).

965. USP 2553770 (1951), Standard Oil Development Co., Inv.: A. R. Kittleson; *C.A.* **45**, 6791i (1951).

966. USP 2867562 (1959), American Cyanamid Co., Inv.: G. Lamb; *C.A.* **53**, 8525i (1959).

967. R. G. Owens, in: D. C. Torgeson, *Fungicides*, Vol. II, p. 147 ff., Academic Press, New York–London (1969).

968. J. G. Horsfall, *Principles of Fungicidal Action*, Chronica Botanica, Press Waltham, Mass. (1956).

969. H. L. Klöpping and G. J. M. van der Kerk, *Rec. Trav. Chim. Pays-Bas* **70**, 917, 949 (1951).

970. H. L. Klöpping, *Chemical Constitution and Antifungal Action of Sulfur Compounds*, Schotanus and Jens, Ultrecht (1951).

971. C. W. Pluijgers, *Direct and Systemic Antifungal Action of Dithiocarbamic Acid Derivatives*, Schotanus and Jens, Utrecht (1959).

972. R. G. Owens and J. H. Rubinstein, *Contrib. Boyce Thompson Inst.* **22**, 241 (1964).

973. R. J. Lukens, *Chemistry of Fungicidal Action*, p. 62, Springer Verlag, Berlin–Heidelberg–New York (1971).

974. J. G. Horsfall and S. Rich, *Indian Phytopathol.* **6**, 1 (1953).

975. P. Allison and G. L. Barnes, *Phytopathology* **46**, 6 (1956).

976. D. E. H. Frear, *Pesticide Index*, 3rd State College/Pa (1965).

977. S. S. Block et al., Abstr. Papers, 140th Meeting, p. 25A, Am. Chem. Soc., Chicago (1961).

978. W. Awe and E. Stoy, *Naturwissenschaften* **37**, 452 (1950).

979. H. Martin, *The Scientific Principles of Crop Protection*, 5th ed., Chapter 7, p. 109, Arnold, London (1964).

980. J. G. Horsfall, *Principles of Fungicidal Action*, p. 185, Chronica Botanica Company, Verlag Waltham, Mass. (1956).

981. Jap. P. 63/11645 (1961), Ihara Agricultural Chemical Co., Inv.: M. Nagarsawa, M. Nakano, and H. Kawada; *C.A.* **59**, 9897d (1963).

982. DRP 583055 (1932), IG Farbenindustrie, Inv.: F. Muth; *Frdl.* **20**, 426.

983. DAS 1227891 (1965), Farbenfabriken Bayer AG, Inv.: H. Schlör and F. Grewe, *C.A.* **66**, 18,592s (1967).

984. L. D. Small, J. H. Bailey, and C. J. Cavallito, *J. Amer. Chem. Soc.* **69**, 1710 (1947); **71**, 3565 (1949).

985. S. S. Block and J. P. Weidner, *Develop. Ind. Microbiol.* **4**, 213 (1963).

986. USP 3249495 (1964), Chemagro Corp., Inv.: P. C. Aichenegg and C. D. Emerson; *C.A.* **65**, 3751b (1966).

987. USP 2959517 (1958), Stauffer Chemical Co., Inv.: J. B. Bowers and R. B. Langford; *C.A.* **55**, 4893g (1961).

988. USP 3268391 (1964), Stauffer Chemical Co., Inv.: D. R. Baker, S. L. Giolito, and D. R. Arneklev; *C.A.* **65**, 19,247h (1966).

989. D. E. H. Frear, *Pesticide Index*, 4th ed., College Science Publishers, State College, Pennsylvania (1969).

990. W. T. Thomson, *Agricultural Chemical, IV – Fungicides*, p. 129, Thomson Publications, Indianapolis, Indiana (1973).

991. DAS 1201115 (1962) ≡ USP 3037907 (1961), Hooker Chemical Corp., Inv.: E. D. Weil; *C.A.* **57**, 6364c (1952).

992. DAS 1094735 (1959) ≡ Brit. P. 910377 (1962), BASF, Inv.: H. Distler; *C.A.* **64**, 2015e (1966).

993. DOS 2047118 (1970, US Prior. 1969), Buckman Laboratories, Inc., Inv.: J. D. Buckman, J. D. Pera, and F. W. Raths; *C.A.* **74**, 140,960r (1971).

994. S. F. Pickard, *Am. Phytopathol. Soc. Monogr.* **26**, 197 (1972).

995. H. Martin, *The Scientific Principles of Crop Protection*, 5th ed., pp. 269–296, Arnold, London (1964).

996. C. W. Pluijgers, *Direct and Systemic Antifungal Action of Dithiocarbamic Acid Derivatives*, Schotanus and Jens, Utrecht (1959).

997. H. L. Klöpping and G. J. van der Kerk, *Rec. Trav. Chim. Pays-Bas* **70**, 917, 949 (1951).

998. S. Rich and J. G. Horsfall, *Science* **120**, 122 (1954).

999. B. G. Benns, B. A. Gingras, and C. H. Bayley, *Appl. Microbiol.* **8**, 353 (1960).

1000. R. G. Owens, *Contrib. Boyce Thompson Inst.* **17**, 473 (1954).

1001. H. W. Gausman, C. L. Rhykerd et al., *Bot. Gaz. (Chicago)* **114**, 292 (1953).

1002. G. D. Thorn and L. T. Richardson, *Meded. Landbouwhogesch. Opzockingssta. Staat Gent* **27**, 1175 (1962).

1003. L. E. Lopatheki and W. Newton, *Can. J. Botany* **30**, 131 (1952).

1004. J. Goksøyr, *Physiol. Plant.* **8**, 719 (1955).

1005. E. C. Gregg and W. P. Tyler, *J. Amer. Chem. Soc.* **32**, 4561 (1950).

1006. R. M. Weed, S. E. A. McCallan, and L. P. Miller, *Contrib. Boyce Thompson Inst.* **17**, 299 (1953).

1007. A. Kaars, Sijpesteijn, *World Rev. Pest Control* **9**, 85 (1970).

1008. R. G. Owens, in: D. C. Torgeson, *Fungicides*, Vol. II, p. 243, Academic Press, New York–London (1969).

1009. A. L. Morehart and D. F. Crossan, *Delaware Univ., Agric. Exp. Sta., Bull.* **357**, (1965).

1010. USP 2766554 (1954), Stauffer Chemical Co., Inv.: S. C. Dorman and A. B. Lindquist; *C.A.* **51**, 4638i (1957).

1011. H. Schlör, in: R. Wegler, *Chemie der Pflanzenschutz- und Schädlingsbekämpfungsmittel, Fungizide*, Vol. II, p. 60, Springer Verlag, Berlin–Heidelberg–New York (1970).

1012. H. Martin, *Guide to the Chemicals used in Crop Protection*, 4th ed., London, Ontario/Canada 1961, with suppl. by E. Y. Spencer (1964).

1013. M. Nagasawa, *Japan Pesticide Information* (Publ.: Japan Plant Protection Association), No. 3, p. 5–10, April 1970 (Kumiai Chem. Ind., Ltd.).

1014. DOS 2036896 (1971, Brit. Prior. 1969), Murphy Chemical Co. Ltd., Inv.: M. Pianka; *C.A.* **74**, 141,826p (1971).

1015. DOS 2308807 (1973, US. Prior. 1972), Buckman Laboratories, Inc., Inv.: J. D. Buckman and J. D. Pera; *C.A.* **80**, 26,786n (1974).

1016. Pesticides to sell in 74, *Farm Chem.* **137**, 1, 36 (1974).

1017. DRP 642532 (1934), IG Farbenindustrie AG, Inv.: H. Günzler, F. Heckmanns, and E. Urbschat, *Frdl.* **23**, 1737.

1018. W. H. Tisdale and A. L. Flenner, *Ind. Eng. Chem.* **34**, 501 (1942).

1019. USP 1972961 (1931), DuPont, Inv.: W. H. Tisdale and I. Williams; *Chem. Zentralblatt* **1935 II**, 557.

1020. USP 2317765 (1941), Röhm & Haas Co., Inv.: W. F. Hester; *C.A.* **37**, 60824 (1943).

1021. J. W. Heuberger and T. F. Manns, *Phytopathology* **33**, 1113 (1943).

1022. USP 2504404 (1946), DuPont, Inv.: A. L. Flenner; *C.A.* **44**, 6076b (1950).

1023. USP 2710822 (1953), DuPont, Inv.: D. R. V. Golding and B. L. Richards; *C.A.* **49**, 13,582i (1955).

1024. G. S. Scheuerer, *Fortschr. Chem. Forsch.* **9**, 254 (1967).

1025. DAS 1057814 (1956), Gebr. Borchers AG, Inv.: H. Reisener and H. Schüler; *C.A.* **55**, 8746f (1961).

1026. DAS 1202266 (1961), BASF, Inv.: H. Windel; *C.A.* **64**, 3102b (1966).

1027. DAS 1226361 (1965), BASF, Inv.: H. Windel and E. H. Pommer; *C.A.* **66**, 1864w (1967).

1028. *Farm Chemicals Handbook* 1972, C 298, Meister Publishing Co., Willoughby/Ohio.

1029. Belg. P. 611960 (1962, DB Prior. 1960), Farbenfabriken Bayer AG, Inv.: H. Lehmann, F. Grewe, and W. Lautenschläger; *C.A.* **57**, 14,236c (1962).

1030. F. Grewe, Rückblick auf 25 Jahre Fungizidforschung, *Pflanzenschutz-Nachr. Bayer* **18**, 45, Sonderheft (1965).

1031. DAS 1163792 (1962), Farbenfabriken Bayer AG, Inv.: A. Frank, G. Spielberger, F. Grewe, and H. Kaspers; *C.A.* **60**, 15,081g (1964).

1032. DAS 1174758 (1960) ≡ Brit. P. 948649 (1964), Farbenfabriken Bayer AG, Inv.: A. Frank and F. Grewe; *C.A.* **60**, 13,150e (1964).

1033. G. D. Thorn and R. A. Ludwig, *Can. J. Bot.* **36**, 389 (1958).

1034. D. E. H. Frear, *Pesticide Index*, College Science Publishers 3rd State College, Pennsylvania (1965).

1035. K. Bodendorf, *J. Prakt. Chem.* [2] **126**, 233 (1930).

1036. DAS 1120801 (1957), Farbenfabriken Bayer AG, Inv.: E. Urbschat and B. Homeyer; *C.A.* **56**, 11,606c (1962).

1037. Jap. P. 66/15578 (1963), Sankyo Co., Inv.: Y. Hamamoto, H. Oda, and Y. Kojitsu; *C.A.* **66**, 10,971v (1967).

1038. S. Wakamori et al., *Agric. Biol.-Chem.* (*Tokyo*) **33**, 1700 (1969).

1039. DOS 1643040 (1967) ≡ S. Afric. P. 68/05172 (1969), Schering AG, Inv.: G. A. Hoyer and E. A. Pieroh; *C.A.* **71**, 112,459d (1969).

1040. Jap. 72/43334 (1971), Sumitomo Chemical Co., Inv.: S. Tanaka, S. Yamamoto, K. Tanaka, and H. Taketa; *C.A.* **78**, 144,283c (1973).

1041. DAS 1272618 (1958, US Prior. 1957), Pfizer & Co., Inv.: P. N. Gordon; *C.A.* **69**, 95,426e (1968).

1042. H. Martin and H. Shaw, *B.I.O.S. Report* 1095 (1946).

1043. B. Nase, *Z. Chem.* **8**, 96 (1968).

1044. DAS 1191628 (1962) ≡ Brit. P. 1000199 (1965, Swiss Prior. 1961), Geigy AG, Inv.: K. Gubler and E. Knüsli; *C.A.* **64**, 17,501a (1966).

1045. H. Schlör, Chemie der Fungizide, in: R. Wegler, *Chemie der Pflanzenschutz- und Schädlingsbekämpfungsmittel*, Vol. 2, p. 82, Springer Verlag, Berlin–Heidelberg–New York (1970).

1046. USP 3154518 (1960), Heyden Newport Chem. Co., Inv.: M. A. Gradsten and T. A. Girard, *C.A.* **62**, 2889g (1965).

1047. H. Schlör, Chemie der Fungizide, in: R. Wegler, *Chemie der Pflanzenschutz- und*

Schädlingsbekämpfungsmittel, Vol. 2, p. 85, Springer Verlag, Berlin–Heidelberg–New York (1970).

1048. W. T. Tomson, *Agricultural Chemicals, IV – Fungicides,* Indianapolis, Indiana (1973).

1049. DOS 1939031 (1968) ≡ S. Afric. P. 69/00714 (1969, US Prior. 1968), Buckman Laboratories, Inc., Inv.: J. D. Buckman, J. D. Pera, and F. W. Rath; *C.A.* **72,** 111,455h (1972).

1050. J. R. Hardison, *Phytopathology* **62,** 605, 611 (1972).

1051. J. C. Walker, S. Morell, and H. H. Foster, *Am. J. Bot.* **24,** 536 (1937).

1052. W. H. Davis, and W. H. Sexton, *Biochem. J.* **40,** 331 (1946).

1053. B. Rathke, *Ber.* **3,** 859 (1870).

1054. J. M. Connolly and G. M. Dyson, *J. Chem. Soc. (London)* **1937,** 827.

1055. A. R. Kittleson, *Science* **155,** 84–86 (1952).

1056. R. Waffler, R. Gasser, A. Margot, and H. Gysin, *Experientia* **11,** 265 (1953).

1057. F. Fischer and O. Wacha, *J. Prakt. Chem.* [4] **12,** 172 (1961).

1058. USP 2553777 (1951), Standard Oil Development Co., Inv.: R. S. Hawley and A. R. Kittleson; *C.A.* **45,** 7742h (1951).

1059. T. P. Johnston et al., *J. Agric. Food Chem.* **5,** 672 (1957).

1060. K. A. Petrov and A. A. Neimysheva, *Zh. Obshch. Khim.* **29,** 3401 (1959); Engl.: 3362.

1061. R. G. Owens and G. Blaak, *Contrib. Boyce Thompson Inst.* **20,** 475 (1960).

1062. R. J. Lukens and H. D. Sisler, *Phytopathology* **48,** 235 (1958).

1063. R. J. Lukens, S. Rich, and J. G. Horsfall, *Phytopathology* **55,** 658 (1965).

1064. D. V. Richmond and E. Somers, *Ann. Appl. Biol.* **50,** 33 (1962).

1065. R. H. Daines, *Am. Fruit Grow.* **73,** 16 (1953).

1066. USP 2553770 (1949), Standard Oil Development Co., Inv.: A. R. Kittleson; *C.A.* **45,** 6791i (1951).

1067. W. D. Thomas, P. H. Eastburg, and M. D. Bankuti, *Phytopathology* **57,** 833 (1967).

1068. Brit. P. 1071996 (1966, US Prior. 1965), Chevron Research Co.; *C.A.* **67,** 63,783m (1967).

1069. Fr.P. 1337286 (1962), Rhône Poulenc S.A., Inv.: J. Metevier and R. Boesch; *C.A.* **60,** 4157d (1964).

1070. G. Matolesy and B. Bordas, *Acta Phyt. Sci. Hung.* **4,** 197 (1969).

1071. F. Grewe, *Pflanzenschutz-Nachr. Bayer* **21,** 147 (1968).

1072. DAS 1193498 (1960) ≡ Belg. P. 609868 (1962), Farbenfabriken Bayer AG, Inv.: E. Klauke, E. Kühle, F. Grewe, H. Kaspers, and R. Wegler; *C.A.* **58,** 9093b (1963).

1073. DBP. 975295 (1955), Farbenfabriken Bayer AG, Inv.: E. Kühle, R. Wegler, and F. Grewe; *C.A.* **56,** 14,166c (1962).

1073a. E. Kühle et al., *Angew. Chem.* **76,** 807 (1964).

1074. Jap. P. 49/13 174 (1971) ≡ USP 3761596, Nihon Noyaku K.K.; *C.A.* **80,** 67411e (1974). See also DOS 2316921 (1973), *C.A.* **80,** 14935x (1974).

1075. USP 3031372 (1959) ≡ Brit. P. 900805 (1962), Hercules Powder Co., Inv.: K. Brack; *C.A.* **61,** 3113g (1964).

1076. Belg. P. 577776 (1959) ≡ Austrian P. 210666 (1959, DB Prior. 1958), E. Merck AG; *Chem. Zentralblatt* **1963,** 6765.

1077. H. G. Bremer, *Z. Pflanzenkr. Pflanzenschutz* **70,** 321 (1963).

1078. R. G. Owens, Organic Sulfur Compounds, in: D. C. Torgeson, *Fungicides,* Vol. 2, p. 288, Academic Press, New York–London (1969).

1079. USP 3260588 (1963), Olin Mathieson Chemical Corp., Inv.: H. J. Schröder; *C.A.* **65**, 12,212f (1966). USP 3260725 (1961) ≡ Belg. P. 624636 (1963), Olin Mathieson Chemical Corp., Inv.: H. J. Schröder and J. H. Reinhart; *C.A.* **59**, 11,508e (1963).

1080. DOS 1953149 (1969, Swiss Prior. 1968), Ciba Ltd., Inv.: S. Janiak; *C.A.* **73**, 25,446v (1970).

1081. DOS 2250077 (1972, US Prior. (1971), Eli Lilly & Co., Inv.: C. J. Paget; *C.A.* **79**, 18,721b (1973).

1082. Jap. P. 7014301 fr. 7.9.1967, Meiji Seika Kaisha Ltd. Inv.: S. Isao; *C.A.* **73**, 45,500m (1970).

1083. USP 2430326 (1947), Shell Development Co., Inv.: H. A. Cheney and S. H. McAllister; *C.A.* **42**, 1960f (1948).

1084. USP 2695859 (1952), Shell Development Co., Inv.: F. B. Hilmer; *C.A.* **49**, 2668c (1955).

1085. USP 3078209 (1960), FMC Corp., Inv.: J. R. Willard, E. G. Maitlen; *C.A.* **58**, 13,075a (1963).

1086. F. C. Stark, *Cornell Univ. Agric. Expt. Sta. Mem.* **1948**, 278.

1087. G. B. Ramsey et al., *Bot. Gaz.* (*Chicago*) **106**, 74 (1944).

1088. H. Yersin et al., *C.R. Acad. Agric. France* **31**, 24 (1945).

1089. H. P. Buschfield and E. E. Storrs, *Contrib. Boyce Thompson Inst.* **19**, 417 (1958).

1090. D. Priest and R. K. S. Wood, *Ann. Appl. Biol.* **49**, 445 (1961).

1091. USP 3134710 (1961), Allied Chemical Corp., Inv.: E. E. Gilbert; *C.A.* **51**, 7641a (1964).

1092. Brit. P. 1019891 (1961), National Research Development Corp., Inv.: R. N. Haszeldine, J. L. Farmer, and H. Goldwhite; *C.A.* **64**, 12,549a (1966).

1093. J. R. Hardison, *Phytopathology* **61**, 936 (1971).

1094. USP 2999118 (1959), Purdue Research Foundation, Inv.: G. B. Bachmand and N. W. Standish; *C.A.* **56**, 2331d (1962).

1095. D. M. Fieldgate et al., *Pestic. Sci.* **2**, 232 (1971).

1096. USP 3265564 (1963) ≡ Neth. Appl. 6402669 (1964), DuPont, Inv.: R. M. Scribner and E. J. Soboczenski; *C.A.* **62**, 12,392b (1965).

1097. P. Rich and J. G. Horsfall, *Phytopathology* **39**, 19 (1949).

1098. R. Sprague, *Rev. Appl. Mycol.* **30**, 475 (1951).

1099. M. Sittig, *Pesticide Production Processes 1967*, Noyes Development Corp., Park Ridge (N.Y./USA) (1967).

1100. L. A. Summers, *Rev. Pure Appl. Chem.* **18**, 1(1968).

1101. H. G. Shirk and H. E. Byrne, *J. Biol. Chem.* **191**, 783 (1951).

1102. H. G. Shirk and H. E. Byrne, *Proc. Soc. Exp. Biol. Med.* **77**, 628 (1951).

1103. T. Geuther, ref. in *Biol. Zbl.* **25**, 879 (1896).

1104. E. Hailer, *Biochem. Z.* **125**, 69 (1921).

1105. USP 3320229 (1962), Stauffer Chemical Co., Inv.: K. Szabo and A. H. Frieberg; *C.A.* **67**, 90521f (1967).

1106. G. A. Carter et al., *Ann. Appl. Biol.* **75**, 49 (1973).

1107. DOS 1901421 (1969, Austrian Prior. 1968), C. H. Boehringer Sohn, Inv.: W. Ost, K. Thomas, D. Jerchel, and K. R. Appel; *C.A.* **72**, 3053s (1970).

1108. Belg. P. 701322 (1968, DB Prior. 1966) ≡ Brit. P. 1123850 (1968), Farbenfabriken Bayer AG, Inv.: H. Malz, F. Grewe, A. Dörken, and H. Kaspers; *C.A.* **70**, 3506a (1969).

1109. DAS 1186467 (1963), Farbenfabriken Bayer AG, Inv.: A. Dörken and G. Schrader; *C.A.* **62**, 16,149a (1965).

1110. L. A. Summers, *Experientia* **31**, 875 (1975).

1111. E. T. Mc Bee et al., *J. Amer. Chem. Soc.* **78**, 1511 (1956).

1112. G. Matolcsy, *Acta Phytopathol. Acad. Sci. Hung.* **4**, 213 (1969).

1113. G. E. W. Wolstenholme and C. M. O'Connor, *Ciba Foundation Symposium on Quinones in Electron Transport*, p. 453, Verlag Churchill, London 1961.

1114. O. Hoffmann-Ostenhof, *Metab. Inhibitors: Comprehensive Treaties* **2**, 145 (1963).

1115. A. L. Smith and R. L. Lester, *Biochim. Biophys. Acta* **48**, 547 (1961).

1116. D. L. Schoene et al., *Agric. Chem.* **4**, 24 (1949).

1117. USP 2349771 (1944), United States Rubber Co., Inv.: P. Ter Horst; *C.A.* **39**, 12464 (1945).

1118. DAS 1233197 (1964, US Prior. 1963) ≡ Neth. Appl. 6413711 (1965), United States Rubber Co., Inv.: B. v. Schmeling; *C.A.* **64**, 4201g (1966).

1119. S. W. Petersen, W. Gauss, and E. Urbschat, *Angew. Chem.* **67**, 217 (1955).

1120. DAS 1231051 (1963) ≡ Brit. P. 999802 (1962), Boots Pure Drug Co., Inv.: A. F. Hams, J. R. Housley, and J. Collyer; *C.A.* **63**, 12,259h (1965).

1121. DAS 1158056 (1960), E. Merck AG, Inv.: H. Hahn, G. Mohr, and A. van Schoor; *C.A.* **60**, 5341f (1964).

1122. R. J. W. Byrde, in: D. C. Torgeson, *Fungicides*, Vol. 2, p. 544, Academic Press, New York–London (1969).

1123. K. Lang, *Arzneim. Forsch.* **10**, 997 (1960).

1124. Neth. Appl. 6610214 (1966, Jap. Prior. 1965), Japan Soda Co.; *C.A.* **67**, 108,253g (1967).

1125. USP 2861919 (1956), Allied Chemical Corp., Inv.: E. E. Gilbert; *C.A.* **53**, 5577f (1959).

1126. P. A. Wolf, *Food Technol.* (*Chicago*) **4**, 294 (1950).

1127. Fr.P. 1397521 (1965, US Prior. 1963), Diamond Alkali Co., Inv.: R. D. Battershell and H. Bluestone; *C.A.* **63**, 4212a (1965).

1128. G. A. White and G. D. Thorn, *Pestic. Biochem. Physiol.* **5**, 380 (1975).

1128a. P. A. Urech, F. Schwinn, and T. Staub, *Proceedings of the 9th British Insecticide and Fungicide Conference, 1977*, p. 623ff.

1128b. DOS 2515091 (1974), Ciba-Geigy AG, Inv.: A. Hubele; *C.A.* **84**, 30,713m (1976).

1128c. DOS 2513732 (1974), Ciba-Geigy AG, Inv.: A. Hubele; *C.A.* **84**, 58,964t (1976).

1128d. F. J. Schwinn, T. Staub, and P. A. Urech, *Mededelingen van de Faculteit van de Landbouwwetenschappen, Rijksuniversiteit Gent*, **1977**, 42.

1128e. USP 3933860 (1975), Chevron Chem. Co., Inv.: D. C. K. Chan; *C.A.* **84**, 135,318q (1976).

1128f. H. Kaspers and J. Reuff, *Mitteilungen aus der Biol. Bundesanstalt für Land- und Forstwirtschaft*, **191**, 240 (1979).

1129. *Anonym, Chem. Week* **107**, (13), 49 (1970).

1130. A. Fujinami et al., *Agric. Biol- Chem.* **36**, 318 (1972).

1131. H. Kaars Sijpesteijn, *World Rev. Pest Control* **9**, 85 (1970).

1132. USP 2867562 (1959), American Cyanamid Co., Inv.: G. Lamb; *C.A.* **53**, 8525i (1959).

1133. D. Kerridge, *J. Gen. Microbiol.* **19**, 427 (1958).

1134. T. Zsolnai, *Biochem. Pharmacol.* **11**, 995 (1962).

1135. W. J. Tolmsoff, Dissertation, University of California, Davis, California; *Dissertation Abstr.* **26**, 5707 (1966).

1136. DAS 1028828 (1956), Farbenfabriken Bayer AG, Inv.: E. Urbschat and P. E. Frohberger; *C.A.* **54**, 13,528h (1960).

1137. W. H. Read, *World Rev. Pest Control* **5**, 45 (1966).

1138. P. L. Thayer, D. H. Ford, and H. R. Hall, *Phytopathology* **57**, 833 (1967).

1139. DOS 2105174 (1971, Swiss. Prior. 1970), Ciba Ltd., Inv.: H. Huber-Emden, A. Hubele, G. Klahre; *C.A.* **76**, 3713g (1972).

1140. Y. Gutter, *Plant Disease Reporter* **53**, 474 (1969).

1141. A. Kreutzberger, *Z. Pflanzenkr. Pflanzenschutz* **80**, 255 (1973).

1141a. F. Schwinn, M. Nakemura, and G. Hadschin, *9th International Congress of Plant Protection, Washington 1979, Abstracts of Papers*, p. 479.

1142. F. Grewe, and K. H. Büchel, Mitteilungen aus der Biol. Bundesanstalt t. Land- u. Forstwirtschaft, 39. Deutsche Pflanzenschutz-Tagung 1.–5.10.1973, Issue 151 (1973).

1142a. H. Buchenauer, *British Crop Protection Conference–Pests and Diseases, Proceedings* **3**, 699 (1977).

1143. G. Clemons and H. Sisler, *Pestic. Biochem. Physiol.* **1**, 32 (1971).

1144. J. R. Decallone et al., *Pestic. Sci.* **6**, 113 (1975).

1145. R. S. Elias, *Nature* **219**, 1160 (1968).

1146. K. Sasse, R. Wegler, G. Unterstenhöfer, and F. Grewe, *Angew. Chem.* **72**, 973 (1960).

1147. DAS 1194631 (1963), Farbenfabriken Bayer AG, Inv.: K. Sasse, R. Wegler, H. Scheinpflug, and H. Jung; *C.A.* **63**, 8381a (1965).

1147a. USP 4085212 (1975), ICI, Inv.: R. A. Burrell and J. M. Cox; *C.A.* **89**, 197,583j (1978).

1148. DOS 1567123 (1965, Jap. Prior. 1964) ≡ Neth. Appl. 6516871 (1966), Sankyo Co., Ltd., Inv.: H. Sumi, Y. Takahi, N. Nakamura, and K. Tomita; *C.A.* **65**, 15,383c (1966).

1149. Fr. P. 1512682 (1967, DB Prior. 1966) ≡ Brit. P. 1120338 (1968), Farbenfabriken Bayer AG, Inv.: E. Kleinheidt and H. Gold; *C.A.* **69**, 96,788m (1968).

1150. S. A. Waksman, *Science* **118**, 259 (1953).

1151. J. R. Dyer and A. W. Todd, *J. Amer. Chem. Soc.* **85**, 3896 (1963).

1152. DAS 1094036 (1959), N.V. Philips' Gloeilampenfabrieken, Inv.: J. C. Overeem and J. D. Bijloo; *C.A.* **55**, 27758i (1961).

1153. Fr. P. 1451606 (1965, Brit. Prior. 1964) ≡ Neth. Appl. 6511395 (1966), Shell Internationale Research Maatschappij N.V.; *C.A.* **65**, 2279d (1966).

1154. DOS 1923939 (1968), Takeda Chemical Industries Ltd., Inv.: O. Wakae, K. Yakushiji, and Y. Okada; *C.A.* **72**, 121,546c (1970).

1154a. Jap. Appl. 73/26213 (1970), Sankyo Co., Inv.: S. Yamazaki et al.; *C.A.* **81**, 10,444j (1974).

1155. E. M. Stoddard and A. E. Dimond, *Phytopathology* **41**, 337 (1951).

Chapter 4.   Herbicides

1156. H. H. Cramer, *Pflanzenschutz und Welternte, Pflanzenschutznachr. Bayer* **20**, 487 (1967).

1157. G. W. Baily and J. L. White, *J. Agric. Food Chem.* **12**, 324 (1964).

1158. P. J. Doherty and G. F. Warren, *Weed Res.* **9**, 20 (1969).

1159. R. R. Schmidt, *Z. Pflanzenkr. Pflanzenschutz, Sonderheft* **VI**, 59 (1972).

1160. D. I. Arnon, F. R. Whatley, and M. B. Allen, *Nature* **180**, 182 (1957).

1161. A. Trebst and G. Hauska, *Naturwissenschaften* **61**, 308 (1974).

1162. J. S. C. Wessels and R. van der Veen, *Biochim. Biophys. Acta* **19**, 548 (1956).

1163. N. E. Good, *Plant Physiol.* **36**, 788 (1961).

1164. D. E. Moreland, *Ann. Rev. Plant Physiol.* **18**, 365 (1967).

1165. A. Trebst and E. Harth, *Z. Naturforsch. Teil C* **29**, 232 (1974).

1166. B. Exer, *Experientia* **14**, 136 (1958).

1167. D. E. Moreland, W. A. Gentner, J. L. Hilton, and K. L. Hill, *Plant Physiol.* **34**, 432 (1959).

1168. B. Exer, *Weed Res.* **1**, 233 (1961).

1169. W. Draber, K. Dickoré, K. H. Büchel, A. Trebst, and E. Pistorius, *Naturwissenschaften* **55**, 445 (1968).

1170. W. Draber, K. Dickoré, K. H. Büchel, A. Trebst, and E. Pistorius, in: H. Metzner, *Progress in Photosynthesis Research*, Vol. III, p. 1789, International Union of Biological Sciences, Tübingen (1969).

1171. A. Trebst and H. Wietoska, *Z. Naturforsch. Teil C* **30**, 499 (1975).

1172. J. L. Hilton, T. J. Monaco, D. E. Moreland, and W. A. Genter, *Weeds* **12**, 129 (1964).

1173. C. E. Hoffmann, J. W. McGahen, and P. B. Sweetser, *Nature* **202**, 577 (1964).

1174. J. L. Hilton, A. L. Scharen, J. B. St. John, D. E. Moreland, and K. H. Norris, *Weed Sci.* **17**, 541 (1969).

1175. R. Frank and C. M. Switzer, *Weed Sci.* **17**, 344 (1969).

1176. M. W. Kerr and R. L. Wain, *Ann. Appl. Biol.* **54**, 447 (1964).

1177. Z. Gromet-Elhanan, *Biochem. Biophys. Res. Commun.* **30**, 28 (1968).

1178. D. E. Moreland and K. L. Hill, *J. Agric. Food Chem.* **7**, 832 (1959).

1179. A. Trebst, E. Pistorius, G. Boroschewski, and H. Schulz, *Z. Naturforsch. Teil B* **23**, 342 (1968).

1180. C. Kötter and F. Arndt, *Z. Pflanzenkr. Pflanzenschutz, Sonderheft* **V**, 81 (1970).

1181. H. Schulz, in: H. Metzner, *Progress in Photosynthesis Research*, Vol. III, p. 1752, International Union of Biological Sciences, Tübingen (1969).

1182. D. E. Moreland and K. L. Hill, *Weeds* **11**, 55 (1962).

1183. B. C. Baldwin in: H. Metzner, *Progress in Photosynthesis Research*, Vol. III, p. 1737, International Union of Biological Sciences, Tübingen (1969).

1184. A. D. Dodge, *Endeavour* **30**, 130 (1971).

1185. C. Hansch and T. Fujita, *J. Amer. Chem. Soc.* **86**, 1616 (1964).

1186. C. Hansch, E. W. Deutsch, and R. N. Smith, *J. Amer. Chem. Soc.* **87**, 2738 (1965).

1187. C. Hansch, A. R. Stewards, and J. Isawa, *Mol. Pharmacol.* **1**, 87 (1965).

1188. C. Hansch and E. W. Deutsch, *J. Med. Chem.* **8**, 705 (1965).

1189. C. Hansch and E. W. Deutsch, *Biochim. Biophys. Acta* **112**, 381 (1966).

1190. P. A. Gabbott, in: H. Metzner, *Progress in Photosynthesis Research*, Vol. III, p. 1712, International Union of Biological Sciences, Tübingen (1969).

1191. K. H. Büchel, W. Draber, A. Trebst, and E. Pistorius, *Z. Naturforsch. Teil B* **21**, 243 (1966).

1192. W. Draber, K. H. Büchel, H. Timler, and A. Trebst, *ACS Symposium Series*, No. 2, 100 (1974).

1193. M. W. Kerr and R. L. Wain, *Ann. Appl. Biol.* **54**, 441 (1964).

1194. D. E. Moreland, W. J. Blackmon, H. G. Todd, and F. S. Farmer, *Weed Sci.* **18**, 636 (1970).

1195. S. Matsunaka, *J. Agric. Food Chem.* **17**, 171 (1969).

1196. J. B. Hanson and F. W. Slife, *Residue Rev.* **25**, 59 (1969).

1197. W. J. van der Woude, C. A. Lembi, and D. J. Morre, *Biochim. Biophys. Res. Commun.* **46**, 245 (1972).

1198. E. Krelle, *Biochem. Physiol. Pflanz.* **161**, 299 (1970).

1199. G. W. Keit and R. A. Baker, *Plant Physiol.* **41**, 1561 (1966).

1200. W. G. Tempelman and W. A. Sexton, *Nature* **156**, 630 (1945).

1201. D. T. Canvin and G. Friesen, *Weeds* **7**, 153 (1959).

1202. G. W. Keitt, *Physiol. Plant.* **20**, 1076 (1967).

1203. J. D. Mann, E. Cota-Robles, K. H. Yung, M. Pu, and H. Haid, *Biochim. Biophys. Acta* **138**, 133 (1967).

1204. E. M. Lignowski and E. G. Scott, *Weed Sci.* **20**, 267 (1972).

1205. J. Bonaly, *C.R. Acad. Sci., Paris, Ser. D* **273**, 150 (1971).

1206. C. T. Chang and D. Smith, *Weed Sci.* **20**, 220 (1972).

1207. E. G. Jaworski, *J. Agric. Food Chem.* **17**, 165 (1969).

1208. M. W. Kerr and R. J. Avery, *Biochem. J.* **128**, 132 P (1972).

1209. D. P. Weeks and R. Baxter, *Biochemistry* **11**, 3060 (1972).

1210. E. R. Burns, G. A. Buchanan, and M. C. Carter, *Plant Physiol.* **47**, 144 (1971).

1211. W. A. Gentner, *Weeds* **14**, 27 (1966).

1212. N. Schmitt, *Z. Pflanzenkr. Pflanzenschutz* **76**, 282 (1969).

1213. USP 2277744 (1940), Du Pont, Inv.: M. E. Cupery, A. P. Tanberg; *C.A.* **36**, 49661 (1942).

1214. D. G. Sturkie, *J. Am. Soc. Agron.* **29**, 803 (1937).

1215. S. Marcovitch, *J. Econ. Entomol.* **38**, 395 (1945).

1216. USP 2733 982 (1952), Western Electro Chemical Co., Inv.: J. C. Schumacher; *C.A.* **50**, 8980c (1956).

1217. USP 2551705 (1951), Union Oil Co. of California, Inv.: R. A. Robers; *C.A.* **45**, 8194g (1951).

1218. USP 2998310 (1958), US Borax and Chemical Corp., Inv.: P. J. O'Brian and G. A. Connel; *C.A.* **56**, 736a (1962).

1219. Brit. P. 882972 (1960), US Borax and Chemical Corp.; *C.A.* **56**, 11,217b (1962).

1220. USP 3189428 (1962), Dow Chemical Co., Inv.: D. R. Mussel; *C.A.* **63**, 7590f (1965).

1221. USP 3419382 (1966), Pennwalt Corp., Inv.: W. H. Culver; *C.A.* **70**, 86,506z (1969).

1222. *Farm Chemicals* **138**, (9), 45 (1975).

1223. Brit. P. 869088 (1960), Shell Research Ltd., Inv.: K. J. Eaton and J. Malone; *C.A.* **55**, 23916h (1961).

1224. USP 2635117 (1949), Allied Chemical and Dye Corp., Inv.: C. Woolf and E. E. Gilbert; *C.A.* **48**, 7051i (1954).

1225. USP 2764479 (1955), Allied Chemical & Dye Corp., Inv.: E. E. Gilbert; *C.A.* **51**, 3912a (1957).

1226. Brit. P. 1026036 (1964) ≡ Neth. Appl. 6410998 (1965, US Prior. 1963), Inv.: W. J. Cunningham; *C.A.* **63**, 8977a (1965).

1227. DOS 2241560 (1972, Jap. Prior. 1971), Nippon Kayaku Co., Inv.: O. Yamada, A. Kurozumi, S. Ishida, F. Futatsuya, K. Ito, and H. Yamamoto; *C.A.* **78**, 147,571f (1973).

1228. G. Zwieg, J. E. Hitt, and D. H. Cho, in: H. Metzner, *Progress in Photosynthesis Research*, Vol. III, p. 1728, International Union of Biological Science, Tübingen (1969).

1229. A. Trebst, E. Harth, and W. Draber, *Z. Naturforsch.* **25b**, 1157 (1970).

1230. DOS 1768447 (1968) ≡ S. Afric. P. (1969), Farbwerke Hoechst AG, Inv.: O. Scherer and H. Röchling; *C.A.* **73**, 14,555d (1970).

1231. Brit. P. 425295 (1933) ≡ Fr. P. 751855 (1932), George Truffaut and Cie., Inv.: G. Truffaut and I. Pastac; *Chem. Zentralblatt* **1934 II**, 831.

1232. A. H. M. Kirby, *World Rev. Pest Control* **5**, 30 (1966).

1233. J. B. St. John and J. L. Hilton, *Weed Sci.* **21**, 477 (1973).

1234. A. S. Crafts, *Science* **101**, 417 (1945).

1235. USP 2192197 (1936), Dow Chemical Co., Inv.: L. E. Mills and B. L. Fayerweather; *C.A.* **34**, 45284 (1940).

1236. DAS 1103072 (1959), Farbwerke Hoechst AG, Inv.: O. Scherer, K. Reichner, and H. Frensch; *C.A.* **55**, 23,916d (1961).

1237. P. Poignant and P. Crisinel, *Proc. 4th COLUMA Conf., Paris*, 196 (1967).

1238. F. Colliot and B. Henrion, *Proc. 6th Brit. Insectic. Fungic. Conf.*, Vol. II, 529 (1971).

1239. Brit. P. 1080282 (1963), Murphy Chemical Co. Ltd., Inv.: M. Pianka and J. Edward; *C.A.* **62**, 16,129f (1965).

1240. USP 2131259 (1937), Dow Chemical Co., Inv.: W. C. Stoesser; *C.A.* **32**, 9102^2 (1938).

1241. K. Carpenter and B. J. Heywood, *Nature* **200**, 28 (1963).

1242. R. L. Wain, *Proc. 7th Brit. Weed Control Conf.*, Vol. I, 1, 306 (1964).

1243. K. Carpenter, H. J. Cottrell, W. H. de Silva, B. J. Heywood, W. Gleeds, K. F. Rivett, and M. L. Soundy, *Weed Res.* **4**, 175 (1964).

1244. Brit. P. 1067031 (1962), May and Baker Ltd., Inv.: B. J. Heywood and W. G. Leeds; *Chem. Zentralblatt* **1969**, 2-2182.

1245. K. Auwers, J. Reis, *Ber.* **29**, 2355 (1896).

1246. Neth. P. 6610042 (1966), May and Baker Ltd., Inv.: R. F. Collins and B. J. Heywood; *C.A.* **67**, 116,728f (1967).

1247. Brit. P. 1096037 (1966) ≡ Neth. Appl. 6600834 (1966, Swiss Prior. 1965), Ciba Ltd., Inv.: A. Hubele; *C.A.* **66**, 28,501a (1967).

1248. USP 2777762 (1953), Dow Chemical Co., Inv.: B. V. Toornman; *C.A.* **51**, 6075a (1957).

1249. DAS 1187852 (1961) ≡ USP 3080225 (1960), Rohm & Haas Co., Inv.: H. F. Wilson and D. H. McRae; *C.A.* **59**, 2114a (1963).

1250. Belg. P. 648282 (1964, Jap. Prior. 1963), Mitsui Kakago Kogyo Kabushiki Kaisha; *C.A.* **63**, 12,249e (1965).

1251. L'Industrie Chimique et le Phosphate Réunis **55**, No. 616, 490 (1968).

1252. Fr. P. 1346122 (1963), Ciba Ltd., Inv.: H. Martin, H. Aebi, and L. Ebner; *C.A.* **60**, 10,602f (1964).

1253. L. Ebner, D. H. Green, and P. Pande, *Proc. 9th Brit. Weed Control Conf.*, Vol. II, 1026 (1968).

1254. USP 3652645 (1969) ≡ DOS 2019821 (1970), Mobil Oil Corp., Inv.: R. J. Theissen; *C.A.* **74**, 30981r (1971).

1255. USP 3776715 (1971) ≡ DOS 2019821 (1970), Mobil Oil Corp., Inv.: R. J. Theissen; *C.A.* **74**, 30,981r (1971).

1255a. DOS 2304006 (1973), Ishihara Sangyo Kaisha Ltd., Inv.: R. Takahashi, K. Fujikawa, I. Yokomichi, and S. Someya; *C.A.* **79**, 115,299c (1973).

1255b. DOS 2311638 (1973) ≡ USP 3798276 (1974), Rohm and Haas, Co., Inv.: H. O. Bayer, C. Swithenbank, and R. Y. Yih; *C.A.* **80**, 3253x (1974).

1256. DRP 833274 (1952, Brit. Prior. 1941), ICI, Inv.: W. A. Sexton and W. G. Templeman.

1257. G. Scheuerer, *Fortschr. Chem. Forsch.* **9**, 254 (1967/68).

1258. D. E. Moreland, in: H. Metzner, *Progress in Photosynthesis Research*, Vol. III, p. 1693, International Union of Biological Sciences, Tübingen (1969).

1259. A. Barth and H.-J. Michel, *Pharmazie* **24**, 11 (1969).

1260. R. A. Herrett and R. V. Berthold, *Science* **149**, 191 (1965).

1261. USP 3140167 (1962), Hercules Powder Co., Inc., Inv.: A. H. Haubein; *C.A.* **61**, 8234e (1964).

1262. Belg. P. 645752 (1964) ≡ Fr. P. 1397525 (1965, US Prior. 1963), Union Carbide Corp., Inv.: R. A. Herrett and R. V. Berthold; *C.A.* **63**, 4207e (1965).

1263. D. K. George, D. H. Moore, W. P. Brian, and J. A. Garman, *J. Agric. Food Chem.* **2**, 356 (1954).

1264. USP 2734911 (1952), Columbia Southern Chemical Corp., Inv.: P. Strein; *C.A.* **50**, 10,769i (1956).

1265. H. C. Bucha and C. W. Todd, *Science* **114**, 493 (1951).

1266. DAS 1137255 (1959), Spencer Chemical Co., Inv.: T. R. Hopkins, J. W. Pullen and P. D. Strickler; *C.A.* **58**, 8373h (1963).

1267. O. L. Hoffmann, *Weeds* **10**, 322 (1962).

1268. DBP 1034912 (1958), BASF, Inv.: H. Windel, A. Fischer, and H. Stummeyer; *C.A.* **54**, 20,060f (1960).

1269. A. Fischer, *Z. Pflanzenkr. Pflanzenschutz* **67**, 577 (1960).

1270. DAS 1195549 (1962) ≡ Belg. P. 612550 (1962, US Prior. 1961), Food Machinery and Chemical Corp., Inv.: J. R. Willard and K. P. Dorschner; *C.A.* **57**, 12,948b (1962).

1271. W. Siefken, *Justus Liebigs Ann. Chem.* **562**, 76 (1949).

1272. Fr. P. 1249434 (1959), Rhône-Poulenc, Inv.: J. Metevier; *C.A.* **56**, 7213e (1962).

1273. G. G. Still, D. G. Rusness, and E. R. Mansager, 167th ACS Nat. Meeting, Los Angeles, April 1974 (Lecture).

1274. DAS 1024746 (1957), Bayer AG, Inv.: E. Kühle, W. Schäfer, L. Eue, and R. Wegler; *C.A.* **54**, 11,367h (1960).

1275. Brit. P. 1040541 (1961) ≡ Belg. P. 622214 (1963), May and Baker, Ltd., Inv.: K. Carpenter, B. J. Heywood, J. Metevier, and R. Boesch; *C.A.* **59**, 12,710b (1963).

1276. H. J. Cottrell and B. J. Heywood, *Nature* **207**, 655 (1965).

1277. DAS 1567151 (1965) ≡ Neth. Appl. 6604363 (1966), Schering AG, Inv.: G. Boroschewski, F. Arndt, and R. Rusch; *C.A.* **66**, 104,813w (1967).

1278. F. Arndt and G. Boroschewsky, 3rd Sympe on New Herbicides, EWRC, Versailles, S. 141 (1969).

1279. H. Kassebeer, *Z. Pflanzenkr. Pflanzenschutz* **78**, 158 (1971).

1280. H. Martin and C. R. Worthing, *Pesticide Manual*, 4th ed. p. 246, British Crop Protection Council, Worcester 1974.

1281. DAS 1031571 (1957), Stauffer Chemical Co., Inv.: H. Tilles and J. Antognini; *C.A.* **54**, 17,416i (1960).

1282. USP 3175897 (1962), Stauffer Chemical Co., Inv.: H. Tilles and J. Antognini; *C.A.* **62**, 16,082g (1965).

1283. USP 3198786 (1964), Stauffer Chemical Co., Inv.: H. Tilles and R. Curtis; *C.A.* **63**, 11,523a (1965).

1284. USP 3330643 (1965), Monsanto Co., Inv.: M. W. Harman and J. J. D'Amico; *C.A.* **67**, 108,271 (1967).

1285. O. Johnson, Pesticides '72, *Chem. Week* **110**(25), 33 (1972).

1285a. DOS 2144700 (1972), Montecatini Edison SpA., Inv.: G. Pellegrini, G. Losco, A. Quattrini, and E. Arsura; *C.A.* **77**, 19,367m (1972).

1286. G. R. Wicks and C. R. Fenster, *Weed Sci.* **19**, 565 (1971).

1287. F. M. Pallos, R. A. Gray, D. R. Arneklev, and M. E. Brokke, *J. Agric. Food Chem.* **23**, 821 (1975).

1288. M. M. Lay, J. P. Hubbel, and J. E. Casida, *Science* **189**, 287 (1975).

1289. J. E. Casida, R. A. Gray, and H. Tilles, *Science* **184**, 573 (1974).

1290. DOS 2350475 (1974, Ital. Prior. 1972), Montecatini Edison S.p.A., Inv.: R. Santi, D. Barton, G. Camaggi; *C.A.* **81**, 25,424b (1974).

1291. C. E. Beste and M. M. Schreiber, *Weed Sci.* **20**, 8 (1972).

1292. T. M. Chen, D. E. Seaman, and F. M. Ashton, *Weed Sci.* **16**, 28 (1968).

1293. R. E. Wilkinson and W. S. Hardcastle, *Weed Sci.* **18**, 125 (1970).

1294. USP 2744898 (1952), Monsanto Chemical Co., Inv.: M. W. Harman and J. J. D'Amico; *C.A.* **51**, 500d (1957).

1295. H. E. Thompson, C. P. Swanson, and A. G. Norman, *Bot. Gaz.* (*Chicago*) **107**, 476 (1946).

1296. DAS 1209801 (1963) ≡ Neth. Appl. 6405138 (1964), Bayer AG, Inv.: H. Hack, W. Schäfer, R. Wegler, L. Eue; *C.A.* **63**, 1170a (1965).

1297. L. Eue, H. Hack, and F. Münz, *Proc. 10th Brit. Weed Control Conf.*, 610 (1970).

1298. H. Hack and R. R. Schmidt, *Proc. 11th Brit. Weed Control Conf.*, 479 (1972).

1299. DBP 1027930 (1956) ≡ Brit. P. 812120 (1959), BASF, Inv.: A. Fischer, G. Scheuerer, O. Schlichtung, and H. Stummeyer; *C.A.* **55**, 9772d (1961).

1300. DAS 1200062 (1964), BASF, Inv.: P. Raff, L. Schuster, A. Fischer, G. Scheuerer, and G. Steinbrunn; *C.A.* **63**, 15,466a (1965).

1301. DBP 935165 (1950), DuPont, Inv.: H. E. Cupery, N. E. Searle, and C. W. Todd; *C.A.* **52**, 20,864h (1958).

1302. USP 2655445 (1952), DuPont, Inv.: C. W. Todd; *C.A.* **48**, 943f (1954).

1303. USP 3134665 (1960) ≡ Brit. P. 914779 (1963; Swiss Prior. 1959), Ciba Ltd., Inv.: H. Martin and H. Aebi; *C.A.* **58**, 7306c (1963).

1304. DOS 2039041 (1970), Farbwerke Hoechst AG, Inv.: G. Hörlein, P. Langelüddeke, and H. Schönowsky; *C.A.* **77**, 15,615t (1972).

1305. Belg. P. 688018 (1966) ≡ Fr.P. 1497867 (1967, Swiss Prior. 1965), Sandoz S.A., Inv.: M. Schuler; *C.A.* **69**, 76,891m (1968).

1306. W. Berg, *Z. Pflanzenkr. Pflanzenschutz, Sonderheft* **IV**, 233 (1968).

1307. DOS 1905598 (1969), Ciba AG, Inv.: H. Martin, S. Janiak, W. Töpfl, D. Dürr, O. Rohr, and G. Pissiotas; *C.A.* **72**, 20,787b (1970).

1308. Neth. Appl. 6503645 (1965), Food Machinery and Chemical Corp., Inv.: K. R. Wilson, K. L. Hill; *C.A.* **64**, 8082d (1966).

1309. DAS 1142251 (1960) ≡ Brit. P. 913383 (1962; Swiss Prior. 1959), Ciba AG, Inv.: H. Martin and H. Aebi; *C.A.* **59**, 7436g (1963).

1310. DOS 2003143 (1970), Bayer AG, Inv.: E. Klauke, E. Kühle, and L. Eue; *C.A.* **75**, 98,343k (1971).

1311. DAS 1028986 (1956), Farbwerke Hoechst AG, Inv.: O. Scherer and P. Heller; *C.A.* **54**, 22,496h (1960).

1312. DAS 1210793 (1962) ≡ Fr.P. 1325926 (1963; Swiss Prior. 1961), Ciba AG, Inv.: H. Martin, H. Aebi, L. Ebner; *C.A.* **59**, 11,512h (1963).

1313. Fr. P. 1392499 (1963), Péchiney-Progil Société, Inv.: D. Pillon and P. Poignant; *C.A.* **63**, 8258h (1965).

1314. DBP 968273 (1950), DuPont, Inv.: C. W. Todd, *C.A.* **54**, 5000g (1960).

1315. DBP 1108977 (1960), BASF, Inv.: A. Fischer, E. Flikkinger, G. Steinbrunn, and H. Stummeyer; *C.A.* **56**, 8631g (1962).

1316. Brit. P. 1085430 (1966, DB Prior. 1965), Bayer AG, Inv.: H. Hack, L. Eue, and W. Schäfer; *C.A.* **68**, 21,132p (1968).

1317. H. Hack, *Z. Pflanzenkr. Pflanzenschutz, Sonderheft* **IV**, 251 (1968).

1318. H. Kubo, R. Sato, I. Hamura, and T. Ohi, *J. Agric. Food Chem.* **18**, 60 (1970).

1318a. DOS 2430467 (1975), Velsicol Chemical Corp., Inv.: J. Krenzer; *C.A.* **82**, 156,322k (1975).

1318b. DOS 2214632 (1973), Schering AG., Inv.: H. Schulz and F. Arndt; *C.A.* **80**, 14,931t (1974).

1319. DOS 1816568 (1968), Bayer AG, Inv.: C. Metzger, L. Eue, and H. Hack; *C.A.* **74**, 13,162g (1971).

1320. L. Eue, C. Metzger, and W. Faust, Compte rendu des Journées d'Etudes sur les Herbicides (7e Conf. du Columa), I, 14 (1973).

1321. DOS 1770467 (1968) ≡ S.Afric.P. 6902910 (1969), Bayer AG, Inv.: D. Rücker, C. Metzger, L. Eue, and H. Hack; *C.A.* **72**, 100,711n (1970).

1322. DOS 1816696 (1968), Agripat S.A., Inv.: K. Hoegerle, P. Rathgeb, H. J. Cellarius, and J. Rumpf; *C.A.* **72**, 12,734s (1970).

1323. DOS 2100057 (1971, US Prior. 1970), Air Products and Chemicals Inc., Inv.: T. Cebalo and J. F. Alderman; *C.A.* **75**, 76,799m (1971).

1324. DOS 2118520 (1971), Gulf Research and Development Co., Inv.: W. C. Doyle, Jr.; *C.A.* **76**, 95,742z (1972).

1325. C. T. Redemann and J. Hamaker, *Weeds* **3**, 387 (1954).

1326. T. R. Kemp, L. P. Stoltz, J. W. Herron, and W. T. Smith, Jr., *Weed Sci.* **17**, 444 (1969).

1327. P. E. Kolattukudy, *Biochemistry* **4**, 1844 (1965).

1328. P. E. Kolattukudy, *Plant Physiol.* **43**, 375 (1968).

1329. USP 2622976 (1950), Boyce Thompson Inst. for Plant Research, Inv.: A. E. Hitchcock and P. W. Zimmerman; *Chem. Zentralblatt* **1954**, 2257.

1330. USP 2642354 (1951), Dow Chemical Co., Inv.: K. C. Barrons; *C.A.* **47**, 8308e (1953).

1331. USP 2634290 (1953), Hooker Electro Chemical Co., Inv.: J. A. Sonia and E. H. Scemin; *C.A.* **48**, 2764c (1954).

1332. F. Kögl, *Ber.* **1935 A**, 16.

1333. W. G. Templeman and J. C. Marmoy, *Ann. Appl. Biol.* **27**, 453 (1940).

1334. P. S. Nutman, H. G. Thornton, and J. H. Quastel, *Nature* **155**, 498 (1945).

1335. R. E. Slade, W. G. Templeman, and W. A. Sexton, *Nature* **155**, 497 (1945).

1336. USP 2977212 (1958), Heydon Newport Chemical Corp., Inv.: N. Tischler; *C.A.* **55**, 13,756g (1961).

1337. USP 3299120 (1965), Hooker Chemical Corp., Inv.: J. S. Newcomer, E. D. Weil, and D. Dorfman; *C.A.* **67**, 31,855v (1967).

1338. Brit. P. 1077194 (1966) ≡ Neth. Appl. 6605103 (1966), Bayer AG, Inv.: L. Eue, H. Hack, K. Westphal, R. Wegler; *C.A.* **66**, 55,246c (1967).

1339. L. Eue, *Z. Pflanzenkr. Pflanzenschutz, Sonderheft* **IV**, 211 (1968).

1340. C. Fedtke, *Weed Res.* **12**, 325 (1972).

1341. T. Schmidt, C. Fedtke, and R. R. Schmidt, *Z. Naturforsch.* **31c**, 252 (1976).

1342. G. Schneider, *Naturwissenschaften* **51**, 416 (1964).

1343. DAS 1301173 (1962) ≡ Belg. P. 637632 (1964), E. Merck AG, Inv.: E. Jacobi, D. Erdmann, S. Lust, G. Schneider, and K. Niethammer; *C.A.* **62**, 11,086d (1965).

1344. G. E. Peterson, *Agric. Hist.* **41**, 243 (1967).

1345. DRP 872206 (1948), I.C.I., Inv.: R. T. Foster.

1346. DBP 915876 (1948) ≡ Swiss P. 262704 (1946), American Chemical Paint Co., Inv.: F. D. Jones; *C.* **1950 II**, 334.

1347. USP 2830083 (1953), Allied Chemical and Dye Corp., Inv.: E. E. Gilbert and S. Tryon; *C.A.* **52**, 14,688g (1958).

1348. C. L. Hamner and H. B. Tukey, *Science* **100**, 154 (1944).

1349. K. D. Courtney, D. W. Gaylor, M. D. Hogan, H. L. Falk, R. R. Bates, and I. Mitchell, *Science* **168**, 864 (1970).

1350. *Chem. Eng. News* **48**, 60 (1970).

1351. USP 2573769 (1951), Union Carbide and Carbon Corp., Inv.: J. A. Lambrech; *C.A.* **46**, 5087c (1952).

1352. DBP 967826 (1954), Henkel and Cie. GmbH, Inv.: W. Gündel and H. Linden; *C.A.* **53**, 16,066i (1959).

1353. DAS 1051277 (1954) ≡ USP 2754324 (1956), Dow Chemical Co., Inv.: H. F. Brust and H. O. Senkbeil; *C.A.* **51**, 15,574i (1957).

1354. USP 2828198 (1955), US Rubber Co., Inv.: W. D. Harris and A. W. Feldmann; *C.A.* **53**, 1624d (1959).

1355. Brit. P. 822199 (1957), Boots Pure Drug Co. Ltd., Inv.: H. A. Stevenson and R. F. Brookes; *C.A.* **55**, 2575i (1961).

1356. C. H. Fawcett, D. J. Osborne, R. L. Waine, and R. D. Walker, *Ann. Appl. Biol.* **40**, 231 (1953).

1357. USP 3076025 (1959), Union Carbide Comp., Inv.: J. N. Hogsett; *C.A.* **59**, 1537d (1963).

1358. M. E. Synerholm and P. W. Zimmerman, *Contrib. Boyce Thompson Inst.* **14**, 369 (1945); *C.A.* **41**, 4132c (1947).

1359. R. L. Wain, *Ann. Appl. Biol.* **42**, 151 (1955).

1359a. DOS 2223894 (1973), Hoechst AG., Inv.: W. Becker, P. Langelüddeke, H. Leditschke, H. Nahm, and F. Schwerdtle; *C.A.* **80**, 70,543e (1974).

1360. H. J. Miller, *Weeds* **1**[2], 185 (1952).

1361. USP 3081162 (1956), Heyden Newport Chemical Corp., Inv.: N. Tischler; *C.A.* **59**, 1040a (1963).

1362. DBP 1019324 (1955), Diamond Alkali Co., Inv.: C. E. Entemann; *C.A.* **54**, 23,168e (1960).

1363. USP 3013054 (1958), Velsicol Chemical Corp., Inv.: S. B. Richter; *C.A.* **56**, 10,049d (1962).

1364. USP 2890243 (1956). DuPont, Inv.: R. L. Brown and J. R. Gobeil; *C.A.* **54**, 424i (1960).

1365. DAS 1133176 (1958) ≡ USP 3027248 (1962, Neth. Prior. 1957), Philips Gloeilampenfabrieken, Inv.: H. Koopman and J. Daams; *C.A.* **57**, 7662a (1962).

1366. H. Koopman and J. Daams, *Nature* **186**, 89 (1960).

1367. H. Sandfort, *Proc. 7th Brit. Weed Control Conf.*, 208 (1964).

1368. Brit. P. 987253 (1965), Shell Research Ltd., Inv.: J. Yates; *C.A.* **62**, 16,150e (1965).

1369. USP 2923634 (1958), Diamond Alkali Co., Inv.: R. F. Lindemann; *C.A.* **54**, 16,732i (1960).

1370. DAS 1115516 (1959) ≡ USP 3014063 (1958), Am. Chem. Products, Inc., Inv.: S. R. McLane, J. R. Bishop, and H. P. Raman; *C.A.* **56**, 8643f (1962).

1371. USP 3234229 (1962) ≡ Belg. P. 628486 (1963), Dow Chemical Co., Inv.: C. T. Redemann; *C.A.* **60**, 15,840d (1964).

1372. USP 3285925 (1962), Dow Chemical Co., Inv.: H. A. Johnston and M. S. Tomita; *C.A.* **66**, 46,338g (1967).

1373. Brit. P. 862226 (1958), Boots Pure Drug Co. Ltd., Inv.: D. H. Godson, E. L. Leafe, G. B. Lush, and H. A. Stevenson; *C.A.* **55**, 19,122g (1961).

1374. R. F. Brookes and E. L. Leafe, *Nature* **198**, 589 (1963).

1375. T. Chapman, D. Jordan, J. M. Moncorge, D. H. Payne, and R. G. Turner, *Proc. 3rd Symposium on Herbicides, EWRC, Paris* **3**, 40 (1969).

1376. Neth. P. 6717715 (1967), Shell International Research, *C.A.* **69**, 106,282h (1968).

1377. E. Haddock, D. Jordan, and A. J. Sampson, *Pestic. Sci.* **6**, 273 (1975).

1378. B. Jeffcoat and W. N. Harries, *Pestic. Sci.* **6**, 283 (1975).

1379. Brit. P. 671153 (1950), US Rubber Co., Inv.: O. L. Hoffmann and A. E. Smith; *C.A.* **46**, 7704b (1952).

1380. O. L. Hoffmann and A. E. Smith, *Science* **109**, 588 (1949).

1381. G. W. Keitt and R. A. Baker, *Plant Physiol.* **20**, 449 (1969).

1382. DAS 1039779 (1957), Bayer AG, Inv.: W. Schäfer, R. Wegler, and L. Eue; *C.A.* **54**, 20,060i (1960).

1383. DAS 1160236 (1959) ≡ Brit. P. 869169 (1961), Food Machinery and Chemical Corp., Inv.: K. P. Dorschner, R. L. Gates, and J. R. Willard; *C.A.* **55**, 23,917b (1961).

1384. DAS 1166547 (1962), Schering AG, Inv.: H. E. Freund, H. F. Arndt, and R. Rusch; *C.A.* **60**, 16,438e (1964).

1385. F. Arndt, *Z. Pflanzenkr. Pflanzenschutz, Sonderheft* **III**, 277 (1965).

1386. DAS 1116469 (1959) ≡ Brit. P. 885043 (1961, US Prior. 1959), Food Machinery and Chemical Corp., Inv.: K. P. Dorschner and J. R. Willard; *C.A.* **57**, 8498e (1962).

1387. R. D. Blackburn and L. W. Weldon, *Weeds* **12**, 295 (1964).

1388. USP 3211544 (1964), Gulf Oil Corp., Inv.: K. P. Dubrovin; *C.A.* **64**, 9635c (1966).

1388a. USP 3639474 (1972), 3M Co., Inv.: J. K. Harrington, D. C. Kvam, A. Mendel, and J. E. Robertson; *C.A.* **72**, 121,183g (1970).

1389. DAS 1014380 (1954), Monsanto Chemical Co., Inv.: P. C. Hamm and A. J. Speziale; *C.A.* **54**, 13,531a (1960).

1390. P. C. Hamm and A. J. Speziale, *J. Agric. Food Chem.* **4**, 518 (1956).

1391. USP 3268584 (1964), Monsanto Chemical Co., Inv.: J. F. Olin; *C.A.* **65**, 19,240b (1966).

1392. Belg. P. 622131 (1962), Monsanto Chemical Co., Inv.: J. F. Olin; *C.A.* **59**, 11,330h (1963).

1392a. DOS 2328340 (1973), Ciba-Geigy AG., Inv.: C. Vogel, and R. Aebi; *C.A.* **80**, 82,440g (1974).

1392b. DOS 2305495 (1973), Ciba-Geigy AG., Inv.: C. Vogel; *C.A.* **79**, 146,248q (1973).

1393. C. C. Still and O. Kazirian, *Nature* **216**, 799 (1967).

1394. G. L. Lamoureux, L. E. Stafford, and F. F. Tanaka, *J. Agric. Food Chem.* **19**, 346 (1971).

1395. USP 3534098 (1967), Rohm & Haas Co., Inv.: B. W. Horrom, A. J. Crovetti, and K. L. Viste; *C.A.* **71**, 91,120u (1969).

1396. K. L. Viste, A. J. Crovetti, and B. W. Horrom, *Science* **167**, 280 (1970).

1397. USP 3120434 (1960), Eli Lilly and Co., Inv.: A. Pohland; *C.A.* **60**, 9208a (1964).

1398. J. Deli and G. F. Warren, *Weed Sci.* **18**, 692 (1970).

1399. Brit. P. 1066606 (1965, US Prior. 1964), Stauffer Chemical Co., Inv.: H. Tilles, D. K. Baker, and C. L. Dewald; *C.A.* **67**, 90,581a (1967).

1400. USP 3396009 (1966), Gulf Research and Co., Inv.: R. P. Neighbors; *C.A.* **69**, 85,643j (1968).

1401. *Farm Chemicals* **131**, 46 (1968).

1402. USP 3111403 (1960), Eli Lilly and Co., Inv.: Q. F. Soper; *C.A.* **60**, 13,184c (1964).

1403. USP 3257190 (1962), Eli Lilly and Co., Inv.: Q. F. Soper; *C.A.* **65**, 13,606e (1966).

1404. USP 3546295 (1968). Esso Research and Engineering Co., Inv.: L. L. Maravetz; *C.A.* **74**, 125,122a (1971).

1405. DOS 2013509 (1970, US Prior. 1969), US Borax and Chemical Corp., Inv.: D. L. Hunter, C. W. Lefevre, W. G. Woods, and J. D. Stone; *C.A.* **73**, 120,287c (1970).

1406. USP 3321292 (1965), Shell Oil Co., Inv.: S. B. Soloway and K. D. Zwahlen; *C.A.* **68**, 49,299x (1968).

1407. D. S. Burgis, *Weeds* **15**, 180 (1967).

1408. USP 3367949 (1966), Eli Lilly and Co., Inv.: Q. F. Soper; *C.A.* **69**, 27,037d (1968).

1408a. DAS 1643719 (1971), BASF AG., Inv.: K. Kiehs, K. H. König, and A. Fischer; *C.A.* **72**, 90,028d (1970).

1408b. DOS 2361464 (1974), Eli Lilly Co., Inv.: C. E. Moore and S. J. Parka; *C.A.* **82**, 27,229m (1975).

1409. DOS 2241408 (1972, US Prior. 1971), American Cyanamid Co., Inv.: A. W. Lutz, R. E. Diehl; *C.A.* **78**, 135,858s (1973).

1410. USP 3272842 (1965), Eli Lilly and Co., Inv.: N. R. Easton and R. D. Dillard; *C.A.* **66**, 2472e (1967).

1411. USP 3280136 (1965), General Electric Co., Inv.: H. L. Finkbeiner; *C.A.* **66**, 28,755m (1967).

1412. DAS 1112855 (1960), Norddeutsche Affinerie, Inv.: W. Perkow; *C.A.* **56**, 11,597h (1962).

1413. Belg. P. 619371 (1962), N. V. Philips' Gloeilampenfabrieken, Inv.: J. Daams and H. Koopman; *C.A.* **59**, 10,091g (1963).

1414. D. E. Burton, A. J. Lambie, J. C. L. Ludgate, G. T. Newbold, A. Percival, and D. T. Saggers, *Nature* **208**, 1166 (1965).

1415. DBP 972405 (1953), Am. Chem. Products, Inc., Inv.: W. W. Allen; *C.A.* **55**, 5855h (1961).

1416. W. F. Donnalley and S. K. Ries, *Science* **145**, 497 (1964).

1417. Brit. P. 1110500 (1968), Rhône-Poulenc; *C.A.* **69**, 52,143t (1968).

1418. W. Furness, *Proc. VII Congr. Intern. Protection Plantes, Paris,* **1970**, 314.

1419. DOS 1695989 (1967) ≡ USP 3437664 (1966), Velsicol Chemical Corp., Inv.: J. Kreuzer; *C.A.* **71**, 38,972x (1969).

1420. Brit. P. 1271659 (1968) ≡ DOS 1926139 (1969), Fisons Ltd., Inv.: P. S. Gates, J. Gillon, and D. T. Saggers; *C.A.* **72**, 100,487u (1970).

1421. DOS 2260485 (1972, US Prior. 1971), American Cyanamid Co., Inv.: B. L. Walworth and E. Klingsberg; *C.A.* **79**, 92,210a (1973).

1422. R. J. Winfield and J. J. B. Caldicott, *Pestic. Sci.* **6**, 297 (1975).

1423. USP 3249419 (1963), Dow Chemical Co., Inv.: R. T. Martin; *C.A.* **65**, 4570g (1966).

1424. Brit. P. 926326 (1959), I.C.I. Ltd., Inv.: A. H. Jubb; *C.A.* **59**, 11,529d (1963).

1425. Belg. P. 626303 (1962, Brit. Prior. 1961), I.C.I. Ltd., Inv.: J. T. Braunholtz; *C.A.* **60**, 8001d (1964).

1426. Brit. P. 785732 (1955), I.C.I. Ltd., Inv.: R. J. Fielden, R. F. Homer, and R. L. Jones; *C.A.* **52**, 6707g (1958).

1427. R. C. Brian, R. F. Homer, J. Stubbs, and R. L. Jones, *Nature* **181**, 446 (1958).

1427a. R. E. Phillips, R. L. Blevins, G. W. Thomas, W. W. Frye, and S. H. Phillips, *Science* **208**, 1108 (1980).

1428. Brit. P. 999585 (1962), I.C.I. Ltd., Inv.: J. E. Downes and D. W. R. Headford; *C.A.* **64**, 3498f (1966).

1429. DBP 1105232 (1958), BASF, Inv.: F. Reicheneder, K. Dury, and A. Fischer; *C.A.* **56**, 12,034b (1962).

1430. A. Fischer, *Weed Res.* **2**, 177 (1962).

1431. S. K. Ries, M. J. Zabik, G. R. Stephenson, and T. M. Chen, *Weed Sci.* **16**, 40 (1968).

1432. USP 3644355 (1969) ≡ Swiss P. 482684 (1967), Sandoz Ltd., Inv.: C. Ebner and M. Schuler; *C.A.* **72**, 121,567k (1970).

1433. USP 2575954 (1950), US Rubber Co., Inv.: W. D. Harris and D. L. Schoene; *C.A.* **46**, 6161g (1952).

1434. D. L. Schoene and O. L. Hoffmann, *Science* **109**, 588 (1949).

1435. DOS 1910620 (1969, Jap. Prior. 1968), Sankyo Co. Ltd., Inv.: T. Jojima; *C.A.* **72**, 12,755z (1970).

1436. T. Jojima and S. Tamura, *Agric. Biol.-Chem.* **30**, 896 (1966).

1437. H. C. Bucha, W. E. Cupery, J. E. Harrod, H. M. Loux, and L. M. Ellis, *Science* **137**, 537 (1962).

1438. USP 3235360 (1962), DuPont, Inv.: E. J. Soboczenski; *C.A.* **64**, 14,196f (1966).

1439. USP 3235357 (1962) ≡ Belg. P. 625897 (1963), DuPont, Inv.: H. M. Loux; *C.A.* **60**, 14,519c (1964).

1440. USP 3352862 (1965), DuPont, Inv.: E. J. Soboczenski; *C.A.* **68**, 105,229p (1968).

1441. DBP 1210242 (1964) ≡ Neth. Appl. 6504991 (1965), BASF, Inv.: G. Scheurer, A. Zeidler, A. Fischer; *C.A.* **64**, 11,226a (1966).

1442. A. Zeidler, A. Fischer, and G. Scheurer, *Z. Naturforsch. Teil B* **24**, 740 (1969).

1443. Belg. P. 702877 (1967) ≡ S. Afric.P. 6705164 (1968, DB Prior. 1966), BASF, Inv.: A. Zeidler, A. Fischer, and G. Weiss; *C.A.* **70**, 37,847c (1969).

1444. A. Gast, E. Knüsli, and H. Gysin, *Experientia* **11**, 107 (1955).

1445. A. Gast, E. Knüsli, and H. Gysin, *Experientia* **12**, 146 (1956).

1446. H. Koopman and J. Daam, *Rec. Trav. Chim. Pays-Bas* **77**, 235 (1958).

1447. DAS 1011904 (1955) ≡ Swiss P. 329277 (1954), J. R. Geigy AG, Inv.: H. Gysin and E. Knüsli; *C.A.* **54**, 20,061h (1960).

1448. USP 3451802 (1966), Gulf Research and Development Co., Inv.: R. P. Neighbors and L. V. Phillips; *C.A.* **71**, 61,426f (1969).

1449. Brit. P. 1132306 (1967; DB Prior. 1966), Degussa; *C.A.* **70**, 37,844z (1969).

1450. Swiss P. 337019 (1955), J. R. Geigy AG, Inv.: H. Gysin and E. Knüsli; *C.A.* **57**, 14,226a (1962).

1451. USP 3145208 (1963) ≡ Fr. P. 1339336 (1963; Swiss Prior. 1961), J. R. Geigy AG, Inv.: E. Knüsli, W. Schäppi, and D. Berrer; *C.A.* **60**, 2985a (1964).

1452. USP 3326914 (1965, Swiss Prior. 1958), J. R. Geigy AG, Inv.: E. Knüsli and H. Gysin; *C.A.* **69**, 43,944c (1968).

1453. USP 3799925 (1968), Ciba-Geigy Corp., Inv.: E. Nikles; *C.A.* **81**, 3973c (1974).

1454. Brit. P. 1093376 (1964) ≡ Neth. Appl. 6413689 (1965, Swiss Prior. 1963), Ciba Ltd.; *C.A.* **64**, 741g (1966).

1455. USP 3303015 (1960), Monsanto Co., Inv.: A. J. Speziale, P. C. Hamm; *C.A.* **67**, 21,944j (1967).

1456. L. S. Jordan, T. Murashige, J. D. Mann, and B. E. Day, *Weeds* **14**, 134 (1966).

1457. O. C. Thompson, B. Truelove, and D. E. Davis, *J. Agric. Food Chem.* **17**, 997 (1969).

1458. S. K. Ries, H. Chmiel, D. R. Dilley, and P. Filner, *Proc. Nat. Acad. Sci. U.S.A.* **58**, 526 (1967).

1459. W. Roth and E. Knüsli, *Experentia* **17**, 312 (1961).

1460. J. L. Hilton, L. L. Jansen, and H. M. Hull, *Ann. Rev. Plant Physiol.* **14**, 353 (1963).

1461. R. H. Shimabukuro, D. S. Frear, H. R. Swanson, and W. C. Walsh, *Plant Physiol.* **47**, 10 (1971).

1462. USP 3902887 (1974), DuPont, Inv.: K. Lin.

1463. Belg. P. 697083 (1967) ≡ Fr. P. 1519180 (1968, DB Prior. 1966), Bayer AG, Inv.: K. Westphal, W. Meiser, L. Eue, and H. Hack; *C.A.* **70**, 106,570w (1969).

1464. A. Dornow, H. Menzel, and P. Marx, *Chem. Ber.* **97**, 2173 (1964).

1465. L. Eue, *Pfanzenschutznachrichten Bayer* **25**, 175 (1972).

1466. H. D. Coble and J. W. Schrader, *Weed Sci.* **21**, 308 (1973).

1467. DAS 1795784 (1966), Bayer AG, Inv.: K. Westphal, W. Meiser, L. Eue, and H. Hack; *C.A.* **70**, 106,570w (1969).

1468. DOS 2107757 (1971), Bayer AG, Inv.: K. Dickore, W. Draber, and L. Eue; *C.A.* **77**, 164,767w (1972).

1469. DOS 2364474 (1973), Bayer AG, Inv.: H. Timmler, W. Draber.

1470. R. R. Schmidt, W. Draber, L. Eue, and H. Timmler, *Pestic. Sci.* **6**, 239 (1975).

1471. USP 2943107 (1959), Chemagro Corp., Inv.: K. H. Rattenbury and J. R. Costello; *C.A.* **54**, 20,876e (1960).

1472. USP 2955803 (1958), Virginia Carolina Chemical Corp., Inv.: L. E. Goyette; *C.A.* **55**, 2991f (1961).

1473. USP 3074790 (1960), Dow Chemical Co., Inv.: J. K. Leasure; *C.A.* **59**, 2112h (1963).

1474. S. Sumida and M. Veda, *ACS Symposium Series No. 2*, 156 (1974).

1475. D. H. Green and L. Ebner, *Proc. 11th Brit. Weed Control Conf.* 822 (1972).

1476. USP 3205253 (1963), Stauffer Chemical Co., Inv.: L. W. Fancher and C. L. Dewald; *C.A.* **64**, 2015c (1966).

1477. DAS 2152826 (1971), Monsanto Co., Inv.: J. E. Franz; *C.A.* **77**, 165,079k (1972).

1477a. USP 3619166 (1971), DuPont, Inv.: B. Quebedeaux; *C.A.* **72**, 90,627y (1970).

USP 3627507 (1971), DuPont, Inv.: W. P. Langsdorf; *C.A.* **72**, 90,627y (1970).

1478. *Bibliography*

R. L. Wain, The Relation of Chemical Structure to Activity for the 2.4-D-Type Herbicide and Plant Growth Regulator, in: R. L. Metcalf, *Adv. in Pest Control Res.*, Vol. II, pp. 263–305, Interscience, New York (1957).

H. Gysin and E. Knüsli, Chemistry and Herbicidal Properties of Triazine Derivatives, in: R. L. Metcalf, *Adv. in Pest Control Res.*, Vol. III, pp. 289–358, Interscience, New York (1957).

C. I. Harris, D. D. Kaufman, T. J. Sheets, R. G. Nash, and P. C. Kearny, Behaviour and Fate of s-Triazines in Soils, in: R. L. Metcalf, *Adv. in Pest Control Res.*, Vol. VIII, pp. 1–55, Interscience, New York (1968).

A. Calderbank, The Bipyridylium Herbicides, in: R. L. Metcalf, *Adv. in Pest Control Res.*, Vol. VIII, pp. 127–235, Interscience, New York (1968).

C. Hansch, Theoretical Considerations of Structure–Activity Relationships in Photosynthesis Inhibitors, in: H. Metzner, *Progress in Photosynthesis Research*, Vol. III, pp. 1685–1692, International Union of Biological Sciences, Tübingen (1969).

D. E. Moreland, Inhibitors of Chloroplast Electron Transport Structure–Activity Relations, in: H. Metzner, *Progress in Photosynthesis Research*, Vol. III, pp. 1693–1711, International Union of Biological Sciences, Tübingen (1969).

A. Barth and H.-J. Michel, Wirkungsmechanismen und Substituenteneinflüsse von Herbiziden des Harnstoff-, Carbamat-, Amid- und Triazintyps, *Pharmazie* **24**, 11–23 (1969).

A. Trebst, Neue Reaktionen und Substanzen im photosynthetischen Elektronentransport, *Ber. Dtsch. Bot. Ges.* **83**, 373–398 (1970).

W. van der Zweep and J. L. P. van Oorschot, Neue Forschungsergebnisse über Wirkungsweise und Verhalten von Herbiziden in Pflanzen, *Z. Pflanzenkr. Pflanzenschutz*, *Sonderheft* **V**, 61–70 (1970).

J. Caseley, Herbicide Activity Involving Light, *Pestic. Sci.* **1**, 28 (1970).

R. Wegler and L. Eue, Herbizide, in: R. Wegler, *Chemie der Pflanzenschutz- und Schädlingsbekämpfungsmittel*, Vol. 2, pp. 165–395, Springer-Verlag, Berlin–Heidelberg–New York (1970).

K. H. Büchel, Mechanisms of Action and Structure Activity Relations of Herbicides that Inhibit Photosynthesis, *Pestic. Sci.* **3**, 89–110 (1972).

F. M. Ashton and A. S. Crafts, *Mode of Action of Herbicides*, John Wiley, New York–London–Sidney–Toronto (1973).

A. Trebst, Energy Conservation in Photosynthetic Electron Transport of Chloroplasts, *Ann. Rev. Plant Physiol.* **25**, 423–458 (1974).

J. R. Corbett, *The Biochemical Mode of Action of Pesticides*, Academic Press, London–New York (1974).

H. Martin and C. R. Worthing, *Pesticide Manual*, 4th ed., British Crop of Protection Council, Worcester (1974).

R. Heitefuß, *Pflanzenschutz*, Georg Thieme Verlag, Stuttgart (1975).

M. B. Green, Polychloroaromatics and Heteroaromatics, in: H. Suschitzky, *Polychloroaromatic Compounds*, pp. 419–473, Plenum Press, London–New York (1975).

Pflanzenschutzmittel-Verzeichnis 1975/76, Teil 1, Biologische Bundesanstalt für Land- und Forstwirtschaft, Braunschweig.

Chapter 5. Plant Growth Regulators

1479. W. Perkow, *Wirksubstanzen der Pflanzenschutz- und Schädlingsbekämpfungsmittel*, P. Parey, Berlin–Hamburg 1971; contrast: D. J. Cook, *World Crops* **21**, 30 (1969).

1480. R. Wegler and L. Eue, in: R. Wegler, *Chemie der Pflanzenschutz- und Schädlingsbekämpfungsmittel*, Vol. 2, p. 268, Springer Verlag, Berlin–Heidelberg–New York (1970).

1481. F. W. Went, *Rec. Trav. Chim. Pays-Bas* **25**, 1 (1929).

1482. F. Kögl, *Ber.* **68A**, 16 (1935).

1483. E. Kurosawa, *Trans. Nat. Hist. Soc. Formosa* **16**, 213 (1926).

1484. J. McMillan, J. C. Seaton, and P. J. Suter, *Adv. Chem. Ser.* **28**, 18 (1961).

1485. P. W. Radley, *Nature* **178**, 1070 (1956).

1486. D. S. Letham, J. S. Shannon, and I. R. McDonald, *Proc. Chem. Soc.* **1964**, 230.

1487. K. Ohkuma, L. J. Lyon, F. T. Addicott, O. E. Smith, and W. E. Thiessen, *Tetrahedron Lett.* **1965**, 2529.

1488. S. P. Burg and E. A. Burg, *Plant Physiol.* **37**, 179 (1962).

1489. S. H. Wittwer, *Farm. Chem.* [2] **131**, 62 (1968).

1490. Anonym, *Farm. Chem.* [9] **138**, 45 (1975).

1491. Anonym, *Farm. Chem.* [3] **138**, 15 (1975).

1492. J. W. Mitchell and G. A. Livingston, Methods of Studying Plant Hormones and Growth Regulating Substances, *Agriculture Handbook*, No. 336, USDA (1968).

1493. K. Steyn, *Documentation East-European Agricult. Lit.* **8**, 1027 (1967).

1494. S. L. Adams, Chemagro, Agricultural Division of Mobay Chemical Corporation, calculation in Sept. 1975.

1495. Figures for 1969: D. W. Young, Chemagro Corporation.

1495a. Figures for 1978: G. Schmahl, Bayer AG, Leverkusen.

1496. J. Jung and P. E. Schott, *Z. Pflanzenkr. Pflanzensch. Sonderheft* **VII**, 333 (1975).

1497. G. Schneider, *Naturwissenschaften* **51**, 416 (1964).

1498. H. Ziegler, D. Köhler, and B. Streitz, *Z. Pflanzenphysiol.* **54**, 118 (1966).

1499. G. Schneider, D. Erdmann, S. Lust, G. Mohr, and K. Niethammer, *Nature* **208**, 1013 (1965).

1500. P. E. Pilet, *Experientia* **26**, 608 (1970).

1501. G. Schneider, *Ann. Rev. Plant Physiol.* **21**, 499 (1970).

1502. J. A. Ridell, H. A. Hageman, C. M. J. Anthony, and W. L. Hubbard, *Science* **136**, 391 (1962).

1503. H. Blömke, *Mitt. Obstversuchsring d. Alt. Landes* **21**, 464 (1966).

1504. S. H. Wittwer, *Outlook on Agriculture* **6**, 205 (1971).

1505. N. Tischler and E. P. Bell, *Proc. N.E. Weed Control Conf.*, 35 (1951).

1506. Anonym, *Chem. Eng. News* **53**(3), 31 (1975).

1507. A. R. Cooke and D. I. Randall, *Nature* **218**, 974 (1968).

1508. R. J. Weaver, *Nature* **181**, 851 (1958).

1509. L. G. Paleg, *Plant Physiol.* **36**, 902 (1960).

1510. L. E. Ahlrichs and C. A. Porter, *Proc. 11th Brit. Weed Contr. Conf. 13.–16.11.1972, Brighton*, p. 1215 (1972).

1511. G. W. Schneider and J. V. Enzie, *Proc. Am. Soc. Hortic. Sci.* **42**, 151 (1943); **44**, 117 (1944).

1512. O. L. Hoffmann and A. E. Smith, *Science* **109**, 588 (1949).

1513. K. Knoevenagel, K. Rolli, and R. Himmelreich, *Naturwissenschaften* **57**, 395 (1970).

1514. I. C. Anderson, *Farm. Chem.* [12] **131**, 29 (1968).

1515. C. W. Huffmann, E. M. Godar, and D. C. Torgeson, *J. Agric. Food Chem.* **15**, 976 (1967).

1516. N. E. Tolbert, *J. Biol. Chem.* **235**, 475 (1960).

1517. N. E. Tolbert, *Plant Physiol.* **35**, 380 (1960).

1518. J. Jung, *Naturwissenschaften* **54**, 356 (1967).

1519. B. G. Coombe, *Nature* **205**, 385 (1965).

1520. W. H. Peston and C. B. Link, *Plant Physiol. (Suppl.)* **33**, 29 (1958).

1521. S. H. Wittwer and N. E. Tolbert, *Plant Physiol.* **35**, 871 (1960).

1522. E. E. Tschabold et al., *Plant Physiol.* **48**, 519 (1971).

1523. A. C. Leopold, *Plant Physiol.* **48**, 537 (1971).

1524. E. C. Kornfeld, R. G. Jones, and T. V. Parke, *J. Amer. Chem. Soc.* **71**, 150 (1949).

1525. J. H. Ford and B. E. Leach, *J. Amer. Chem. Soc.* **70**, 1223 (1948).

1526. H. B. Curier, A. S. Crafts, and A. E. Smith, *Science* **111**, 152 (1950).

1527. S. H. Wittwer and R. C. Sharma, *Science* **112**, 597 (1950).

1528. W. C. Wilson, *Abstr. Pap. 170th ACS National Meeting 25.–29.8.1975*, Pest 56.

1528a. P. F. Bocion, W. H. deSilva, H. Walter, and H. R. Graf, *Gartenbau und Gartenwirtschaft* **27**, 634 (1977).

W. H. deSilva, H. R. Graf, and H. Walter, *Proc. British Crop Protection Conf. 1976*, p. 349.

P. F. Bocion and W. H. deSilva, *Plant Growth Regulation–9th Intern. Conf. on Plant Growth Substances*, 1977, p. 189, Springer Verlag, Berlin–Heidelberg–New York (1977).

R. M. Sachs, H. Hield, and J. deBie, *Hort. Science* **10**, 367 (1975).

H. L. Malstrom and J. L. McMeans, *Hort. Science* **12**, 68 (1977).

1529. *Technical Data Bulletin of 3 M-Company*, 1.1.1971.

1530. R. D. Gruenhagen, T. C. Hargroder, J. W. Matteson, J. W. O'Malley, D. R. Pauly, and F. L. Selman, *Abstr. 1975 PGRW Meeting, 27.8.1975, Chicago*, p. 3.

1531. H. M. Cathey and G. L. Steffens, in: Plant Growth Regulators, *SCI Monograph* **31**, 224 (1968).

1532. C. DePauw, *23. Intern. Symposium on Crop Protection, Gent*, 4.5.1971, F4.

1533. A. Busschots, *Meded. Rijksfac. Landbouwwet., Gent* **34**, 474 (1969).

1534. Brochure "PRB 8" from Poudreries Reunies de Belgique S.A., Dez. 1970.

1535. J. C. Vendrig, *Nature* **234**, 557 (1971).

1536. J. R. Bearder, F. G. Dennis, J. Macmillan, G. C. Martin, and B. O. Phinney, *Tetrahedron Lett.* **1975**, 669.

1537. R. L. Palmer, L. N. Lewis, H. Z. Hield, and J. Kumamoto, *Nature* **216**, 1216 (1967).

1538. P. W. Morgan and W. C. Hall, *Physiol. Plant.* **15**, 420 (1962); *Nature* **201**, 99 (1964).

1539. McLeish, *Heredity* **6**, Suppl., 125 (1952).

1540. J. Bonaly, *C. R. Acad. Sci., Paris, Ser. D* **273**, 150 (1971).

1541. H. M. Cathey and G. L. Steffens, in: Plant Growth Regulators, *SCI Monograph No. 31*, 225 (1968).

1542. D. W. Jones and C. L. Fox, *Weed Sci.* **19**, 595 (1971).

1543. N. N. Ragsdale and H. D. Sisler, *Pestic. Biochem. Physiol.* **3**, 20 (1973).

1544. E. G. Jaworski, *J. Agric. Food Chem.* **20**, 1195 (1972).

1545. *Bibliography*. Only includes books and reviews from 1970; for older literature consult the corresponding works by W. Draber and L. J. Audus.

   W. Draber and R. Wegler, Natürliche Pflanzenwuchsstoffe–Phytohormone (Literature review up to 1969), in: R. Wegler, *Chemie der Pflanzenschutz- und Schädlingsbekämpfungsmittel*, Vol. 2, pp. 400–401, Springer Verlag, Berlin–Heidelberg–New York (1970).

   D. J. Carr, *Plant Growth Substances 1970*, Springer Verlag, Berlin–Heidelberg–New York (1972).

   L. J. Audus, *Plant Growth Substances*, 3rd ed., Vol. 1: *Chemistry and Physiology*, Leonard Hill, London (1972).

   F. Skoog and D. J. Armstrong, Cytokinins, *Ann. Rev. Plant Physiol.* **21**, 359 (1970).

   J. van Overbeek, Growth Regulators and Their Potential, *Proc. 10th Brit. Weed Control Conf.* **3**, 1022 (1970).

   S. H. Wittwer, Growth Regulants in Agriculture, *Outlook on Agriculture* **6**, 205 (1971).

   W. J. Fleming and M. E. H. Howden, The Chemistry of Naturally Occurring Plant Growth Regulators, *Rev. Pure Appl. Chem.* **22**, 67 (1972).

   K. V. Thimann, L. G. Paleg, C. A. West, F. Skoog, and R. Schmitz, The Natural Plant Hormones, in: F. C. Steward, *Plant Physiology, A Treatise*, Vol. 6 C, pp. 3–332, Academic Press, New York (1972).

   G. Mohr, Pflanzliche Wuchsstoffe und Wachstumsregulatoren, *Chem. Z.* **97**, 409 (1973).

   D. Hess, Phytohormone–interzelluläre Regulation bei höheren Pflanzen, *Naturwiss. Rundsch.* **26**, 284 (1973).

   T. H. Tudor, An Assessment of the Value of Plant Growth Regulators and Herbicides for Crops Grown for the Processed Food Industries, *Pestic. Sci.* **5**, 87 (1974).

   Anonym, Entering the Age of Plant Growth Regulators, *Farm. Chem.* [3] **138**, 15 (1975).

   K. H. Neumann, Phytohormone und die Entwicklung der höheren Pflanze, *Z. Pflanzenkr. Pflanzenschutz, Sonderheft* **VII**, 309 (1975).

   J. Jung and P. E. Schott, Möglichkeiten der Anwendung von Wachstumsregulatoren in der Pflanzenproduktion, *Z. Pflanzenkr. Pflanzenschutz, Sonderheft* **VII**, 333 (1975).

   R. L. Wain, Some Developments in Research on Plant Growth Inhibitors, *Proc. Roy. Soc., Ser. B* **191**, 335 (1975).

Chapter 6.   Formulation Aids

1546. R. Heusch and H. Niessen, Tenside in Pflanzenschutz- und Schädlingsbekämpfungsmitteln (In preparation).

1547. O. Telle, Handelsformen von Pflanzenschutz- und Schädlingsbekämpfungsmitteln, in: R. Wegler, *Chemie der Pflanzenschutz- und Schädlingsbekämpfungsmittel*, Vol. 1, pp. 41–54, Springer Verlag, Berlin–Heidelberg–New York (1970).

1548. H. Niessen, Möglichkeiten und Grenzen der Formulierung von Pflanzenschutzmitteln, *Pflanzenschutz-Nachr. Bayer* **27**, 76–90 (1974).

1549. H. Niessen, Importance of Storage Stability Studies in the Development of Pesticide Formulations, *Pestic. Sci.* **6**, 181–188 (1975).

1550. Anonym, Exemptions from Tolerances, *The Pesticide Chemical News Guide*, Issue 1.4.75, pp. 107–120.

1551. T. C. Watkins and L. B. Norton, *Handbook of Insecticide Dust Diluents and Carriers*, 2nd ed., Caldwell, New Jersey (1955).

1552. K. Ullrich, *Chem. Technik* **16**, 263 (1964).

1553. E. Wagner and H. Brümer, Aerosil, Herstellung, Eigenschaften und Verhalten in organischen Flüssigkeiten, *Angew. Chem.* **72**, 744–750 (1960).

1554. New Zealand Pat. Appl. No. 154193.

1555. *Biologische Bundesanstalt für Land- und Forstwirtschaft, Pflanzenschutzmittel-Verzeichnis 1975/76*, 24th ed., Braunschweig 1975, Teil 1, p. 163 and Teil 2, p. 174.

Index

See and *see also* references may appear in any of the four index sections, parts A through D.

A. General Subjects

B. Classes of Agents

C. Individual Compounds, Trivial Names

D. Trade Names, Common Names, etc.